Student Solutions Manual

to accompany

College Physics

With an Integrated Approach to Forces and Kinematics

Fourth Edition

Alan Giambattista

Cornell University

Betty McCarthy Richardson

Cornell University

Robert C. Richardson

Cornell University

Prepared by

William Fellers

Fellers Math & Science.

The McGraw-Hill Companies

Connect
Learn
Succeed™

Student Solutions Manual to accompany
COLLEGE PHYSICS: WITH AN INTEGRATED APPROACH TO FORCES AND KINEMATICS, FOURTH EDITION
ALAN GIAMBATTISTA, BETTY MCCARTHY RICHARDSON, ROBERT C. RICHARDSON

Published by McGraw-Hill Higher Education, an imprint of The McGraw-Hill Companies, Inc., 1221 Avenue of the Americas, New York, NY 10020. Copyright © 2013, 2010, 2008 and 2006 by The McGraw-Hill Companies, Inc. All rights reserved. Printed in the United States of America.

1 2 3 4 5 6 7 8 9 0 QDB/QDB 1 0 9 8 7 6 5 4 3 2

ISBN: 978-0-07-743788-6
MHID: 0-07-743788-8

www.mhhe.com

Student's Solutions Manual
to accompany
COLLEGE PHYSICS

Table of Contents

Chapter 1

INTRODUCTION

Conceptual Questions

1. Knowledge of physics is important for a full understanding of many scientific disciplines, such as chemistry, biology, and geology. Furthermore, much of our current technology can only be understood with knowledge of the underlying laws of physics. In the search for more efficient and environmentally safe sources of energy, for example, physics is essential. Also, many study physics for the sense of fulfillment that comes with learning about the world we inhabit.

5. Scientific notation eliminates the need to write many zeros in very large or small numbers. Also, the appropriate number of significant digits is unambiguous when written this way.

9. The kilogram, meter, and second are three of the base units used in the SI system.

13. Trends in a set of data are often the most interesting aspect of the outcome of an experiment. Such trends are more apparent when data is plotted graphically rather than listed in numerical tables.

Problems

1. **Strategy** The new fence will be $100\% + 37\% = 137\%$ of the height of the old fence.

   Solution Find the height of the new fence.
   $$1.37 \times 1.8 \text{ m} = \boxed{2.5 \text{ m}}$$

3. **Strategy** Relate the surface area S to the radius r using $S = 4\pi r^2$.

   Solution Find the ratio of the new radius to the old.
   $$S_1 = 4\pi r_1^2 \text{ and } S_2 = 4\pi r_2^2 = 1.160 S_1 = 1.160(4\pi r_1^2).$$
   $$4\pi r_2^2 = 1.160(4\pi r_1^2)$$
   $$r_2^2 = 1.160 r_1^2$$
   $$\left(\frac{r_2}{r_1}\right)^2 = 1.160$$
   $$\frac{r_2}{r_1} = \sqrt{1.160} = 1.077$$

   The radius of the balloon increases by $\boxed{7.7\%}$.

5. **Strategy** To find the factor by which the metabolic rate of a 70 kg human exceeds that of a 5.0 kg cat use a ratio.

   Solution Find the factor.
   $$\left(\frac{m_h}{m_c}\right)^{3/4} = \left(\frac{70}{5.0}\right)^{3/4} = \boxed{7.2}$$

9. **Strategy** Use a proportion.

   Solution Find Jupiter's orbital period.
   $$T^2 \propto R^3, \text{ so } \frac{T_J^2}{T_E^2} = \frac{R_J^3}{R_E^3} = 5.19^3. \text{ Thus, } T_J = 5.19^{3/2} T_E = \boxed{11.8 \text{ yr}}.$$

11. **Strategy** The area of the poster is given by $A = \ell w$. Let the original and final areas be $A_1 = \ell_1 w_1$ and $A_2 = \ell_2 w_2$, respectively.

    Solution Calculate the percentage reduction of the area.
    $$A_2 = \ell_2 w_2 = (0.800\ell_1)(0.800 w_1) = 0.640 \ell_1 w_1 = 0.640 A_1$$
    $$\frac{A_1 - A_2}{A_1} \times 100\% = \frac{A_1 - 0.640 A_1}{A_1} \times 100\% = \boxed{36.0\%}$$

13. **Strategy** Assuming that the cross section of the artery is a circle, we use the area of a circle, $A = \pi r^2$.

    Solution
    $$A_1 = \pi r_1^2 \text{ and } A_2 = \pi r_2^2 = \pi(2.0 r_1)^2 = 4.0 \pi r_1^2.$$
    Form a proportion.
    $$\frac{A_2}{A_1} = \frac{4.0 \pi r_1^2}{\pi r_1^2} = 4.0$$
    The cross-sectional area of the artery increases by a factor of $\boxed{4.0}$.

17. **(a) Strategy** Rewrite the numbers so that the power of 10 is the same for each. Then add and give the answer with the number of significant figures determined by the less precise of the two numbers.

    Solution Perform the operation with the appropriate number of significant figures.
    $$3.783 \times 10^6 \text{ kg} + 1.25 \times 10^8 \text{ kg} = 0.03783 \times 10^8 \text{ kg} + 1.25 \times 10^8 \text{ kg} = \boxed{1.29 \times 10^8 \text{ kg}}$$

    (b) Strategy Find the quotient and give the answer with the number of significant figures determined by the number with the fewest significant figures.

    Solution Perform the operation with the appropriate number of significant figures.
    $$(3.783 \times 10^6 \text{ m}) \div (3.0 \times 10^{-2} \text{ s}) = \boxed{1.3 \times 10^8 \text{ m/s}}$$

21. **Strategy** Multiply and give the answer in scientific notation with the number of significant figures determined by the number with the fewest significant figures.

    Solution Solve the problem.
    $$(3.2 \text{ m}) \times (4.0 \times 10^{-3} \text{ m}) \times (1.3 \times 10^{-8} \text{ m}) = \boxed{1.7 \times 10^{-10} \text{ m}^3}$$

25. Strategy Use the rules for determining significant figures and for writing numbers in scientific notation.

Solution

(a) 0.00574 kg has three significant figures, 5, 7, and 4. The zeros are not significant, since they are used only to place the decimal point. To write this measurement in scientific notation, we move the decimal point three places to the right and multiply by 10^{-3}.

(b) 2 m has one significant figure, 2. This measurement is already written in scientific notation

(c) 0.450×10^{-2} m has three significant figures, 4, 5, and the 0 to the right of 5. The zero is significant, since it comes after the decimal point and is not used to place the decimal point. To write this measurement in scientific notation, we move the decimal point one place to the right and multiply by 10^{-1}.

(d) 45.0 kg has three significant figures, 4, 5, and 0. The zero is significant, since it comes after the decimal point and is not used to place the decimal point. To write this measurement in scientific notation, we move the decimal point one place to the left and multiply by 10^{1}.

(e) 10.09×10^{4} s has four significant figures, 1, 9, and the two zeros. The zeros are significant, since they are between two significant figures. To write this measurement in scientific notation, we move the decimal point one place to the left and multiply by 10^{1}.

(f) 0.09500×10^{5} mL has four significant figures, 9, 5, and the two zeros to the right of 5. The zeros are significant, since they come after the decimal point and are not used to place the decimal point. To write this measurement in scientific notation, we move the decimal point two places to the right and multiply by 10^{-2}.

The results of parts (a) through (f) are shown in the table below.

	Measurement	Significant Figures	Scientific Notation
(a)	0.00574 kg	3	5.74×10^{-3} kg
(b)	2 m	1	2 m
(c)	0.450×10^{-2} m	3	4.50×10^{-3} m
(d)	45.0 kg	3	4.50×10^{1} kg
(e)	10.09×10^{4} s	4	1.009×10^{5} s
(f)	0.09500×10^{5} mL	4	9.500×10^{3} mL

29. Strategy There are approximately 39.37 inches per meter.

Solution Find the thickness of the cell membrane in inches.

$$7.0 \times 10^{-9} \text{ m} \times 39.37 \text{ inches/m} = \boxed{2.8 \times 10^{-7} \text{ inches}}$$

33. **Strategy** For (a), convert meters per second to miles per hour using the conversion 1 mi/h = 0.4470 m/s. For (b), convert meters per second to centimeters per millisecond using the conversions 1 m = 100 cm and 1 s = 1000 ms.

Solution Convert each speed.

(a) $80 \text{ m/s} \times \dfrac{1 \text{ mi/h}}{0.4470 \text{ m/s}} = \boxed{180 \text{ mi/h}}$

(b) $80 \text{ m/s} \times \dfrac{10^2 \text{ cm}}{1 \text{ m}} \times \dfrac{1 \text{ s}}{10^3 \text{ ms}} = \boxed{8.0 \text{ cm/ms}}$

37. **Strategy** Convert the radius to centimeters; then use the conversions $1 \text{ L} = 10^3 \text{ cm}^3$ and $60 \text{ s} = 1 \text{ min}$.

Solution Find the volume rate of blood flow

$$\textit{volume rate of blood flow} = \pi r^2 v = \pi (1.2 \text{ cm})^2 (18 \text{ cm/s}) \times \dfrac{1 \text{ L}}{10^3 \text{ cm}^3} \times \dfrac{60 \text{ s}}{1 \text{ min}} = \boxed{4.9 \text{ L/min}}$$

41. (a) **Strategy** There are 12 inches in one foot, 2.54 centimeters in one inch, and 60 seconds in one minute.

Solution Express the snail's speed in feet per second.

$$\dfrac{5.0 \text{ cm}}{1 \text{ min}} \times \dfrac{1 \text{ min}}{60 \text{ s}} \times \dfrac{1 \text{ in}}{2.54 \text{ cm}} \times \dfrac{1 \text{ ft}}{12 \text{ in}} = \boxed{2.7 \times 10^{-3} \text{ ft/s}}$$

(b) **Strategy** There are 5280 feet in one mile, 12 inches in one foot, 2.54 centimeters in one inch, and 60 minutes in one hour.

Solution Express the snail's speed in miles per hour.

$$\dfrac{5.0 \text{ cm}}{1 \text{ min}} \times \dfrac{60 \text{ min}}{1 \text{ h}} \times \dfrac{1 \text{ in}}{2.54 \text{ cm}} \times \dfrac{1 \text{ ft}}{12 \text{ in}} \times \dfrac{1 \text{ mi}}{5280 \text{ ft}} = \boxed{1.9 \times 10^{-3} \text{ mi/h}}$$

45. **Strategy** Replace each quantity in $T^2 = 4\pi^2 r^3/(GM)$ with its dimensions.

Solution Show that the equation is dimensionally correct.

$$T^2 \text{ has dimensions } [T]^2 \text{ and } \dfrac{4\pi^2 r^3}{GM} \text{ has dimensions } \dfrac{[L]^3}{\dfrac{[L]^3}{[M][T]^2} \times [M]} = \dfrac{[L]^3}{[M]} \times \dfrac{[M][T]^2}{[L]^3} = [T]^2.$$

Since $\boxed{[T]^2 = [T]^2}$, the equation is dimensionally correct.

49. **Strategy** Approximate the distance from your eyes to a book held at your normal reading distance.

Solution The normal reading distance is about 30–40 cm, so the approximate distance from your eyes to a book you are reading is $\boxed{30\text{–}40 \text{ cm}}$.

53. **Strategy** One story is about 3 m high.

Solution Find the order of magnitude of the height in meters of a 40-story building.

$(3 \text{ m})(40) \sim \boxed{100 \text{ m}}$

57. Strategy Put the equation that describes the line in slope-intercept form, $y = mx + b$.

$$at = v - v_0$$
$$v = at + v_0$$

Solution

(a) v is the dependent variable and t is the independent variable, so $\boxed{a}$ is the slope of the line.

(b) The slope-intercept form is $y = mx + b$. Find the vertical-axis intercept.
$v \leftrightarrow y$, $t \leftrightarrow x$, $a \leftrightarrow m$, so $v_0 \leftrightarrow b$.

Thus, $\boxed{+v_0}$ is the vertical-axis intercept of the line.

61. Strategy Use the slope-intercept form, $y = mx + b$.

Solution Since x is on the vertical axis, it corresponds to y. Since t^4 is on the horizontal axis, it corresponds to x (in $y = mx + b$). So, the equation for x as a function of t is $\boxed{x = (25 \ \mathrm{m/s^4})t^4 + 3 \ \mathrm{m}}$.

65. Strategy Replace v, r, ω, and m with their dimensions. Then use dimensional analysis to determine how v depends upon some or all of the other quantities.

Solution v, r, ω, and m have dimensions $\dfrac{[\mathrm{L}]}{[\mathrm{T}]}$, $[\mathrm{L}]$, $\dfrac{1}{[\mathrm{T}]}$, and $[\mathrm{M}]$, respectively. No combination of r, ω, and m gives dimensions without $[\mathrm{M}]$, so v does not depend upon m. Since $[\mathrm{L}] \times \dfrac{1}{[\mathrm{T}]} = \dfrac{[\mathrm{L}]}{[\mathrm{T}]}$ and there is no dimensionless constant involved in the relation, v is equal to the product of ω and r, or $\boxed{v = \omega r}$.

69. Strategy There are about 10^3 hairs in a one-square-inch area of the average human head. An order-of-magnitude estimate of the area of the average human head is 10^2 square inches.

Solution Calculate the estimate.
$$10^3 \ \mathrm{hairs/in^2} \times 10^2 \ \mathrm{in^2} = \boxed{10^5 \ \mathrm{hairs}}$$

73. (a) Strategy There are 3.28 feet in one meter.

Solution Find the length in meters of the largest recorded blue whale.
$$1.10 \times 10^2 \ \mathrm{ft} \times \frac{1 \ \mathrm{m}}{3.28 \ \mathrm{ft}} = \boxed{33.5 \ \mathrm{m}}$$

(b) Strategy Divide the length of the largest recorded blue whale by the length of a double-decker London bus.

Solution Find the length of the blue whale in double-decker-bus lengths.
$$\frac{1.10 \times 10^2 \ \mathrm{ft}}{8.0 \ \frac{\mathrm{m}}{\mathrm{bus \ length}}} \times \frac{1 \ \mathrm{m}}{3.28 \ \mathrm{ft}} = \boxed{4.2 \ \mathrm{bus \ lengths}}$$

77. Strategy If s is the speed of the molecule, then $s \propto \sqrt{T}$ where T is the temperature.

Solution Form a proportion.

$$\frac{s_{\text{cold}}}{s_{\text{warm}}} = \frac{\sqrt{T_{\text{cold}}}}{\sqrt{T_{\text{warm}}}}$$

Find s_{cold}.

$$s_{\text{cold}} = s_{\text{warm}}\sqrt{\frac{T_{\text{cold}}}{T_{\text{warm}}}} = (475 \text{ m/s})\sqrt{\frac{250.0 \text{ K}}{300.0 \text{ K}}} = \boxed{434 \text{ m/s}}$$

81. Strategy The SI base unit for mass is kg. Replace each quantity in $W = mg$ with its SI base units.

Solution Find the SI unit for weight.

$$\text{kg} \cdot \frac{\text{m}}{\text{s}^2} = \boxed{\frac{\text{kg} \cdot \text{m}}{\text{s}^2}}$$

85. (a) Strategy There are 7.0 leagues in one pace and 4.8 kilometers in one league.

Solution Find your speed in kilometers per hour.

$$\frac{120 \text{ paces}}{1 \text{ min}} \times \frac{7.0 \text{ leagues}}{1 \text{ pace}} \times \frac{4.8 \text{ km}}{1 \text{ league}} \times \frac{60 \text{ min}}{1 \text{ h}} = \boxed{2.4 \times 10^5 \text{ km/h}}$$

(b) Strategy The circumference of the earth is approximately 40,000 km. The time it takes to march around the Earth is found by dividing the distance by the speed.

Solution Find the time of travel.

$$40,000 \text{ km} \times \frac{1 \text{ h}}{2.4 \times 10^5 \text{ km}} \times \frac{60 \text{ min}}{1 \text{ h}} = \boxed{10 \text{ min}}$$

89. Strategy Since there are about 3×10^8 people in the U.S., a reasonable estimate of the number of automobiles is 1.5×10^8. There are 365 days per year. A reasonable estimate for the average volume of gasoline used per day per car is greater than 1 gal, but less than 10 gal; for a rough estimate, let's guess 2 gallons per day.

Solution Calculate the estimate.

$$1.5 \times 10^8 \text{ cars} \times 365 \text{ days} \times 2 \frac{\text{gal}}{\text{car} \cdot \text{day}} \approx \boxed{10^{11} \text{ gal}}$$

93. Strategy The dimensions of k and m are mass per time squared and mass, respectively. Dividing either quantity by the other will eliminate the mass dimension.

Solution The square root of k/m has dimensions of inverse time, which is correct for frequency. So, $f = \sqrt{k/m}$. Find k.

$$f_1 = \sqrt{\frac{k}{m_1}}, \text{ so } f_1^2 = \frac{k}{m_1}, \text{ or } k = m_1 f_1^2.$$

Find the frequency of the chair with the 75-kg astronaut.

$$f_2 = \sqrt{\frac{k}{m_2}} = \sqrt{\frac{m_1 f_1^2}{m_2}} = f_1\sqrt{\frac{m_1}{m_2}} = (0.50 \text{ s}^{-1})\sqrt{\frac{62 \text{ kg} + 10.0 \text{ kg}}{75 \text{ kg} + 10.0 \text{ kg}}} = \boxed{0.46 \text{ s}^{-1}}$$

Chapter 2

FORCE

Conceptual Questions

1. In an automobile accident the force due to the collision changes the motion of the car, but the driver and passengers continue to move in accordance with Newton's first law. Seat belts supply the force necessary to change their motion and slow them down. Without seat belts people would collide with the steering wheel or windshield, for example, and stand a greater risk of injury.

5. When the handle hits the board and stops abruptly, Newton's first law says that the steel head will continue to move for a short distance, resisting changes in velocity, until the force of friction between the head and the handle has brought it to rest. It will have then moved down some to where the handle is a little wider, resulting in a tightening of the head onto the handle.

9. (a) The reading of the scale is the magnitude of the normal force pushing up on you. This equals your weight as long as the normal force and the force of gravity are the only forces acting on you, and you are at rest or moving with a constant velocity.

   (b) If you were standing on the scale in a swimming pool for example, there would be a buoyant force from the water pushing up on you, and the scale would read a smaller apparent weight. Also, the scale would not read your weight if you were accelerating—for example standing on the scale in an elevator as it was moving upward with increasing speed.

13. The key is that the equal and opposite forces of Newton's third law are acting on two different objects—one on the wagon and the other on you. The wagon can therefore experience a non-zero net force, which causes it to accelerate forward.

17. Newton's third law tells us that if the person on the raft walks away from the pier, the raft will in turn move toward the pier. Thus, after walking the length of the raft, it should be possible for the person with the hook on the pier to grab the raft and reel it in. Without the person on the pier to hold the raft, this technique would be of no use, as the raft would move back away from the pier on the return walk.

21. Yes, as long as the y-axis is perpendicular to the chosen x-axis. This will often simplify a problem.

Problems

1. **Strategy** Determine the forces *not* acting on the scale.

   Solution The scale is in contact with the floor, so a contact force due to the floor is exerted on the scale. The scale is in contact with the person's feet, so a contact force due to the person's feet is exerted on the scale. The scale is in the proximity of a very large mass (Earth), so the weight of the scale is a force exerted on the scale. The weight of the person is a force exerted on the person due to the very large mass, so it is not a force exerted on the scale.

5. **Strategy** There are 0.2248 pounds per newton.

   Solution Find the weight of the astronaut in newtons.
   $$175 \text{ lb} \times \frac{1 \text{ N}}{0.2248 \text{ lb}} = \boxed{778 \text{ N}}$$

9. **Strategy** Use the fact that $|\vec{A}| = |\vec{B}|$ and symmetry to determine the direction of $\vec{C}$; then sketch $\vec{C}$.

   Solution By symmetry, we find that $\vec{C}$ points downward; the horizontal components cancel when $\vec{A}$ and $\vec{B}$ are added. The downward components of each vector have the same magnitude, about 0.7 N. So, the magnitude of $\vec{C}$ is about 1.4 N. The sketch is shown:

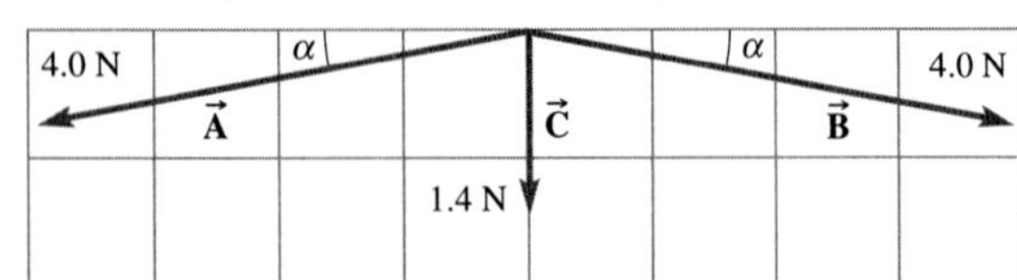

13. **Strategy** Represent 10 N as a length of 1 cm. Then, 30 N is represented by 3 cm and 40 N is represented by 4 cm.

   Solution Use graph paper, ruler, and protractor to find the magnitude and direction of the vector sum of the two forces.

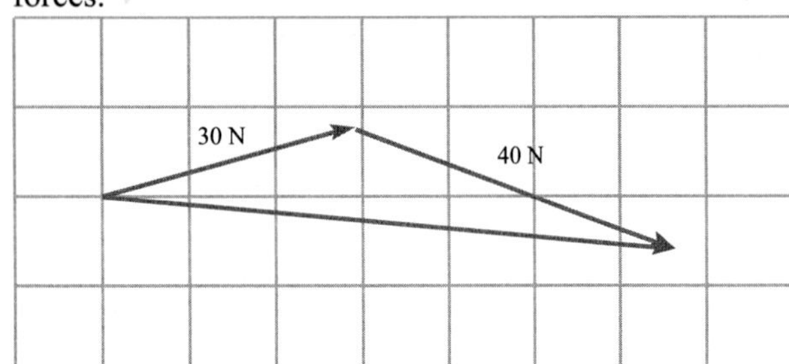

   Using the ruler, we find that the magnitude of the vector sum of the forces is about 70 N. Using the protractor, we find that the direction of the vector sum of the forces is about 5° below the horizontal.

17. **Strategy** Use graph paper to draw a diagram.

   Solution Find the vector sum of the vectors.

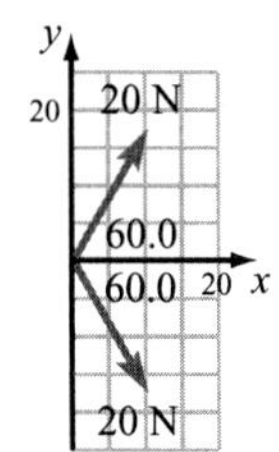

   Because of symmetry, the y-components of the vectors cancel. The x-components look to be about 10 N, so the vector sum is $10\text{ N} + 10\text{ N} = 20\text{ N}$, or 20 N in the positive x-direction.

19. **Strategy** Graph the vectors and their sum. Use the graph and the grid to estimate the components of the vectors. Then use the component method to find the vector sum.

   Solution Find the vector sum.
   x-comp: $A_x + B_x + C_x = 3(2\text{ N}) + 0(2\text{ N}) - 2(2\text{ N}) = 2\text{ N}$
   y-comp: $A_y + B_y + C_y = 4(2\text{ N}) - 4(2\text{ N}) + 0(2\text{ N}) = 0$

   The vector sum of the forces is 2 N in the positive x-direction or 2 N to the east.

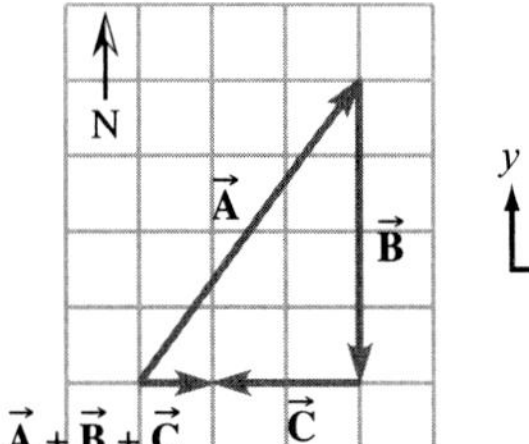

21. Strategy The components of $\vec{v}$ are given. Since the x-component is positive and the y-component is negative, the vector lies in the fourth quadrant. Give the angle with respect to the axes.

Solution Find the magnitude and direction of $\vec{v}$.

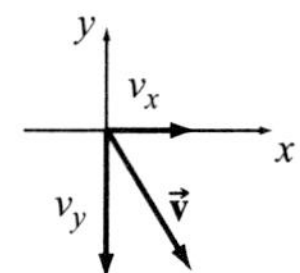

(a) $v = \sqrt{v_x{}^2 + v_y{}^2} = \sqrt{(16.4 \text{ m/s})^2 + (-26.3 \text{ m/s})^2} = \boxed{31.0 \text{ m/s}}$

(b) $\theta = \tan^{-1} \dfrac{-26.3}{16.4} = \boxed{58.1° \text{ with the } +x\text{-axis and } 31.9° \text{ with the } -y\text{-axis}}$

25. Strategy The components of $\vec{a}$ are given. Since the x-component is negative and the y-component is positive, the vector lies in the second quadrant. Give the angle with respect to the axis to which it lies closest.

Solution Find the magnitude and direction of $\vec{a}$.

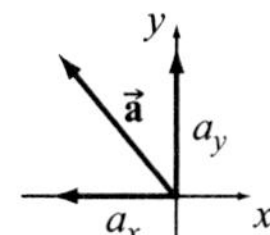

(a) $a = \sqrt{a_x{}^2 + a_y{}^2} = \sqrt{(-3.0 \text{ m/s}^2)^2 + (4.0 \text{ m/s}^2)^2} = \boxed{5.0 \text{ m/s}^2}$

(b) $\theta = \tan^{-1} \dfrac{4.0}{-3.0} = \boxed{37° \text{ CCW from the } +y\text{-axis}}$

27. Strategy and Solution Multiplying a vector by a scalar is equivalent to multiplying the vector's components by that scalar value. Multiplying a vector by a positive scalar—other than 1—changes the magnitude of the vector and its components but not the direction. Therefore, doubling the magnitude of the vector $\boxed{\text{doubles both}}$ $\boxed{\text{components, without changing their signs}}$.

29. Strategy Use the Pythagorean theorem to find the magnitude of each vector. Sketch a right triangle to find the direction angle. Give the angle with respect to the axis to which it lies closest.

Solution Find the magnitude and direction of each vector.

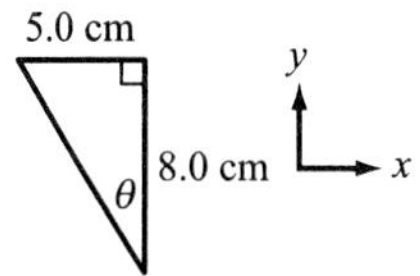

(a) $r = \sqrt{(-5.0 \text{ cm})^2 + (8.0 \text{ cm})^2} = \boxed{9.4 \text{ cm}}$ and

$\theta = \tan^{-1} \dfrac{5.0}{8.0} = \boxed{32° \text{ CCW from the } +y\text{-axis}}$.

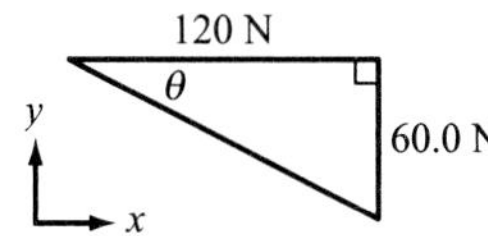

(b) $F = \sqrt{(120 \text{ N})^2 + (-60.0 \text{ N})^2} = \boxed{130 \text{ N}}$ and

$\theta = \tan^{-1} \dfrac{60.0}{120} = \boxed{27° \text{ CW from the } +x\text{-axis}}$.

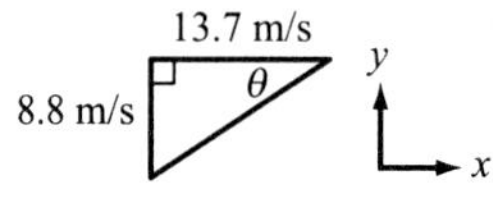

(c) $v = \sqrt{(-13.7 \text{ m/s})^2 + (-8.8 \text{ m/s})^2} = \boxed{16.3 \text{ m/s}}$ and

$\theta = \tan^{-1} \dfrac{8.8}{13.7} = \boxed{33° \text{ CCW from the } -x\text{-axis}}$.

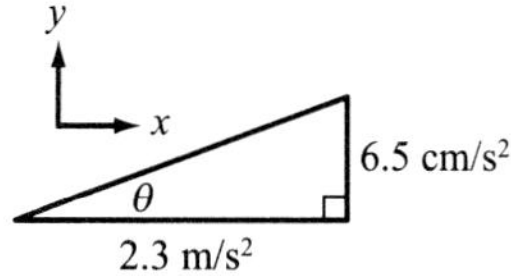

(d) $a = \sqrt{(2.3 \text{ m/s}^2)^2 + (6.5 \times 10^{-2} \text{ m/s}^2)^2} = \boxed{2.3 \text{ m/s}^2}$ and

$\theta = \tan^{-1} \dfrac{0.065}{2.3} = \boxed{1.6° \text{ CCW from the } +x\text{-axis}}$.

33. Strategy Since the mattress is neither moving upward nor downward, the net force must be zero in the vertical direction.

Solution So that the net force is zero, the upward force of the water must be equal to the combined weight of the man and the air mattress, or $\boxed{806 \text{ N}}$.

35. Strategy The force of the lake on the boat must be equal in magnitude and opposite in direction to the weight of the boat. The force of the wind on the boat must be equal in magnitude and opposite in direction to that of the line. Let the subscripts be the following:
s = sailboat   e = Earth   w = wind   l = lake   m = mooring line

Solution The free-body diagram is shown.

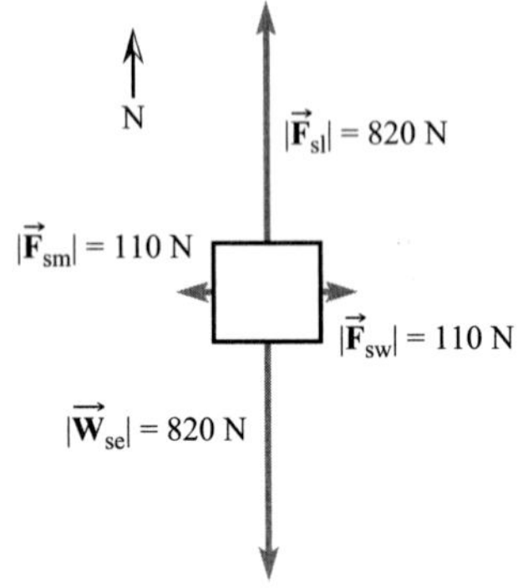

37. Strategy Make a scale drawing to determine which net force is greater.

Solution The scale drawing is shown.

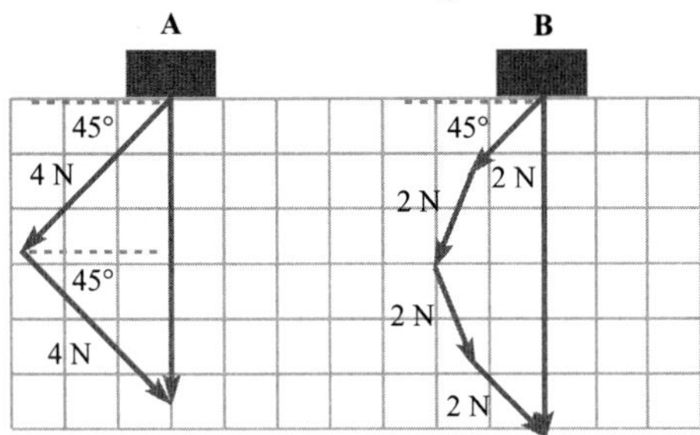

The net force magnitude on object B is greater than that on object A because two of the forces acting on B are directed at an angle greater than 45° with respect to the horizontal and contribute more to the downward-directed net force.

39. Strategy Draw a free-body diagram and add the force to find the net force on the truck.

Solution The vertically directed forces balance, so the net force is due to the difference in the east-west forces.
7 kN east + 5 kN west = 7 kN east − 5 kN east = $\boxed{2 \text{ kN east}}$

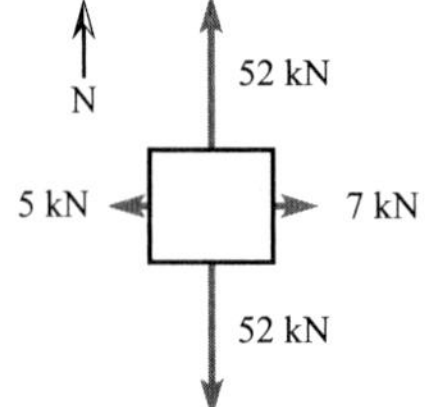

41. Strategy Use Newton's laws of motion. Let the y-direction be perpendicular to the canal and the $+x$-direction be parallel to the center line in the direction of motion.

Solution Find the sum of the two forces on the barge.

$\sum F_y = T\sin 15° - T\sin 15° = 0$ and

$\sum F_x = T\cos 15° + T\cos 15° = 2T\cos 15° = 2(560\text{ N})\cos 15° = 1.1\text{ kN}.$

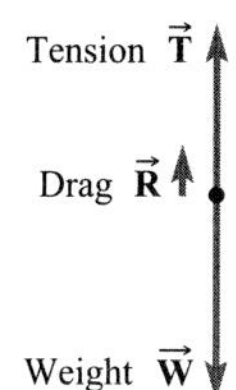

So, $\vec{\mathbf{F}} = \boxed{1.1\text{ kN forward (along the center line)}}$.

$\boxed{\text{No; there are other forces acting on the barge, which are not included.}}$

45. Strategy Consider forces acting on the fish suspended by the line.

Solution

$\boxed{\begin{array}{l}\text{One force acting on the fish is an upward force on the fish by the line; its interaction partner is a}\\ \text{downward force on the line by the fish. A second force acting on the fish is the downward gravitational}\\ \text{force on the fish; its interaction partner is the upward gravitational force on the Earth by the fish.}\end{array}}$

49. (a) Strategy Identify each force acting on the skydiver.

Solution

$\boxed{\begin{array}{l}\text{Gravitational force exerted on the skydiver by Earth; drag exerted on the}\\ \text{skydiver by the air; tension exerted on the skydiver by the parachute.}\end{array}}$

(b) Strategy Draw an FBD using the force information in part (a).

Solution The FBD for the forces exerted on the skydiver is shown at right.

Tension $\vec{\mathbf{T}}$

Drag $\vec{\mathbf{R}}$

Weight $\vec{\mathbf{W}}$

(c) Strategy Determine the magnitude of the force exerted by the air using the force of the parachute and the weight of the skydiver.

Solution Both the upward tension force exerted by the parachute and the upward drag force exerted by the air act to oppose the downward force due to gravity exerted on the skydiver (the weight). Since the skydiver is falling at constant speed, the net force on the skydiver is zero. Thus, the sum of the magnitudes of the upward forces must be equal to that of the skydiver's weight. So, $F_{\text{air}} + 620\text{ N} = 650\text{ N}$, or $F_{\text{air}} = \boxed{30\text{ N}}$.

(d) Strategy Use Newton's laws to identify the interaction partners of each force acting on the skydiver.

Solution

$\boxed{\begin{array}{l}\text{Gravitational force exerted on Earth by the skydiver, 650 N upward; drag exerted on the air by the}\\ \text{skydiver, 30 N downward; tension exerted on the parachute by the skydiver, 620 N downward.}\end{array}}$

53. Strategy Analyze the forces due to and on the three interacting objects: the woman, the chair, and the floor.

Solution

(a) The weight of the woman is directed downward. The forces on the woman due to the seat and armrests are directed upward and total $25\text{ N} + 25\text{ N} + 500\text{ N} = 550\text{ N}$. The chair and floor must support her entire weight, so the balance of her weight to support is $600\text{ N} - 2(25\text{ N}) - 500\text{ N} = 50\text{ N}$. Thus, the floor exerts a force on the woman's feet of $\boxed{50.0\text{ N upward}}$.

(b) The force exerted by the floor on the chair must be equal to the weight of the chair plus the weight of the woman supported by the chair, or $600.0\text{ N} + 100.0\text{ N} - 50.0\text{ N} = 650.0\text{ N}$. Thus, the floor exerts a force on the chair of $\boxed{650.0\text{ N upward}}$.

(c) The two forces acting on the woman and chair system are the upward force due to the floor and the downward gravitational force due to the Earth. Let the subscripts be the following: s = woman and chair system, e = Earth, f = floor.

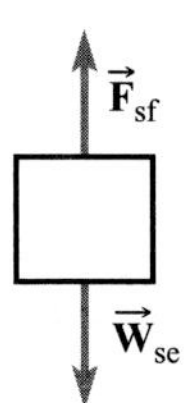

55. Strategy Use the conversion factor for pounds to newtons, $0.2248\text{ lb} = 1\text{ N}$.

Solution

(a) Find the weight of the girl in newtons.
$$W = mg = (40.0\text{ kg})(9.80\text{ N/kg}) = \boxed{392\text{ N}}$$

(b) Find the weight of the girl in pounds.
$$(392\text{ N})(0.2248\text{ lb/N}) = \boxed{88.1\text{ lb}}$$

57. Strategy Use Newton's universal law of gravitation. The Voyager spacecraft are approximately 17 billion kilometers from the Sun. Use this value for the distance between the Earth and the spacecraft.

Solution Find the approximate magnitude of the gravitational force between Earth and the spacecraft.
$$F = \frac{Gm_1m_2}{r^2} = \frac{(6.674\times10^{-11}\text{ N}\cdot\text{m}^2/\text{kg}^2)(825\text{ kg})(5.974\times10^{24}\text{ kg})}{(17\times10^{12}\text{ m})^2} = \boxed{1.1\times10^{-9}\text{ N}}$$

61. Strategy Gravitational field strength is given by $g = GM/R^2$. Find $H = R_2 - R_1$, where R_1 and R_2 are the distances from the center of the Earth to the surface of the Earth and the location of the balloon, respectively.

Solution Find the height above sea level of the balloon, H.
$$H = R_2 - R_1 = \sqrt{\frac{GM}{g_2}} - \sqrt{\frac{GM}{g_1}} = \sqrt{GM}\left(g_2^{-1/2} - g_1^{-1/2}\right)$$
$$= \sqrt{(6.674\times10^{-11}\text{ N}\cdot\text{m}^2/\text{kg}^2)(5.974\times10^{24}\text{ kg})}\left[(9.792\text{ N/kg})^{-1/2} - (9.803\text{ N/kg})^{-1/2}\right] = \boxed{4\text{ km}}$$

65. Strategy Use Newton's universal law of gravitation.

Solution Find the ratio.
$$\frac{F_1}{F_2} = \frac{Gm_1m_2}{r_1^2} \div \frac{Gm_1m_2}{r_2^2} = \left(\frac{r_2}{r_1}\right)^2 = \left(\frac{r_1 + 6.00\times10^6\text{ m}}{r_1}\right)^2 = \left(\frac{6.371\times10^6\text{ m} + 6.00\times10^6\text{ m}}{6.371\times10^6\text{ m}}\right)^2 = \boxed{3.770}$$

69. Strategy Draw free-body diagrams for each situation. Let the subscripts be the following:
b = book   t = table   e = Earth   h = hand

Solution The diagrams are shown.

(a) 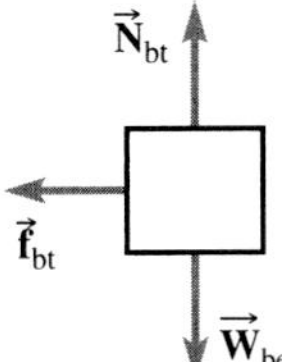   **(b)** 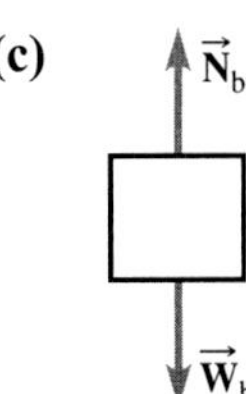   **(c)**

(d) **Strategy and Solution** In cases (a) and (b), the book is accelerating; so in these cases, the net force is not zero.

(e) **Strategy and Solution** The normal force on the book is equal to its weight, $(0.50 \text{ kg})(9.80 \text{ m/s}^2) = 4.9 \text{ N}$. The net force acting on the book in part (b) is equal to the force of kinetic friction. The force of kinetic friction is opposite the direction of motion. The magnitude is $\mu_k N = 0.40(4.9 \text{ N}) = 2.0 \text{ N}$. Thus, the net force on the book is $\boxed{2.0 \text{ N opposite the direction of motion}}$.

(f) **Strategy and Solution** The free-body diagram would look $\boxed{\text{just like the diagram for part (c) and the book}}$ $\boxed{\text{would not slow down because there is no net force on the book}}$ (friction is zero).

73. Strategy The crate is sliding down the ramp. The force of kinetic friction must oppose the motion of the crate parallel to the ramp. The normal force must oppose the force of gravity perpendicular to the ramp.

Solution Find the magnitude of the normal force and the magnitude and direction of the force of kinetic friction.

$$\Sigma F_y = N - mg\cos\theta = 0, \text{ so}$$

$$N = mg\cos\theta = (18.0 \text{ kg})(9.80 \text{ N/kg})\cos 30° = \boxed{150 \text{ N}}.$$

$$\Sigma F_x = f_k = \mu_k N = \mu_k mg\cos\theta, \text{ so}$$

$$f_k = 0.40(18.0 \text{ kg})(9.80 \text{ N/kg})\cos 30° = 61 \text{ N}.$$

The force of kinetic friction is $\boxed{61 \text{ N up the ramp}}$.

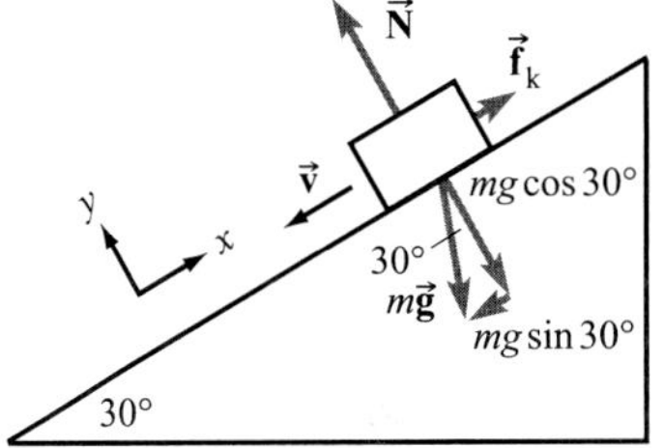

75. Strategy The crate is moving at constant speed up the ramp on the conveyor belt (without sliding), so the net force on it is zero. The force of static friction must oppose the force of gravity parallel to the ramp (and conveyor belt). The normal force must oppose the force of gravity perpendicular to the ramp (and conveyor belt).

Solution Find the magnitude of the normal force and the magnitude and direction of the force of static friction.

$$\Sigma F_y = N - mg\cos\theta = 0, \text{ so}$$

$$N = mg\cos\theta = (18.0 \text{ kg})(9.80 \text{ N/kg})\cos 30° = \boxed{150 \text{ N}}.$$

$$\Sigma F_x = f_s - mg\sin\theta = 0, \text{ so}$$

$$f_s = mg\sin\theta = (18.0 \text{ kg})(9.80 \text{ N/kg})\sin 30° = 88 \text{ N}.$$

The force of static friction is $\boxed{88 \text{ N up the ramp}}$.

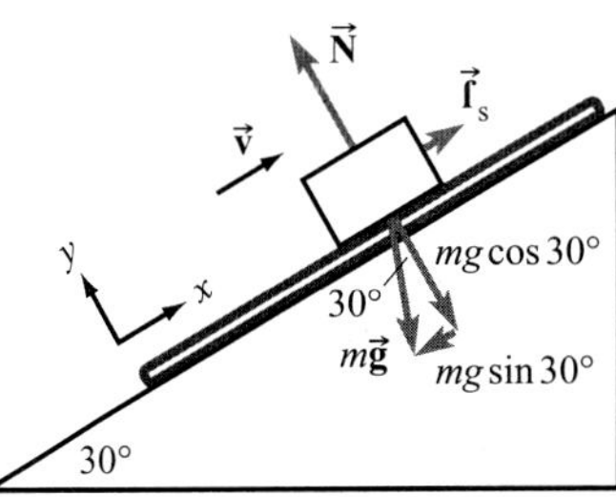

77. (a) Strategy To just get the block to move, the force must be equal to the maximum force of static friction.

Solution Solve for μ_s.

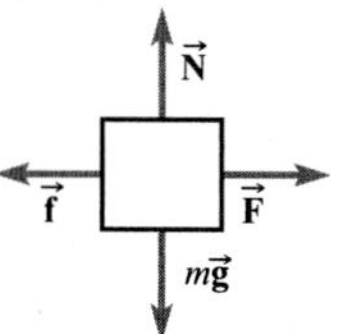

$$F = f_{\max} = \mu_s N = \mu_s mg, \text{ so } \mu_s = \frac{F}{mg} = \frac{12.0\ \text{N}}{(3.0\ \text{kg})(9.80\ \text{N/kg})} = \boxed{0.41}.$$

(b) Strategy The maximum static frictional force is now proportional to the total mass of the two blocks. The free-body diagram is the same as before, except that the mass m is now the sum of the masses of both blocks.

Solution Find the magnitude F of the force required to make the two blocks start to move.

$$F = \mu_s mg = 0.41(3.0\ \text{kg} + 7.0\ \text{kg})(9.80\ \text{N/kg}) = \boxed{40\ \text{N}}$$

79. Strategy Since the block moves with constant speed, there is no net force on the block. Draw the free-body diagram using this information. Let the subscripts be the following:
b = block　　　　B = Brenda　　　　w = wall　　　　e = Earth

Solution Find the coefficient of kinetic friction between the wall and the block.

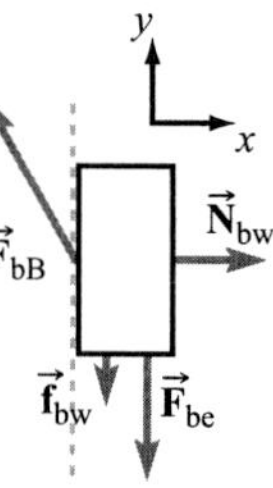

$$\Sigma F_x = N_{\text{bw}} - F_{\text{bB}} \sin\theta = 0, \text{ so } N_{\text{bw}} = F_{\text{bB}} \sin\theta.$$
$$\Sigma F_y = F_{\text{bB}} \cos\theta - F_{\text{be}} - f_{\text{bw}} = 0, \text{ so } f_{\text{bw}} = F_{\text{bB}} \cos\theta - F_{\text{be}}.$$

Since $f_{\text{bw}} = \mu_k N_{\text{bw}}$, we have

$$\mu_k = \frac{F_{\text{bB}} \cos\theta - F_{\text{be}}}{N_{\text{bw}}} = \frac{F_{\text{bB}} \cos\theta - F_{\text{be}}}{F_{\text{bB}} \sin\theta} = \cot\theta - \frac{F_{\text{be}}}{F_{\text{bB}}} \csc\theta$$

$$= \cot 30.0^\circ - \frac{2.0\ \text{N}}{3.0\ \text{N}} \csc 30.0^\circ = \boxed{0.4}$$

81. (a) Strategy and Solution Since the apples neither slide nor roll as they move up the incline, the apples are not moving relative to the belt, so the belt exerts forces of static friction on the apples.

(b) Strategy and Solution The apples are in equilibrium; therefore, the force of static friction is equal to 0.40 N, and we cannot conclude anything about the coefficient of kinetic friction. Find the coefficient of static friction.

$$f_s \le \mu_s N, \text{ so } \mu_s \ge \frac{f_s}{N} = \frac{0.40\ \text{N}}{1.0\ \text{N}} = \boxed{0.40}.$$

85. (a) Strategy Draw a diagram and use Newton's laws of motion.

Solution According to the first law, since the skier is moving with constant velocity, the net force on the skier is zero. Calculate the force of kinetic friction.

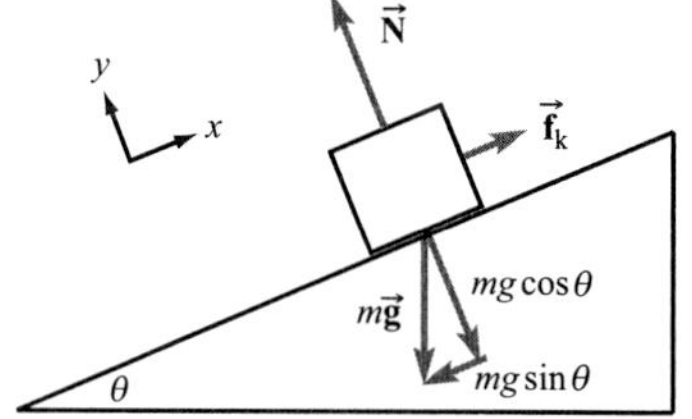

$$\Sigma F_x = f_k - mg \sin\theta = 0, \text{ so}$$
$$f_k = mg \sin\theta = (85\ \text{kg})(9.80\ \text{N/kg}) \sin 11^\circ = 160\ \text{N}.$$

The force of kinetic friction is $\boxed{160\ \text{N up the slope}}$.

(b) Strategy Use the diagram and results from part (a).

Solution Find the normal force.

$$\Sigma F_y = N - mg \cos\theta = 0, \text{ so } N = mg \cos\theta. \text{ Since } f_k = \mu_k N,$$

$$\mu_k = \frac{f_k}{N} = \frac{mg \sin\theta}{mg \cos\theta} = \tan\theta = \tan 11^\circ = \boxed{0.19}.$$

87. Strategy Recall that the tension in the rope is the same along its length.

Solution The tension is equal to the weight at the end of the rope, 120 N. Therefore, scale A reads 120 N.

There are two forces pulling downward on the pulley due to the tension of 120 N in each part of the rope. Therefore, $T_B = T_A + T_A = 2T_A = 240$ N. Scale B reads 240 N, since it supports the pulley.

89. Strategy Use Newton's laws of motion.

Solution The Earth exerts a force on the mass, which then exerts a force on the scale, which then exerts a force on the hook, which then exerts a force on the ceiling. All these forces are equal (assuming that the masses of the spring scale and hook are negligible). In addition, each body, on which a force is exerted, exerts an equal and opposite force on the other object. So, the ceiling exerts a force on the hook, the hook on the scale, etc. One person replaces the force on the mass due to the Earth, and the other person replaces the force on the scale due to the hook. So, each person must exert a force of 98 N.

93. Strategy Use Newton's laws of motion. Let $+y$ be down and $+x$ to the right.

Solution Find the force $\vec{F}$ applied to the front tooth.
$\sum F_x = T \sin\theta - T \sin\theta = 0$ and $\sum F_y = T \cos\theta + T \cos\theta - F = 0$. So, we have

$F = 2T \cos\theta = 2(12\text{ N})\cos 37.5° = 19$ N. By symmetry, the force is directed toward the back of the mouth, so

$\vec{F} = \boxed{19\text{ N toward the back of the mouth}}$.

97. Strategy Use Newton's laws of motion. Draw a free-body diagram.

Solution

(a) Find the tension in the rope from which the pulley hangs.
$\sum F_y = T_1 \sin\theta - Mg = 0$ and $\sum F_x = T_1 \cos\theta - T_2 = 0$.

The tension in T_2 is due to the mass M, so $T_2 = Mg$.

Thus, $T_1 \cos\theta = Mg$ and $T_1 \sin\theta = Mg$.

According to these equations, $\cos\theta = \sin\theta$, which is true only if

$\theta = 45°$ for $0° \le \theta \le 90°$.

Therefore, $T_1 = Mg/\cos 45° = \boxed{\sqrt{2}\,Mg}$.

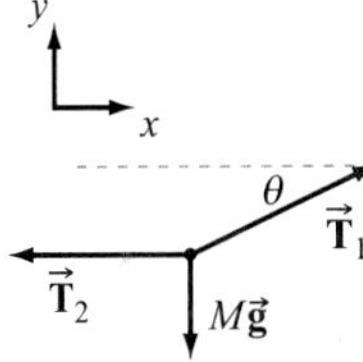

(b) As found in part (a), $\theta = \boxed{45°}$.

101. Strategy Consider the ranges and natures of the fundamental forces.

Solution Of all of the fundamental forces, the weak force has the shortest range (about 10^{-17} m). In the Sun, the weak interaction enables thermonuclear reactions to occur, without which there would be no sunlight.

105. Strategy Draw a diagram and use Newton's laws of motion.

Solution

(a) The magnitude of the force of static friction on block A due to the floor must be equal to the magnitude of the tension in the cord, so $T = f_{sA} = \mu_A N = \mu_A mg$.

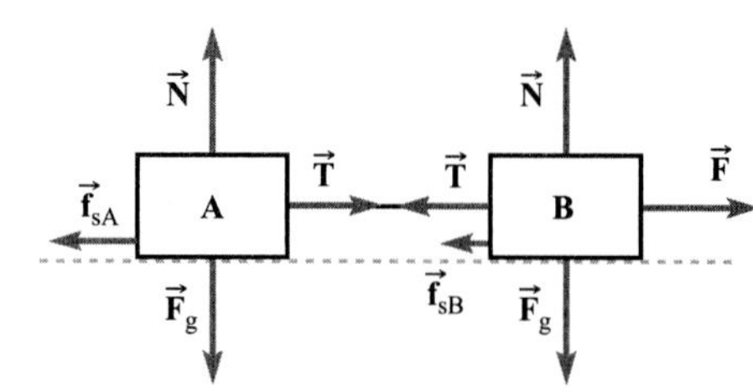

The magnitude of the applied force must be equal to the magnitude of the tension in the cord plus the magnitude of the force of static friction on block B due to the floor. Thus,

$$F = T + f_{sB} = \mu_A mg + \mu_B mg = mg(\mu_A + \mu_B)$$
$$= (2.0 \text{ kg})(9.80 \text{ N/kg})(0.45 + 0.30) = \boxed{15 \text{ N}}.$$

(b) $T = \mu_A mg = 0.45(2.0 \text{ kg})(9.80 \text{ N/kg}) = \boxed{8.8 \text{ N}}$

109. Strategy Let the $+y$-direction be in the direction of the 360.0-N force (F). Draw a diagram and use Newton's laws of motion.

Solution Find the force exerted on the poplar tree.
$\sum F_y = F - 2T \sin \theta = 0$ before the poplar is cut through.

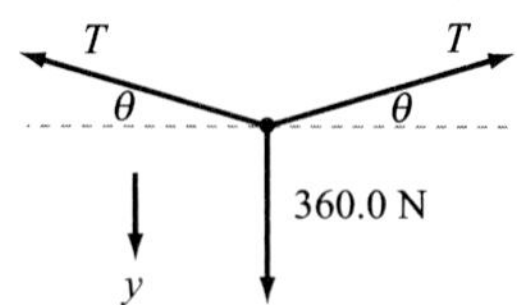

So, $F = 2T \sin \theta$. The force exerted on the poplar is the tension T, so $T = \dfrac{F}{2 \sin \theta}$.

Refer to the figure in the text to find $\sin \theta$.

$$\sin \theta = \frac{\text{displacement}}{\text{half length of rope}} = \frac{2.00 \text{ m}}{\sqrt{(20.0 \text{ m})^2 + (2.00 \text{ m})^2}} = 0.0995$$

Thus, $T = \dfrac{360.0 \text{ N}}{2(0.0995)} = \boxed{1810 \text{ N}}$. Compare the forces.

$$\frac{1810 \text{ N}}{360.0 \text{ N}} \approx \boxed{5 \text{ times the force with which Yoojin pulls}}$$

The values for the two situations are different because $\boxed{\text{the oak tree supplies additional force}}$.

113. Strategy Use Newton's laws of motion.

Solution

(a) Since the airplane is cruising in a horizontal level flight (straight line) at constant velocity, it is in equilibrium and the net force is $\boxed{\text{zero}}$.

(b) The air pushes upward with a force equal to the weight of the airplane: $\boxed{2.6 \times 10^4 \text{ N}}$.

117. Strategy Use Newton's laws of motion.

Solution

(a) Since the train is moving at constant speed, and air resistance and friction are negligible, the readings on the three scales are $\boxed{\text{all 0}}$.

(b) Air resistance and friction are not considered negligible this time. The engine pulls the cars against these forces. Since the cars are identical, each car contributes about one-third of the total frictional and drag forces. Each spring scale will measure the net force due to the cars behind it, so the relative readings on the three spring scales are $\boxed{A > B > C}$. The free-body diagram is shown.

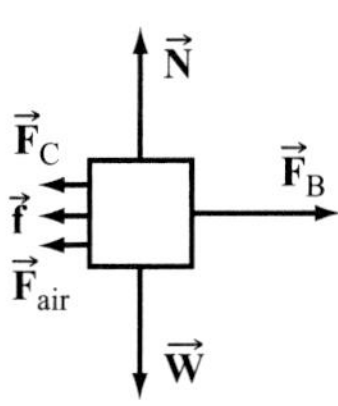

(c) Spring A measures the forces on all 3 cars. Spring B measures the forces on the latter 2 cars. Spring C measures the forces on the final 1 car.
$A = 5.5\text{ N} + 5.5\text{ N} + 5.5\text{ N} = \boxed{16.5\text{ N}}$; $B = 5.5\text{ N} + 5.5\text{ N} = \boxed{11.0\text{ N}}$; $C = \boxed{5.5\text{ N}}$

121. Strategy Let the subscripts be the following:
$i = $ ice $e = $ Earth $s = $ stone $o = $ opponent's stone

Solution

(a) The only forces on the stone are gravity due to the Earth and the normal force due to the ice.

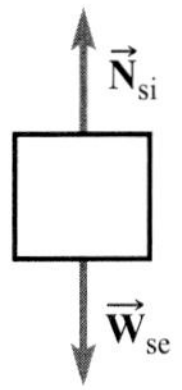

(b) As the stone slides down the rink, it experiences a force of kinetic friction opposite to its motion.

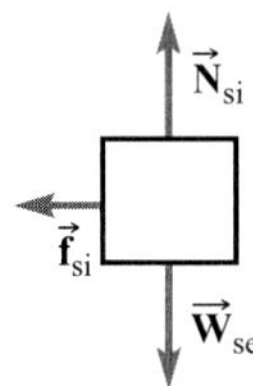

(c) The additional force is that due to the opponent's stone.

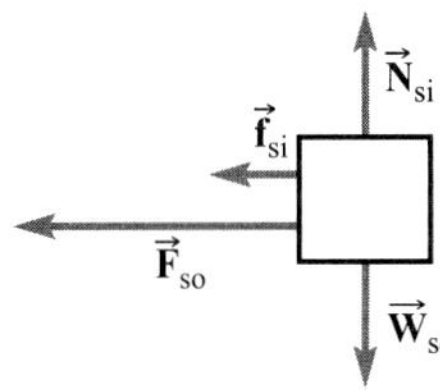

123. (a) Strategy Let the subscripts be the following:
$c = $ computer $d = $ desk $e = $ Earth

Solution The only forces on the computer are gravity due to the Earth and the normal force due to the desk. The free-body diagram is shown.

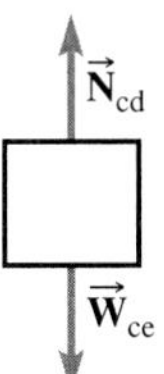

(b) Strategy Consider the nature of friction forces.

Solution Since the only forces acting on the computer are in the vertical direction, the friction force is $\boxed{\text{zero}}$.

(c) Strategy Find the maximum force of static friction on the computer due to the desk; this is the horizontal force necessary to make it begin to slide.

Solution $F = f_\text{s} = \mu_\text{s} N = \mu_\text{s} W = 0.60(87\text{ N}) = \boxed{52\text{ N}}$

125. (a) Strategy Scale A measures the weight of both masses. Scale B only measures the weight of the 4.0-kg mass.

Solution Find the readings of the two scales if the masses of the scales are negligible.

Scale A $= (10.0 \text{ kg} + 4.0 \text{ kg})(9.80 \text{ N/kg}) = \boxed{137 \text{ N}}$ and Scale B $= (4.0 \text{ kg})(9.80 \text{ N/kg}) = \boxed{39 \text{ N}}$.

(b) Strategy Scale A measures the weight of both masses and scale B. Scale B only measures the weight of the 4.0-kg mass.

Solution Find the readings if each scale has a mass of 1.0 kg.

Scale A $= (10.0 \text{ kg} + 4.0 \text{ kg} + 1.0 \text{ kg})(9.80 \text{ N/kg}) = \boxed{147 \text{ N}}$ and Scale B $= \boxed{39 \text{ N}}$.

129. Strategy Use Newton's laws of motion and draw a free-body diagram.

Solution Find the tension in the cable.

$$\Sigma F_y = T \cos\theta - mg = 0, \text{ so } T = \boxed{\dfrac{mg}{\cos\theta}}.$$

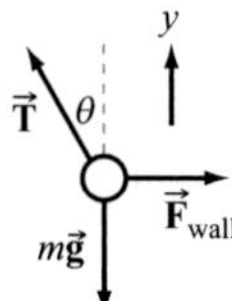

131. Strategy Set the magnitudes of the forces on the spaceship due to the Earth and the Moon equal. (The forces are along the same line.)

Solution Find the distance from the Earth expressed as a percentage of the distance between the centers of the Earth and the Moon.

$$F_{sE} = \frac{GM_E m}{r_E^2} = F_{sM} = \frac{GM_M m}{r_M^2}, \text{ so } r_E = r_M \sqrt{\frac{M_E}{0.0123 M_E}} = 9.02 r_M.$$

Find the percentage.

$$\frac{r_E}{r_E + r_M} = \frac{9.02 r_M}{9.02 r_M + r_M} = \frac{9.02}{10.02} = 0.900$$

The distance from the Earth is $\boxed{90.0\% \text{ of the Earth-Moon distance}}$.

133. (a) Strategy Set the magnitudes of the forces on the spacecraft due to the Earth and the Sun equal.

Solution Find the distance of the spacecraft from Earth.

$$F_{sS} = \frac{GM_S m}{r_S^2} = F_{sE} = \frac{GM_E m}{r_E^2}, \text{ so } \frac{r_E}{r_S} = \sqrt{\frac{M_E}{M_S}}.$$

This is the ratio of the Earth-spacecraft distance to the Sun-spacecraft distance. If this is multiplied by the Earth-Sun mean distance, the product is the distance of the spacecraft from the Earth.

$$(1.50 \times 10^{11} \text{ m}) \sqrt{\frac{5.974 \times 10^{24} \text{ kg}}{1.987 \times 10^{30} \text{ kg}}} = \boxed{2.60 \times 10^8 \text{ m from Earth}}$$

(b) Strategy Imagine the spacecraft is a small distance d closer to the Earth and find out which gravitational force is stronger, the Earth's or the Sun's.

Solution At the equilibrium point the net gravitational force is zero. If the spacecraft is closer to the Earth than the equilibrium point distance from the Earth, then the force due to the Earth is greater than that due to the Sun. If the spacecraft is closer to the Sun than the equilibrium point distance from the Sun, then the force due to the Sun is greater than that due to the Earth. So, if the spacecraft is close to, but not at, the equilibrium point, the net force tends to pull it $\boxed{\text{away from}}$ the equilibrium point.

Chapter 3

ACCELERATION AND NEWTON'S SECOND LAW OF MOTION

Conceptual Questions

1. Distance traveled is a scalar quantity equal to the total length of the path taken in moving from one point to another. Displacement is a vector quantity directed from the initial point towards the final point with a magnitude equal to the straight line distance between the two points. The magnitude of the displacement is always less than or equal to the total distance traveled.

5. The area under the curve of an a_x versus time graph is equal to the change in the x-component of the velocity.

9. The average speed and the magnitude of the average velocity of an object are equal if and only if the object travels along a straight line path without changing direction. In all other cases, the average speed is greater than the magnitude of the average velocity because the total distance traveled must be greater than the straight-line distance between the starting and ending points.

13. Newton's second law states that the vector sum of all external forces acting on an object is equal to the product of the mass of the object and its acceleration. From this law, we conclude that objects connected to opposite ends of a negligibly massive cord must exert a negligible net force—the second law would otherwise predict an infinite acceleration. The forces acting on each side of the cord must therefore be equal in magnitude and opposite in direction—and from Newton's third law, the forces exerted on the objects by the cord must also be equal in magnitude.

17. **(a)** $a_x > 0$ and $v_x < 0$ means you are moving south and slowing down.

    (b) $a_x = 0$ and $v_x < 0$ means you are moving south at a constant speed.

    (c) $a_x < 0$ and $v_x = 0$ means you are momentarily at rest but speeding up in a southward direction.

    (d) $a_x < 0$ and $v_x < 0$ means you are moving south and speeding up.

    (e) As can be seen from our answers above, it is not a good idea to use the term "negative acceleration" to mean slowing down. In parts (c) and (d), the acceleration is negative, but the bicycle is speeding up. Also, in part (a), the acceleration is positive, but the bicycle is slowing down.

21. The following are sample examples:

    (a) pushing a crate up a ramp with increasing speed

    (b) a tennis ball as it hits the wall

    (c) a rocket coasting in empty space

    (d) an orbiting satellite

Problems

1. **Strategy** Use the fact that $|\vec{A}| = |\vec{B}|$ and symmetry to determine the direction of $\vec{D}$.

   Solution The vertical components cancel when $\vec{B}$ is subtracted from $\vec{A}$. The direction of the horizontal component of $\vec{B}$ is reversed due to the subtraction, and so the vector resulting from the subtraction is in the direction of the horizontal component of $\vec{A}$; that is, to the left. The horizontal components of each vector have the same magnitude, which is about 3.9 cm; so the magnitude of $\vec{D}$ is $\boxed{\text{about 7.9 cm}}$. The sketch is shown:

   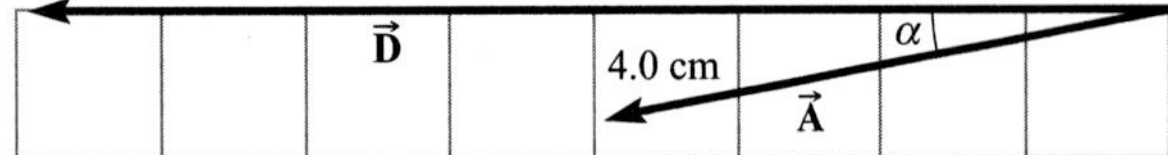

5. **Strategy** Add the displacement from Jerry's dorm to the fitness center to the displacement from Cindy's apartment to Jerry's dorm to find the total displacement from Cindy's apartment to the fitness center.

   Solution Add the displacements.
   1.50 mi east + 2.00 mi north + 3.00 mi east = 4.50 mi east + 2.00 mi north
   Let north be along the $+y$-axis and east be along the $+x$-axis. Then, the components of the total displacement are $\Delta x = 4.50$ mi and $\Delta y = 2.00$ mi.
   Find the magnitude.

   $$\Delta r = \sqrt{(\Delta x)^2 + (\Delta y)^2} = \sqrt{(4.50\text{ mi})^2 + (2.00\text{ mi})^2} = \boxed{4.92\text{ mi}}$$

   Find the direction.

   $$\theta = \tan^{-1}\frac{2.00}{4.50} = \boxed{24.0^\circ \text{ north of east}}$$

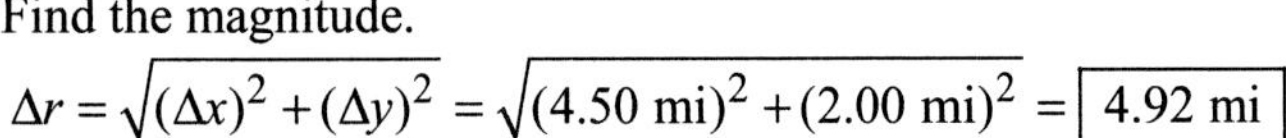

9. **Strategy** Let east be the $+x$-direction and north be the $+y$-direction.

   Solution Compute the direction of travel and the distance walked.
   $$\Delta r_{1x} = 1.2\text{ km}$$
   $$\Delta r_{1y} = 0$$
   $$\Delta r_{2x} = (2.7\text{ km})\cos 135^\circ = -1.9\text{ km}$$
   $$\Delta r_{2y} = (2.7\text{ km})\sin 135^\circ = 1.9\text{ km}$$
   $$\Delta r_x = \Delta r_{1x} + \Delta r_{2x} = 1.2\text{ km} - 1.9\text{ km} = -0.7\text{ km}$$
   $$\Delta r_y = \Delta r_{1y} + \Delta r_{2y} = 1.9\text{ km}$$
   $$|\Delta\vec{r}| = \sqrt{(\Delta r_x)^2 + (\Delta r_y)^2} = \sqrt{(-0.7\text{ km})^2 + (1.9\text{ km})^2} = 2.0\text{ km}$$
   The direction of the return trip is opposite the displacement vector found.
   $\Delta r_x' = 0.7$ km and $\Delta r_y' = -1.9$ km.
   $$\theta = \tan^{-1}\frac{-1.9}{0.7} = 20^\circ \text{ east of south}$$
   So, they must travel $\boxed{2.0\text{ km at }20^\circ \text{ east of south}}$.

13. Strategy Use the definition of average velocity.

Solution Find the average velocity of the cyclist in meters per second.

$$\vec{v}_{av} = \frac{\Delta \vec{r}}{\Delta t} = \frac{10.0 \text{ km east}}{11 \text{ min } 40 \text{ s}} = \frac{10.0 \times 10^3 \text{ m east}}{700 \text{ s}} = \boxed{14.3 \text{ m/s east}}$$

15. Strategy Use the definition of average speed.

Solution Find the time it took the ball to get to home plate.

$$v_{av} = \frac{\Delta r}{\Delta t}, \text{ so } \Delta t = \frac{\Delta r}{v_{av}} = \frac{18.4 \text{ m}}{45.1 \text{ m/s}} = \boxed{0.408 \text{ s}}.$$

17. (a) Strategy From Problem 7, the distance Michaela traveled between Killarney to Cork via Mallow was 61 km.

Solution Find Michaela's average speed.

$$v = \frac{\text{distance traveled}}{\text{total time}} = \frac{61 \text{ km}}{48 \text{ min}} \times \frac{1000 \text{ m}}{1 \text{ km}} \times \frac{1 \text{ min}}{60 \text{ s}} = \boxed{21 \text{ m/s}}$$

(b) Strategy From Problem 7, the magnitude of Michaela's displacement was 45 km.

Solution Find the magnitude of Michaela's average velocity.

$$\left| \vec{v}_{av} \right| = \frac{\Delta r}{\Delta t} = \frac{45 \text{ km}}{48 \text{ min}} \times \frac{1000 \text{ m}}{1 \text{ km}} \times \frac{1 \text{ min}}{60 \text{ s}} = \boxed{16 \text{ m/s}}$$

21. (a) Strategy Find the distance traveled during the first part of the trip; then compute the speed required to travel the remaining distance in the time required.

Solution There is 48.0 min = 0.800 h left to complete the trip. The harpsichordist has traveled $(55.0 \text{ mi/h})(1.20 \text{ h}) = 66.0 \text{ mi}$, so he has $122 \text{ mi} - 66 \text{ mi} = 56 \text{ mi}$ to go. To get to the concert on time, he must travel at a speed of $(56 \text{ mi})/(0.800 \text{ h}) = \boxed{70 \text{ mi/h}}$.

(b) Strategy Use the definition of average velocity. Use components for the displacements.

Solution Let the $+x$-direction be west and the $+y$-direction be south. Then, $\Delta x = 66.0 \text{ mi} + (56 \text{ mi}) \cos 30.0° = 114.5 \text{ mi}$ and $\Delta y = (56 \text{ mi}) \sin 30.0° = 28 \text{ mi}$.

The magnitude of the average velocity is $v = \dfrac{\sqrt{(114.5 \text{ mi})^2 + (28 \text{ mi})^2}}{2.00 \text{ h}} = 59 \text{ mi/h}$.

The direction of the average velocity is $\theta = \tan^{-1} \dfrac{28}{114.5} = 14°$.

So, the average velocity for the entire trip was $\boxed{59 \text{ mi/h at } 14° \text{ south of west}}$.

23. Strategy Use the area under the curve to find the displacement of the car.

Solution The displacement of the car is given by the area under the v_x vs. t curve. Under the curve, there are 16 squares and each square represents $(5 \text{ m/s})(2 \text{ s}) = 10 \text{ m}$. Therefore, the car moves $16(10 \text{ m}) = \boxed{160 \text{ m}}$. Counting squares at 2-s intervals gives the motion diagram.

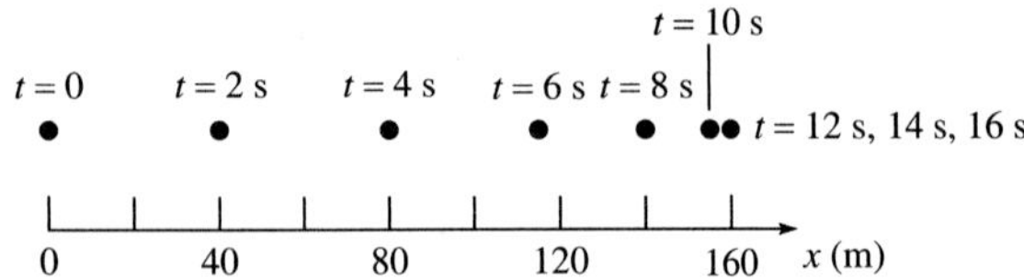

Using the motion diagram for position gives the graph of $x(t)$.

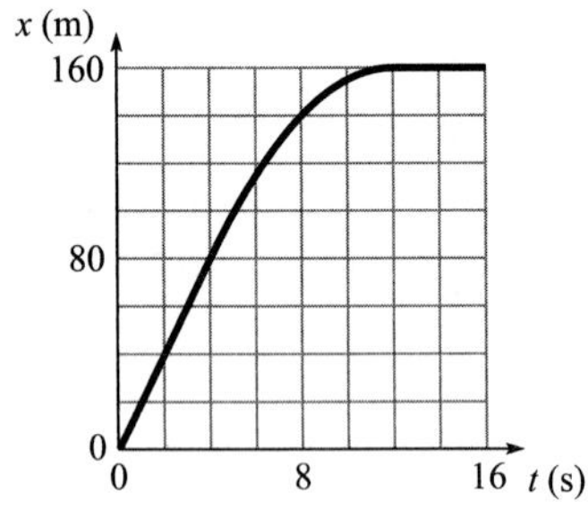

25. Strategy Use the definition of average speed.

Solution Find the average speeds of the skater.

(a) $v_{\text{av},x} = \dfrac{\Delta x}{\Delta t} = \dfrac{6.0 \text{ m} - 0}{4.0 \text{ s}} = \boxed{1.5 \text{ m/s}}$

(b) $v_{\text{av},x} = \dfrac{6.0 \text{ m} - 0}{5.0 \text{ s}} = \boxed{1.2 \text{ m/s}}$

29. Strategy Determine the maximum time allowed to complete the run.

Solution Massimo must take no more than $1000 \text{ m} \div 4.0 \text{ m/s} = 250 \text{ s}$ to complete the run. Since he ran the first 900 m is 250 s, $\boxed{\text{he cannot pass the test because he would have to run the last 100 m in 0 s}}$.

31. (a) Strategy Find the average speed by dividing the total distance traveled by the total time.

Solution Since the time traveled at each speed is the same, we can simply add the speeds and divide by 2.0.

$$v_{\text{av}} = \frac{96 \text{ km/h} + 128 \text{ km/h}}{2.0} = \frac{224 \text{ km/h}}{2.0} = \boxed{110 \text{ km/h}}$$

(b) Strategy Use the definition of average velocity. Draw a diagram.

Solution Let east be the $+x$-direction and north be the $+y$-direction.

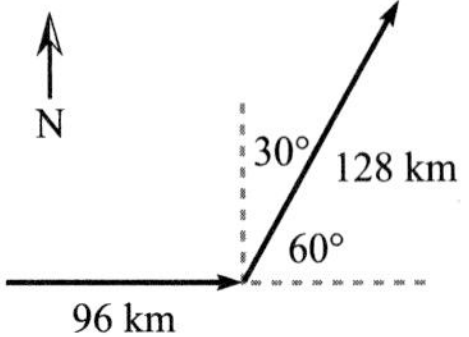

Find the magnitude of $\Delta \vec{\mathbf{r}}$.

$$|\Delta \vec{r}| = \sqrt{[96 \ \text{km/h} + (128 \ \text{km/h})\cos 60^\circ]^2 (1.0 \ \text{h})^2 + [(128 \ \text{km/h})\sin 60^\circ]^2 (1.0 \ \text{h})^2} = 195 \ \text{km}$$

Find the direction of $\Delta \vec{r}$.

$$\theta = \tan^{-1} \frac{110.85 \ \text{km}}{160 \ \text{km}} = 35^\circ \ \text{north of east}$$

So, $|\vec{v}_{av}| = \dfrac{\Delta r}{\Delta t} = \dfrac{194.65 \ \text{km}}{2.0 \ \text{h}} = 97 \ \text{km/h}$ and $\vec{v}_{av} = \dfrac{\Delta \vec{r}}{\Delta t} = \boxed{97 \ \text{km/h at } 35^\circ \text{ north of east}}$.

33. Strategy Use vector subtraction to find the change in velocity.

Solution Find the change in velocity of the scooter.

$$\Delta \vec{v} = \vec{v}_f - \vec{v}_i = 15 \ \text{m/s west} - 12 \ \text{m/s east} = 15 \ \text{m/s west} - (-12 \ \text{m/s west}) = \boxed{27 \ \text{m/s west}}$$

37. (a) Strategy Use $\Delta v = a \Delta t$ and solve for Δt.

Solution

$$\Delta t = \frac{\Delta v}{a} = \frac{22 \ \text{m/s}}{1.7 \ \text{m/s}^2} = \boxed{13 \ \text{s}}$$

(b) Strategy The antelope's hooves push backward on the ground. The force of friction on the hooves opposes this force, propelling the antelope forward. Use Newton's second law.

Solution Find the antelope's mass.

$$f = ma, \ \text{so} \ m = \frac{f}{a} = \frac{78 \ \text{N}}{1.7 \ \text{m/s}^2} = \boxed{46 \ \text{kg}}.$$

41. Strategy The acceleration a_x is equal to the slope of the v_x versus t graph. The displacement is equal to the area under the curve.

Solution

(a) a_x is the slope of the graph at $t = 11$ s.

$$a_x = \frac{\Delta v_x}{\Delta t} = \frac{10.0 \ \text{m/s} - 30.0 \ \text{m/s}}{12.0 \ \text{s} - 10.0 \ \text{s}} = \boxed{-10 \ \text{m/s}^2}$$

(b) Since v_x is constant, $a_x = \boxed{0}$ at $t = 3$ s.

(c) Sketch the acceleration using the v_x versus t graph.

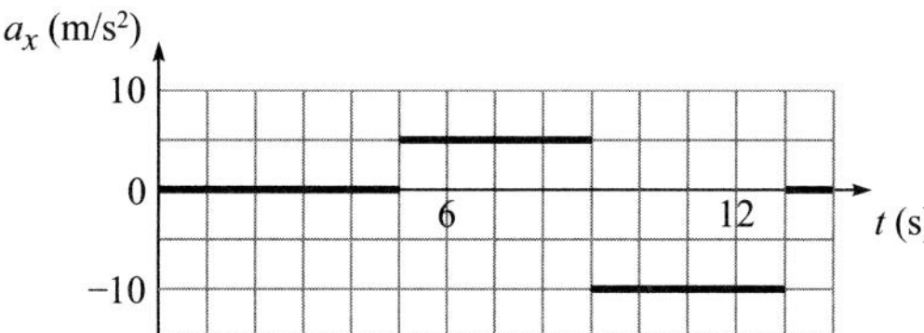

(d) The area under the v_x vs. t curve from $t = 12$ s to $t = 14$ s represents the displacement of the body. Each square represents $(10.0 \ \text{m/s})(1.0 \ \text{s}) = 1.0 \times 10^1$ m, and there is $1/2$ square under the curve for $t = 12$ s to $t = 14$ s, so the car travels $\boxed{5.0 \ \text{m}}$.

45. Strategy Use Newton's second law of motion.

Solution Find the average force on the airplane.

$$\Sigma F = ma = m\frac{\Delta v}{\Delta t} = (1100 \text{ kg})\frac{35 \text{ m/s}}{8.0 \text{ s}} = \boxed{4.8 \text{ kN}}$$

47. (a) Strategy Use the definition of average velocity. Draw a diagram.

Solution Let the center of the circle be the origin, then
$$\Delta\vec{r} = \vec{r}_f - \vec{r}_i = 20.0 \text{ m east} - 20.0 \text{ m south.}$$

$$\left|\Delta\vec{r}\right| = \sqrt{(20.0 \text{ m})^2 + (-20.0 \text{ m})^2} = 28.3 \text{ m}$$

Let east be the $+x$-direction and north the $+y$-direction.

$$\theta = \tan^{-1}\frac{20.0}{20.0} = 45.0° \text{ north of east}$$

So, $\left|\vec{v}_{av}\right| = \dfrac{\Delta r}{\Delta t} = \dfrac{28.3 \text{ m}}{3.0 \text{ s}} = 9.4 \text{ m/s}$ and $\vec{v}_{av} = \dfrac{\Delta\vec{r}}{\Delta t} = \boxed{9.4 \text{ m/s at } 45° \text{ north of east}}$.

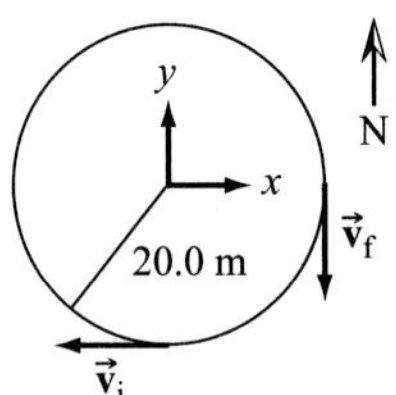

(b) Strategy Use the definition of average acceleration and the fact that $C = 2\pi r$.

Solution Find the average acceleration of the car.

$$\left|\vec{v}_f\right| = \left|\vec{v}_i\right| = \frac{2\pi r(3/4)}{\Delta t} = \frac{3\pi r}{2\Delta t}, \text{ so } \vec{a}_{av} = \frac{3\pi r}{2(\Delta t)^2}(\text{south} - \text{west}).$$

$$\left|\vec{a}_{av}\right| = \frac{3\pi r}{2(\Delta t)^2}\sqrt{1^2 + (-1)^2} = \frac{3\pi r}{\sqrt{2}(\Delta t)^2}$$

$$\theta = \tan^{-1}\frac{-1}{1} = 45° \text{ south of east}$$

$$\vec{a}_{av} = \frac{\Delta\vec{v}}{\Delta t} = \frac{3\pi(20.0 \text{ m})}{\sqrt{2}(3.0 \text{ s})^2} \text{ at } 45° \text{ south of east} = \boxed{15 \text{ m/s}^2 \text{ at } 45° \text{ south of east}}$$

(c) Strategy Consider Newton's first law of motion.

Solution Although the magnitude of the velocity is constant, its direction must change continuously for the car to travel in a circle; $\boxed{\text{changing the direction of the velocity requires an acceleration}}$.

49. (a) Strategy Let east be in the $+x$-direction and north be in the $+y$-direction.

Solution See the figure.

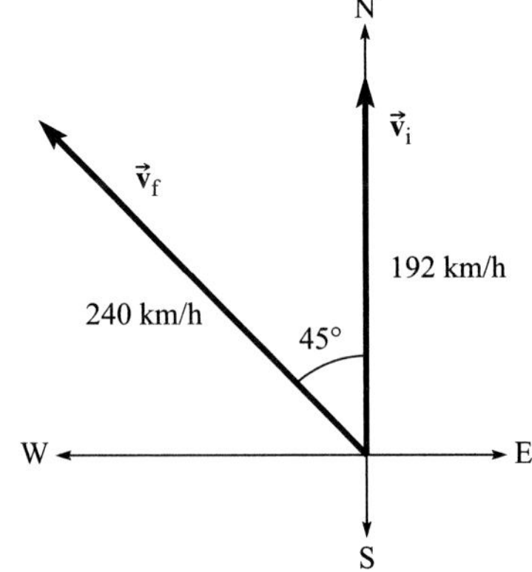

(b) Strategy Use the component method to subtract the initial velocity vector from the final velocity vector.

Solution

$\Delta \vec{v} = \vec{v}_f - \vec{v}_i = 240$ km/h NW $- 192$ km/h N, so

$$\left|\Delta \vec{v}\right| = \sqrt{\left[(240 \text{ km/h})\cos 135°\right]^2 + \left[-192 \text{ km/h} + (240 \text{ km/h})\sin 135°\right]^2} = 170 \text{ km/h}.$$

$$\theta = \tan^{-1} \frac{-192 \text{ km/h} + (240 \text{ km/h})\sin 135°}{(240 \text{ km/h})\cos 135°} = 7° \text{ south of west}$$

So, $\Delta \vec{v} = \boxed{170 \text{ km/h at } 7° \text{ south of west}}$.

(c) Strategy Use the definition of average acceleration.

Solution

$$\vec{a}_{av} = \frac{\Delta \vec{v}}{\Delta t} = \frac{170 \text{ km/h at } 7° \text{ south of west}}{3.0 \text{ h}} = \boxed{57 \text{ km/h}^2 \text{ at } 7° \text{ south of west}}$$

51. Strategy Since the particle is moving to the east and is accelerated to the south, its velocity in 8.00 s will be between east and south. Use the component method. Let north be in the $+y$-direction and east be in the $+x$-direction.

Solution

$v_x = 40.0$ m/s and $v_y = a_y \Delta t = (-2.50 \text{ m/s}^2)(8.00 \text{ s}) = -20.0$ m/s.

$$\left|\vec{v}\right| = \sqrt{v_x^2 + v_y^2} = \sqrt{(40.0 \text{ m/s})^2 + (-20.0 \text{ m/s})^2} = 44.7 \text{ m/s}$$

$$\theta = \tan^{-1} \frac{v_y}{v_x} = \tan^{-1} \frac{-20.0 \text{ m/s}}{40.0 \text{ m/s}} = 26.6° \text{ south of east}$$

So, $\vec{v} = \boxed{44.7 \text{ m/s at } 26.6° \text{ south of east}}$.

53. Strategy Use the definition of average acceleration and Newton's second law.

Solution Calculate the average acceleration.

$$\vec{a}_{av} = \frac{\Delta \vec{v}}{\Delta t} = \frac{50.2 \text{ m/s} - (-47.5 \text{ m/s}) \text{ away from the racquet}}{0.00360 \text{ s}} = 27,100 \text{ m/s}^2 \text{ away from the racquet}$$

Compute the average force on the ball due to the racquet.

$$\vec{F}_{av} = m\vec{a}_{av} = (0.0570 \text{ kg})(27,140 \text{ m/s}^2 \text{ away from the racquet}) = \boxed{1550 \text{ N away from the racquet}}$$

57. Strategy Use Newton's second law and $\Delta v = a\Delta t$.

Solution Compute the speeds at the indicated times.

(a) $F = ma$, so $v_a = at_1 = \frac{F}{m} t_1$; (b) $2F = 2ma$, so $v_b = at_1 = \frac{2F}{2m} t_1 = \frac{F}{m} t_1$;

(c) $F = ma$, so $v_c = a(2t_1) = \frac{F}{m}(2t_1) = 2\frac{F}{m} t_1$; (d) $2F = ma$, so $v_d = at_1 = \frac{2F}{m} t_1 = 2\frac{F}{m} t_1$;

(e) $F = 2ma$, so $v_e = a(2t_1) = \frac{F}{2m}(2t_1) = \frac{F}{m} t_1$

Ranking the objects according to their speeds, from largest to smallest, gives $\boxed{(c) = (d), (a) = (b) = (e)}$.

61. Strategy Use Newton's second law for the vertical direction.

Solution Draw a free-body diagram. Find the tension.
$\Sigma F_y = T - mg = ma_y$, so

$T = m(a_y + g) = (2010 \text{ kg})(1.50 \text{ m/s}^2 + 9.80 \text{ m/s}^2) = 22.7 \text{ kN}.$

The tension in the cable is $\boxed{22.7 \text{ kN upward}}$.

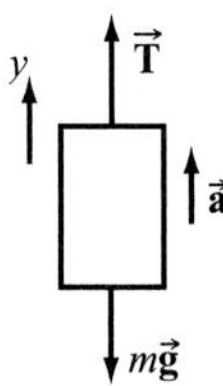

65. Strategy Use the expressions for a_y and T found in Example 3.11.

Solution

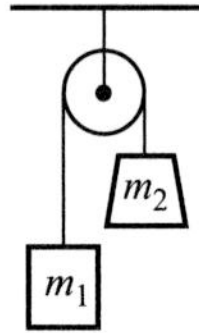

(a) $a_y = \dfrac{(m_2 - m_1)g}{m_2 + m_1} = \dfrac{(5.0 \text{ kg} - 3.0 \text{ kg})(9.80 \text{ m/s}^2)}{5.0 \text{ kg} + 3.0 \text{ kg}} = 2.5 \text{ m/s}^2$

Since $m_2 > m_1$, $\boxed{\vec{a}_1 = 2.5 \text{ m/s}^2 \text{ up and } \vec{a}_2 = 2.5 \text{ m/s}^2 \text{ down}}$.

(b) $T = \dfrac{2m_1 m_2}{m_1 + m_2} g = \dfrac{2(3.0 \text{ kg})(5.0 \text{ kg})}{3.0 \text{ kg} + 5.0 \text{ kg}}(9.80 \text{ m/s}^2) = \boxed{37 \text{ N}}$

69. Strategy Use Newton's second law to evaluate the situation.

Solution Since $F = ma$, the maximum acceleration is $a_{\text{max}} = \dfrac{F}{m} = \dfrac{T}{m_{\text{car}}} = \dfrac{2500 \text{ N}}{1400 \text{ kg}} = \boxed{1.8 \text{ m/s}^2}$. With this

acceleration, the truck could reach 30 mph in about $\Delta t = \dfrac{v}{a} = \dfrac{30 \text{ mph}}{1.8 \text{ m/s}^2} = 7.5$ s. This is certainly possible. So, $\boxed{\text{yes}}$,

the truck driver should be concerned about the rope breaking, particularly when friction is also considered.

71. Strategy Use Newton's laws of motion.

Solution The stone is lifted with constant velocity, so the net force on the stone in the vertical direction is zero. The force is equal and opposite to the force of gravity on the stone, which is mg downward.

$F = mg = (2.0 \text{ kg})(9.80 \text{ m/s}^2) = 20 \text{ N}$

The force exerted by the stone on the man's hand is equal in magnitude to the force of gravity on the stone. So, the magnitude of the total force is $\boxed{20 \text{ N}}$.

73. (a) Strategy and Solution The intersection of the two curves indicates when the motorcycle and the police car are moving at the same speed. According to the graph, they are moving at the same speed at $\boxed{t = 11 \text{ s}}$.

  (b) Strategy and Solution The displacement of each vehicle is represented by the area under each curve. The answer is $\boxed{\text{no; the area under the police car curve is less than the area under the motorcycle curve.}}$

  (c) Strategy and Solution The velocity of the motorcycle with respect to the police car is the difference of the speeds and the relative directions. At $t = 5$ s, the motorcycle is moving faster than the police car by 36 m/s – 24 m/s = 12 m/s, so the relative speed is 12 m/s. At $t = 10$ s, the relative speed is 36 m/s – 34 m/s = 2 m/s. The relative velocities are $\boxed{\text{12 m/s and 2m/s, both in the same direction as the motorcycle's velocity with respect to the highway}}$.

75. **Strategy** Consider the relative motion of the two vehicles.

Solution Let north be in the $+x$-direction.

$v_{\mathrm{JR}x}$ = the velocity of the Jeep relative to the road $= 82$ km/h

$v_{\mathrm{RF}x}$ = the velocity of the road relative to the Ford $= -v_{\mathrm{FR}x} = 48$ km/h

$v_{\mathrm{JF}x}$ = the velocity of the Jeep relative to the (observer in the) Ford $= v_{\mathrm{JR}x} + v_{\mathrm{RF}x} = 82$ km/h $+ 48$ km/h

$\quad = 130$ km/h

So, $\vec{\mathbf{v}}_{\mathrm{JF}x} = \boxed{130 \text{ km/h north}}$.

77. **Strategy** Consider the motion of the person relative to the escalator.

Solution

v_{w} = the speed of the person walking on the stalled escalator

v_{r} = the speed of the person riding on the escalator without walking

v_{wr} = the speed of the person walking while riding $= v_{\mathrm{w}} + v_{\mathrm{r}}$ and x = the distance traveled $= v_{\mathrm{w}} t_{\mathrm{w}} = v_{\mathrm{r}} t_{\mathrm{r}} = v_{\mathrm{wr}} t$,

so $t = \dfrac{x}{v_{\mathrm{wr}}} = \dfrac{x}{v_{\mathrm{w}} + v_{\mathrm{r}}} = \dfrac{v_{\mathrm{w}} t_{\mathrm{w}}}{v_{\mathrm{w}} + v_{\mathrm{w}} \dfrac{t_{\mathrm{w}}}{t_{\mathrm{r}}}} = \dfrac{94 \text{ s}}{1 + \dfrac{94 \text{ s}}{66 \text{ s}}} = \boxed{39 \text{ s}}$.

81. **Strategy** Consider the relative motion of the two vehicles. Use the component method.

Solution Let the $+y$-direction be north and the $+x$-direction be east.

$\vec{\mathbf{v}}_{\mathrm{ps}}$ = the velocity of the Pierce Arrow relative to the Stanley Steamer

$\vec{\mathbf{v}}_{\mathrm{pg}}$ = the velocity of the Pierce Arrow relative to the ground

$\vec{\mathbf{v}}_{\mathrm{sg}}$ = the velocity of the Stanley Steamer relative to the ground

Compute the components of the velocity of the Pierce Arrow relative to the observer riding
in the Stanley Steamer.

$v_{\mathrm{ps}x} = v_{\mathrm{pg}x} + v_{\mathrm{gs}x} = 50$ km/h $+ 0$, so $\boxed{v_x = 50 \text{ km/h east}}$.

$v_{\mathrm{ps}y} = v_{\mathrm{pg}y} + v_{\mathrm{gs}y} = v_{\mathrm{pg}y} - v_{\mathrm{sg}y} = 0 - 40$ km/h and -40 km/h north $= 40$ km/h south, so

$\boxed{v_y = 40 \text{ km/h south}}$.

83. (a) **Strategy** Consider the relative motion of the boy and the water.

Solution

$\Delta t = \dfrac{d_{\mathrm{across}}}{v_{\mathrm{boy}}}$ and $v_{\mathrm{water}} = \dfrac{d_{\mathrm{downstream}}}{\Delta t}$, so

$v_{\mathrm{water}} = \dfrac{d_{\mathrm{downstream}}}{d_{\mathrm{across}}/v_{\mathrm{boy}}} = \dfrac{d_{\mathrm{downstream}}}{d_{\mathrm{across}}} v_{\mathrm{boy}} = \dfrac{50.0 \text{ m}}{25.0 \text{ m}} (0.500 \text{ m/s}) = \boxed{1.00 \text{ m/s}}$

(b) **Strategy** Use the Pythagorean theorem.

Solution Find the speed of the boy relative to the friend.

$v_{\mathrm{bf}} = \sqrt{(0.500 \text{ m/s})^2 + (1.00 \text{ m/s})^2} = \boxed{1.12 \text{ m/s}}$

85. Strategy Consider the relative motion of the water (w) and Sheena (s). Let the $+y$-direction be upstream and the $+x$-direction be toward the opposite bank (b).

Solution

(a) Find the x-component.

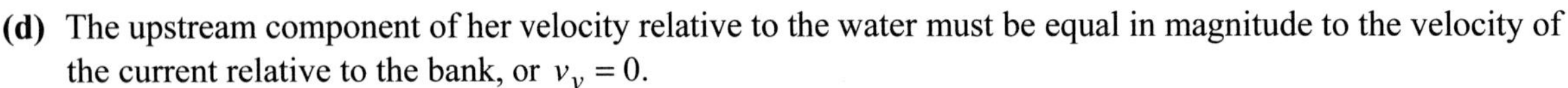

$$v_x = (3.00\ \text{mi/h})\cos 60.0^\circ = 1.50\ \text{mi/h}$$

The y-component is $v_y = v_{Sby} = v_{Swy} + v_{wby}$.

$$v_y = (3.00\ \text{mi/h})\sin 60.0^\circ - 1.60\ \text{mi/h} = 1.00\ \text{mi/h}$$

Use the Pythagorean theorem.

$$v_{Sb} = \sqrt{(1.50\ \text{mi/h})^2 + (1.00\ \text{mi/h})^2} = \boxed{1.80\ \text{mi/h}}$$

(b) $\Delta t = \dfrac{\Delta x}{v_x} = \dfrac{1.20\ \text{mi}}{1.50\ \text{mi/h}} = (0.800\ \text{h})\left(\dfrac{60\ \text{min}}{\text{h}}\right) = \boxed{48.0\ \text{min}}$

(c) $\Delta y = v_y \Delta t = (1.00\ \text{mi/h})(0.800\ \text{h}) = \boxed{0.800\ \text{mi upstream}}$

(d) The upstream component of her velocity relative to the water must be equal in magnitude to the velocity of the current relative to the bank, or $v_y = 0$.

$$(3.00\ \text{mi/h})\sin\theta - 1.60\ \text{mi/h} = 0,\ \text{so}\ \theta = \sin^{-1}\dfrac{1.60}{3.00} = \boxed{32.2^\circ\ \text{upstream}}.$$

89. Strategy The cutter must move with the moving glass to cut perpendicularly to the direction of motion of the conveyor belt. Thus, the cutter must be set at some angle with respect to the width of the belt and toward the direction of motion of the belt. Draw a diagram.

Solution Let d be the distance that the sheet of glass travels in the time t that it takes the cutter to cut it. Then, $d = (15.0\ \text{cm/s})t$. Since the cutter moves across the width at a speed of $24.0\ \text{cm/s}$ and the width is 72.0 cm, the time t is given by

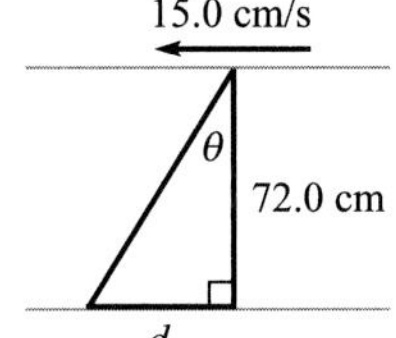

$$t = \dfrac{72.0\ \text{cm}}{24.0\ \text{cm/s}} = 3.00\ \text{s. Solve for } \theta.$$

$$\tan\theta = \dfrac{d}{72.0\ \text{cm}} = \dfrac{(15.0\ \text{cm/s})t}{72.0\ \text{cm}} = \dfrac{(15.0\ \text{cm/s})(3.00\ \text{s})}{72.0\ \text{cm}} = 0.625,\ \text{so}$$

$$\theta = \tan^{-1} 0.625 = \boxed{32.0^\circ}.$$

93. Strategy Use Newton's second law.

Solution

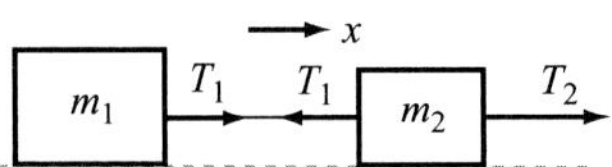

For m_2: $\sum F_x = T_2 - T_1 = m_2 a$

For m_1: $\sum F_x = T_1 = m_1 a$

Find T_2.

$T_2 - T_1 = m_2 a$, so $T_2 = T_1 + m_2 a = m_1 a + m_2 a = (m_1 + m_2)a.$

Find T_1 / T_2.

$$\dfrac{T_1}{T_2} = \dfrac{m_1 a}{(m_1 + m_2)a} = \boxed{\dfrac{m_1}{m_1 + m_2}}$$

97. (a) Strategy Draw a diagram and use vector addition.

Solution Find the magnitude of the displacement.

$$|\Delta \vec{r}| = \sqrt{[600.0 \text{ km} + (300.0 \text{ km})\cos(-30.0°)]^2 + [(300.0 \text{ km})\sin(-30.0°)]^2}$$
$$= \boxed{873 \text{ km}}$$

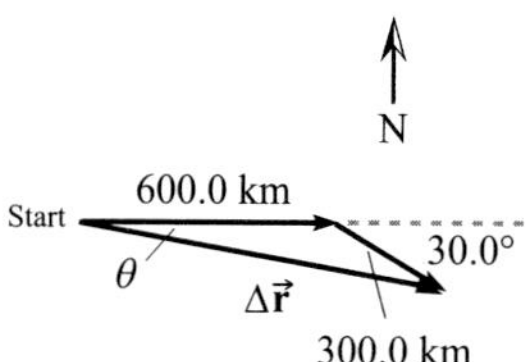

(b) Strategy Refer to the diagram in part (a). Find the angle between the initial displacement vector and $\Delta \vec{r}$.

Solution Find the direction of the displacement.

$$\theta = \tan^{-1} \frac{(300.0 \text{ km})\sin(-30.0°)}{600.0 \text{ km} + (300.0 \text{ km})\cos(-30.0°)} = \boxed{9.90° \text{ south of east}}$$

(c) Strategy The flight time is given by the quotient of the distance traveled and the speed of the jetliner.

Solution

$$\Delta t = \frac{d}{v} = \frac{600.0 \text{ km} + 300.0 \text{ km}}{400.0 \text{ km/h}} = \boxed{2.250 \text{ h}}$$

(d) Strategy The direct flight time is given by the quotient of the magnitude of the displacement and the speed of the jetliner.

Solution

$$\Delta t = \frac{|\Delta \vec{r}|}{v} = \frac{873 \text{ km}}{400.0 \text{ km/h}} = \boxed{2.18 \text{ h}}$$

101. Strategy The slope of the $x(t)$ curve for any interval in the graph is equal to the velocity component v_x.

Solution The intervals with positive slope are AB, CD, and EF. The slopes of AB and CD are the same, while that of EF is greater. The interval BC has zero slope, which is between positive and negative. The only interval with negative slope is DE. Ranking the intervals in order of the velocity component v_x from greatest positive to greatest negative, we have $\boxed{\text{EF, AB} = \text{CD, BC, DE}}$.

103. Strategy Use Newton's second law. Let T be the maximum tension.

Solution

(a) Find the acceleration of the block plus cart system.

$$\sum F_x = T = (m_1 + m_2)a_x, \text{ so } a_x = \frac{T}{m_1 + m_2}.$$

Now, switch to the block system.

$$\sum F_y = N - m_2 g = 0, \text{ so } N = m_2 g. \quad \sum F_x = f_{\max} = m_2 a_x, \text{ since}$$

the block must not slide.

$$m_2 a_x = m_2 \left(\frac{T}{m_1 + m_2} \right) = f_{\max} = \mu N = \mu m_2 g, \text{ so}$$

$$T = \boxed{(m_1 + m_2)\mu g}.$$

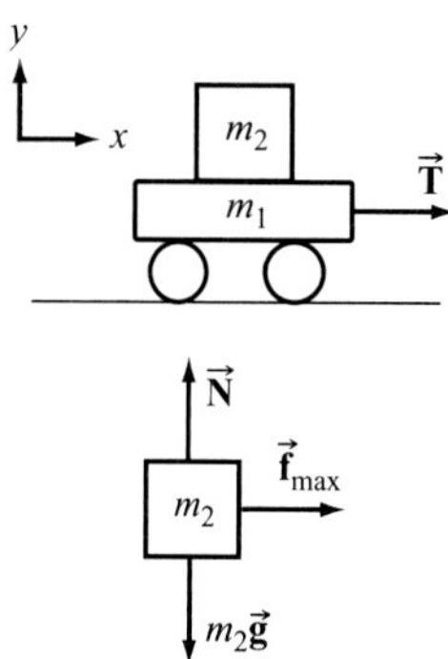

(b) Find the acceleration of the block plus cart system.

$$\sum F_x = T - (m_1 + m_2)g\sin\theta = (m_1 + m_2)a_x, \text{ so}$$

$$a_x = \frac{T}{m_1 + m_2} - g\sin\theta.$$

Now, switch to the block system.

$$\sum F_y = N - m_2 g\cos\theta = 0, \text{ so } N = m_2 g\cos\theta.$$

$$\sum F_x = f_{max} = m_2 a_x, \text{ since the block must not slide.}$$

$$m_2 a_x = m_2\left(\frac{T}{m_1 + m_2} - g\sin\theta\right) = f_{max} = \mu N = \mu m_2 g\cos\theta, \text{ so}$$

$$T = \boxed{(m_1 + m_2)g(\mu\cos\theta + \sin\theta)}.$$

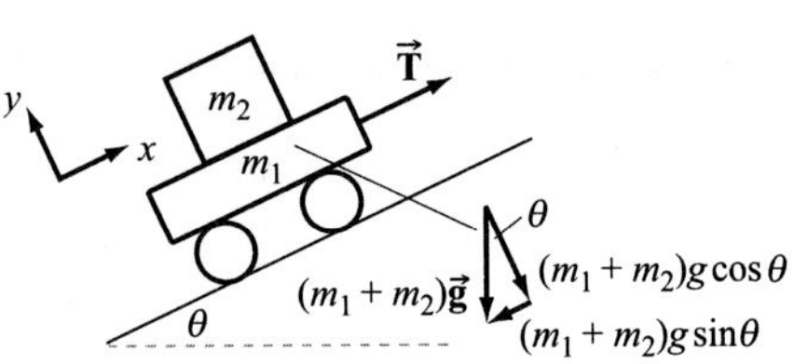
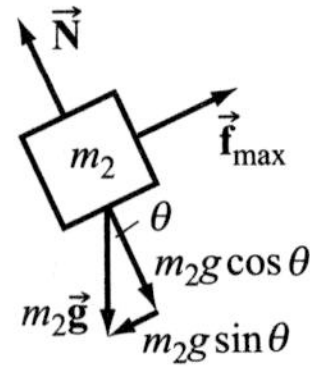

105. (a) Strategy Use the definition of average speed.

Solution

$$v_{av} = \frac{\Delta x}{\Delta t} = \frac{100\times10^{-9}\text{ m}}{0.10\times10^{-3}\text{ s}} = \boxed{1.0\text{ mm/s}}$$

(b) Strategy Find the time it takes the pain signal to travel the length of a 1.0-m long neuron. Then, add the times of travel across synapses and neurons.

Solution

$$t_n = \frac{x}{v} = \frac{1.0\text{ m}}{100\text{ m/s}} = 10\text{ ms}$$

Find the total time to reach the brain.

$$t_n + t_{syn} + t_n + t_{syn} = 2t_n + 2t_{syn} = 2(t_n + t_{syn}) = 2(10\text{ ms} + 0.10\text{ ms}) = \boxed{20\text{ ms}}$$

(c) Strategy Use the definition of average speed.

Solution

$$v_{av} = \frac{\Delta x}{\Delta t} = \frac{2.0\text{ m} + 2(100\times10^{-9}\text{ m})}{20\times10^{-3}\text{ s}} = \boxed{100\text{ m/s}}$$

109. (a) Strategy Since $mg = (51\text{ kg})(9.80\text{ m/s}^2) = 500\text{ N} > 408\text{ N}$, the woman feels less than her normal weight, so the elevator is accelerating downward. Use Newton's second law.

Solution Let the $+y$-direction be up.

$$\sum F_y = 408\text{ N} - mg = ma_y, \text{ so } a_y = \frac{408\text{ N} - mg}{m} = \frac{408\text{ N}}{51\text{ kg}} - 9.80\text{ m/s}^2 = -1.8\text{ m/s}^2.$$

Thus, $\vec{a} = \boxed{1.8\text{ m/s}^2 \text{ down}}$.

(b) Strategy Find the change in speed of the elevator after 4.0 s at the acceleration found in part (a).

Solution Let down be positive. Find the speed of the elevator.

$$\Delta v = v_f - v_i = a_y\Delta t, \text{ so } v_f = v_i + a_y\Delta t = 1.5\text{ m/s} + (1.8\text{ m/s}^2)(4.0\text{ s}) = \boxed{8.7\text{ m/s}}.$$

113. (a) Strategy Use the definition of average velocity. Draw a diagram.

Solution Since the pilot has traveled 15 km due west in 1.0 h, the average velocity of the wind must be $\boxed{15 \text{ km/h due west}}$.

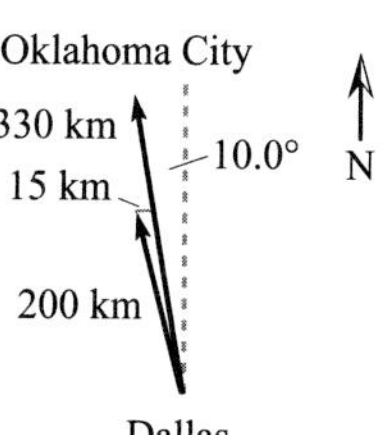

(b) Strategy The angle θ is the angle with respect to the vertical (north) in which the pilot should have headed his plane to get directly to Oklahoma city from Dallas without being blown off course. Draw a vector diagram. Use the Law of Sines and Eq. (3-17).

Solution Let the velocities of the plane with respect to the ground and to the air be $\vec{v}_{pg}$ and $\vec{v}_{pa}$, respectively. Let the velocity of the air with respect to the ground be $\vec{v}_{ag}$. Then, we have $\vec{v}_{pg} = \vec{v}_{pa} + \vec{v}_{ag}$.

From the diagram, we see that $\theta = 10.0° - \alpha$. We need to find α to find θ. The vectors form a triangle. We know two sides of the triangle, and the angles α and β are opposite those sides. If we can find the angle β, we can use the Law of Sines to find α. Since $\vec{v}_{pg}$ is 10.0° from the vertical (west of north) and $\vec{v}_{ag}$ is horizontal (due west), we see that $\beta = 90.0° - 10.0° = 80.0°$. Find α.

$$\frac{\sin \alpha}{15 \text{ km/h}} = \frac{\sin \beta}{200 \text{ km/h}} = \frac{\sin 80.0°}{200 \text{ km/h}}, \text{ so } \alpha = \sin^{-1}\frac{15 \sin 80.0°}{200} = 4.2°.$$

Thus, the direction the pilot should have headed his plane is $\theta = 10.0° - 4.2° = \boxed{5.8° \text{ west of north}}$.

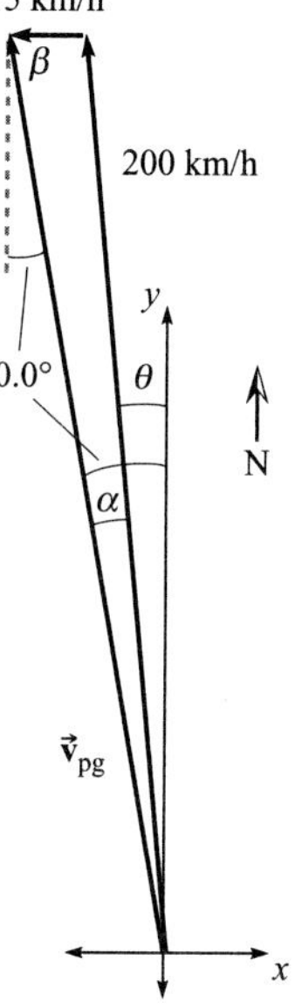

117. (a) Strategy Since the downward speed is decreasing at a rate of $0.10g$, the acceleration of the truck is $0.10g$ upwards. Use Newton's second law.

Solution

$\sum F_x = 0$ and $\sum F_y = T - mg = ma_y = m(0.10g)$, so $T = \boxed{1.10mg}$.

(b) Strategy and Solution Although the motion of the helicopter has changed, the acceleration of the truck is the same as in part (a), so the tension is the same, $\boxed{1.10mg}$.

Chapter 4

MOTION WITH CONSTANT ACCELERATION

Conceptual Questions

1. Neglecting air resistance, the trajectory of a bullet that has exited the muzzle of a rifle is solely influenced by gravity. The force due to gravity causes the bullet to accelerate downward toward the Earth but does not influence its horizontal motion. Thus, to hit a target, the muzzle must be aimed above the target by a distance equal to the amount that the bullet will fall in the course of its travel. If aimed at the target instead of above it, the bullet will miss low.

5. The terminal velocity of an object depends linearly on its mass and inversely on a parameter determined by its size and shape. The mass of a feather and brick differ by several orders of magnitude. Therefore, if the difference in the drag parameters of the two objects is much smaller than their mass differences, the brick must have a higher terminal velocity. Because the brick is traveling at a higher velocity for the majority of its journey, it arrives at the ground first. The density of the atmosphere on the surface of the Moon is much less than the density of the atmosphere on Earth. Thus, the effects of drag are reduced on the Moon and both objects will hit the ground at nearly the same instant—however, only in a perfect vacuum would the objects fall at exactly identical rates.

9. The forces of gravity and air resistance act upon the parachutist. Descending to Earth with a constant velocity, the parachutist has zero acceleration.

13. The only force is gravity.

17. The instantaneous velocity is zero at the high point. The acceleration is constant and directed vertically downward throughout the motion, including at the high point. See Fig. 4.19.

Problems

1. **Strategy** Since the time intervals are the same, the greater the change in distance between each successive pairs of dots indicates a greater magnitude acceleration. If there is no change in distance between dots, the acceleration is zero.

   Solution The distance between dots in (b) and (c) is constant; therefore, the accelerations are zero. The distance between dots in (a) and (d) is increasing, indicating a positive acceleration. The increase is greater for (a) than for (d); therefore, (a) represents a greater magnitude acceleration than (d). Ranking the motion diagrams in order of the magnitude of the acceleration, from greatest to lest, we have $\boxed{\text{(a), (d), (b) = (c)}}$.

3. **Strategy** Refer to the figure. Each square represents $(10 \text{ m/s})(1 \text{ s}) = 10$ m. Count squares to determine the distance traveled at each time.

   Solution Sketch the motion diagram and describe the motion in words.

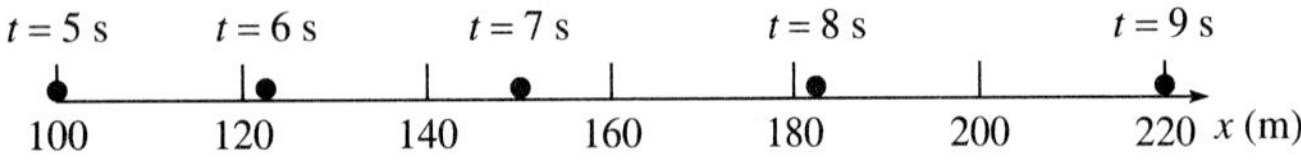

$\boxed{\text{The } x\text{-coordinates are determined by choosing } x = 0 \text{ at } t = 0. \text{ At } t = 5.0 \text{ s and } x = 100 \text{ m, the object's speed is } 20 \text{ m/s. From } t = 5.0 \text{ s to } t = 9.0 \text{ s, the speed of the object increases at a constant rate until it reaches 40 m/s at } x = 220 \text{ m.}}$

Find the slope of the graph to find the acceleration.

$$a_{\text{av}, x} = \frac{40 \text{ m/s} - 20 \text{ m/s}}{9.0 \text{ s} - 5.0 \text{ s}} = 5.0 \text{ m/s}^2$$

The acceleration is $\boxed{5.0 \text{ m/s}^2 \text{ in the } +x\text{-direction}}$.

5. (a) Strategy The graph will be a line with a slope of 1.20 m/s^2.

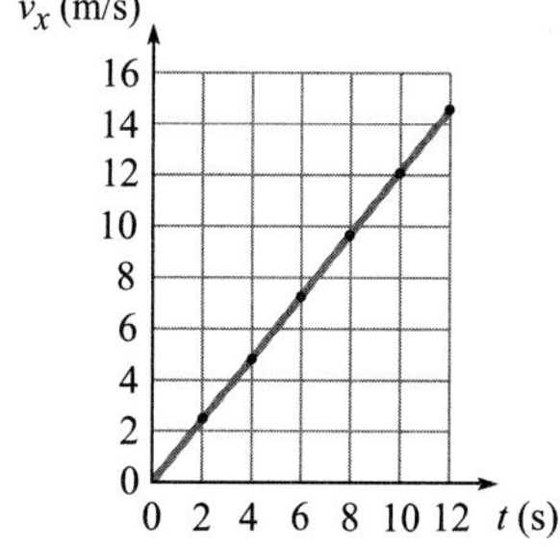

Solution $v_x = 0$ when $t = 0$. The graph is shown.

(b) Strategy Use Eq. (4-4).

Solution Find the distance the train traveled.

$$\Delta x = v_{\text{ix}}\Delta t + \frac{1}{2}a_x(\Delta t)^2 = (0)\Delta t + \frac{1}{2}a_x(\Delta t)^2 = \frac{1}{2}a_x(\Delta t)^2$$
$$= \frac{1}{2}(1.20 \text{ m/s}^2)(12.0 \text{ s})^2 = \boxed{86.4 \text{ m}}$$

(c) Strategy Use Eq. (4-1).

Solution Find the final speed of the train.

$$v_{\text{fx}} - v_{\text{ix}} = v_{\text{fx}} - 0 = a_x\Delta t, \text{ so } v_{\text{fx}} = a_x\Delta t = (1.20 \text{ m/s}^2)(12.0 \text{ s}) = \boxed{14.4 \text{ m/s}}.$$

(d) Strategy Refer to Figure 4.10, which shows a motion diagram.

Solution The motion diagram is shown.

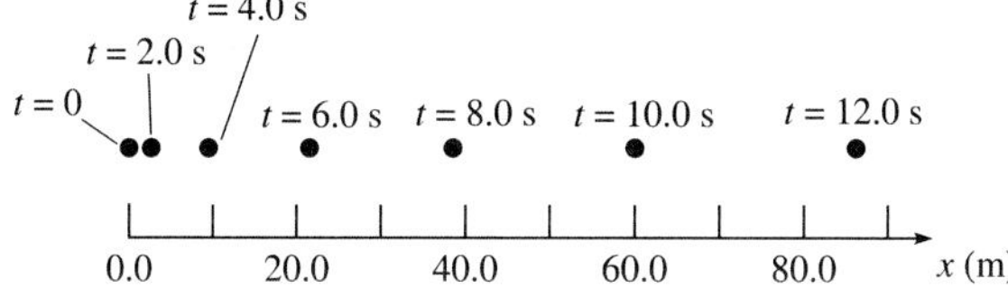

9. Strategy Relate the acceleration, speed, and distance using Eq. (4-5). Let southwest be the positive direction. Use the acceleration for the slope of the $v_x(t)$ curve.

Solution Find the constant acceleration required to stop the airplane. The acceleration must be opposite to the direction of motion of the airplane, so the direction of the acceleration is $-$southwest = northeast.

$$v_{\text{fx}}^2 - v_{\text{ix}}^2 = 2a_x\Delta x, \text{ so } a_x = \frac{v_{\text{fx}}^2 - v_{\text{ix}}^2}{2\Delta x} = \frac{0 - (55 \text{ m/s})^2}{2(1.0\times10^3 \text{ m})} = -1.5 \text{ m/s}^2.$$

Thus, the acceleration is $\boxed{1.5 \text{ m/s}^2 \text{ northeast}}$. The slope is -1.5 m/s^2. The v_x-intercept is 55 m/s. Sketch the graph.

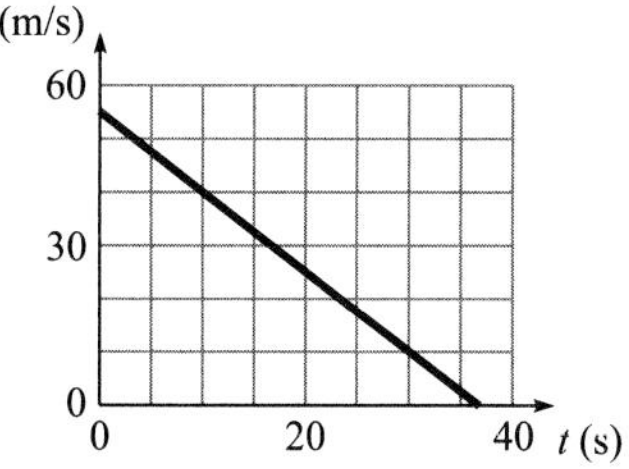

13. (a) Strategy Use the definition of average acceleration.

Solution Find the magnitude of the acceleration.
$$a = \frac{\Delta v}{\Delta t} = \frac{24 \text{ m/s}}{2.0 \text{ s}} = \boxed{12 \text{ m/s}^2}$$

(b) Strategy Relate the distance traveled, acceleration, and time using Eq. (4-4).

Solution Find the distance traveled.
$$\Delta x = \frac{1}{2} a (\Delta t)^2 = \frac{1}{2}(12 \text{ m/s}^2)(2.0 \text{ s})^2 = \boxed{24 \text{ m}}$$

(c) Strategy Refer to part (a).

Solution The magnitude of the acceleration of the runner is
$$a = \frac{\Delta v}{\Delta t} = \frac{6.0 \text{ m/s}}{2.0 \text{ s}} = \boxed{3.0 \text{ m/s}^2}.$$
Find a_c/a_r.
$$\frac{a_\text{c}}{a_\text{r}} = \frac{12 \text{ m/s}^2}{3.0 \text{ m/s}^2} = \boxed{4.0}$$

17. (a) Strategy Draw a diagram and use Newton's second law to find the acceleration of the skier. Then, relate the acceleration, speed, and distance using Eq. (4-5).

Solution Find the acceleration.
$$\Sigma F_x = mg \sin \theta = ma_x, \text{ so } a_x = g \sin \theta.$$
Find the speed at the bottom of the slope.
$$v_{\text{f}x}^2 - v_{\text{i}x}^2 = 2a_x \Delta x$$
$$v_{\text{f}x}^2 - 0 = 2g \sin \theta \Delta x$$
$$v_{\text{f}x} = \sqrt{2g \sin \theta \Delta x} = \sqrt{2(9.80 \text{ m/s}^2)(\sin 32°)(50 \text{ m})} = \boxed{23 \text{ m/s}}$$

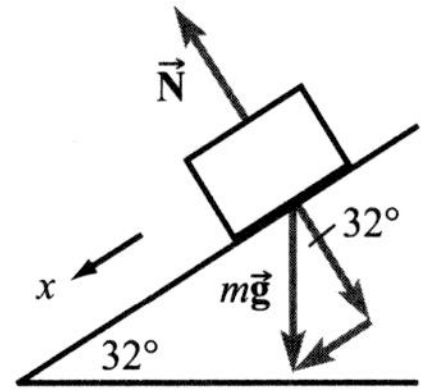

(b) Strategy Find the acceleration using Eq. (4-5). Then, use Newton's second law.

Solution Find a_x.
$$v_{\text{f}x}^2 - v_{\text{i}x}^2 = 2a_x \Delta x, \text{ so}$$
$$a_x = \frac{v_{\text{f}x}^2 - v_{\text{i}x}^2}{2\Delta x} = \frac{0 - \left(\sqrt{2g \sin \theta \Delta x_{\text{slope}}}\right)^2}{2\Delta x} = -\frac{2g \sin \theta \Delta x_{\text{slope}}}{2\Delta x} = -\frac{g \sin \theta \Delta x_{\text{slope}}}{\Delta x}.$$
Find μ_k.

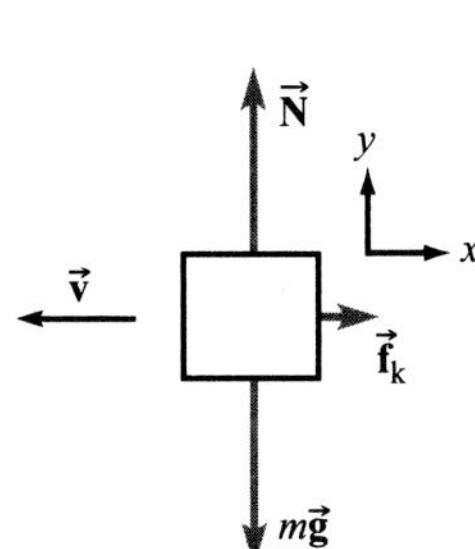

$$\Sigma F_x = -f_\text{k} = ma_x, \text{ so } f_\text{k} = \mu_\text{k} N = -ma_x \text{ and } \mu_\text{k} = -\frac{ma_x}{N}.$$

$$\Sigma F_y = N - mg = 0, \text{ so } N = mg. \text{ Thus, we have}$$

$$\mu_\text{k} = -\frac{ma_x}{N} = -\frac{ma_x}{mg} = -\frac{a_x}{g} = \frac{\sin \theta \Delta x_{\text{slope}}}{\Delta x} = \frac{(\sin 32°)(50 \text{ m})}{140 \text{ m}} = \boxed{0.19}.$$

21. (a) Strategy Use Newton's second law. Since the pumpkin and the watermelon are attached by the cord, they must have the same magnitude of acceleration.

Solution For the pumpkin:

$$\sum F_x = m_p g \sin 53.0° - T = m_p a \quad (1)$$

For the watermelon:

$$\sum F_x = T - m_w g \sin 30.0° = m_w a \quad (2)$$

Adding (1) and (2) gives

$$g(m_p \sin 53.0° - m_w \sin 30.0°) = (m_p + m_w)a.$$

Solving for a, we have

$$a = \frac{g(m_p \sin 53.0° - m_w \sin 30.0°)}{m_p + m_w} = \frac{(9.80 \text{ m/s}^2)[(7.00 \text{ kg}) \sin 53.0° - (10.0 \text{ kg}) \sin 30.0°]}{7.00 \text{ kg} + 10.0 \text{ kg}} = 0.34 \text{ m/s}^2.$$

The acceleration is positive, so the watermelon slides up the ramp and the pumpkin slides down. Therefore, the acceleration is $\vec{\mathbf{a}} = \boxed{0.34 \text{ m/s}^2 \text{, where the watermelon moves up and to the left}}$.

(b) Strategy The pumpkin will travel down the ramp with the acceleration found in part (a).

Solution Use Eq. (4-4).

$$\Delta x = \frac{1}{2} a (\Delta t)^2 = \frac{1}{2}(0.34 \text{ m/s}^2)(0.30 \text{ s})^2 = \boxed{1.5 \text{ cm}}$$

(c) Strategy Use Eq. (4-1).

Solution

$$v = a\Delta t = (0.34 \text{ m/s}^2)(0.20 \text{ s}) = \boxed{6.8 \text{ cm/s}}$$

25. Strategy The final speed is zero. Use Eq. (4-10).

Solution Find the initial speed.

$$v_{fy}^2 - v_{iy}^2 = 0 - v_{iy}^2 = -2g\Delta y, \text{ so } v_{iy} = \sqrt{2g\Delta y} = \sqrt{2(9.80 \text{ m/s}^2)(1.3 \text{ m} - 0)} = \boxed{5.0 \text{ m/s}}.$$

27. Strategy Ignoring air resistance, the golf ball is in free fall. Use Eq. (4-9).

Solution

(a) Find the time it takes the golf ball to fall 12.0 m.

$$v_{iy} = 0, \text{ so } \Delta y = -\frac{1}{2} g (\Delta t)^2 \text{ and } \Delta t = \sqrt{-\frac{2\Delta y}{g}} = \sqrt{\frac{-2(0 - 12.0 \text{ m})}{9.80 \text{ m/s}^2}} = \boxed{1.6 \text{ s}}.$$

(b) Find how far the golf ball would fall in $2\sqrt{\dfrac{-2(0 - 12.0 \text{ m})}{9.80 \text{ m/s}^2}} = 3.13 \text{ s}.$

$$\Delta y = -\frac{1}{2} g (\Delta t)^2 = -\frac{1}{2}(9.80 \text{ m/s}^2)(3.13 \text{ s})^2 = -48 \text{ m}$$

The golf ball would fall $\boxed{48 \text{ m}}$.

29. Strategy Use Eq. (4-10).

Solution Find the sandbag's speed when it hits the ground.

$$v_{fy}^2 - v_{iy}^2 = -2g\Delta y, \text{ so } v_{fy} = \sqrt{v_{iy}^2 - 2g\Delta y} = \sqrt{(10.0 \text{ m/s})^2 - 2(9.80 \text{ m/s}^2)(-40.8 \text{ m})} = \boxed{30.0 \text{ m/s}}.$$

33. Strategy Use Eq. (4-9) to find the time it takes for the coin to reach the water. Then, find the time it takes the sound to reach Glenda's ear. Add these two times. Let $h = 7.00$ m.

Solution Find the time elapsed between the release of the coin and the hearing of the splash.

$$h = v_{iy}\Delta t + \frac{1}{2}a_y(\Delta t)^2 = 0 + \frac{1}{2}g(\Delta t_1)^2, \text{ so } \Delta t_1 = \sqrt{\frac{2h}{g}}. \quad h = v_s\Delta t_2, \text{ so } \Delta t_2 = \frac{h}{v_s}.$$

Therefore, the time elapsed is $\Delta t = \Delta t_1 + \Delta t_2 = \sqrt{\dfrac{2h}{g}} + \dfrac{h}{v_s} = \sqrt{\dfrac{2(7.00 \text{ m})}{9.80 \text{ m/s}^2}} + \dfrac{7.00 \text{ m}}{343 \text{ m/s}} = \boxed{1.22 \text{ s}}.$

37. Strategy Find and subtract the time it took for the sound of the rock hitting the bottom of the well to reach your ears from the total time (3.20 s) to find the time it took the rock to reach the bottom. Then, use Eq. (4-9) to determine the depth of the well.

Solution The time it took for the sound of the rock hitting the bottom to reach you is $\Delta t_{\text{sound}} = d/v_s$, where d is the depth of the well. So, $\Delta t_{\text{fall}} = \Delta t_{\text{total}} - d/v_s$. Find d.

$$d = \frac{1}{2}g(\Delta t)^2 = \frac{1}{2}g\left(\Delta t_{\text{total}} - \frac{d}{v_s}\right)^2 = \frac{1}{2}g\left[(\Delta t_{\text{total}})^2 - \frac{2\Delta t_{\text{total}}}{v_s}d + \frac{1}{v_s^2}d^2\right], \text{ so}$$

$$\frac{2}{g}d = (\Delta t_{\text{total}})^2 - \frac{2\Delta t_{\text{total}}}{v_s}d + \frac{1}{v_s^2}d^2$$

$$\frac{2v_s^2}{g}d = (\Delta t_{\text{total}})^2 v_s^2 - 2v_s\Delta t_{\text{total}}d + d^2$$

$$0 = d^2 - 2v_s\left(\Delta t_{\text{total}} + \frac{v_s}{g}\right)d + (\Delta t_{\text{total}})^2 v_s^2$$

Use the quadratic formula with $a = 1$, $b = -2v_s\left(\Delta t_{\text{total}} + \dfrac{v_s}{g}\right)$, and $c = (\Delta t_{\text{total}})^2 v_s^2$ to solve for d.

$$d = \frac{2v_s\left(\Delta t_{\text{total}} + \dfrac{v_s}{g}\right) \pm \sqrt{\left[-2v_s\left(\Delta t_{\text{total}} + \dfrac{v_s}{g}\right)\right]^2 - 4(1)(\Delta t_{\text{total}})^2 v_s^2}}{2(1)}$$

$$= \frac{2(343 \text{ m/s})\left(3.20 \text{ s} + \dfrac{343 \text{ m/s}}{9.80 \text{ m/s}^2}\right) \pm \sqrt{\left[-2(343 \text{ m/s})\left(3.20 \text{ s} + \dfrac{343 \text{ m/s}}{9.80 \text{ m/s}^2}\right)\right]^2 - 4(3.20 \text{ s})^2(343 \text{ m/s})^2}}{2}$$

$$= \boxed{46 \text{ m}} \quad (26{,}000 \text{ m is extraneous.})$$

39. Strategy Use equations of motion with constant acceleration to determine the vertical and horizontal positions of the clay after 1.50 s have elapsed.

Solution Find the position of the clay.

$$x_f = v_{ix}\Delta t = (20.0 \text{ m/s})(1.50 \text{ s}) = 30.0 \text{ m}$$

$$y_f = y_i + v_{iy}\Delta t - \frac{1}{2}g(\Delta t)^2 = 8.50 \text{ m} + 0 - \frac{1}{2}(9.80 \text{ m/s}^2)(1.50 \text{ s})^2 = -2.53 \text{ m}$$

The clay cannot pass through the ground, so it hit and stuck prior to 1.50 s. Find the time it took for the clay to land.

$$y_f = 0 = y_i + v_{iy}\Delta t - \frac{1}{2}g(\Delta t)^2 = y_i - \frac{1}{2}g(\Delta t)^2, \text{ so}$$

$$\Delta t = \sqrt{\frac{2y_i}{g}} = \sqrt{\frac{2(8.50 \text{ m})}{9.80 \text{ m/s}^2}} = 1.317 \text{ s, and } x_f = v_{ix}\Delta t = (20.0 \text{ m/s})(1.317 \text{ s}) = 26.3 \text{ m}.$$

The clay hits and $\boxed{\text{it is on the ground after 1.32 s, so the horizontal distance along the ground is 26.3 m}}$.

41. Strategy Use Eqs. (4-10) and (4-11). Set $v_{fy} = 0$, since the vertical component of the velocity is zero at the maximum height.

Solution

(a) Find the maximum height.

$$v_{fy}^2 - v_{iy}^2 = 0 - v_{iy}^2 = -2g\Delta y, \text{ so } \Delta y = \frac{v_{iy}^2}{2g} = y_f - y_i \text{ and}$$

$$y_f = \frac{v_i^2 \sin^2\theta}{2g} + y_i = \frac{(19.6 \text{ m/s})^2 \sin^2 30.0°}{2(9.80 \text{ m/s}^2)} + 1.0 \text{ m} = \boxed{5.9 \text{ m}}.$$

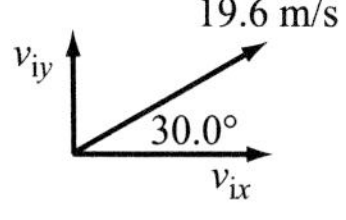

(b) At the ball's highest point, $v_{fy} = 0$, so the speed v equals v_x.

$$v = v_x = v_{ix} = v_i \cos\theta = (19.6 \text{ m/s})\cos 30.0° = \boxed{17.0 \text{ m/s}}$$

45. Strategy The skater must be up the ramp far enough for their speed at the end of the horizontal section to be just great enough so that the skater travels a horizontal distance of 7.00 m while falling 3.00 m. Draw a diagram of the skater on the ramp to find the acceleration of the skater caused by the force of gravity.

Solution According to Newton's second law, $\Sigma F = mg \sin 15.0° = ma$, so $a = g \sin 15.0°$ along the surface of the ramp. Use Eq. (4-5) to relate the distance up the ramp to the speed of the skater at the end of the ramp.

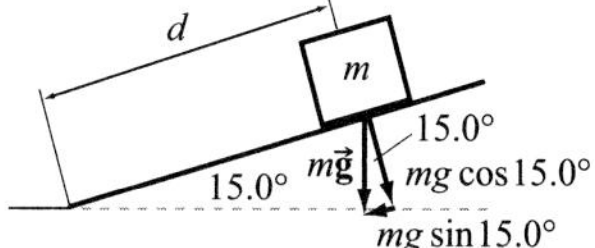

$$v_f^2 - v_i^2 = v_f^2 - 0 = 2a\Delta x = 2(g\sin 15.0°)d, \text{ so } d = \frac{v_f^2}{2g\sin 15.0°}.$$

Since the ramp is frictionless, the velocity of the skater at the end of the horizontal part of the ramp is in the x-direction with magnitude equal to v_f. So, the components of the displacement are $\Delta x = v_f\Delta t$ (1) and

$$\Delta y = v_{iy}\Delta t + \frac{1}{2}a_y(\Delta t)^2 = 0 - \frac{1}{2}g(\Delta t)^2 = -\frac{1}{2}g(\Delta t)^2 \text{ (2)}. \text{ Solving for } \Delta t \text{ in (1) and substituting into (2) gives}$$

$$\Delta y = -\frac{1}{2}g\left(\frac{\Delta x}{v_f}\right)^2, \text{ or } v_f^2 = -\frac{g(\Delta x)^2}{2\Delta y}. \text{ Substitute this result into the equation for } d.$$

$$d = \frac{v_f^2}{2g\sin 15.0°} = \frac{-\dfrac{g(\Delta x)^2}{2\Delta y}}{2g\sin 15.0°} = -\frac{(\Delta x)^2}{4\Delta y\sin 15.0°} = -\frac{(7.00 \text{ m})^2}{4(-3.00 \text{ m})\sin 15.0°} = \boxed{15.8 \text{ m}}$$

49. Strategy Solve $\Delta x = v_x \Delta t$ for the time and substitute the result into Eq. (4-9). Then, solve for Δx to find the required distance from the cannon.

Solution $\Delta x = v_x \Delta t = (v_i \cos \theta)\Delta t$, so $\Delta t = \Delta x / (v_i \cos \theta)$. Substitute.

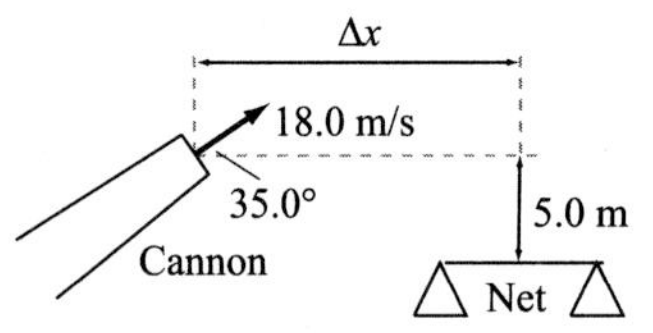

$$\Delta y = y_f - y_i = v_{iy}\Delta t + \frac{1}{2}a_y(\Delta t)^2$$

$$= (v_i \sin \theta)\frac{\Delta x}{v_i \cos \theta} - \frac{1}{2}g\frac{(\Delta x)^2}{v_i^2 \cos^2 \theta} = \Delta x \tan \theta - \frac{g(\Delta x)^2}{2v_i^2 \cos^2 \theta}, \text{ so}$$

$$0 = \frac{g}{2v_i^2 \cos^2 \theta}(\Delta x)^2 - (\tan \theta)\Delta x + \Delta y.$$

Use the quadratic formula.

$$\Delta x = \frac{\tan \theta \pm \sqrt{\tan^2 \theta - 4\left(\frac{g}{2v_i^2 \cos^2 \theta}\right)\Delta y}}{2\left(\frac{g}{2v_i^2 \cos^2 \theta}\right)} = \frac{\tan 35.0° \pm \sqrt{\tan^2 35.0° - \frac{2(9.80 \text{ m/s}^2)(-5.0 \text{ m})}{(18.0 \text{ m/s})^2 \cos^2 35.0°}}}{\frac{9.80 \text{ m/s}^2}{(18.0 \text{ m/s})^2 \cos^2 35.0°}} = 37.1 \text{ m or } -6.0 \text{ m}$$

Since the cannon won't fire backward, -6.0 m is extraneous. So, you tell the ringmaster to place the net such that its center is $\boxed{37.1 \text{ m}}$ in front of the cannon.

53. Strategy Use $\Delta y = (v_i \sin \theta)\Delta t - \frac{1}{2}g(\Delta t)^2$ and $\Delta x = (v_i \cos \theta)\Delta t$ for the change in the projectile's position.

Solution When a projectile returns to its original height, $\Delta y = 0$.

$$0 = (v_i \sin \theta)\Delta t - \frac{1}{2}g(\Delta t)^2 = v_i \sin \theta - \frac{1}{2}g\Delta t, \text{ so } \frac{1}{2}g\Delta t = v_i \sin \theta \text{ or } \Delta t = \frac{2v_i \sin \theta}{g}.$$

Substitute this value for Δt into $R = \Delta x = (v_i \cos \theta)\Delta t$.

$$R = (v_i \cos \theta)\Delta t = (v_i \cos \theta)\left(\frac{2v_i \sin \theta}{g}\right) = \frac{2v_i^2 \sin \theta \cos \theta}{g}$$

Substitute the given values.

$$R = \frac{2(1.46 \times 10^3 \text{ m/s})^2 \sin 55° \cos 55°}{9.80 \text{ m/s}^2} = \boxed{200 \text{ km}}$$

57. Strategy Let the positive y-direction be up. Use Newton's second law. When the elevator is moving with constant speed, the only acceleration is due to gravity and the scale reading is equal to the person's weight. When the elevator is accelerating, the scale reading will differ from the weight depending on the direction and magnitude of the elevator's acceleration.

Solution In scenarios (b) and (c), the elevator is moving with constant speed—zero acceleration—and the scale reading is equal to the weight W. Use Newton's second law to determine the scale readings in (a), (d), and (e). The scale reading is equal to the normal force.

$$\sum F_y = N - W = ma_y, \text{ so } N = W + ma_y = W + \frac{W}{g}a_y = W\left(1 + \frac{a_y}{g}\right).$$

(a) $N = \left(1 + \frac{1.0 \text{ m/s}^2}{9.8 \text{ m/s}^2}\right)W = 1.10W$; (d) $N = \left(1 + \frac{-2.0 \text{ m/s}^2}{9.8 \text{ m/s}^2}\right)W = 0.80W$; (e) $N = \left(1 + \frac{-2.0 \text{ m/s}^2}{9.8 \text{ m/s}^2}\right)W = 0.80W$

Ranking the scale readings from highest to lowest, we have $\boxed{\text{(a), (b) = (c), (d) = (e)}}$.

61. Strategy When Jaden is on the ground, his weight is equal to $mg = 600$ N. While on the accelerating elevator, his apparent weight is 550 N. Since 550 N < 600 N, the acceleration must be downward.

Solution According to Newton's second law, $\Sigma F_y = N - W = ma_y$, so

$$a_y = \frac{N - W}{m} = \frac{W' - W}{W/g} = g\left(\frac{W'}{W} - 1\right) = (9.80 \text{ m/s}^2)\left(\frac{550 \text{ N}}{600 \text{ N}} - 1\right) = -0.8 \text{ m/s}^2, \text{ or}$$

$$\vec{\mathbf{a}} = \boxed{0.8 \text{ m/s}^2 \text{ downward}}.$$

63. Strategy Refer to Example 4.12.

Solution

(a) The elevator is accelerating downward, so $a_y = -0.50 \text{ m/s}^2$.

$$W' = \frac{W}{g}(g + a_y) = W\left(1 + \frac{a_y}{g}\right) = (598 \text{ N})\left(1 + \frac{-0.50 \text{ m/s}^2}{9.80 \text{ m/s}^2}\right) = \boxed{567 \text{ N}}$$

(b) Since the elevator is moving downward and slowing down, it is accelerating upward, so $a_y = 0.50 \text{ m/s}^2$.

$$W' = W\left(1 + \frac{a_y}{g}\right) = (598 \text{ N})\left(1 + \frac{0.50 \text{ m/s}^2}{9.80 \text{ m/s}^2}\right) = \boxed{629 \text{ N}}$$

65. Strategy The apparent weight is given by $W' = W(1 + a_y/g)$.

Solution Up is the positive direction. Find Felipe's actual weight.

$$W' = W\left(1 + \frac{a_y}{g}\right), \text{ so } W = \frac{W'}{1 + \frac{a_y}{g}} = \frac{750 \text{ N}}{1 + \frac{2.0 \text{ m/s}^2}{9.80 \text{ m/s}^2}} = \boxed{620 \text{ N}}.$$

69. Strategy $v_{1fx}^2 - v_{1ix}^2 = 2a_1 d_1$, where $a_1 = 10.0 \text{ ft/s}^2$ and d_1 is the distance to the point of no return.
$v_{2fx}^2 - v_{2ix}^2 = 2a_2 d_2$, where $a_2 = -7.00 \text{ ft/s}^2$ and d_2 is the distance from the point of no return to the end of the runway. The initial speed v_{1ix} and the final speed v_{2fx} are zero. The speed at the point of no return is $v_{1fx} = v_{2ix}$. Let $v_{1fx} = v_{2ix} = v$ for simplicity. Also, $d = d_1 + d_2$ is the length of the runway.

Solution From the setup, we have $v^2 = 2a_1 d_1$ and $-v^2 = 2a_2 d_2 = 2a_2(d - d_1)$.

Eliminate v^2.

$$2a_1 d_1 = 2a_2(d_1 - d)$$
$$a_1 d_1 = a_2 d_1 - a_2 d$$
$$(a_1 - a_2)d_1 = -a_2 d$$

$$d_1 = \frac{a_2}{a_2 - a_1}d = \frac{-7.00 \text{ ft/s}^2}{-7.00 \text{ ft/s}^2 - 10.0 \text{ ft/s}^2}(1.50 \text{ mi})\left(\frac{5280 \text{ ft}}{1 \text{ mi}}\right) = \boxed{3260 \text{ ft}}$$

Find the time to d_1 using Eq. (4-4).

$$d_1 = \frac{1}{2}a_1(\Delta t)^2, \text{ so } \Delta t = \sqrt{\frac{2d_1}{a_1}} = \sqrt{\frac{2(3260 \text{ ft})}{10.0 \text{ ft/s}^2}} = \boxed{25.5 \text{ s}}.$$

73. (a) Strategy Find the time it takes the ball to reach the ground using Eq. (4-9).

Solution Solve for Δt. The initial velocity in the vertical direction is zero.

$$\Delta y = v_{iy}\Delta t + \frac{1}{2}a_y(\Delta t)^2 = (0)\Delta t - \frac{1}{2}g(\Delta t)^2, \text{ so } \Delta t = \sqrt{-\frac{2\Delta y}{g}} = \sqrt{-\frac{2(-20.0 \text{ m})}{9.80 \text{ m/s}^2}} = \boxed{2.02 \text{ s}}.$$

(b) Strategy and Solution It would still take $\boxed{2.02 \text{ s}}$ for the ball to fall to the ground, since $v_y = 0$ for both cases.

(c) Strategy Use Eq. (4-9) and the quadratic formula to find the time.

Solution

$$\Delta y = (v_i \sin\theta)\Delta t - \frac{1}{2}g(\Delta t)^2$$

$$0 = \frac{1}{2}g(\Delta t)^2 - (v_i \sin\theta)\Delta t + \Delta y$$

$$0 = (4.90 \text{ m/s}^2)(\Delta t)^2 + (20.0 \text{ m/s})\sin 18°\Delta t - 20.0 \text{ m}$$

Solve for Δt using the quadratic formula.

$$\Delta t = \frac{-(20.0 \text{ m/s})\sin 18° \pm \sqrt{(20.0 \text{ m/s})^2 \sin^2 18° - 4(4.90 \text{ m/s}^2)(-20.0 \text{ m})}}{2(4.90 \text{ m/s}^2)} = 1.5 \text{ s or } -2.7 \text{ s}$$

$\Delta t > 0$, so $\Delta t = \boxed{1.5 \text{ s}}$.

77. Strategy Use the graph to answer the questions. The slope of the graph represents the acceleration of the ball.

Solution

(a) The ball reaches its maximum height the first time $v_y = 0$, or at $t = \boxed{0.30 \text{ s}}$.

(b) The time it takes for the ball to make the transition from its negative-most velocity to its positive-most velocity is the time that the ball is in contact with the floor.

$0.65 \text{ s} - 0.60 \text{ s} = \boxed{0.05 \text{ s}}$

(c) Using Eq. (4-9) and the definition of average acceleration, we find that the maximum height of the ball is

$$\Delta y = v_{iy}\Delta t + \frac{1}{2}a_y(\Delta t)^2 = v_{iy}\Delta t + \frac{\Delta v_y}{2\Delta t}(\Delta t)^2 = (3.0 \text{ m/s})(0.30 \text{ s}) + \frac{0 - 3.0 \text{ m/s}}{2}(0.30 \text{ s}) = \boxed{0.45 \text{ m}}.$$

(d) $a_y = \dfrac{\Delta v_y}{\Delta t} = \dfrac{0 - 3.0 \text{ m/s}}{0.30 \text{ s}} = -10 \text{ m/s}^2$, so the acceleration is $\boxed{10 \text{ m/s}^2 \text{ down}}$.

(e) $a_{av} = \dfrac{\Delta v}{\Delta t} = \dfrac{3.0 \text{ m/s} - (-3.0 \text{ m/s})}{0.05 \text{ s}} = 120 \text{ m/s}^2$, so the acceleration is $\boxed{120 \text{ m/s}^2 \text{ up}}$.

81. Strategy Use Eqs. (3-3), (4-1), and (4-9).

Solution Find the time of flight in terms of h and v_i.

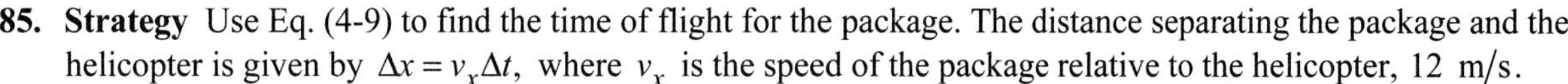

$\Delta x = v_i \Delta t = h$, so $\Delta t = \dfrac{h}{v_i}$.

Find the time of flight in terms of v_i and g.

$$h = \frac{1}{2}g(\Delta t)^2 = \frac{1}{2}g\left(\frac{h}{v_i}\right)^2 = \frac{gh^2}{2v_i^2}, \text{ so } h = \frac{2v_i^2}{g} \text{ and } \Delta t = \frac{h}{v_i} = \frac{1}{v_i}\left(\frac{2v_i^2}{g}\right) = \frac{2v_i}{g}.$$

Find the components of $\bar{\mathbf{v}}$.

$$v_x = v_i \text{ and } v_y = -g\Delta t = -g\left(\frac{2v_i}{g}\right) = -2v_i, \text{ so } \theta = \tan^{-1}\frac{v_y}{v_x} = \tan^{-1}\frac{-2v_i}{v_i} = \tan^{-1}(-2) = -63^\circ,$$

or $\boxed{63^\circ \text{ below the horizontal}}$.

85. Strategy Use Eq. (4-9) to find the time of flight for the package. The distance separating the package and the helicopter is given by $\Delta x = v_x \Delta t$, where v_x is the speed of the package relative to the helicopter, 12 m/s.

Solution Find the time the package takes to reach the ground.

$$y_f = y_i + v_{iy}\Delta t + \frac{1}{2}a_y(\Delta t)^2 = y_i + (0)\Delta t - \frac{1}{2}g(\Delta t)^2 = 0, \text{ so } \frac{1}{2}g(\Delta t)^2 = y_i, \text{ or } \Delta t = \sqrt{\frac{2y_i}{g}}.$$

Compute the horizontal distance.

$$\Delta x = v_x \Delta t = v_x \sqrt{\frac{2y_i}{g}} = (12 \text{ m/s})\sqrt{\frac{2(18 \text{ m})}{9.80 \text{ m/s}^2}} = \boxed{23 \text{ m}}$$

89. Strategy and Solution Since the acceleration is constant, the speed of the glider at the gate is its average speed as it passes through.

$$v_f = v_{\text{gate, av}} = \frac{8.0 \text{ cm}}{0.333 \text{ s}} = 24 \text{ cm/s}$$

Also, since the acceleration is constant, the displacement of the glider equals the average speed times the time of travel; $v_{\text{av}} = v_f/2 = 12 \text{ cm/s}$. Find Δt.

$$\Delta x = v_{\text{av}}\Delta t, \text{ so } \Delta t = \frac{\Delta x}{v_{\text{av}}}.$$

Find a_{av} from its definition; $a_{\text{av}} = a$.

$$a = \frac{\Delta v}{\Delta t} = \frac{\Delta v}{\Delta x/v_{\text{av}}} = \frac{v_{\text{av}}\Delta v}{\Delta x} = \frac{(12 \text{ cm/s})(24 \text{ cm/s} - 0)}{96 \text{ cm}} = 3.0 \text{ cm/s}^2. \text{ So, } \bar{\mathbf{a}} = \boxed{3.0 \text{ cm/s}^2 \text{ parallel to the velocity.}}$$

93. (a) Strategy Find h such that the final speed of the water is 5.0 m/s.

Solution Use Eq. (4-10) to find $h = -\Delta y$.

$$v_{fy}^2 - v_{iy}^2 = v_{fy}^2 - 0 = 2a_y \Delta y = -2g\Delta y = 2gh, \text{ so } v_{fy}^2 = 2gh, \text{ or } h = \frac{v_{fy}^2}{2g} = \frac{(5.0 \text{ m/s})^2}{2(9.80 \text{ m/s}^2)} = \boxed{1.3 \text{ m}}.$$

(b) Strategy and Solution Since the salmon can only swim through water that has fallen 1.3 m or less, the salmon must jump $1.5 \text{ m} - 1.3 \text{ m} = \boxed{0.2 \text{ m}}$.

(c) Strategy and Solution From part (a), we know that $v_{fy}^2 = 2gh$, so

$$v_{fy} = \sqrt{2gh} = \sqrt{2(9.80 \text{ m/s}^2)(0.2 \text{ m})} = \boxed{2 \text{ m/s}}.$$

(d) Strategy Consider the motion of the salmon and water relative to the ground.

Solution

$$v_{sgy} = v_{swy} + v_{wgy} = v_{swy} - \sqrt{2gh} = 5.0 \text{ m/s} - \sqrt{2(9.80 \text{ m/s}^2)(1.0 \text{ m})} = \boxed{0.6 \text{ m/s}}$$

Chapter 5

CIRCULAR MOTION

Conceptual Questions

1. Depressing the gas pedal is not the only way to make the car accelerate. The driver can also apply the brakes or turn the car to make it accelerate.

5. In uniform circular motion, the acceleration always points toward the center of the circle. Hence it remains perpendicular to the velocity the whole time. When a projectile is launched horizontally, the acceleration is initially perpendicular to the velocity, but does not remain so.

9. When the roller coaster turns hard to the right, the inertia of a rider's upper body keeps it moving in a straight line until it runs into the wall of the car. The wall exerts a normal force on the upper body that causes it to accelerate radially with the car. Thus, no force pushes the rider to the left as they enter a turn—the rider's inertia simply carries them forward while the car moves to the right.

Problems

1. **Strategy** Find the arc length swept out by the carnival swing.

   Solution Use Eq. (5-4).
   $$s = r\theta = (8.0 \text{ m})(120°)\left(\frac{2\pi \text{ rad}}{360°}\right) = \boxed{17 \text{ m}}$$

5. **(a) Strategy** Use the definition of average angular velocity.

   Solution In 1.0 s, the wheel rotates
   $$\Delta\theta = \omega\Delta t = \left(\frac{2.0 \text{ rev}}{0.080 \text{ s}}\right)\left(\frac{2\pi \text{ rad}}{\text{rev}}\right)(1.0 \text{ s}) = \boxed{160 \text{ rad}}.$$

   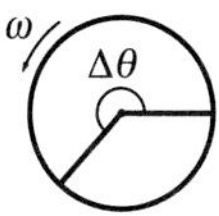

   (b) Strategy Use Eq. (5-7) to find the linear speed of a point on the wheel's rim.

   Solution
   $$v = r\omega = (30 \text{ cm})\left(\frac{2.0 \text{ rev}}{0.080 \text{ s}}\right)\left(\frac{2\pi \text{ rad}}{\text{rev}}\right) = \boxed{4700 \text{ cm/s}}$$

   (c) Strategy Use Eq. (5-8) to find the frequency of rotation of the wheel.

   Solution
   $$f = \frac{1}{T} = \frac{2.0 \text{ rev}}{0.080 \text{ s}} = \boxed{25 \text{ Hz}}$$

9. **Strategy** Use the conversion factor between degrees and radians and $s = r\theta$, where $s = 100.0$ ft, $\theta = 1.5°$, and r is the radius of curvature.

   Solution Find the radius of curvature of a "1.5° curve".
   $$r = \frac{s}{\theta} = \frac{100.0 \text{ ft}}{1.5°}\left(\frac{360°}{2\pi \text{ rad}}\right) = \boxed{3800 \text{ ft}}$$

13. **Strategy** Use the relationship between angular speed and radial acceleration.

   Solution The number of seconds in one day is 86,400, so the angular speed of the Earth (and baobab) is $|\omega| = 2\pi$ rad/86,400 s. Compute the radial acceleration.

   $$a_r = \omega^2 r = \omega^2 R_{\text{Earth}} = \left(\frac{2\pi \text{ rad}}{86,400 \text{ s}}\right)^2 (6.371 \times 10^6 \text{ m}) = \boxed{3.37 \text{ cm/s}^2}$$

17. **Strategy** Use dimensional analysis.

   Solution The dimensions are $[L]/[T]$ for v, $1/[T]$ for ω, and $[L]$ for r.

   $$v\omega: \quad \frac{[L]}{[T]} \cdot \frac{1}{[T]} = [L]/[T]^2$$

   $$\frac{v^2}{r}: \quad \left(\frac{[L]}{[T]}\right)^2 \cdot \frac{1}{[L]} = \frac{[L]^2}{[T]^2} \cdot \frac{1}{[L]} = [L]/[T]^2$$

   $$\omega^2 r: \quad \left(\frac{1}{[T]}\right)^2 \cdot [L] = \frac{1}{[T]^2} \cdot [L] = [L]/[T]^2$$

   Since the dimensions of acceleration are $[L]/[T]^2$, all three expressions are verified.

19. **(a) Strategy and Solution** Let the tensions in strings A and B be T_A and T_B, respectively. Draw the diagram.

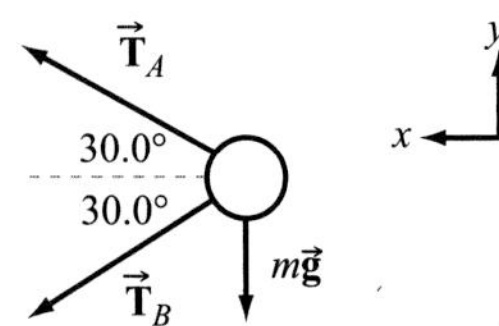

   (b) Strategy Use Newton's second law and the relationship between radial acceleration and angular speed.

   Solution The net force must point (horizontally) toward the pole, since the ball is in uniform circular motion.

   $$\Sigma F_y = T_A \sin 30.0° - T_B \sin 30.0° - mg = 0, \text{ so } T_A - T_B = \frac{mg}{\sin 30.0°} \quad (1).$$

   $$\Sigma F_x = T_A \cos 30.0° + T_B \cos 30.0° = ma_x = m\omega^2 r, \text{ so } T_A + T_B = \frac{m\omega^2 r}{\cos 30.0°} \quad (2).$$

   Add (1) and (2) to find T_A. Note that $r = (15.0 \text{ cm}) \cos 30.0°$.

   $$2T_A = \frac{mg}{\sin 30.0°} + \frac{m\omega^2 r}{\cos 30.0°}, \text{ so}$$

   $$T_A = \frac{m}{2}\left(\frac{g}{\sin 30.0°} + \frac{\omega^2 r}{\cos 30.0°}\right) = \frac{0.100 \text{ kg}}{2}\left[\frac{9.80 \text{ m/s}^2}{\sin 30.0°} + \frac{(6.00\pi \text{ rad/s})^2 (0.150 \text{ m}) \cos 30.0°}{\cos 30.0°}\right] = \boxed{3.64 \text{ N}}.$$

   For T_B, we have $T_B = T_A - \frac{mg}{\sin 30.0°} = 3.64 \text{ N} - \frac{(0.100 \text{ kg})(9.80 \text{ m/s}^2)}{\sin 30.0°} = \boxed{1.68 \text{ N}}$.

21. **(a) Strategy** Use Newton's second law and the relationship between linear speed and radial acceleration.

   Solution According to Newton's second law,

   $$\Sigma F_r = T = ma_r = m\frac{v^2}{r} = m\frac{v^2}{L}, \text{ thus, } T = \boxed{\frac{mv^2}{L}}.$$

(b) Strategy Draw a free-body diagram for the rock. Use Newton's second law.

Solution Decompose the force into vertical (y) and radial (r) components.

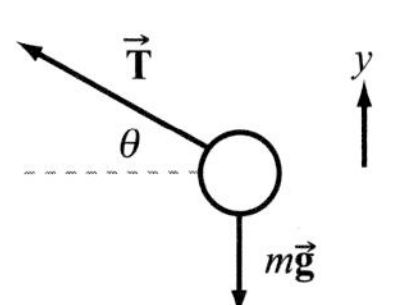

$$\Sigma F_{\mathrm{r}} = T_{\mathrm{r}} = \frac{mv^2}{r} = \frac{mv^2}{L\cos\theta} \text{ and } \Sigma F_y = T_y - mg = 0.$$

Find the magnitude of the tension.

$$T = \sqrt{T_{\mathrm{r}}^2 + T_y^2} = \sqrt{\left(\frac{mv^2}{L\cos\theta}\right)^2 + (mg)^2}$$

$$= \boxed{\; m\sqrt{g^2 + \left(\frac{v^2}{L\cos\theta}\right)^2} \;}$$

25. Strategy Let the x-axis point toward the center of curvature and the y-axis point upward. Draw a free-body diagram. Use Newton's second law and the relationship between radial acceleration and linear speed.

Solution

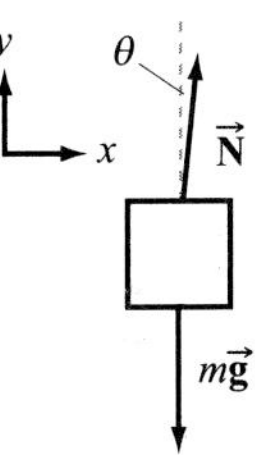

$$\Sigma F_y = N\cos\theta - mg = 0, \text{ so } N\cos\theta = mg, \text{ and } \Sigma F_x = N\sin\theta = ma_{\mathrm{r}} = m\frac{v^2}{r}.$$

Solve for v.

$$\frac{N\sin\theta}{N\cos\theta} = \frac{m\frac{v^2}{r}}{mg}, \text{ so } v^2 = rg\tan\theta, \text{ or}$$

$$v = \sqrt{rg\tan\theta} = \sqrt{(120\ \mathrm{m})(9.80\ \mathrm{m/s^2})\tan 3.0°} = \boxed{7.9\ \mathrm{m/s}}.$$

29. Strategy Let the x-axis point toward the center of curvature and the y-axis point upward. Draw a free-body diagram. Use Newton's second law and the relationship between radial acceleration and linear speed.

Solution

(a) $\Sigma F_y = N\cos\theta - mg - f\sin\theta = 0$ and $\Sigma F_x = N\sin\theta + f\cos\theta = mv^2/r$.

Solve for N in the first equation and substitute into the second.

$$N = \frac{f\sin\theta + mg}{\cos\theta}, \text{ so}$$

$$\frac{f\sin\theta + mg}{\cos\theta}\sin\theta + f\cos\theta = m\frac{v^2}{r}$$

$$f\sin^2\theta + mg\sin\theta + f\cos^2\theta = m\frac{v^2}{r}\cos\theta$$

$$f(\sin^2\theta + \cos^2\theta) = m\frac{v^2}{r}\cos\theta - mg\sin\theta$$

$$f(1) = m\left(\frac{v^2}{r}\cos\theta - g\sin\theta\right)$$

$$f = (1400\ \mathrm{kg})\left[\frac{(32\ \mathrm{m/s})^2}{410\ \mathrm{m}}\cos 5.0° - (9.80\ \mathrm{m/s^2})\sin 5.0°\right] = \boxed{2300\ \mathrm{N}}$$

(b) Set the expression found for the force of friction equal to zero.

$$f = 0 = m\left(\frac{v^2}{r}\cos\theta - g\sin\theta\right), \text{ so } \frac{v^2}{r}\cos\theta = g\sin\theta, \text{ or}$$

$$v = \sqrt{gr\tan\theta} = \sqrt{(9.80 \text{ m/s}^2)(410 \text{ m})\tan 5.0°} = \boxed{19 \text{ m/s}}.$$

33. Strategy Use $v = r\omega$ and $\omega = 2\pi/T$, where the radius is the average Earth-Sun distance and the period is one year.

Solution

$$v = r\omega = r\left(\frac{2\pi}{T}\right) = \frac{2\pi r}{T} = \frac{2\pi(1.50\times10^{11} \text{ m})}{1 \text{ y}}\left(\frac{1 \text{ y}}{3.156\times10^7 \text{ s}}\right) = \boxed{2.99\times10^4 \text{ m/s}}$$

35. Strategy According to Kepler's third law, $r^3 \propto T^2$. Form a proportion.

Solution Find the orbital period of the second satellite.

$$\left(\frac{4.0r}{r}\right)^3 = 64 = \left(\frac{T_{4.0}}{T}\right)^2 = \frac{T_{4.0}^{\;2}}{T^2}, \text{ so } T_{4.0}^{\;2} = 64T^2, \text{ or } T_{4.0} = 8.0T = 8.0(16 \text{ h}) = \boxed{130 \text{ h}}.$$

37. Strategy Use Eq. (5-14) with the mass of Jupiter in place of the mass of the Sun.

Solution Solve for r.

$$\frac{4\pi^2}{GM_J}r^3 = T^2, \text{ so } r^3 = \frac{GM_J}{4\pi^2}T^2, \text{ or } r = \sqrt[3]{\frac{GM_J}{4\pi^2}T^2}.$$

Compute the distance from the center of Jupiter for each satellite.

$$r_{Io} = \sqrt[3]{\frac{GM_J}{4\pi^2}T_{Io}^{\;2}} = \sqrt[3]{\frac{(6.674\times10^{-11} \text{ N}\cdot\text{m}^2/\text{kg}^2)(1.9\times10^{27} \text{ kg})}{4\pi^2}(1.77 \text{ d})^2\left(\frac{86,400 \text{ s}}{1 \text{ d}}\right)^2} = \boxed{420,000 \text{ km}}$$

$$r_{Europa} = \sqrt[3]{\frac{GM_J}{4\pi^2}T_{Europa}^{\;2}} = \sqrt[3]{\frac{(6.674\times10^{-11} \text{ N}\cdot\text{m}^2/\text{kg}^2)(1.9\times10^{27} \text{ kg})}{4\pi^2}(3.54 \text{ d})^2\left(\frac{86,400 \text{ s}}{1 \text{ d}}\right)^2} = \boxed{670,000 \text{ km}}$$

41. (a) Stratey Draw a diagram. Use Newton's second law and the relationship between radial acceleration and linear speed.

Solution Find the tension.

$$\Sigma F_r = T - mg = ma_r, \text{ so}$$

$$T = m(g + a_r) = m\left(g + \frac{v^2}{r}\right) = (1.0 \text{ kg})\left[9.80 \text{ m/s}^2 + \frac{(1.6 \text{ m/s})^2}{0.80 \text{ m}}\right] = \boxed{13 \text{ N}}.$$

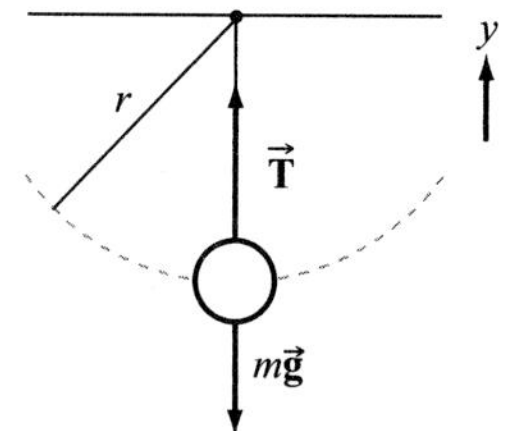

(b) Strategy and Solution If the bob were at rest, the tension would be equal to the weight of the bob. Since the bob is moving and its velocity is changing—its vertical component is increasing from zero—the tension must be greater than the weight. Thus, $\boxed{\text{the bob has an upward acceleration, so the net } F_y \text{ must be upward and greater than the weight of the bob}}$.

45. Strategy Use Eq. (5-20) to find the constant angular acceleration.

Solution Since the cyclist starts from rest, the initial angular velocity is zero.

$$\Delta\theta = \omega_i \Delta t + \frac{1}{2}\alpha(\Delta t)^2 = (0)\Delta t + \frac{1}{2}\alpha(\Delta t)^2 = \frac{1}{2}\alpha(\Delta t)^2, \text{ so } \alpha = \frac{2\Delta\theta}{(\Delta t)^2} = \frac{2(8.0 \text{ rev})\left(\frac{2\pi \text{ rad}}{\text{rev}}\right)}{(5.0 \text{ s})^2} = \boxed{4.0 \text{ rad/s}^2}.$$

49. (a) Strategy Since the car moves with constant acceleration, use Eq. (4-5) and $C = 2\pi r$ to find the speed of the car.

Solution The distance traveled by the car is $\Delta x = \dfrac{C}{4} = \dfrac{2\pi r}{4} = \dfrac{\pi r}{2}$. Find the final speed.

$$v_{\text{fx}}^2 - v_{\text{ix}}^2 = v_{\text{fx}}^2 - 0 = v_t^2 = 2a_t \Delta x = 2a_t\left(\frac{\pi r}{2}\right) = a_t\pi r, \text{ so}$$

$$v_t = \sqrt{a_t\pi r} = \sqrt{(2.00 \text{ m/s}^2)\pi(50.0 \text{ m})} = \boxed{17.7 \text{ m/s}}.$$

(b) Strategy Use the relationship between radial acceleration and linear speed.

Solution

$$a_r = \frac{v_t^2}{r} = \frac{a_t\pi r}{r} = \pi a_t = \pi(2.00 \text{ m/s}^2) = \boxed{6.28 \text{ m/s}^2}$$

(c) Strategy Draw a diagram. Use the Pythagorean Theorem to find the magnitude of the total acceleration. Then use trigonometry to find its direction.

Solution Find the total acceleration.

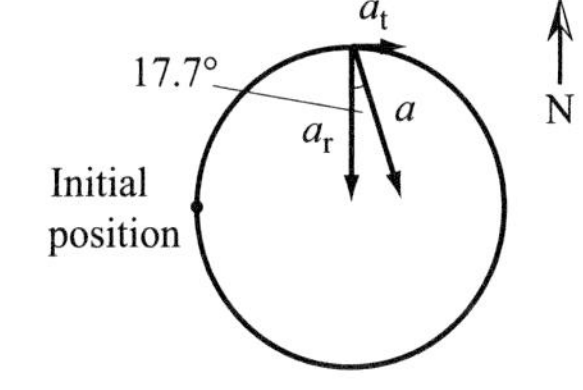

$$a = \sqrt{a_r^2 + a_t^2} = \sqrt{\pi^2 a_t^2 + a_t^2} = a_t\sqrt{\pi^2 + 1} = (2.00 \text{ m/s}^2)\sqrt{\pi^2 + 1}$$
$$= 6.59 \text{ m/s}^2$$

Let $+y$ be north and $+x$ be east. Then, $a_x = a_t$ and $a_y = -a_r$.

$$\theta = \tan^{-1}\frac{a_y}{a_x} = \tan^{-1}\frac{-a_r}{a_t} = \tan^{-1}\frac{-\pi a_t}{a_t} = \tan^{-1}(-\pi)$$
$$= -72.3° \text{ or } 17.7° \text{ east of south}$$

Thus, $\vec{a} = \boxed{6.59 \text{ m/s}^2 \text{ at an angle of } 17.7° \text{ east of south}}$.

53. (a) Strategy Use Eq. (5-18) to find the time it takes for the rotor to come to rest.

Solution The acceleration is opposite the rotation of the rotor, so it is negative.

$$\Delta t = \frac{\omega_f - \omega_i}{\alpha} = \frac{0 - 5.0\times10^5 \text{ rad/s}}{-0.40 \text{ rad/s}^2} = \boxed{1.3\times10^6 \text{ s}}$$

(b) Strategy Use Eq. (5-21) to find the number of revolutions the rotor spun before it stopped.

Solution

$$\Delta\theta = \frac{\omega_f^2 - \omega_i^2}{2\alpha}$$

$$\left(\frac{1 \text{ rev}}{2\pi \text{ rad}}\right)\Delta\theta = \frac{0 - (5.0\times10^5 \text{ rad/s})^2}{2(-0.40 \text{ rad/s}^2)}\left(\frac{1 \text{ rev}}{2\pi \text{ rad}}\right) = \boxed{5.0\times10^{10} \text{ rev}}$$

55. Strategy Draw a free-body diagram for the bob. Use Newton's second law and the relationship between radial acceleration and linear speed.

Solution Refer to the figure.

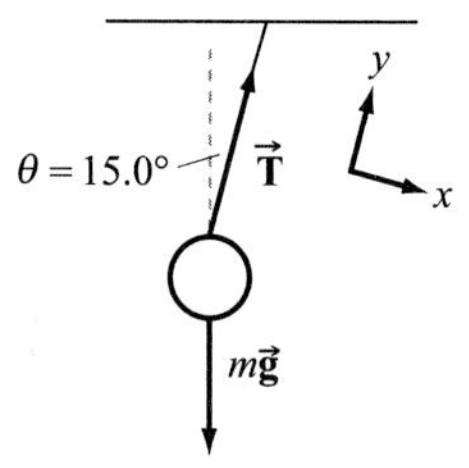

$\Sigma F_y = T - mg\cos\theta = ma_r$ and $\Sigma F_x = mg\sin\theta = ma_t$, so

$a_t = g\sin\theta = (9.80 \text{ m/s}^2)\sin 15.0° = \boxed{2.54 \text{ m/s}^2}$ and

$a_r = \dfrac{v^2}{r} = \dfrac{(1.40 \text{ m/s})^2}{0.800 \text{ m}} = \boxed{2.45 \text{ m/s}^2}$. The tension is

$T = m(a_r + g\cos\theta) = (1.00 \text{ kg})\left[2.45 \text{ m/s}^2 + (9.80 \text{ m/s}^2)\cos 15.0°\right] = \boxed{11.9 \text{ N}}$.

57. Strategy The strength of the artificial gravity is equal to the radial acceleration.

Solution Compute the magnitude of the radial acceleration.

$$a_r = \omega^2 r = \left(\frac{\omega^2 r}{g}\right)g = \frac{(4.0 \text{ rev/s})^2 \left(\frac{2\pi \text{ rad}}{\text{rev}}\right)^2 (0.25 \text{ m})}{9.80 \text{ m/s}^2}g = \boxed{16g}$$

59. Strategy Use $a_r = \omega^2 r$ to find the angular speed required.

Solution The magnitude of the radial acceleration must be the same as the magnitude of the gravitational field strength.

$$a_r = \omega^2 r = g, \text{ so } \omega = \sqrt{\frac{g}{r}} = \sqrt{\frac{9.80 \text{ m/s}^2}{0.20 \text{ m}}} = \boxed{7.0 \text{ rad/s}}.$$

61. (a) Strategy Earth rotates once every 24 hours. Use $a_r = \omega^2 r$.

Solution The radius r is the radius of Earth.

$$a_r = \omega^2 r = \left(\frac{2\pi \text{ rad}}{24 \text{ h}}\right)^2 \left(\frac{1 \text{ h}}{3600 \text{ s}}\right)^2 (6.371\times10^6 \text{ m}) = \boxed{0.034 \text{ m/s}^2}$$

(b) Strategy and Solution Since the object is on the outside of the Earth, the rotation of the Earth seemingly "pushes outward" on the object. (This is sometimes referred to as the fictitious "centrifugal force.") So, since $g - a < g$ for $a > 0$, the object's apparent weight is $\boxed{\text{less}}$ than its true weight.

(c) Strategy Compare the radial acceleration to the gravitational field strength.

Solution Divide the radial acceleration by g.

$$(100\%)\frac{a_r}{g} = (100\%)\frac{0.0337}{9.80} = 0.34\%$$

The actual weight is reduced by this amount, so the apparent weight is $\boxed{0.34\% \text{ smaller}}$ than the actual weight.

(d) Strategy and Solution The rotation of the Earth $\boxed{\text{at the poles}}$ has no effect on the reading of a bathroom scale, so the actual weight is measured.

63. (a) Strategy At the top, $\vec{g}$ and $\vec{a}$ are both directed downward. Draw a free-body diagram. Use Newton's second law.

Solution

$\Sigma F_{\mathrm{r}} = N - mg = ma_{\mathrm{r}} = m(-a_y)$, so $W' = N = mg - ma_y$.

The apparent weight is less than the true weight by ma_y. Thus, the lower weight,

$\boxed{518.5\ \mathrm{N}}$, is measured at the top.

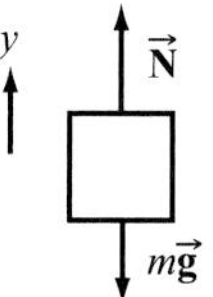

(b) Strategy At the bottom, $\vec{g}$ is directed downward and $\vec{a}$ is directed upward. Draw a free-body diagram.

Solution

$\Sigma F_{\mathrm{r}} = N - mg = ma_{\mathrm{r}} = m(a_y)$, so $W' = N = mg + ma_y$.

The apparent weight is greater than the true weight by ma_y. Thus, the higher weight,

$\boxed{521.5\ \mathrm{N}}$, is measured at the bottom.

(c) Strategy The apparent weight at the top is given by $W'_{\mathrm{top}} = W(1 - a_y/g)$, where $W = mg$. Use

$a_{\mathrm{r}} = \omega^2 r = a_y$ to find the radius.

Solution Solve for the radius r.

$$W'_{\mathrm{top}} = W\left(1 - \frac{a_y}{g}\right) = W\left(1 - \frac{\omega^2 r}{g}\right),\ \text{so}\ r = \frac{g}{\omega^2}\left(1 - \frac{W'_{\mathrm{top}}}{W}\right) = \frac{9.80\ \mathrm{m/s^2}}{(0.025\ \mathrm{rad/s})^2}\left(1 - \frac{518.5\ \mathrm{N}}{520.0\ \mathrm{N}}\right) = \boxed{45\ \mathrm{m}}.$$

65. Strategy Use Eq. (5-18).

Solution After one minute:

$\omega_{\mathrm{f}} - \omega_{\mathrm{i}} = 0.80\omega - \omega = -0.20\omega = \alpha\Delta t$

After three minutes:

$\omega_{\mathrm{f}} - \omega_{\mathrm{i}} = \omega_{\mathrm{f}} - \omega = \alpha(3\Delta t) = 3\alpha\Delta t = 3(-0.20\omega)$, so $\omega_{\mathrm{f}} = -0.60\omega + \omega = \boxed{0.40\omega}$.

69. Strategy Use the relationship between radial acceleration and linear speed. Recall that the circumference of a circle is given by $C = 2\pi r$.

Solution The distance is $C/2 = \pi r = v\Delta t$ and $a_{\mathrm{r}} = v^2/r$, so $v = \sqrt{a_{\mathrm{r}} r}$. The time to complete the U-turn is given by $\Delta t = \pi r/v = \pi r/\sqrt{a_{\mathrm{r}} r} = \pi\sqrt{r/a_{\mathrm{r}}}$, so the larger the radius the greater the time to complete the U-turn. Therefore, the $\boxed{\text{smallest}}$ possible radius should be used to make the turn. Calculating the minimum time required to complete the U-turn, we find that $\Delta t_{\mathrm{min}} = \pi\sqrt{(5.0\ \mathrm{m})/(3.0\ \mathrm{m/s^2})} = \boxed{4.1\ \mathrm{s}}$.

71. Strategy Earth rotates once per 24.0 hours.

Solution Compute the tangential speed of Mt. Kilimanjaro.

$$v = \omega r = \left(\frac{2\pi\ \mathrm{rad}}{24.0\ \mathrm{h}}\right)\left(\frac{1\ \mathrm{h}}{3600\ \mathrm{s}}\right)(6.378\times 10^6\ \mathrm{m} + 5895\ \mathrm{m}) = \boxed{464\ \mathrm{m/s}}$$

73. Strategy Use the definition of average angular velocity.

Solution Compute the number of degrees that the drill rotates.

$$\omega_{\text{av}} = \frac{\Delta\theta}{\Delta t}, \text{ so } \Delta\theta = \omega_{\text{av}}\Delta t = (3.14\times10^4 \text{ rad/s})\left(\frac{360°}{2\pi \text{ rad}}\right)(1.00 \text{ s}) = \boxed{1.80\times10^6 \text{ degrees}}.$$

77. Strategy Use the relationship between linear and angular speed.

Solution

$$v = r|\omega| = r\left(\frac{2\pi}{T}\right) = \frac{2\pi r}{T} = \frac{2\pi(2\times10^{17} \text{ km})}{200\times10^6 \text{ y}}\left(\frac{1 \text{ y}}{3.156\times10^7 \text{ s}}\right) = \boxed{200 \text{ km/s}}$$

81. Strategy For each revolution of the flagellum, the bacterium moves the distance of the pitch.

Solution Compute the speed of the bacterium.

$$v = (1.0 \text{ μm/rev})(110 \text{ rev/s}) = \boxed{110 \text{ μm/s}}$$

85. Strategy Draw a free-body diagram. Use the relationship between radial acceleration and angular speed and Newton's second law.

Solution

$$\Sigma F_y = T\cos\theta - mg = 0, \text{ so } T = \frac{mg}{\cos\theta}, \text{ and } \Sigma F_x = T\sin\theta = ma_r.$$

Use these results to find the angular speed.

$$T\sin\theta = ma_r$$

$$\frac{mg}{\cos\theta}\sin\theta = ma_r$$

$$g\tan\theta = \omega^2 r$$

$$\omega = \sqrt{\frac{g\tan\theta}{r}} = \sqrt{\frac{(9.80 \text{ m/s}^2)\tan 45.0°}{6.00 \text{ m} + (4.25 \text{ m})\sin 45.0°}} = \boxed{1.04 \text{ rad/s}}$$

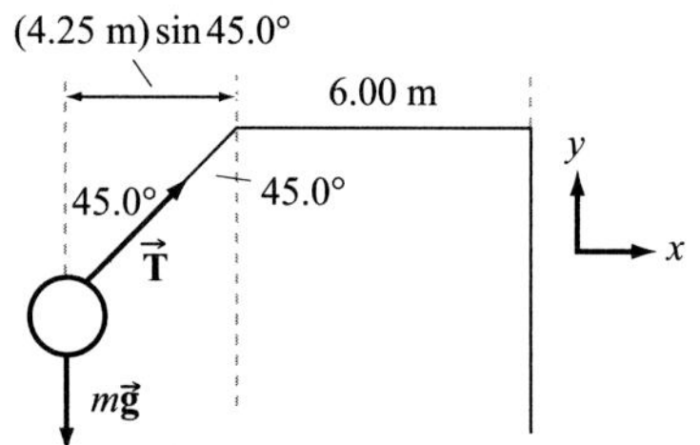

89. Strategy and Solution The cutting tool moves one inch in the time $\Delta t = \dfrac{d}{v} = \dfrac{1 \text{ in}}{0.080 \text{ in/s}}$. The lathe chuck must complete 18 revolutions in the time Δt. Thus, the rotational speed must be $\dfrac{18 \text{ rev}}{\frac{1 \text{ in}}{0.080 \text{ in/s}}} = \boxed{1.4 \text{ rev/s}}$.

REVIEW AND SYNTHESIS: CHAPTERS 1–5

Review Exercises

1. **Strategy** Replace the quantities with their units.

   Solution Find the units of the spring constant k.
   $$F = kx, \text{ so } k = \frac{F}{x}, \text{ and the units of } k \text{ are } \boxed{\text{N/m}} = \frac{\text{kg} \cdot \text{m/s}^2}{\text{m}} = \boxed{\text{kg/s}^2}.$$

5. **(a) Strategy** Find the total distance traveled and the time of travel. Then divide the distance by the time to obtain the average speed.

   Solution Find Mike's average speed.
   $$\Delta x = 50.0 \text{ m} + 34.0 \text{ m} = 84.0 \text{ m and } \Delta t = \frac{50.0 \text{ m}}{1.84 \text{ m/s}} + \frac{34.0 \text{ m}}{1.62 \text{ m/s}} = 48.2 \text{ s, so the average speed is}$$
   $$v_{\text{av}} = \frac{\Delta x}{\Delta t} = \frac{84.0 \text{ m}}{48.2 \text{ s}} = \boxed{1.74 \text{ m/s}}.$$

   (b) Strategy Find Mike's total displacement and divide it by the time found in part (a) to obtain his average velocity. Let his initial direction be positive.

   Solution Mike's total displacement is $\Delta \vec{r} = 50.0 \text{ m forward} - 34.0 \text{ m back} = 16.0 \text{ m forward}$. So, his average velocity is $\vec{v}_{\text{av}} = \frac{\Delta \vec{r}}{\Delta t} = \frac{16.0 \text{ m forward}}{48.2 \text{ s}} = 0.332 \text{ m/s forward, or}$
   $$\boxed{0.332 \text{ m/s in his original direction of motion}}.$$

9. **Strategy** Let north be up. Using a ruler and a protractor, draw the force vectors to scale; then, find the sum of the force vectors graphically.

   Solution Draw the diagram and measure the length and angle of the sum of the force vectors.

   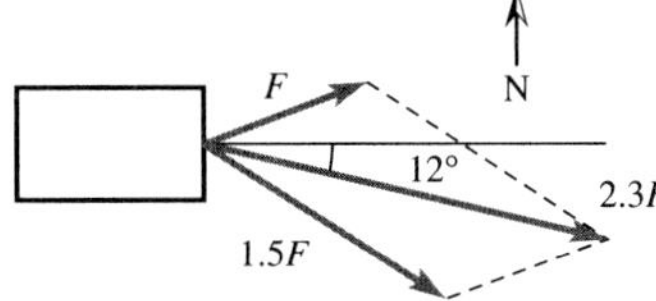

   The net force has a magnitude of about 2.3F, where F is the magnitude of the force with which Sandy pulls. The net force is at an angle of about 12° south of east. So, $\boxed{\text{the cart will go off the road toward south}}$.

13. **Strategy** Use Newton's second law to find the acceleration due to friction. Then use the acceleration and distance the plate must travel to determine the necessary initial speed.

   Solution According to Newton's second law, $\Sigma F_y = N - mg = 0$ and $\Sigma F_x = -f_k = ma$.

   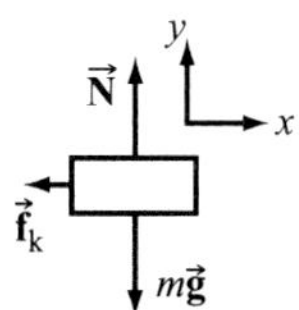

   So, $a = -\frac{f_k}{m} = -\frac{\mu_k N}{m} = -\frac{\mu_k mg}{m} = -\mu_k g$. Find the initial speed.
   $$v_f^2 - v_i^2 = 0 - v_i^2 = 2a\Delta x = -2\mu_k g \Delta x, \text{ so}$$
   $$v_i = \sqrt{2\mu_k g \Delta x} = \sqrt{2(0.32)(9.80 \text{ m/s}^2)(0.44 \text{ m})} = \boxed{1.7 \text{ m/s}}.$$

17. **Strategy and Solution** Consider a cord attached to a wall at one end and pulled by one of the boys at the other end. The cord does not accelerate when the boy pulls it; thus, the force on the cord from the wall must be equal in magnitude to the pulling force. This situation is identical to the one in which the two boys pull from opposite ends of the cord—the tension in the cord is the same as the case when only one boy is pulling. However, if both pull from one end, the tension is doubled, so Stefan's plan is superior and thus more likely to work.

21. **(a) Strategy** Use Eq. (4-10). Let the $+y$-direction be down and neglect air resistance.

    Solution Let the initial speed for all three rocks be v_i and the vertical distance from the cliff to the ground be h. For the first rock (thrown straight down):

$$v_{fy}^2 - v_{iy}^2 = v_f^2 - v_i^2 = 2gh, \text{ so } v_f = \sqrt{v_i^2 + 2gh}.$$

For the second rock (thrown straight up):

$$v_{fy}^2 - v_{iy}^2 = v_f^2 - v_i^2 = 2gh, \text{ so } v_f = \sqrt{v_i^2 + 2gh}.$$

For the third rock (thrown horizontally):

$$v_{fy}^2 - v_{iy}^2 = v_{fy}^2 - 0 = v_{fy}^2 = 2gh \text{ and } v_{fx} = v_{ix} = v_i, \text{ so } v_f = \sqrt{v_{fx}^2 + v_{fy}^2} = \sqrt{v_i^2 + 2gh}.$$

Therefore, just before the rocks hit the ground at the bottom of the cliff, all three have the same final speed.

    (b) Strategy Use the result obtained for the final speed in part (a).

    Solution Compute the final speed.

$$v_f = \sqrt{v_i^2 + 2gh} = \sqrt{(10.0 \text{ m/s})^2 + 2(9.80 \text{ m/s}^2)(15.00 \text{ m})} = \boxed{19.8 \text{ m/s}}$$

25. **Strategy** Draw a free-body diagram for the package and use Newton's second law.

    Solution Find the maximum acceleration the truck can have without the package falling of the back.

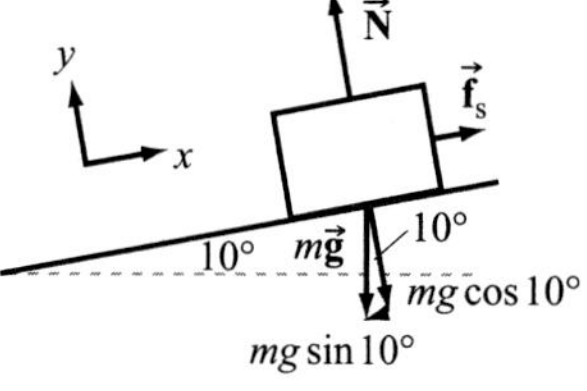

$$\Sigma F_x = f_s - mg \sin 10° = \mu_s N - mg \sin 10° = ma_x \text{ and}$$
$$\Sigma F_y = N - mg \cos 10° = 0, \text{ so } N = mg \cos 10°.$$

The maximum acceleration allowed for the truck is
$$a_x = g(\mu_s \cos 10° - \sin 10°)$$
$$= (9.80 \text{ m/s}^2)(0.380 \cos 10° - \sin 10°) = \boxed{2.0 \text{ m/s}^2}.$$

29. **(a) Strategy** Use the equations for circular motion.

    Solution Find the angular acceleration.

$$\Delta \omega = \omega_f - \omega_i = \omega_f - 0 = \alpha \Delta t, \text{ so } \alpha = \frac{\omega_f}{\Delta t}.$$

Find the tangential acceleration.

$$a = r\alpha = r\frac{\omega_f}{\Delta t} = (0.100 \text{ m})\frac{1.00 \text{ Hz}}{0.800 \text{ s}}\left(\frac{2\pi \text{ rad}}{\text{cycle}}\right) = \boxed{0.785 \text{ m/s}^2}$$

    (b) Strategy While the hamster is running, the forces on it are due to gravity and the normal force due to the wheel. Use Newton's second law.

    Solution Find the normal force on the hamster.

$$\Sigma F_r = N - mg = ma_r = m\frac{v^2}{r} = m\frac{(r\omega)^2}{r} = mr\omega^2, \text{ so}$$
$$N = m(g + r\omega^2) = (0.100 \text{ kg})[9.80 \text{ m/s}^2 + (0.100 \text{ m})(1.00 \text{ Hz})^2(2\pi \text{ rad/cycle})^2] = \boxed{1.37 \text{ N}}.$$

33. (a) Strategy Let $+y$ be down for m_1 and up for m_2, since $m_1 \gg m_2$. Use Newton's second law.

Solution Find the acceleration of each block.

For m_1: $\Sigma F_y = m_1 g - T = m_1 a_y$, so $T = m_1 g - m_1 a_y$.

For m_2: $\Sigma F_y = T - m_2 g = m_2 a_y$, so $T = m_2 g + m_2 a_y$.

T and a_y are identical in these two equations. Eliminate T.

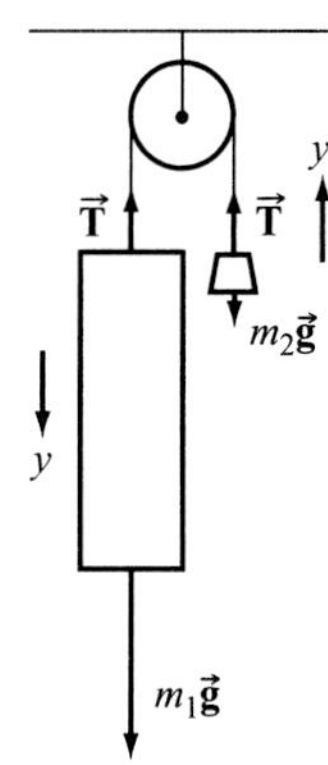

$$m_1 g - m_1 a_y = m_2 g + m_2 a_y$$
$$m_2 a_y + m_1 a_y = m_1 g - m_2 g$$
$$(m_1 + m_2) a_y = (m_1 - m_2) g$$
$$a_y = \frac{m_1 - m_2}{m_1 + m_2} g$$

Since $m_1 \gg m_2$, $m_1 - m_2 \approx m_1$ and $m_1 + m_2 \approx m_1$, so $a_y \approx \boxed{g}$.

(b) Strategy Use the results from part (a) for the tension and the vertical component of the acceleration.

Solution Find the tension.

$$T = m_2 g + m_2 a_y = m_2 g + m_2 \frac{m_1 - m_2}{m_1 + m_2} g = m_2 g \left(1 + \frac{m_1 - m_2}{m_1 + m_2}\right) = m_2 g \frac{m_1 + m_2 + m_1 - m_2}{m_1 + m_2} = \frac{2 m_1 m_2}{m_1 + m_2} g$$

Since $m_1 \gg m_2$, $m_1 + m_2 \approx m_1$, so $T \approx \frac{2 m_1 m_2}{m_1} g = \boxed{2 m_2 g}$.

MCAT Review

1. Strategy and Solution Gravity contributes an acceleration of $-g$. Air resistance is always opposite an object's direction of motion, so the vertical component of the acceleration contributed by air resistance is negative as well. According to Newton's second law, $F = ma$, so the magnitude of the acceleration due to air resistance is $a_R = F_R/m = bv^2/m$. Since we want the vertical component of acceleration, the correct answer is $\boxed{D}$, $-g - (bvv_y)/(0.5 \text{ kg})$.

2. Strategy Use the result for the range derived in Problem 4.48b.

Solution Assuming air resistance is negligible, the horizontal distance the projectile travels before returning to the elevation from which it was launched is $R = \dfrac{v_i^2 \sin 2\theta}{g} = \dfrac{(30 \text{ m/s})^2 \sin[2(40°)]}{9.80 \text{ m/s}^2} = 90$ m. Thus, the correct answer is $\boxed{C}$.

3. Strategy and Solution The magnitude of the horizontal component of air resistance is $F_R \cos\theta = bv^2 \cos\theta = bv(v\cos\theta) = bvv_x$. Thus, the correct answer is $\boxed{D}$.

4. **Strategy** Use Newton's second law to analyze each case. For simplicity, consider only vertical motion.

   Solution Let the positive y-direction be up.
   On the way up:

   $$\Sigma F_y = -mg - bv^2 = ma_y, \text{ so } a_y = -g - \frac{bv^2}{m}.$$

   On the way down:

   $$\Sigma F_y = -mg + bv^2 = ma_y, \text{ so } a_y = -g + \frac{bv^2}{m}.$$

   The magnitude of the acceleration is greater on the way up than on the way down. On the way up, the magnitude of the acceleration is never less than g. On the way down, it may be as small as zero. The projectile must travel the same distance in each case. So, when a projectile is rising, it begins with an initial speed which is reduced to zero relatively quickly due to the relatively large negative acceleration it experiences. When a projectile is falling, it begins with zero speed and is accelerated toward the ground by a smaller acceleration relative to when it is rising. Thus, it must take the projectile longer to reach the ground than to reach its maximum height; therefore, the correct answer is $\boxed{C}$.

5. **Strategy** Find the time it takes to cross the river. Use this time and the speed of the river to find how far downstream the raft travels while crossing. Then use the Pythagorean theorem to find the total distance traveled.

   Solution Let x be the width of the river and y be the distance traveled down the river during the crossing. The raft takes the time $\Delta t = x/v_{\text{raft}}$ to cross the river. During this time, the raft travels the distance $y = v_{\text{river}} \Delta t = v_{\text{river}}(x/v_{\text{raft}})$ down the river. Compute the distance traveled.

   $$\sqrt{x^2 + y^2} = \sqrt{x^2 + \left(\frac{v_{\text{river}}}{v_{\text{raft}}} x\right)^2} = x\sqrt{1 + \left(\frac{v_{\text{river}}}{v_{\text{raft}}}\right)^2} = (200 \text{ m})\sqrt{1 + \left(\frac{2 \text{ m/s}}{2 \text{ m/s}}\right)^2} = 283 \text{ m}$$

   The correct answer is $\boxed{C}$.

6. **Strategy** To row directly across the river, the component of the raft's velocity that is antiparallel to the current of the river must equal the speed of the current, 2 m/s.

   Solution Since the angle is relative to the shore, the antiparallel component of the raft's velocity is $(3 \text{ m/s})\cos\theta$. Set this equal to the speed of the current and solve for θ.

   $$(3 \text{ m/s})\cos\theta = 2 \text{ m/s, so } \cos\theta = \frac{2}{3}, \text{ or } \theta = \cos^{-1}\frac{2}{3}. \text{ The correct answer is } \boxed{D}.$$

7. **Strategy** Use Eq. (4-9).

   Solution Find the time it takes the rock to reach the ground.

   $$\Delta y = v_{iy}\Delta t + \frac{1}{2}a_y(\Delta t)^2 = (0)\Delta t - \frac{1}{2}g(\Delta t)^2, \text{ so } \Delta t = \sqrt{-\frac{2\Delta y}{g}} = \sqrt{-\frac{2(0 - 100 \text{ m})}{10 \text{ m/s}^2}} = 4.5 \text{ s}.$$

   The correct answer is $\boxed{A}$.

Chapter 6

CONSERVATION OF ENERGY

Conceptual Questions

1. Assuming the object can be treated as a point particle, the total work done on it by external forces is equal to the change in its kinetic energy. An object moving in a circle may be changing its speed as it goes around, so the total work done on it is not necessarily zero.

5. Yes, static friction can do work. As an example, imagine a book on a conveyor belt that carries it up an incline. The force of static friction on the book is directed upward along the surface of the belt and has a component that is parallel to the book's displacement. Thus, the force of static friction does positive work on the book. (At the same time, the work done on the book by gravity is negative and the total work done on the book is zero.)

9. The bicyclist requires a minimum amount of energy to climb the hill. This quantity is independent of the means that the bicyclist employs to acquire the energy and is solely a function of the height of the hill (the energy required is also affected by the work done by frictional and drag forces—the magnitude of this effect is approximately equal for any method used by the bicyclist to climb the hill and therefore doesn't affect our reasoning). After beginning the ascent, a component of the gravitational force acts in the direction opposite to the displacement thereby increasing the amount of negative work done on the rider with respect to the amount done while riding on flat land. Therefore, the rate at which the rider must do work to acquire the necessary energy is greater when pedaling uphill than when on flat land. It is thus advantageous to acquire as much energy as possible before the ascent when the amount of kinetic energy gained per amount of work done by the rider is greatest.

13. Zorba is correct. You get to a top speed sooner on the first slide, so it takes less time to get to the bottom, but the final speeds are the same from $mgh = \frac{1}{2}mv^2$.

Problems

1. **Strategy** Use Eq. (6-1).

   Solution Find the work done by Denise dragging her basket of laundry.
   $$W = F\Delta r \cos\theta = (30.0 \text{ N})(5.0 \text{ m})\cos 60.0° = \boxed{75 \text{ J}}$$

   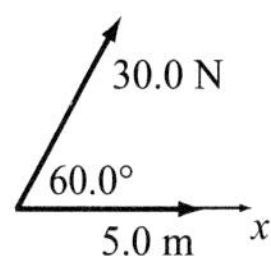

5. **Strategy** Use Newton's second law and Eq. (6-2).

   Solution Find the net force on the barge.
   $$\Sigma F_y = T\sin\theta - T\sin\theta = 0 \text{ and } \Sigma F_x = T\cos\theta + T\cos\theta = F_x.$$
   Find the work done on the barge.
   $$W = F_x\Delta x = (2T\cos\theta)\Delta x = 2(1.0 \text{ kN})\cos 45°(150 \text{ m}) = \boxed{210 \text{ kJ}}$$

   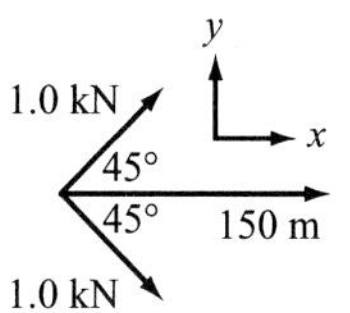

9. **Strategy** Use Eq. (6-2). Let the x-axis point in the direction of motion.

   Solution The force of friction is opposite the motion of the box. Dirk's horizontal force is in the direction of motion.
   $$W = F_x\Delta x = (F - f_k)\Delta x = (66.0 \text{ N} - 4.80 \text{ N})(2.50 \text{ m}) = \boxed{153 \text{ J}}$$

11. Strategy Use Eq. (6-6).

Solution Compute the kinetic energy of the automobile.
$$K = \frac{1}{2}mv^2 = \frac{1}{2}(1600 \text{ kg})(30.0 \text{ m/s})^2 = \boxed{720 \text{ kJ}}$$

13. Strategy Use Eq. (6-6) for each case.

Solution Compute the kinetic energies.

The kinetic energy of Murphy and his bike was $K = \frac{1}{2}mv^2 = \frac{1}{2}(70.5 \text{ kg})(27.8 \text{ m/s})^2 = \boxed{27.2 \text{ kJ}}$.

For Howard and his bike, the kinetic energy was $K = \frac{1}{2}mv^2 = \frac{1}{2}(70.5 \text{ kg})(68.04 \text{ m/s})^2 = \boxed{163 \text{ kJ}}$.

15. Strategy Use Eqs. (6-2), (6-6), and (6-7).

Solution Find the magnitude of the force.
$$W = F_x \Delta x = \Delta K = \frac{1}{2}m(v_f^2 - v_i^2) = \frac{1}{2}mv^2 - 0, \text{ so } F_x = \frac{mv^2}{2\Delta x} = \frac{(12 \text{ kg})(0.40 \text{ m/s})^2}{2(8.0 \text{ m})} = \boxed{0.12 \text{ N}}.$$

17. Strategy The sum of the work done on Jim and his skateboard by gravity and that of friction is equal to the change in kinetic energy of Jim and his skateboard. Use Eqs. (6-6) and (6-7).

Solution Find the work done by friction on Jim and his skateboard. Let the y-axis point upward.
$$W_{\text{total}} = W_{\text{gravity}} + W_{\text{friction}} = -mg\Delta y + W_{\text{friction}} = \Delta K = K_f - K_i = \frac{1}{2}mv^2 - 0 = \frac{1}{2}mv^2, \text{ so}$$
$$W_{\text{friction}} = \frac{1}{2}mv^2 + mg\Delta y = (65.0 \text{ kg})[(9.00 \text{ m/s})^2/2 + (9.80 \text{ m/s}^2)(0 - 5.00 \text{ m})] = \boxed{-550 \text{ J}}.$$

21. Strategy Since $U = 0$ at ground level, the potential energy of Sean and the parachute at the top of the tower is equal to the negative of the work done by gravity as Sean climbed the tower.

Solution Find the potential energy of Sean and the parachute at the top of the tower.
$$U = mgh_{\text{tower}} = (68.0 \text{ kg})(9.80 \text{ m/s}^2)(82.3 \text{ m}) = \boxed{54.8 \text{ kJ}}.$$

25. Strategy and Solution Mechanical energy is the sum of the kinetic and potential energies. Assuming no friction or air resistance, there is no change in mechanical energy; therefore, the mechanical energy will be the same at every point. Ranking the points in order of mechanical energy, from greatest to least, we have $\boxed{A = B = C = D = E}$.

29. Strategy Use Eq. (6-9).

Solution

(a) Since the orange returns to its original position $(\Delta y = 0)$ and air resistance is ignored, the change in its potential energy is $\boxed{0}$.

(b) Let the y-axis point upward and the initial position be $y = 0$.
$$\Delta U_{\text{grav}} = mg\Delta y = (0.30 \text{ kg})(9.80 \text{ m/s}^2)(-1.0 \text{ m} - 0) = \boxed{-2.9 \text{ J}}$$

33. (a) Strategy Use conservation of energy.

Solution Find the speed of the cart as it passes point 3.

$$E_1 = \frac{1}{2}mv_1^2 \text{ if } y_1 = 0 \text{ and } E_3 = \frac{1}{2}mv_3^2 + mgy_3.\ E_f = E_3 = \frac{1}{2}mv_3^2 + mgy_3 = E_i = E_1 = \frac{1}{2}mv_1^2,\ \text{so}$$

$$v_3 = \sqrt{v_1^2 - 2gy_3} = \sqrt{(20.0\ \text{m/s})^2 - 2(9.81\ \text{m/s}^2)(10.0\ \text{m})} = \boxed{14.3\ \text{m/s}}.$$

(b) Strategy Use the result of part (a), replacing 3 with 4. If the result is real—the argument of the square root is nonnegative—the cart will reach position 4.

Solution Compute the speed of the cart at position 4.

$$v_4 = \sqrt{v_1^2 - 2gy_4} = \sqrt{(20.0\ \text{m/s})^2 - 2(9.81\ \text{m/s}^2)(20.0\ \text{m})} = 3\ \text{m/s}$$

The answer is $\boxed{\text{yes; the cart will reach position 4}}$.

35. Strategy The initial height of the rope is $l\cos\theta$ where l is the length of the rope and θ is the angle it makes with the vertical. Then $\Delta y = l\cos\theta - l = l(\cos\theta - 1)$. Use conservation of energy.

Solution Find Bruce's speed at the bottom of the swing.

$$\Delta K = \frac{1}{2}mv^2 - 0 = \frac{1}{2}mv^2 = -\Delta U = -mg\Delta y = mgl(1 - \cos\theta),\ \text{so}$$

$$v = \sqrt{2gl(1 - \cos\theta)} = \sqrt{2(9.80\ \text{m/s}^2)(20.0\ \text{m})(1 - \cos 35.0°)} = \boxed{8.42\ \text{m/s}}.$$

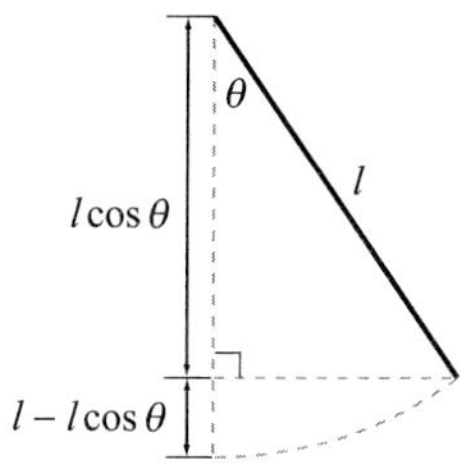

37. Strategy Use Eq. (6-10) to find the nonconservative work.

Solution Calculate the work done by friction and air resistance during the run.

$$W_{\text{total}} = W_c + W_{\text{nc}} = \Delta K = \frac{1}{2}mv_f^2,\ \text{so}$$

$$W_{\text{nc}} = \frac{1}{2}mv_f^2 - W_c = \frac{1}{2}mv_f^2 - mgh = \frac{1}{2}(75\ \text{kg})(12\ \text{m/s})^2 - (75\ \text{kg})(9.80\ \text{m/s}^2)(78\ \text{m}) = \boxed{-52\ \text{kJ}}.$$

41. (a) Strategy Since the gravitational field is uniform, the work done by gravity is $W_{\text{grav}} = F_y\Delta y = -mg\Delta y$, where the y-axis points up.

Solution Note that the slope is inclined at $15.0°$ to the horizontal.

$$W_{\text{grav}} = -mg\Delta y = -(75.0\ \text{kg})(9.80\ \text{m/s}^2)[0 - (32.0\ \text{m})\sin 15.0°]$$

$$= \boxed{6.09\ \text{kJ}}$$

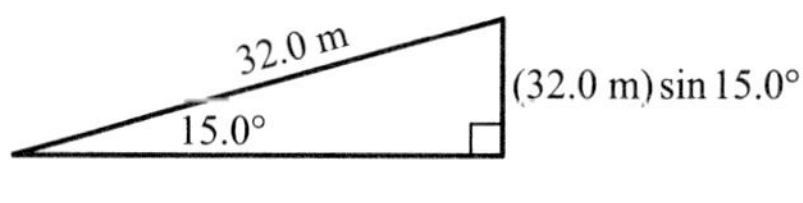

The normal force is perpendicular to the motion of the skier, so the work done by the normal force is $\boxed{0\ \text{J}}$.

(b) Strategy Refer to part (a). Use conservation of energy and Newton's second law.

Solution The work done by gravity is the same as found in part (a), $\boxed{6.09\ \text{kJ}}$. As before, the normal force is $\boxed{0\ \text{J}}$. The total work done on the skier is equal to the sum of the work done by gravity and the work done by friction. We use Eqs. (6-6) and (6-7) to find the work done by friction.

$$W_{\text{total}} = W_{\text{gravity}} + W_{\text{friction}} = -mg\Delta y + W_{\text{friction}} = \Delta K = K_f - K_i = \frac{1}{2}mv^2 - 0 = \frac{1}{2}mv^2,\ \text{so}$$

$$W_{\text{friction}} = \frac{1}{2}mv^2 + mg\Delta y$$

$$= \frac{1}{2}(75.0 \text{ kg})(10.0 \text{ m/s})^2 + (75.0 \text{ kg})(9.80 \text{ m/s}^2)[0 - (32.0 \text{ m})\sin 15.0°] = \boxed{-2.34 \text{ kJ}}.$$

Now that we know the work done by friction, we use Eq. (6-2) to find the force of friction.

$$W = F_x \Delta x = f_k \Delta x = W_{\text{friction}}, \text{ so } f_k = \frac{W_{\text{friction}}}{\Delta x} = \frac{-2337 \text{ J}}{32.0 \text{ m}} = -73.0 \text{ N}.$$

The force of friction is $\boxed{73.0 \text{ N opposite the direction of motion}}$. To find the coefficient of kinetic friction, we draw a diagram and use Newton's second law.

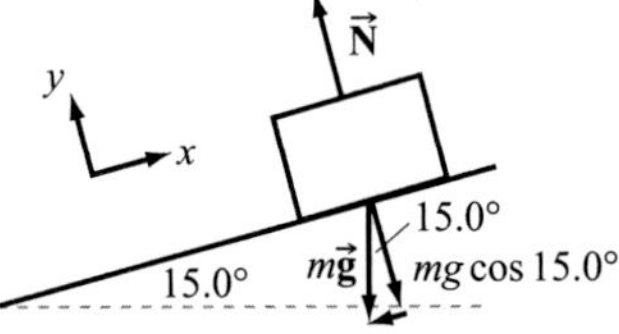

$$\Sigma F_y = N - mg\cos 15.0° = 0, \text{ so } N = mg\cos 15.0°. \text{ Since } f_k = \mu_k N, \text{ the coefficient of kinetic friction is}$$

$$\mu_k = \frac{f_k}{N} = \frac{f_k}{mg\cos 15.0°} = \frac{73.0 \text{ N}}{(75.0 \text{ kg})(9.80 \text{ m/s}^2)\cos 15.0°} = \boxed{0.103}.$$

45. **Strategy** In the equation for the escape speed found in Example 6.8, replace the values for Earth with appropriate values for the fictional planet. Use proportional reasoning and the relationship between the volume of a sphere and its radius to relate the mass and radius of the planet with those of Earth.

Solution Find the escape speed.
Earth:

$$v_{\text{esc}} = \sqrt{\frac{2GM_E}{R_E}} \text{ and } \rho_E = \text{density} = \frac{M_E}{\frac{4}{3}\pi R_E^3}.$$

Planet:

$$v_{\text{esc}} = \sqrt{\frac{2GM}{R}} = \sqrt{\frac{2G}{R}\rho_E V} = \sqrt{\frac{2G}{2R_E}\left(\frac{M_E}{\frac{4}{3}\pi R_E^3}\right)\left[\frac{4}{3}\pi(2R_E)^3\right]} = \sqrt{\frac{8GM_E}{R_E}}$$

$$= \sqrt{\frac{8(6.674\times 10^{-11} \text{ N}\cdot\text{m}^2/\text{kg}^2)(5.974\times 10^{24} \text{ kg})}{6.37\times 10^6 \text{ m}}} = \boxed{22.4 \text{ km/s}}$$

49. **Strategy** Use conservation of energy.

Solution Find the comet's speed at perihelion.

$$K_p + U_p = \frac{1}{2}mv_p^2 - \frac{GmM}{r_p} = K_a + U_a = \frac{1}{2}mv_a^2 - \frac{GmM}{r_a}, \text{ so}$$

$$v_p = \sqrt{v_a^2 + 2GM\left(\frac{1}{r_p} - \frac{1}{r_a}\right)}$$

$$= \sqrt{(10.0\times 10^3 \text{ m/s})^2 + 2(6.674\times 10^{-11} \text{ N}\cdot\text{m}^2/\text{kg}^2)(1.987\times 10^{30} \text{ kg})\left(\frac{1}{8.9\times 10^{10} \text{ m}} - \frac{1}{5.3\times 10^{12} \text{ m}}\right)}$$

$$= \boxed{55 \text{ km/s}}$$

53. Strategy The work done on the spring is negative the work done by the spring. Use the relationship between work and the extension or compression of a spring.

Solution Find the work done to stretch the spring.

$$W = \frac{1}{2}kx^2 = \frac{1}{2}(20.0\ \text{N/m})(0.40\ \text{m})^2 = \boxed{1.6\ \text{J}}$$

57. (a) Strategy Use Hooke's law and form a proportion.

Solution Form the proportion.

$$k = \frac{F_1}{x_1} = \frac{F_2}{x_2}$$

Solve for x_2 to find the amount that the spring stretches.

$$x_2 = \frac{F_2}{F_1}x_1 = \frac{7.0\ \text{N}}{5.0\ \text{N}}(3.5\ \text{cm}) = \boxed{4.9\ \text{cm}}$$

(b) Strategy Substitute known values for F_1 and x_1 to find k.

Solution Compute the spring constant.

$$k = \frac{F_1}{x_1} = \frac{5.0\ \text{N}}{3.5\ \text{cm}} = \boxed{1.4\ \text{N/cm}}$$

(c) Strategy The triangular area under a forces on the spring vs. the stretch of the spring graph is equal to the work done by the forces.

Solution Compute the work done by the forces on the spring.

$$W = \frac{1}{2}Fx = \frac{1}{2}(5.0\ \text{N})(0.035\ \text{m}) = \boxed{88\ \text{mJ}}$$

61. Strategy and Solution Since the displacement of the model airplane is zero, $\boxed{\text{zero}}$ work has been done on it by the string.

65. Strategy The mechanical energy is constant, so we set the elastic potential energy of the spring on the toy gun equal to the gravitational potential energy of the rubber ball. Assume $x \ll h$.

Solution $mgh = (1/2)kx^2$, so $h \propto x^2$. Thus, the height reached by the ball is proportional to the square of the compression of the spring. Form a proportion.

$$\frac{h_2}{h_1} = \frac{x_2^{\,2}}{x_1^{\,2}},\ \text{so}\ h_2 = \frac{x_2^{\,2}}{x_1^{\,2}}h_1 = \left(\frac{2x}{x}\right)^2 h = \boxed{4h}.$$

69. (a) Strategy The increase in kinetic energy of the block is equal to the decrease in its potential energy. Let the potential energy be zero at $y = 0.25$ m.

Solution To find the speed of the block, set Eqs. (6-6) and (6-13) equal and solve for v.

$$\frac{1}{2}mv^2 = mgy,\ \text{so}\ v = \sqrt{2gy} = \sqrt{2(9.80\ \text{m/s}^2)(0.25\ \text{m})} = \boxed{2.2\ \text{m/s}}.$$

(b) Strategy The elastic potential energy increase of the spring is equal to the decrease in gravitational potential energy of the block. Let the potential energy be zero at $y = 0$ m.

Solution To find the compression of the spring, set Eqs. (6-24) and (6-13) equal and solve for x.

$$\frac{1}{2}kx^2 = mgy, \text{ so } x = \sqrt{\frac{2mgy}{k}} = \sqrt{\frac{2(2.0 \text{ kg})(9.80 \text{ m/s}^2)(0.50 \text{ m})}{450 \text{ N/m}}} = \boxed{0.21 \text{ m}}.$$

(c) Strategy and Solution Since the surface is frictionless, no nonconservative forces do work on the block. So, the block will return to its previous height, or $\boxed{0.50 \text{ m}}$.

73. Strategy Use the definition of average power and the potential energy in a uniform gravitational field.

Solution Find the minimum time required for the man to lift the boxes.

$$P_{\text{av}} = \frac{\Delta E}{\Delta t} \text{ and } \Delta E = \Delta U = m_{\text{total}}gh, \text{ so}$$

$$\Delta t = \frac{\Delta E}{P_{\text{av}}} = \frac{50mgh}{P_{\text{av}}} = \frac{50(10.0 \text{ kg})(9.80 \text{ m/s}^2)(2.00 \text{ m})}{40.0 \text{ W}}\left(\frac{1 \text{ min}}{60 \text{ s}}\right) = \boxed{4.08 \text{ min}}.$$

77. Strategy Use $P = Fv$ for the average mechanical power output of the heart.

Solution Find the average speed of blood leaving the heart.

$$v_{\text{av}} = \frac{\text{volume flow rate}}{\text{cross sectional area}} = \frac{(5.0 \text{ L/min})(10^{-3} \text{ m}^3/\text{L})\frac{1 \text{ min}}{60 \text{ s}}}{\pi(0.0090 \text{ m})^2} = 0.327 \text{ m/s}$$

Find the average mechanical power output.

$$P_{\text{av}} = Fv_{\text{av}} = (16 \text{ N})(0.327 \text{ m/s}) = \boxed{5.2 \text{ W}}$$

81. Strategy The instantaneous power is given by Eq. (6-27), where $\theta = 0°$. Obtain the necessary values of the force from the graph.

Solution

(a) Compute the instantaneous power.

$$P_{\text{av}} = Fv\cos\theta = F_x v_x = (800 \text{ N})(11 \text{ m/s}) = \boxed{8.8 \text{ kW}}$$

(b) As in part (a), we have $P_{\text{av}} = Fv\cos\theta = F_x v_x = (400 \text{ N})(16 \text{ m/s}) = \boxed{6.4 \text{ kW}}$.

85. Strategy Convert the speed limit in mi/h to m/s using the conversion 1 mi/h = 0.4470 m/s. Use expressions for work, friction, and kinetic energy to find the initial speed of travel.

Solution Convert the speed limit.

$$(25 \text{ mi/h})\frac{0.4470 \text{ m/s}}{1 \text{ mi/h}} = 11 \text{ m/s}$$

Find the speed of the car before the brakes were applied.

$$W_{\text{friction}} = -f_k\Delta x = -\mu_k N\Delta x = -\mu_k mg\Delta x = \Delta K = 0 - \frac{1}{2}mv_i^2, \text{ so}$$

$$v_i = \sqrt{2\mu_k g\Delta x} = \sqrt{2(0.60)(9.8 \text{ m/s}^2)(9.0 \text{ m})} = 10 \text{ m/s}.$$

The answer is $\boxed{\text{no}}$; the speed limit is 11 m/s, but you were traveling at about 10 m/s.

89. (a) Strategy Let $+x$ be up the incline. Use Newton's second law and Eq. (6-27).

Solution Compute the power the engine must deliver.

$\Sigma F_x = F_{\text{air}} - mg \sin \phi = 0$ at terminal speed.

The rate at which air resistance dissipates energy is $P_{\text{air}} = F_{\text{air}} v \cos 180° = -F_{\text{air}} v = -mg \sin \phi v$.

(We use cos 180° since the force of air resistance is opposite the car's velocity.)
The power the engine must deliver to drive the car on level ground is

$P_{\text{engine}} = -P_{\text{air}} = mg \sin \phi v = (1500 \text{ kg})(9.80 \text{ m/s}^2)(20.0 \text{ m/s}) \sin 2.0° = \boxed{10 \text{ kW}}$.

(b) Strategy The power available to climb the hill is the power delivered by the engine minus the dissipating power of air resistance.

Solution From part (a), for a slope of ϕ:

$P = mgv \sin \phi$, so $\phi = \sin^{-1} \dfrac{P}{mgv} = \sin^{-1} \dfrac{40.0 \times 10^3 \text{ W} - 10.26 \times 10^3 \text{ W}}{(1500 \text{ kg})(9.80 \text{ m/s}^2)(20.0 \text{ m/s})} = \boxed{5.8°}$.

93. Strategy Use conservation of energy.

Solution Find the maximum height of the pole-vaulter's center of gravity.

$K_i + U_i = \dfrac{1}{2} mv^2 + mgh_i = K_f + U_f = 0 + mgh_f$, so $\dfrac{1}{2} mv^2 = mg(h_f - h_i)$, or

$h_f = \dfrac{v^2}{2g} + h_i = \dfrac{(10.0 \text{ m/s})^2}{2(9.80 \text{ m/s}^2)} + 1.0 \text{ m} = \boxed{6.1 \text{ m}}$.

97. Strategy Use conservation of energy.

Solution The elastic potential energy of the spring is converted to gravitational potential energy, so

$\dfrac{1}{2} kx^2 = mgh = mg(l \sin \theta)$ where l is the distance the object travels up the incline.

Thus, $l = \dfrac{kx^2}{2mg \sin \theta} = \dfrac{(40.0 \text{ N/m})(0.20 \text{ m})^2}{2(0.50 \text{ kg})(9.80 \text{ N/kg}) \sin 30.0°} = \boxed{0.33 \text{ m}}$.

101. Strategy The basal metabolic rate is equal to the amount of food energy per day required by a person resting under standard conditions.

Solution

(a) Compute Jermaine's basal metabolic rate.

$\text{BMR} = \left(\dfrac{1 \text{ kcal}}{0.010 \text{ mol}} \right) \left(\dfrac{0.015 \text{ mol}}{\text{min}} \right) \left(\dfrac{1440 \text{ min}}{\text{day}} \right) = \boxed{2200 \text{ kcal/day}}$

(b) Find the mass of fat lost.

$\dfrac{2160 \text{ kcal/day}}{9.3 \text{ kcal/g}} \left(\dfrac{2.2 \text{ lb}}{10^3 \text{ g}} \right) = 0.51 \text{ lb/day}$

Since Jermaine is not resting the entire time, he loses $\boxed{\text{more than 0.51 lb}}$.

105. (a) Strategy Use Newton's second law and Eq. (6-10).

Solution Find the speed at the bottom of the incline.

$$\Delta K = \frac{1}{2}mv^2 - 0 = W_c + W_{nc} = mgh - fd \text{ so } v = \sqrt{2gh - \frac{2fd}{m}}.$$

Use Newton's second law with $+y$ perpendicular to the incline and $+x$ down the incline.

$$\Sigma F_y = N - mg\cos\theta = 0, \text{ so } N = mg\cos\theta.$$

Now, $d = 0.85$ m, $h = d\sin\theta$, and $f = \mu N = \mu mg\cos\theta.$

Substitute.

$$v = \sqrt{2gd\sin\theta - \frac{2\mu mgd\cos\theta}{m}} = \sqrt{2gd(\sin\theta - \mu\cos\theta)}$$

Find the maximum compression.

$$K_f + U_f = 0 + \frac{1}{2}kx^2 = K_i + U_i = \frac{1}{2}mv^2 + 0, \text{ so}$$

$$x = v\sqrt{\frac{m}{k}} = \sqrt{\frac{2mgd}{k}(\sin\theta - \mu\cos\theta)} = \sqrt{\frac{2(0.50 \text{ kg})(9.80 \text{ m/s}^2)(0.85 \text{ m})}{35 \text{ N/m}}(\sin 30.0° - 0.25\cos 30.0°)}$$

$$= \boxed{26 \text{ cm}}.$$

(b) Strategy When the block is accelerated by the spring, it attains its previous kinetic energy and speed.

Solution Find the distance along the incline, d'.

$$\Delta K = 0 - \frac{1}{2}mv^2 = W_c + W_{nc} = -mgh - fd' = -mgd'\sin\theta - \mu mgd'\cos\theta = -d'[mg(\sin\theta + \mu\cos\theta)], \text{ so}$$

$$d' = \frac{v^2}{2g(\sin\theta + \mu\cos\theta)} = \frac{2gd(\sin\theta - \mu\cos\theta)}{2g(\sin\theta + \mu\cos\theta)} = (85 \text{ cm})\frac{\sin 30.0° - 0.25\cos 30.0°}{\sin 30.0° + 0.25\cos 30.0°} = \boxed{34 \text{ cm}}.$$

109. Strategy Use Hooke's law and Newton's laws.

Solution

(a) The mass connected to the lower spring exerts a force on the lower spring equal to its weight, W. The spring stretches an amount $x = F/k = W/k$. The lower spring exerts a force on the upper spring equal to $F = W$, and causes it to stretch by $x = F/k = W/k$. Thinking of the two springs as a single spring:

$$2x = \frac{F}{k} + \frac{F}{k} = \frac{2F}{k} = x', \text{ so } F = \frac{k}{2}x' = k'x'. \text{ Therefore, } \boxed{\frac{k}{2}} = k', \text{ the effective spring constant.}$$

(b) Sum the forces on the mass.

$$F + F - W = kx + kx - W = 2kx - W = 0, \text{ so } W = 2kx = k'x.$$

Therefore, $\boxed{2k} = k'$, the effective spring constant.

113. Strategy Draw a diagram. Then, use Newton's second law and conservation of energy. Let the positive direction be away from the slope.

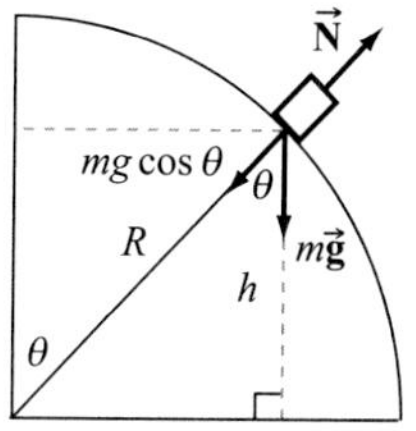

Solution According to Newton's second law, $\Sigma F_r = N - mg\cos\theta = ma_r = -mv^2/R$.

When the normal force becomes zero, we have $mg\cos\theta = mv^2/R$, or $mgR\cos\theta = mv^2$. When this condition is true, the skier leaves the surface of the ice. Note from the figure that $h = R\cos\theta$, thus, the condition becomes $mgh = mv^2$. Now, mgh is the final gravitational potential energy $(U_i = mgR)$ and mv^2 is twice the final kinetic energy $(K_i = 0)$.

So, the condition becomes $U_f = 2K_f$, or $K_f = U_f/2$. Use conservation of energy to find h in terms of R.

$$0 = \Delta K + \Delta U = K_f - K_i + U_f - U_i = \frac{1}{2}U_f - 0 + U_f - U_i = \frac{3}{2}U_f - U_i, \text{ so } \frac{3}{2}mgh - mgR = \frac{3}{2}h - R = 0, \text{ or}$$

$$h = \boxed{\frac{2}{3}R}.$$

115. (a) Strategy Use Hooke's law and Newton's laws of motion.

Solution According to Hooke's law, $F_1 = k_1 x_1$ and $F_2 = k_2 x_2$.

Imagine that the springs are suspended from a ceiling such that the bottom of each is at the same height. Then a mass m is attached to the bottom of both, the springs stretch, and the system comes to equilibrium. Assume that the masses of the springs are negligible. Sum the vertical forces.

$F_1 + F_2 - W = 0$, so $W = F_1 + F_2 = k_1 x_1 + k_2 x_2$.

Assuming the springs are attached to the same point on the top of the mass, $x_1 = x_2 = x$.

$W = k_1 x_1 + k_2 x_2 = k_1 x + k_2 x = (k_1 + k_2)x = kx = W$

So, in response to a force that stretches the springs (W, in this case), the springs act like one spring with a spring constant $\boxed{k = k_1 + k_2}$.

(b) Strategy Use the result from part (a) and Eq. (6-24).

Solution Compute the potential energy stored in the spring.

$$U = \frac{1}{2}kx^2 = \frac{1}{2}(k_1 + k_2)x^2 = \frac{1}{2}(500 \text{ N/m} + 300 \text{ N/m})(0.020 \text{ m})^2 = \boxed{0.16 \text{ J}}$$

117. (a) Strategy The work is represented by the area under the curve. Estimate the work done during stretching and contraction; then, find the total work done.

Solution Estimate the work.

$$W_{\text{stretch}} \approx \frac{1}{2}(0.34 \text{ m})(18 \text{ N}) = 3.1 \text{ J and } W_{\text{contract}} \approx \frac{1}{2}(0.34 \text{ m} - 0.02 \text{ m})(16 \text{ N}) = 2.6 \text{ J}.$$

Therefore, the total work $= 3.1 \text{ J} - 2.6 \text{ J} = \boxed{0.5 \text{ J}}$

(b) Strategy and Solution Hooke's Law is a conservative force, so the total work to stretch and contract the rubber band would be $\boxed{\text{zero}}$.

(c) Strategy and Solution The work done on the rubber band does not all go into increasing its elastic potential energy; $\boxed{\text{some of the energy is dissipated as heat}}$.

Chapter 7

LINEAR MOMENTUM

Conceptual Questions

1. The likelihood of injury resulting from jumping from a second floor window is primarily determined by the average force acting to decelerate the body.

5. The law of the conservation of linear momentum states that in the absence of external interactions, the linear momentum of a closed system is constant. Floating in free space, the astronaut and the wrench form a closed system free from interactions with other bodies. If the astronaut throws the wrench in the direction opposite the ship, conservation of momentum dictates that he must in turn move toward the ship.

9. First law: The momentum of an object is constant unless acted upon by an external force. Second law: The net force acting on an object is equal to the rate of change of the object's momentum. Third law: When two objects interact, the changes in momentum that each imparts to the other are equal in magnitude and opposite in direction.

13. An impulse must be supplied to the egg to change its momentum and bring it to rest. A good strategy is to make the time interval over which the stopping force is applied as large as possible. This will reduce the magnitude of the force required to stop the egg. One should therefore attempt to catch the egg with a swinging motion, moving the hand backwards as it is being caught, to bring it to rest as slowly and gently as possible.

Problems

1. **Strategy** Use the definition of linear momentum.

   Solution Find the magnitude of the total momentum of the system.
   $$\vec{p}_{total} = \vec{p}_1 + \vec{p}_2 = m\vec{v}_1 + m\vec{v}_2 = m(\vec{v}_1 + \vec{v}_2) = m[\vec{v}_1 + (-\vec{v}_1)] = 0, \text{ so the magnitude is } \boxed{0}.$$

5. **Strategy** Add the momenta of the three particles.

   Solution Find the total momentum of the system.
   $$\vec{p}_{tot} = \vec{p}_1 + \vec{p}_2 + \vec{p}_3 = m_1\vec{v}_1 + m_2\vec{v}_2 + m_3\vec{v}_3 = m_1 v_1 \text{ north} + m_2 v_2 \text{ south} + m_3 v_3 \text{ north}$$
   $$= (m_1 v_1 - m_2 v_2 + m_3 v_3) \text{ north} = \left[(3.0 \text{ kg})(3.0 \text{ m/s}) - (4.0 \text{ kg})(5.0 \text{ m/s}) + (7.0 \text{ kg})(2.0 \text{ m/s})\right] \text{ north}$$
   $$= \boxed{3 \text{ kg} \cdot \text{m/s north}}$$

9. **Strategy** The initial momentum is toward the wall and the final momentum is away from the wall.

   Solution Find the change in momentum.
   $$\Delta p = p_f - p_i = mv_f - mv_i = m(v_f - v_i) = (5.0 \text{ kg})(-2.0 \text{ m/s} - 2.0 \text{ m/s}) = -20 \text{ kg} \cdot \text{m/s, so}$$
   $$\Delta\vec{p} = \boxed{20 \text{ kg} \cdot \text{m/s in the } -x\text{-direction}}.$$

13. Strategy Since $\Delta p = F\Delta t$, and the braking force F is a constant and the final momentum for each car is zero, ranking the cars in order of their initial momentum magnitudes is the same as ranking them in order of their times to stop.

Solution Compute the initial momentum for each car.
(a) $p = mv = (1500\ \text{kg})(30\ \text{m/s}) = 45{,}000\ \text{kg}\cdot\text{m/s}$; (b) $(1500\ \text{kg})(20\ \text{m/s}) = 30{,}000\ \text{kg}\cdot\text{m/s}$;
(c) $(1000\ \text{kg})(30\ \text{m/s}) = 30{,}000\ \text{kg}\cdot\text{m/s}$; (d) $(1000\ \text{kg})(20\ \text{m/s}) = 20{,}000\ \text{kg}\cdot\text{m/s}$;
(e) $(2000\ \text{kg})(40\ \text{m/s}) = 80{,}000\ \text{kg}\cdot\text{m/s}$
Ranking the initial momentum magnitudes—and times to stop—from smallest to largest, we have
$\boxed{\text{(d), (b) = (c), (a), (e)}}$.

17. Strategy Use the impulse-momentum theorem. Let the positive direction be in the direction of motion.

Solution Find the average horizontal force exerted on the automobile during breaking.
$$F_{\text{av}} = \frac{\Delta p}{\Delta t} = \frac{m(v_{\text{f}} - v_{\text{i}})}{\Delta t} = \frac{(1.0\times10^3\ \text{kg})(0-30.0\ \text{m/s})}{5.0\ \text{s}} = -6.0\times10^3\ \text{N}$$
So, $\vec{\mathbf{F}}_{\text{av}} = \boxed{6.0\times10^3\ \text{N opposite the car's direction of motion}}$.

21. Strategy Use conservation of momentum.

Solution Find the recoil speed of the frog and lily pad.
$$\vec{\mathbf{P}}_{\text{tf}} + \vec{\mathbf{P}}_{\text{ff}} = m_{\text{t}}\vec{\mathbf{v}}_{\text{tf}} + m_{\text{f}}\vec{\mathbf{v}}_{\text{ff}} = \vec{\mathbf{P}}_{\text{ti}} + \vec{\mathbf{P}}_{\text{fi}} = 0+0,\ \text{so}\ \left|\vec{\mathbf{v}}_{\text{ff}}\right| = \frac{m_{\text{t}}}{m_{\text{f}}}\left|-\vec{\mathbf{v}}_{\text{tf}}\right| = \frac{0.41\ \text{g}}{12.5\ \text{g}}(3.7\ \text{m/s}) = \boxed{0.12\ \text{m/s}}.$$

25. Strategy Use conservation of momentum.

Solution Find the recoil speed of the submarine.
$$\vec{\mathbf{P}}_{\text{sf}} + \vec{\mathbf{P}}_{\text{tf}} = m_{\text{s}}\vec{\mathbf{v}}_{\text{sf}} + m_{\text{t}}\vec{\mathbf{v}}_{\text{tf}} = \vec{\mathbf{P}}_{\text{si}} + \vec{\mathbf{P}}_{\text{ti}} = 0+0,\ \text{so}\ \left|\vec{\mathbf{v}}_{\text{sf}}\right| = \frac{m_{\text{t}}}{m_{\text{s}}}\left|-\vec{\mathbf{v}}_{\text{tf}}\right| = \frac{250\ \text{kg}}{2.5\times10^6\ \text{kg}}(100.0\ \text{m/s}) = \boxed{0.010\ \text{m/s}}.$$

29. Strategy Use conservation of momentum.

Solution Find the recoil speed of the railroad car.
$\vec{\mathbf{p}}_{\text{i}} = 0 = -\vec{\mathbf{p}}_{\text{f}}$, and since we are only concerned with the horizontal direction, we have:
$$m_{\text{c}}v_{\text{cx}} = m_{\text{s}}v_{\text{sx}},\ \text{so}\ v_{\text{cx}} = \frac{m_{\text{s}}}{m_{\text{c}}}v_{\text{sx}} = \frac{98\ \text{kg}}{5.0\times10^4\ \text{kg}}(105\ \text{m/s})\cos 60.0° = \boxed{0.10\ \text{m/s}}.$$

31. Strategy Use the component form of the definition of center of mass.

Solution Find the location of particle B.
Find x_{CM}.

$$x_{\text{CM}} = \frac{m_{\text{A}}x_{\text{A}} + m_{\text{B}}x_{\text{B}}}{m_{\text{A}} + m_{\text{B}}} = \frac{0 + m_{\text{B}}x_{\text{B}}}{m_{\text{A}} + m_{\text{B}}},\ \text{so}$$

$$x_{\text{B}} = \frac{m_{\text{A}} + m_{\text{B}}}{m_{\text{B}}}x_{\text{CM}} = \frac{30.0\ \text{g} + 10.0\ \text{g}}{10.0\ \text{g}}(2.0\ \text{cm}) = 8.0\ \text{cm}.$$

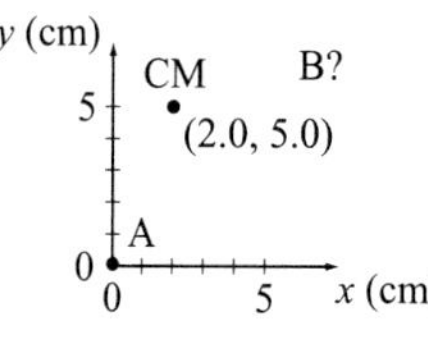

Similarly,
$$y_{\text{B}} = \frac{30.0\ \text{g} + 10.0\ \text{g}}{10.0\ \text{g}}(5.0\ \text{cm}) = 20\ \text{cm}.$$

The coordinates of particle B are $(x_{\text{B}}, y_{\text{B}}) = \boxed{(8.0\ \text{cm},\ 20\ \text{cm})}$.

33. Strategy Use the component form of the definition of center of mass.

Solution Find the distance in front of the woman's hips to the new horizontal component of her center of mass.

$$x_{CM} = \frac{mx + m_{gained}x_{gained}}{m + m_{gained}} = \frac{(68 \text{ kg})(0) + (8.0 \text{ kg})(18 \text{ cm})}{68 \text{ kg} + 8.0 \text{ kg}} = \boxed{1.9 \text{ cm}}$$

37. Strategy Use symmetry and the component form of the definition of center of mass to determine the center of mass of each object with respect to the origin at the top left corner of the sculpture.

Solution The centers of mass are as follows:
rectangle: (1.0 m, -0.25 m); circle: (0 m, -2.5 m); square: (1.4 m, -1.9 m); octagon: (2.0 m, -3.0 m)
Find the components of the center of mass of the entire sculpture.

$$x_{CM} = \frac{m_r x_r + m_c x_c + m_s x_s + m_o x_o}{m_r + m_c + m_s + m_o}$$

$$= \frac{(2.0 \text{ kg})(1.0 \text{ m}) + (5.0 \text{ kg})(0 \text{ m}) + (2.0 \text{ kg})(1.4 \text{ m}) + (3.0 \text{ kg})(2.0 \text{ m})}{2.0 \text{ kg} + 5.0 \text{ kg} + 2.0 \text{ kg} + 3.0 \text{ kg}} = 0.900 \text{ m}$$

$$y_{CM} = \frac{m_r y_r + m_c y_c + m_s y_s + m_o y_o}{m_r + m_c + m_s + m_o}$$

$$= \frac{(2.0 \text{ kg})(-0.25 \text{ m}) + (5.0 \text{ kg})(-2.5 \text{ m}) + (2.0 \text{ kg})(-1.9 \text{ m}) + (3.0 \text{ kg})(-3.0 \text{ m})}{2.0 \text{ kg} + 5.0 \text{ kg} + 2.0 \text{ kg} + 3.0 \text{ kg}} = -2.15 \text{ m}$$

The center of mass of the sculpture is $\boxed{(0.900 \text{ m}, -2.15 \text{ m})}$.

41. Strategy The total momentum of the system is equal to the total mass of the system times the velocity of the center of mass. Let east be in the positive direction.

Solution Find the total momentum.

$$\vec{p} = M\vec{v}_{CM} = m_A \vec{v}_A + m_B \vec{v}_B \text{ since } \vec{p} = \vec{p}_A + \vec{p}_B. \text{ Thus, } \vec{v}_{CM} = \frac{m_A \vec{v}_A + m_B \vec{v}_B}{m_A + m_B}.$$

Find the velocity of the center of mass.

$$v_{CM} = \frac{(5.0 \text{ kg})(10 \text{ m/s}) + (15 \text{ kg})(-10 \text{ m/s})}{5.0 \text{ kg} + 15 \text{ kg}} = -5.0 \text{ m/s, so } \vec{v}_{CM} = \boxed{5.0 \text{ m/s west}}.$$

45. Strategy Linear momentum is conserved, so $p_f = p_i$.

Solution Find the change in speed of the car.

$$p_f = (m_{car} + m_{clay})v_f = p_i = m_{car}v_i, \text{ so } v_f = \frac{m_{car}}{m_{car} + m_{clay}}v_i, \text{ and}$$

$$\Delta v = v_f - v_i = \frac{m_{car}}{m_{car} + m_{clay}}v_i - v_i = \left(\frac{m_{car}}{m_{car} + m_{clay}} - 1\right)v_i = \left(\frac{120 \text{ g}}{120 \text{ g} + 30.0 \text{ g}} - 1\right)(0.75 \text{ m/s}) = \boxed{-0.15 \text{ m/s}}.$$

47. Strategy Use conservation of momentum. The block is initially at rest, so $v_{2i} = 0$. Let east be in the $+x$-direction.

Solution Find the final velocity of the block.
$$m_1 v_{1f} + m_2 v_{2f} = m_1 v_{1i} + m_2 v_{2i} = m_1 v_{1i} + m_2(0), \text{ so}$$
$$v_{2f} = \frac{m_1(v_{1i} - v_{1f})}{m_2} = \frac{0.020 \text{ kg}}{2.0 \text{ kg}}[200.0 \text{ m/s} - (-100.0 \text{ m/s})] = 3.0 \text{ m/s}.$$

Thus, $\vec{v}_{block} = \boxed{3.0 \text{ m/s east}}$.

49. Strategy The collision is perfectly inelastic since the bullet embeds in the wood. Friction does negative work on the block and bullet combination. Use Newton's second law, Eq. (4-5), and conservation of momentum.

Solution Let the $+x$-direction be in the direction of motion. Find the acceleration of the block and bullet due to friction.

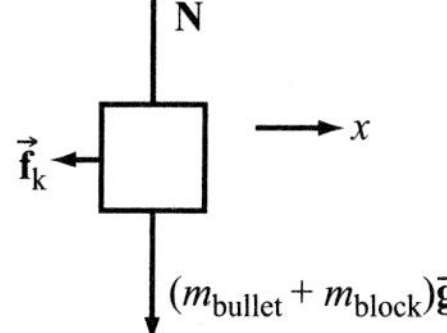

$$\Sigma F_y = N - (m_{bullet} + m_{block})g = 0, \text{ so } N = (m_{bullet} + m_{block})g.$$
$$\Sigma F_x = -f_k = -\mu_k N = -\mu_k(m_{bullet} + m_{block})g = (m_{bullet} + m_{block})a_x, \text{ so } a_x = -\mu_k g.$$

Find the initial speed of the block and bullet (just after the collision).
$$v_{fx}^2 - v_{ix}^2 = 0 - v_{ix}^2 = 2a_x \Delta x = -2\mu_k g \Delta x, \text{ so } v_{ix} = \sqrt{2\mu_k g \Delta x} = v.$$

Use conservation of momentum to find the speed of the bullet just before its collision with the block.
$$m_{bullet} v_{bullet} = (m_{bullet} + m_{block})v = (m_{bullet} + m_{block})\sqrt{2\mu_k g \Delta x}, \text{ so}$$
$$v_{bullet} = \frac{(m_{bullet} + m_{block})\sqrt{2\mu_k g \Delta x}}{m_{bullet}} = \frac{(2.02 \text{ kg})\sqrt{2(0.400)(9.80 \text{ m/s}^2)(1.50 \text{ m})}}{0.020 \text{ kg}} = \boxed{350 \text{ m/s}}.$$

53. Strategy Use conservation of linear momentum. Since the collision is perfectly elastic, kinetic energy is conserved.

Solution The 100-g ball is (1) and the 300-g ball is (2). Note that $m_2 = 3m_1$.

$$m_1 v_{1i} + m_2 v_{2i} = m_1 v_{1i} + m_2(0) = m_1 v_{1i} = m_1 v_{1f} + m_2 v_{2f}, \text{ so } v_{1i} = v_{1f} + \frac{m_2}{m_1}v_{2f} = v_{1f} + 3v_{2f}.$$

$$\frac{1}{2}m_1 v_{1i}^2 + \frac{1}{2}m_2(0)^2 = \frac{1}{2}m_1 v_{1i}^2 = \frac{1}{2}m_1 v_{1f}^2 + \frac{1}{2}m_2 v_{2f}^2 = \frac{1}{2}m_1 v_{1f}^2 + \frac{1}{2}(3m_1)v_{2f}^2, \text{ so } v_{1i}^2 = v_{1f}^2 + 3v_{2f}^2.$$

Substitute for v_{1i}.

$$(v_{1f} + 3v_{2f})^2 = v_{1f}^2 + 6v_{1f}v_{2f} + 9v_{2f}^2 = v_{1f}^2 + 3v_{2f}^2, \text{ so } 6v_{1f}v_{2f} = -6v_{2f}^2, \text{ or } v_{1f} = -v_{2f}.$$

Find the final velocities of each ball.

$$v_{1i} = v_{1f} + 3v_{2f} = -v_{2f} + 3v_{2f} = 2v_{2f}, \text{ so } v_{2f} = \frac{1}{2}v_{1i} = \frac{1}{2}(5.00 \text{ m/s}) = 2.50 \text{ m/s}.$$

Since $v_{1f} = -v_{2f}$, $v_{1f} = -2.50 \text{ m/s}$. So, the 300-g ball moves at $\boxed{2.50 \text{ m/s in the } +x\text{-direction}}$ and the 100-g ball moves at $\boxed{2.50 \text{ m/s in the } -x\text{-direction}}$.

57. Strategy The collision is perfectly inelastic, so $v_{1f} = v_{2f} = v$. The block is initially at rest, so $v_{2i} = 0$ and $v_{1i} = v_i$. Use conservation of momentum and Eq. (4-9).

Solution Find the speed of the bullet and block system.

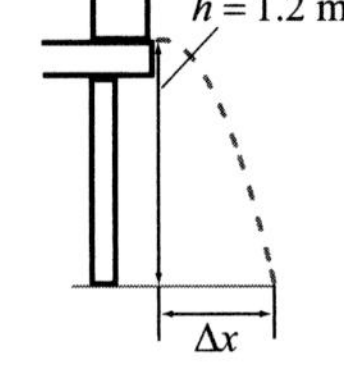

$$(m_{bul} + m_{blk})v = m_{bul}v_i + m_{blk}(0), \text{ so } v = \frac{m_{bul}}{m_{bul} + m_{blk}}v_i.$$

Determine the time it takes the system to hit the floor.

$$\Delta y = -h = v_{iy}\Delta t - \frac{1}{2}g(\Delta t)^2 = 0 - \frac{1}{2}g(\Delta t)^2, \text{ so } \Delta t = \sqrt{\frac{2h}{g}}.$$

Find the horizontal distance traveled.

$$\Delta x = v_{ix}\Delta t = v\Delta t = \frac{m_{bul}}{m_{bul} + m_{blk}}v_i\sqrt{\frac{2h}{g}} = \frac{0.010 \text{ kg}}{0.010 \text{ kg} + 4.0 \text{ kg}}(400.0 \text{ m/s})\sqrt{\frac{2(1.2 \text{ m})}{9.80 \text{ m/s}^2}} = \boxed{0.49 \text{ m}}$$

61. Strategy Use conservation of momentum. Let each of the first two pieces be $45°$ from the positive x-axis (one CW, one CCW).

Solution Find the speed of the third piece.
Find v_{3x}.

$$p_{1x} + p_{2x} + p_{3x} = mv_{1x} + mv_{2x} + mv_{3x} = 0, \text{ so } v_{3x} = -v_{1x} - v_{2x} = -v\cos 45° - v\cos(-45°) = -\frac{v}{\sqrt{2}} - \frac{v}{\sqrt{2}} = -v\sqrt{2}.$$

Similarly,

$$v_{3y} = -v_{1y} - v_{2y} = -v\sin 45° - v\sin(-45°) = -\frac{v}{\sqrt{2}} + \frac{v}{\sqrt{2}} = 0, \text{ so } v_3 = |v_{3x}| = v\sqrt{2} = (120 \text{ m/s})\sqrt{2} = \boxed{170 \text{ m/s}}.$$

65. Strategy Use conservation of momentum.

Solution Find v_{2f} in terms of v_{1f}.

$$mv_{1fy} + mv_{2fy} = mv_{1f}\sin\theta_1 + mv_{2f}\sin\theta_2 = mv_{1iy} + mv_{2iy} = 0 + 0, \text{ so}$$

$$v_{2f} = \frac{-\sin\theta_1}{\sin\theta_2}v_{1f} = \frac{-\sin 60.0°}{\sin(-30.0°)}v_{1f} = \boxed{1.73v_{1f}}.$$

69. Strategy The collision is perfectly inelastic, so the final velocities of the cars are identical. Use conservation of momentum.

Solution Let the 1700-kg car be (1) and the 1300-kg car be (2).

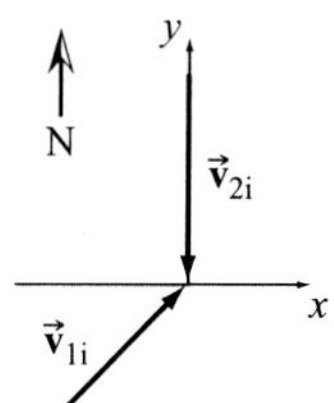

$$p_{ix} = m_1 v_{1ix} + m_2 v_{2ix} = m_1 v_{1ix} + 0 = p_{fx} = (m_1 + m_2)v_{fx}, \text{ so } v_{fx} = \frac{m_1}{m_1 + m_2} v_{1ix}.$$

$$p_{iy} = m_1 v_{1iy} + m_2 v_{2iy} = p_{fy} = (m_1 + m_2)v_{fy}, \text{ so } v_{fy} = \frac{m_1 v_{1iy} + m_2 v_{2iy}}{m_1 + m_2}.$$

Compute the final speed and the direction.

$$v_f = \sqrt{v_{fx}^2 + v_{fy}^2} = \sqrt{\left(\frac{m_1 v_{1ix}}{m_1 + m_2}\right)^2 + \left(\frac{m_1 v_{1iy} + m_2 v_{2iy}}{m_1 + m_2}\right)^2}$$

$$= \frac{\sqrt{[(1700 \text{ kg})(14 \text{ m/s})\cos 45°]^2 + [(1700 \text{ kg})(14 \text{ m/s})\sin 45° + (1300 \text{ kg})(-18 \text{ m/s})]^2}}{1700 \text{ kg} + 1300 \text{ kg}} = 6.0 \text{ m/s}$$

$$\theta = \tan^{-1}\frac{v_{fy}}{v_{fx}} = \tan^{-1}\frac{(1700 \text{ kg})(14 \text{ m/s})\sin 45° + (1300 \text{ kg})(-18 \text{ m/s})}{(1700 \text{ kg})(14 \text{ m/s})\cos 45°} = -21°$$

Thus, the final velocity of the cars is $\boxed{6.0 \text{ m/s at } 21° \text{ S of E}}$.

73. Strategy Use conservation of momentum.

Solution Let swallow 1 and its coconut be (1) and swallow 2 and its coconut be (2) (before the collision). After the collision, let swallow 1's coconut be (3), swallow 2's coconut be (4), and the tangled-up swallows be (5).

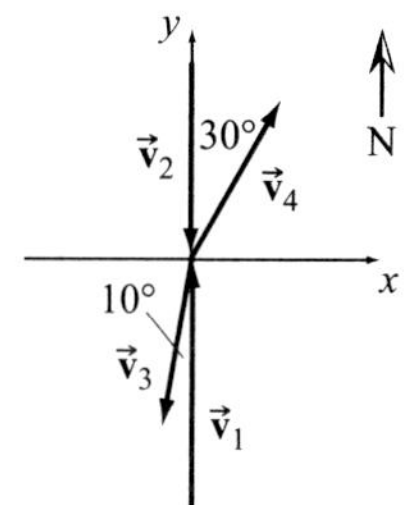

$$p_{ix} = m_1 v_{1x} + m_2 v_{2x} = 0 + 0 = p_{fx} = m_3 v_{3x} + m_4 v_{4x} + m_5 v_{5x}, \text{ so } v_{5x} = -\frac{m_3 v_{3x} + m_4 v_{4x}}{m_5}.$$

$$p_{iy} = m_1 v_{1y} + m_2 v_{2y} = m_1 v_1 + m_2 v_2 = p_{fy} = m_3 v_{3y} + m_4 v_{4y} + m_5 v_{5y}, \text{ so}$$

$$v_{5y} = \frac{m_1 v_1 + m_2 v_2 - m_3 v_{3y} - m_4 v_{4y}}{m_5}.$$

Compute the final speed of the tangled swallows, v_5. Then, compute the direction.

$$v_5 = \sqrt{v_{5x}^2 + v_{5y}^2} = \frac{1}{m_5}\sqrt{[-(m_3 v_{3x} + m_4 v_{4x})]^2 + (m_1 v_1 + m_2 v_2 - m_3 v_{3y} - m_4 v_{4y})^2}$$

$$= \frac{\sqrt{\begin{array}{l}[(0.80 \text{ kg})(13 \text{ m/s})\cos 260° + (0.70 \text{ kg})(14 \text{ m/s})\cos 60°]^2 \\ + [(1.07 \text{ kg})(20 \text{ m/s}) + (0.92 \text{ kg})(-15 \text{ m/s}) - (0.80 \text{ kg})(13 \text{ m/s})\sin 260° - (0.70 \text{ kg})(14 \text{ m/s})\sin 60°]^2\end{array}}}{0.270 \text{ kg} + 0.220 \text{ kg}}$$

$$= 20 \text{ m/s}$$

$$\theta = \tan^{-1}\frac{v_{5y}}{v_{5x}}$$

$$= \tan^{-1}\frac{(1.07 \text{ kg})(20 \text{ m/s}) + (0.92 \text{ kg})(-15 \text{ m/s}) - (0.80 \text{ kg})(13 \text{ m/s})\sin 260° - (0.70 \text{ kg})(14 \text{ m/s})\sin 60°}{-(0.80 \text{ kg})(13 \text{ m/s})\cos 260° - (0.70 \text{ kg})(14 \text{ m/s})\cos 60°}$$

$$= -72°$$

Since $v_{5x} < 0$ and $v_{5y} > 0$, the velocity vector is located in the second quadrant, so the angle is $180° - 72° = 108°$ from the positive x-axis or $18°$ west of north. Thus, the velocity of the birds immediately after the collision is $\boxed{20 \text{ m/s at } 18° \text{ W of N}}$.

77. (a) Strategy Use conservation of momentum. Let the $+x$-direction be to the right.

Solution Find the final velocity of the other glider.
$$mv_{1f} + mv_{2f} = mv_i + mv_i, \text{ so } v_{2f} = v_i + v_i - v_{1f} = 0.50 \text{ m/s} + 0.50 \text{ m/s} - 1.30 \text{ m/s} = -0.30 \text{ m/s}.$$

So, the velocity of the other glider is $\boxed{0.30 \text{ m/s to the left}}$.

(b) Strategy Form a ratio of the final to the initial kinetic energies.

Solution Compute the ratio.
$$\frac{K_f}{K_i} = \frac{\frac{1}{2}mv_{1f}^2 + \frac{1}{2}mv_{2f}^2}{2\left(\frac{1}{2}mv_i^2\right)} = \frac{v_{1f}^2 + v_{2f}^2}{2v_i^2} = \frac{(1.30 \text{ m/s})^2 + (0.30 \text{ m/s})^2}{2(0.50 \text{ m/s})^2} = 3.6$$

> The final kinetic energy is greater than the initial kinetic energy. The extra kinetic energy comes from the elastic potential energy stored in the spring.

79. Strategy Use conservation of momentum. The collision is perfectly inelastic, so $v_{1f} = v_{2f} = v_f$.

Solution Find the speed of the sled once the book is on it.
$$m_1 v_{1f} + m_2 v_{2f} = (m_1 + m_2)v_f = m_1 v_{1i} + m_2 v_{2i} = m_1 v_i + 0, \text{ so}$$
$$v_f = \frac{m_1}{m_1 + m_2} v_i = \frac{5.0 \text{ kg}}{5.0 \text{ kg} + 1.0 \text{ kg}}(1.0 \text{ m/s}) = \boxed{0.83 \text{ m/s}}.$$

81. Strategy Use Eqs. (7-10) and (7-11).

Solution Find the velocity of the center of mass of the system.
$$\vec{p}_{CM} = M\vec{v}_{CM} = m_1\vec{v}_1 + m_2\vec{v}_2 + m_3\vec{v}_3, \text{ so}$$
$$v_{CMx} = \frac{(3.0 \text{ kg})(290 \text{ m/s}) + (5.0 \text{ kg})(-120 \text{ m/s}) + (2.0 \text{ kg})(52 \text{ m/s})}{3.0 \text{ kg} + 5.0 \text{ kg} + 2.0 \text{ kg}} = 37 \text{ m/s}$$

$v_{CMy} = 0$, so $\vec{v}_{CM} = \boxed{37 \text{ m/s in the } +x\text{-direction}}$.

83. Strategy Use the definition of linear momentum and the impulse-momentum theorem.

Solution

(a) Compute the magnitude of the change in momentum of the ball.
$$\Delta p = p_f - p_i = mv_f - mv_i = m(v_f - v_i) = (0.145 \text{ kg})\left[37 \text{ m/s} - (-41 \text{ m/s})\right] = \boxed{11 \text{ kg}\cdot\text{m/s}}$$

(b) Compute the impulse delivered to the ball by the bat.
$$\text{Impulse} = \Delta p = \boxed{11 \text{ kg}\cdot\text{m/s}}$$

(c) Compute the magnitude of the average force exerted on the ball by the bat.
$$F_{av} = \frac{\Delta p}{\Delta t} = \frac{11.31 \text{ kg}\cdot\text{m/s}}{3.0\times10^{-3} \text{ s}} = \boxed{3.8 \text{ kN}}$$

85. Strategy The center of each length is its center of mass. Use the component form of the definition of center of mass.

Solution Find the location of the center of mass of the rod.

$$x_{CM} = \frac{\frac{m}{3}x_1 + \frac{m}{3}x_2 + \frac{m}{3}x_3}{m} = \frac{1}{3}(0 + 5.0 \text{ cm} + 10.0 \text{ cm}) = 5.00 \text{ cm}$$

$$y_{CM} = \frac{\frac{m}{3}y_1 + \frac{m}{3}y_2 + \frac{m}{3}y_3}{m} = \frac{1}{3}(5.0 \text{ cm} + 10.0 \text{ cm} + 5.0 \text{ cm}) = 6.67 \text{ cm}$$

Thus, $(x_{CM}, y_{CM}) = \boxed{(5.00 \text{ cm}, \ 6.67 \text{ cm})}$.

89. Strategy The fly splatters on the windshield, so the collision is perfectly inelastic $(v_{\text{fly, final}} = v_{\text{car, final}} = v_f)$. Use conservation of momentum. Let the positive direction be along the velocity of the automobile.

Solution

(a) Compute the change in momentum.

$$\Delta p_{\text{car}} = -\Delta p_{\text{fly}} = -m_{\text{fly}}(v_{\text{fly, f}} - v_{\text{fly, i}}) \approx -m_{\text{fly}}(v_{\text{car, i}} - 0) = -(0.1 \times 10^{-3} \text{ kg})(100 \text{ km/h}) = -0.01 \text{ kg} \cdot \text{km/h}$$

So, the change in the car's momentum due to the fly is $\boxed{0.01 \text{ kg} \cdot \text{km/h} \text{ opposite the car's motion}}$.

(b) Compute the change in momentum.

$$\Delta p_{\text{fly}} = -\Delta p_{\text{car}} = 0.01 \text{ kg} \cdot \text{km/h}, \text{ or } \boxed{0.01 \text{ kg} \cdot \text{km/h} \text{ along the car's velocity}}.$$

(c) Compute the number of flies N required to slow the car.

$$N\Delta p_{\text{fly}} = -m_{\text{car}}\Delta v_{\text{car}}, \text{ so } N = -\frac{m_{\text{car}}\Delta v_{\text{car}}}{\Delta p_{\text{fly}}} = -\frac{(1000 \text{ kg})(-1 \text{ km/h})}{0.01 \text{ kg} \cdot \text{km/h}} = \boxed{10^5 \text{ flies}}.$$

93. Strategy Use conservation of momentum and the definition of center of mass. Let the pier be to the left of the raft and woman at $x = 0$.

Solution

(a) Since $\Delta \bar{\mathbf{p}}_{CM} = 0$, as the woman walks toward the pier, the raft moves away from the pier, and the center of mass does not change. So, $x_{CM} = \dfrac{m_w x_{wi} + m_r x_{ri}}{m_w + m_r} = \dfrac{m_w x_{wf} + m_r x_{rf}}{m_w + m_r}$.

Initially, x_{CM} is to the right of x_{ri}. When the woman has walked to the other end of the raft, x_{CM} is to the left of x_{rf}. By symmetry, the distance $x_{CM} - x_{ri}$ equals the distance $x_{rf} - x_{CM}$, thus $x_{rf} - x_{CM} = x_{CM} - x_{ri}$, so $x_{rf} = 2x_{CM} - x_{ri}$. The final distance of the raft from the dock, d_f, is equal to the difference between x_{rf} and half its length, 3.0 m.

$$d_f = x_{rf} - 3.0 \text{ m} = 2x_{CM} - x_{ri} - 3.0 \text{ m} = 2x_{CM} - (3.0 \text{ m} + 0.50 \text{ m}) - 3.0 \text{ m} = 2x_{CM} - 6.5 \text{ m}$$

Calculate x_{CM}.

$$x_{CM} = \frac{(60.0 \text{ kg})(6.5 \text{ m}) + (120 \text{ kg})(3.5 \text{ m})}{60.0 \text{ kg} + 120 \text{ kg}} = 4.5 \text{ m, so } d_f = 2(4.5 \text{ m}) - 6.5 \text{ m} = \boxed{2.5 \text{ m}}.$$

(b) Find the distance the woman walked relative to the pier.

$$|\Delta x_w| = |x_{wf} - x_{wi}| = |d_f - x_{wi}| = |2.5 \text{ m} - 6.5 \text{ m}| = \boxed{4.0 \text{ m}}$$

97. Strategy Use conservation of energy and momentum. Let $2m = m_B = 2m_A$.

Solution Find the maximum kinetic energy of A alone and, thus, its speed just before it strikes B.

$$\Delta K = \frac{1}{2}mv_1^2 - 0 = -\Delta U = mgh - 0, \text{ so } v_1 = \sqrt{2gh}.$$

Use conservation of momentum to find the speed of the combined bobs just after impact. The collision is perfectly inelastic, so $v_{Af} = v_{Bf} = v_2$.

$$m_A v_{Af} + m_B v_{Bf} = (m + 2m)v_2 = m_A v_{Ai} + m_B v_{Bi} = mv_1 + 0, \text{ so } v_2 = \frac{1}{3}v_1.$$

Find the maximum height.

$$\Delta K = 0 - \frac{1}{2}mv_2^2 = -\frac{1}{2}m\left(\frac{1}{3}\sqrt{2gh}\right)^2 = -\Delta U = 0 - mgh_2, \text{ so } h_2 = \boxed{\frac{1}{9}h}.$$

101. Strategy Use conservation of momentum and Eq. (6-6) for the kinetic energies. Since the radium nucleus is at rest, $\vec{p}_i = \vec{p}_{Ra} = 0$.

Solution

(a) Find the ratio of the speed of the alpha particle to the speed of the radon nucleus.

$$p_f = m_{Rn}v_{Rn} + m_\alpha v_\alpha = p_i = 0, \text{ so } m_\alpha v_\alpha = -m_{Rn}v_{Rn}. \text{ Therefore,}$$

$$\frac{v_\alpha}{v_{Rn}} = \frac{m_{Rn}}{m_\alpha} = \frac{222 \text{ u}}{4 \text{ u}} = \frac{222}{4} = \boxed{\frac{111}{2}}, \text{ where the negative was dropped because speed is nonnegative.}$$

(b) Since the initial momentum is zero, $\vec{p}_{Rn} = -\vec{p}_\alpha$; therefore, $\dfrac{|\vec{p}_\alpha|}{|\vec{p}_{Rn}|} = \dfrac{p_\alpha}{p_{Rn}} = \boxed{1}$.

(c) Find the ratio of the kinetic energies.

$$\frac{K_\alpha}{K_{Rn}} = \frac{\frac{1}{2}m_\alpha v_\alpha^2}{\frac{1}{2}m_{Rn}v_{Rn}^2} = \frac{m_\alpha}{m_{Rn}}\left(\frac{v_\alpha}{v_{Rn}}\right)^2 = \frac{4 \text{ u}}{222 \text{ u}}\left(\frac{111}{2}\right)^2 = \boxed{\frac{111}{2}}$$

Chapter 8

TORQUE AND ANGULAR MOMENTUM

Conceptual Questions

1. To maximize the torque, locate it as far as possible from the rotation axis: along the lower edge.

5. For a body to be in equilibrium, both the net force and the net torque acting on it must equal zero. To satisfy the first requirement, the two forces must be equal in magnitude and opposite in direction. To satisfy the second requirement, the two forces must act along the same line—a net torque would otherwise act to rotate the object.

9. An object's moment of inertia depends on how its mass is distributed with respect to the axis of rotation. The farther the mass is from the axis, the greater the object's moment of inertia. When animals have leg muscles that are concentrated close to the hip joint, their legs have relatively small moments of inertia. This makes it easier for them to rotate their legs, allowing them to run faster.

13. The vertical component of the angular momentum of the system (merry-go-round and child) is conserved throughout this process, since there are no external torques about the vertical axis of the merry-go-round. When the child moves out to the rim, the rotational inertia of the system increases, because the child is located farther from the axis. To conserve angular momentum, the angular velocity must therefore decrease. Noting that the rotational kinetic energy can be written as $L^2/(2I)$ and that L remains constant while I increases, we see that the rotational kinetic energy of the system decreases.

17. The astronaut and satellite constitute an isolated system. The initial angular momentum of the system is zero. When the astronaut tries to remove the bolt, both he and the satellite will rotate. They will rotate in opposite directions so that the total angular momentum of the system remains zero. To put it another way, when the astronaut applies a torque to a part of the satellite, the satellite applies an equal and opposite torque to him. The astronaut must anchor the satellite and himself somehow before trying to remove the bolt.

21. The melting of Earth's polar ice caps would distribute some of its mass from locations near its rotation axis to locations that are on average farther from its rotation axis. The rotational inertia of a sphere is greater if its mass is distributed farther from its axis of rotation—the Earth's moment of inertia would therefore increase. Angular momentum conservation requires that the product of the Earth's rotational inertia and its angular velocity be constant. A larger moment of inertia must be accompanied by a smaller angular velocity—the melting of the caps would therefore increase the length of the day.

Problems

1. **Strategy and Solution** I has units $\text{kg} \cdot \text{m}^2$. ω^2 has units $(\text{rad/s})^2$. So, $\frac{1}{2}I\omega^2$ has units

$$\text{kg} \cdot \text{m}^2 \cdot \text{rad}^2/\text{s}^2 = \text{kg} \cdot \text{m}^2/\text{s}^2 = \text{J, which is a unit of energy.}$$

3. **Strategy** $I = \frac{2}{5}MR^2$ for a solid sphere and mass density is $\rho = M/V$.

Solution

(a) $M = \rho V = \rho \frac{4}{3}\pi R^3$ for a solid sphere. Form a proportion.

$$\frac{M_{\text{child}}}{M_{\text{adult}}} = \left(\frac{R_{\text{child}}}{R_{\text{adult}}}\right)^3 = \left(\frac{1}{2}\right)^3 = \frac{1}{8}, \text{ so the mass is } \boxed{\text{reduced by a factor of 8}}.$$

(b) Form a proportion.

$$\frac{I_{\text{child}}}{I_{\text{adult}}} = \frac{1}{8}\left(\frac{R_{\text{child}}}{R_{\text{adult}}}\right)^2 = \frac{1}{8}\left(\frac{1}{2}\right)^2 = \frac{1}{32}$$

The rotational inertia is $\boxed{\text{reduced by a factor of 32}}$.

5. **Strategy** Rotational inertia depends upon the location of the rotation axis. In each situation, the mass and the distribution of the mass is the same; only the location of the rotation axis differs. The greater the distance a point mass is from the rotation axis, the greater its contribution to the rotational inertia.

    Solution In arrangements (a) and (b), two of the point masses are located along the axis of rotation; therefore, these point masses do not contribute to the rotational inertia. In (a), the two point masses not along the axis are farther from the axis than the two in (b); thus, arrangement (a) has the greater rotational inertia. In (c), none of the point masses are located along the axis; thus, the contribution to the rotational inertia due to mass is four times that in either (a) or (b). All four masses in (c) are the same distance from the rotation axis as the two contributing masses in (b); thus the rotational inertia of (c) is twice that of (b). In (c), the distance of each mass from the axis is a leg of an isosceles right triangle; thus, the distance is the length of the hypotenuse—a side of the square, s— divided by the square root of two. Squaring each distance and multiplying by four such distances gives $2s^2$. In (a), squaring the two distances, s, and multiplying by two such distances gives $2s^2$; thus, the rotational inertia for arrangements (a) and (c) is given by $2ms^2$—they are equal. Ranking the three arrangements in increasing order of the rotational inertia gives $\boxed{\text{(b), (a) = (c)}}$.

9. **Strategy** Use Eq. (8-1) and form a proportion.

    Solution Find the fraction of the total kinetic energy that is rotational.

$$\frac{K_{\text{rot}}}{K_{\text{total}}} = \frac{2\left(\frac{1}{2}I\omega^2\right)}{2\left(\frac{1}{2}I\omega^2\right)+\frac{1}{2}Mv^2} = \frac{2}{2+\frac{Mv^2}{I\omega^2}} = \frac{2}{2+\frac{Mv^2}{I(v^2/R^2)}} = \frac{2}{2+\frac{MR^2}{I}} = \frac{2}{2+\frac{(79\text{ kg})(0.32\text{ m})^2}{0.080\text{ kg·m}^2}} = \boxed{0.019}$$

13. **Strategy** Use Eq. (8-3).

    Solution Find the magnitude of the torque applied to the drum.

$$|\tau| = rF_\perp = (0.0600\text{ m})(75\text{ N}) = \boxed{4.5\text{ N·m}}$$

15. **Strategy** Use Eq. (8-3).

    Solution Find the magnitude of the torque.

$$|\tau| = F_\perp r = mgr = (0.124\text{ kg})(9.80\text{ N/kg})(0.25\text{ m}) = \boxed{0.30\text{ N·m}}$$

17. **Strategy** In each situation, calculate the torque using Eq. (8-3).

    Solution Calculate the torques.
    (a) $|\tau| = rF_\perp = (50\text{ cm})(20\text{ N}) = 1000\text{ N·cm}$; (b) $(25\text{ cm})(40\text{ N}) = 1000\text{ N·cm}$;
    (c) $(25\text{ cm})(80\text{ N})\sin 60° = 1700\text{ N·cm}$; (d) $(25\text{ cm})(80\text{ N})\sin 30° = 1000\text{ N·cm}$; (e) $(50\text{ cm})(40\text{ N})\sin 0° = 0$
    Ranking the situations in order of the magnitude of the torque applied to the handle, from smallest to largest, we have $\boxed{\text{(e), (a) = (b) = (d), (c)}}$.

21. Strategy Use Eq. (8-3) to compute the torque in each case.

Solution

(a) The force is applied perpendicularly to the door, so $\tau = rF = (1.26 \text{ m})(46.4 \text{ N}) = \boxed{58.5 \text{ N} \cdot \text{m}}$.

(b) The force is applied at $43.0°$ from the door's surface, so
$$|\tau| = rF_\perp = rF \sin \theta = (1.26 \text{ m})(46.4 \text{ N}) \sin 43.0° = \boxed{39.9 \text{ N} \cdot \text{m}}.$$

(c) Since the force is applied such that its line of action passes through the axis of the door hinges—the axis of rotation—there is no perpendicular component of the force and the torque is $\boxed{0}$.

25. Strategy The center of gravity is at the center of mass of the plate. Imagine that the plate consists of a rectangular plate (on the left) and a square (on the right). The mass is proportional to the area for a uniform distribution.

Solution Find the center of gravity.
$$x_{\text{CM}} = \frac{A_1 x_1 + A_2 x_2}{A_1 + A_2} = \frac{0.50s^2 \left(\frac{0.50s}{2} \right) + 0.50^2 s^2 \left(0.50s + \frac{0.50s}{2} \right)}{0.50s^2 + 0.50^2 s^2} = 0.42s$$

$$y_{\text{CM}} = \frac{0.50s^2 (0.50s) + 0.50^2 s^2 \left(0.50s + \frac{0.50s}{2} \right)}{0.50s^2 + 0.50^2 s^2} = 0.58s$$

So, the center of gravity is located at $\boxed{(0.42s,\, 0.58s)}$.

29. (a) Strategy The rotational inertia of a hoop is MR^2. Use the work-kinetic energy theorem and Eq. (8-1).

Solution Find the work.
$$W = \Delta K = \frac{1}{2} I(\omega_f^2 - \omega_i^2) = \frac{1}{2}(MR^2)(\omega_f^2 - 0)$$
$$= \frac{1}{2}(1.90 \times 10^6 \text{ kg})(67.5 \text{ m})^2 (3.50 \times 10^{-3} \text{ rad/s})^2 = \boxed{53.0 \text{ kJ}}$$

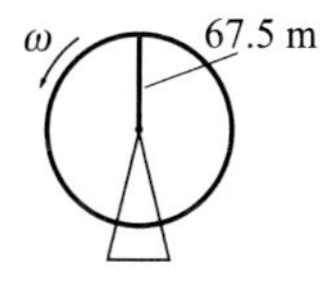

(b) Strategy Constant torque implies constant angular acceleration, so $\Delta\theta = \omega_{\text{av}}\Delta t$. Use Eq. (8-6).

Solution Find the torque.
$$W = \tau\Delta\theta = \tau\omega_{\text{av}}\Delta t = \tau\left(\frac{\omega_f + \omega_i}{2} \right)\Delta t = \tau\left(\frac{\omega_f + 0}{2} \right)\Delta t, \text{ so}$$

$$\tau = \frac{2W}{\omega_f \Delta t} = \frac{2(53.0 \times 10^3 \text{ J})}{(3.50 \times 10^{-3} \text{ rad/s})(20.0 \text{ s})} = \boxed{1.51 \text{ MN} \cdot \text{m}}.$$

31. Strategy Choose the axis of rotation at the fulcrum. Use Eqs. (8-8).

Solution Find F.
$$\Sigma\tau = 0 = -F(3.0 \text{ m}) + (1200 \text{ N})(0.50 \text{ m}), \text{ so } F = \frac{(1200 \text{ N})(0.50 \text{ m})}{3.0 \text{ m}} = \boxed{200 \text{ N}}.$$

33. **(a)** **Strategy** Choose the axis of rotation at the point at which the right-hand cable connects to the platform. Let $m_1 = 75$ kg and $m_2 = 20.0$ kg. Let $l = 5.0$ m. The system is in equilibrium.

Solution Find the force exerted by the left-hand cable.

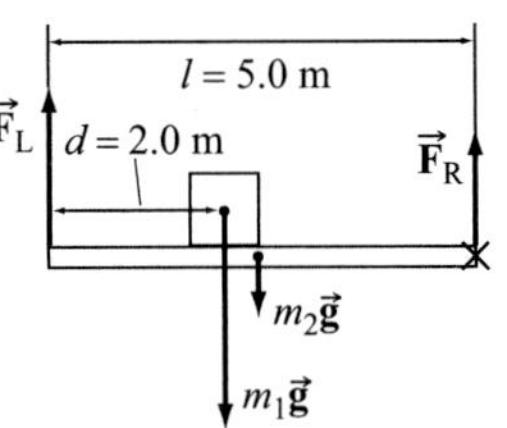

$$\Sigma\tau = 0 = -F_L l + m_1 g(l-d) + m_2 g\left(\frac{l}{2}\right), \text{ so}$$

$$F_L = g\left[m_1\left(1-\frac{d}{l}\right) + \frac{m_2}{2}\right]$$

$$= (9.80 \text{ N/kg})\left[(75 \text{ kg})\left(1-\frac{2.0 \text{ m}}{5.0 \text{ m}}\right) + \frac{20.0 \text{ kg}}{2}\right] = \boxed{540 \text{ N}}.$$

(b) **Strategy** Use Newton's second law.

Solution Find the force exerted by the right-hand cable.
$$\Sigma F = 0 = -m_1 g - m_2 g + F_L + F_R, \text{ so}$$

$$F_R = (m_1 + m_2)g - F_L = (75 \text{ kg} + 20.0 \text{ kg})(9.80 \text{ N/kg}) - 539 \text{ N} = \boxed{390 \text{ N}}.$$

37. **Strategy** Use Eqs. (8-8).

Solution

(a) Choose the axis of rotation at the point of contact between the vertical wall and the climber's feet.
$$\Sigma\tau = T\cos\theta(1.06 \text{ m}) - W_c(0.91 \text{ m}) = 0, \text{ so } T = \frac{(0.91 \text{ m})W_c}{(1.06 \text{ m})\cos\theta} = \frac{(0.91 \text{ m})(770 \text{ N})}{(1.06 \text{ m})\cos 25°} = \boxed{730 \text{ N}}.$$

(b) $\Sigma F_x = 0 = F_x - T\sin\theta$ and $\Sigma F_y = 0 = F_y + T\cos\theta - W_c$, so $F_x = T\sin\theta$ and $F_y = W_c - T\cos\theta$.
Find the magnitude of the force.
$$F = \sqrt{F_x^2 + F_y^2} = \sqrt{T^2\sin^2\theta + W_c^2 + T^2\cos^2\theta - 2W_c T\cos\theta} = \sqrt{T^2 + W_c^2 - 2W_c T\cos\theta}$$

$$= \sqrt{(730 \text{ N})^2 + (770 \text{ N})^2 - 2(770 \text{ N})(730 \text{ N})\cos 25°} = 330 \text{ N}$$

Find the direction.
$$\theta = \tan^{-1}\frac{F_y}{F_x} = \tan^{-1}\frac{W_c - T\cos\theta}{T\sin\theta} = \tan^{-1}\frac{770 \text{ N} - (730 \text{ N})\cos 25°}{(730 \text{ N})\sin 25°} = 19°$$

Thus, $\vec{\mathbf{F}} = \boxed{330 \text{ N at } 19° \text{ above the horizontal}}$.

39. **Strategy** Use Eqs. (8-8).

Solution Choose the axis of rotation at the hinge.
$$\Sigma\tau = 0 = Wl\cos\theta - Tl\sin\theta + mg\frac{l}{2}\cos\theta, \text{ so } T = \boxed{\dfrac{mg/2 + W}{\tan\theta}}.$$

$$\boxed{\text{For } \theta = 0, T \to \infty, \text{ and for } \theta = 90°, T \to 0.}$$

41. Strategy Use the results from Problem 40.

Solution We must add one term (for the cat) and substitute for the angle in the summation of the torques. Let d be the distance between the store and the center of mass of the cat.

$$\Sigma\tau = 0 = (417\ \text{N})\sin 33.8°(1.50\ \text{m}) - (50.0\ \text{N})(0.75\ \text{m}) - (200.0\ \text{N})(1.00\ \text{m}) - (8.7\ \text{kg})(9.80\ \text{m/s}^2)d,\ \text{so}$$

$$d = \frac{(417\ \text{N})\sin 33.8°(1.50\ \text{m}) - (50.0\ \text{N})(0.75\ \text{m}) - (200.0\ \text{N})(1.00\ \text{m})}{(8.7\ \text{kg})(9.80\ \text{m/s}^2)} = \boxed{1.3\ \text{m}}.$$

45. Strategy Use Eqs. (8-8). Choose the axis of rotation at the shoulder joint. One arm supports half of the person's weight, so $F_{\text{p}} = \frac{1}{2}(700\ \text{N}) = 350\ \text{N}$.

Solution Find the force each muscle exerts.
$$\Sigma\tau = 0 = F_{\text{m}}(12\ \text{cm})\sin 15° - F_{\text{g}}(27.5\ \text{cm}) - F_{\text{p}}(60\ \text{cm}),\ \text{so}$$

$$F_{\text{m}} = \frac{F_{\text{g}}(27.5\ \text{cm}) + F_{\text{p}}(60\ \text{cm})}{(12\ \text{cm})\sin 15°} = \frac{(30.0\ \text{N})(27.5\ \text{cm}) + (350\ \text{N})(60\ \text{cm})}{(12\ \text{cm})\sin 15°} = \boxed{7.0\ \text{kN}}.$$

47. Strategy Use Eqs. (8-8). Choose the axis of rotation at the knee.

Solution Find the forces exerted by the patellar tendon.

(a) $\Sigma\tau = 0 = F_{\text{p}}(10.0\ \text{cm})\sin 20.0° - F_{\text{w}}(41\ \text{cm})\sin 30.0° - F_{\text{L}}(22\ \text{cm})\sin 30.0°,\ \text{so}$

$$F_{\text{p}} = \frac{g\sin 30.0°[m_{\text{w}}(41\ \text{cm}) + m_{\text{L}}(22\ \text{cm})]}{(10.0\ \text{cm})\sin 20.0°} = \frac{(9.80\ \text{N/kg})\sin 30.0°[(3.0\ \text{kg})(41\ \text{cm}) + (5.0\ \text{kg})(22\ \text{cm})]}{(10.0\ \text{cm})\sin 20.0°}$$

$$= \boxed{330\ \text{N}}.$$

(b) $\Sigma\tau = 0 = F_{\text{q}}(10.0\ \text{cm})\sin 20.0° - F_{\text{w}}(41\ \text{cm})\sin 90.0° - F_{\text{L}}(22\ \text{cm})\sin 90.0°,\ \text{so}$

$$F_{\text{q}} = \frac{g[m_{\text{w}}(41\ \text{cm}) + m_{\text{L}}(22\ \text{cm})]}{(10.0\ \text{cm})\sin 20.0°} = \frac{(9.80\ \text{N/kg})[(3.0\ \text{kg})(41\ \text{cm}) + (5.0\ \text{kg})(22\ \text{cm})]}{(10.0\ \text{cm})\sin 20.0°} = \boxed{670\ \text{N}}.$$

49. Strategy and Solution Torque has units $\text{N}\cdot\text{m} = \text{kg}\cdot\text{m}\cdot\text{s}^{-2}\cdot\text{m} = \text{kg}\cdot\text{m}^2\cdot\text{s}^{-2}$. Inertia times angular acceleration has units $\text{kg}\cdot\text{m}^2\cdot\text{s}^{-2} = \text{N}\cdot\text{m}$. Thus, the units are consistent.

51. Strategy Use the rotational form of Newton's second law and Eq. (5-21).

Solution Find the torque that the motor must deliver.

$I = \frac{1}{2}MR^2$ for a uniform disk, so

$$\Sigma\tau = I\alpha = \frac{1}{2}MR^2\left(\frac{\omega_{\text{f}}^2 - \omega_{\text{i}}^2}{2\Delta\theta}\right) = \frac{MR^2\omega_{\text{f}}^2}{4\Delta\theta} = \frac{(0.22\ \text{kg})\left(\frac{0.305\ \text{m}}{2}\right)^2(3.49\ \text{rad/s})^2}{4(2.0\ \text{rev})(2\pi\ \text{rad/rev})} = \boxed{0.0012\ \text{N}\cdot\text{m}}.$$

53. (a) Strategy Use the definition of average angular speed.

Solution Find the average angular speed of the discus just before release.
$$\omega_{\text{ave}} = \frac{\Delta\theta}{\Delta t} = \frac{\omega_{\text{i}} + \omega_{\text{f}}}{2} = \frac{0 + \omega_{\text{f}}}{2},\ \text{so}\ \omega_{\text{f}} = 2\frac{\Delta\theta}{\Delta t} = \frac{2(1.5\ \text{rev})(2\pi\ \text{rad/rev})}{1.4\ \text{s}} = \boxed{13\ \text{rad/s}}.$$

(b) Strategy Treating the discus as a point mass, compute the rotational inertia using MR^2. Use Eq. (8-9) to find the torque.

Solution Find the torque applied to the discus by the athlete.

$$\tau = I\alpha = MR^2 \frac{\Delta\omega}{\Delta t} = MR^2 \frac{\omega_f - \omega_i}{\Delta t} = MR^2 \frac{\omega_f - 0}{\Delta t} = MR^2 \frac{2\Delta\theta/\Delta t}{\Delta t} = \frac{2MR^2\Delta\theta}{(\Delta t)^2}, \text{ so}$$

$$\tau = \frac{2(2.0 \text{ kg})(0.90 \text{ m})^2[(1.5 \text{ rev})(2\pi \text{ rad/rev})]}{(1.4 \text{ s})^2} = \boxed{16 \text{ N} \cdot \text{m}}.$$

(c) Strategy Use $v = R\omega$ to find the initial speed of the discus as it is released by the athlete. Then, use the expression for the range of a projectile—given in Problem 4-54—to estimate the distance traveled by the discus.

Solution Compute the initial speed of the discus.

$$v_i = R\omega_f = \frac{2R\Delta\theta}{\Delta t} = \frac{2(0.90 \text{ m})(1.5 \text{ rev})(2\pi \text{ rad/rev})}{1.4 \text{ s}} = 12 \text{ m/s}$$

Estimate the distance.

$$\Delta x = \frac{2v_i^2 \sin\theta\cos\theta}{g} = \frac{2(12 \text{ m/s})^2 \sin 45°\cos 45°}{9.8 \text{ m/s}^2} = 15 \text{ m}$$

The discus travels $\boxed{15 \text{ m to the same height}}$ as it was released, $\boxed{\text{plus about another meter if released 1 m above the ground.}}$

57. (a) Strategy The rotational inertia of the merry-go-round is $I = \frac{1}{2}MR^2$ and that of the children is $I = 2MR^2$. Use the rotational form of Newton's second law.

Solution Find the torque on the merry-go-round.

$$\Sigma\tau = I\alpha = \left(\frac{1}{2}MR^2 + 2mR^2\right)\frac{\Delta\omega}{\Delta t}$$

$$= \left[\frac{1}{2}(350.0 \text{ kg})(1.25 \text{ m})^2 + 2(30.0 \text{ kg})(1.25 \text{ m})^2\right]\left(\frac{25 \text{ rpm}}{20.0 \text{ s}}\right)\left(\frac{2\pi \text{ rad}}{\text{rev}}\right)\left(\frac{1 \text{ min}}{60 \text{ s}}\right) = \boxed{48 \text{ N} \cdot \text{m}}$$

(b) Strategy Let F be the magnitude of the tangential force with which each child must push the rim.

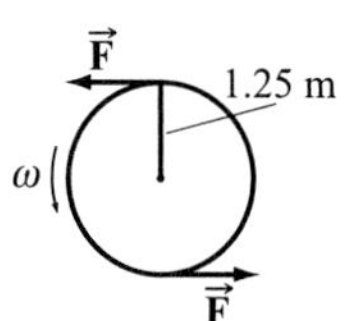

Solution Find F.

$$FR + FR = \Sigma\tau, \text{ so } F = \frac{\Sigma\tau}{2R} = \frac{48 \text{ N} \cdot \text{m}}{2(1.25 \text{ m})} = \boxed{19 \text{ N}}.$$

61. Strategy The rotational inertia of a uniform solid sphere is $\frac{2}{5}MR^2$. Use the expression for the acceleration found in Example 8.13.

Solution Find the acceleration of the solid sphere.

$$a_{CM} = \frac{g\sin\theta}{1 + I/(MR^2)} = \frac{g\sin\theta}{1 + \frac{2}{5}MR^2/(MR^2)} = \frac{g\sin\theta}{1 + 2/5} = \frac{(9.80 \text{ m/s}^2)\sin 35°}{1 + 2/5}$$

$$= \boxed{4.0 \text{ m/s}^2}$$

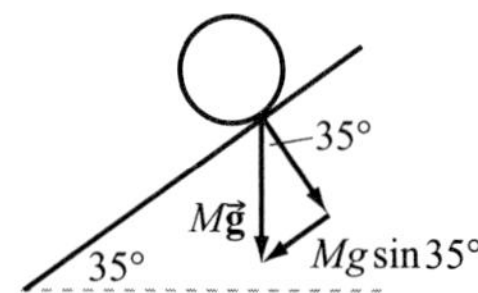

65. (a) Strategy Use conservation of energy. The rotational inertia of a uniform solid cylinder is $\frac{1}{2}MR^2$.

Solution Let $h = 0.80$ m, m be the mass of the bucket, and M be the mass of the cylinder. The tangential speed of the cylinder is the same as the linear speed of the bucket, since they are attached by a rope.

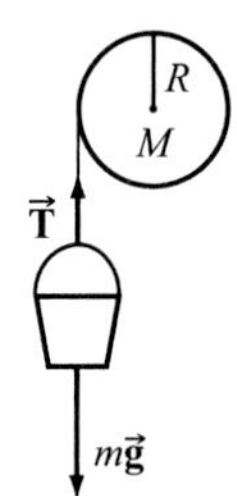

$$\Delta K = \frac{1}{2}mv^2 + \frac{1}{2}I\omega^2 = \frac{1}{2}mv^2 + \frac{1}{2}\left(\frac{1}{2}MR^2\right)\left(\frac{v}{R}\right)^2 = \frac{1}{2}mv^2 + \frac{1}{4}Mv^2 = -\Delta U = mgh, \text{ so}$$

$$v = \sqrt{\frac{4mgh}{2m+M}} = \sqrt{\frac{4(2.0\text{ kg})(9.80\text{ m/s}^2)(0.80\text{ m})}{2(2.0\text{ kg})+3.0\text{ kg}}} = \boxed{3.0\text{ m/s}}.$$

(b) Strategy Use the work-kinetic energy theorem.

Solution Find the tension T in the rope as the bucket falls a distance h.

$$W_{\text{total}} = \Delta K = \frac{1}{2}mv^2 = W_{\text{rope}} + W_{\text{grav}} = -Th + mgh, \text{ so}$$

$$T = m\left(g - \frac{v^2}{2h}\right) = (2.0\text{ kg})\left[9.80\text{ m/s}^2 - \frac{8.96\text{ m}^2/\text{s}^2}{2(0.80\text{ m})}\right] = \boxed{8.4\text{ N}}.$$

(c) Strategy Use Newton's second law.

Solution Find the acceleration of the bucket as it falls.

$$\Sigma F_y = T - mg = ma_y, \text{ so } a_y = -g + \frac{T}{m} = -9.80\text{ m/s}^2 + \frac{8.4\text{ N}}{2.0\text{ kg}} = -5.6\text{ m/s}^2, \text{ or } \boxed{5.6\text{ m/s}^2 \text{ down}}.$$

67. (a) Strategy Use conservation of energy and the relationship between speed and radial acceleration.

Solution At the top of the loop, the sphere's speed must be at least the speed that results in a radial acceleration of g.

$$\frac{v^2}{r} = g, \text{ so } v^2 = gr.$$

The sphere's kinetic energy is $\frac{1}{2}mv^2 = \frac{1}{2}mgr$, and it must equal the potential energy difference $mgh - mg(2r)$. Thus, $\frac{1}{2}r = h - 2r$ or $h = \boxed{\frac{5}{2}r}$.

(b) Strategy The rotational inertia of a uniform solid sphere is $\frac{2}{5}mr^2$. Use conservation of energy.

Solution Find the kinetic energy of the sphere.

$$K = \frac{1}{2}mv^2 + \frac{1}{2}\left(\frac{2}{5}mr^2\right)\left(\frac{v^2}{r^2}\right) = \frac{7}{10}mv^2 = \frac{7}{10}mgr$$

Find h.

$$\Delta K = \frac{7}{10}mgr = -\Delta U = mgh - mg(2r), \text{ so } h = \boxed{\frac{27}{10}r}.$$

69. Strategy Consider the rotational inertia of each object. Use conservation of energy and the relationship between speed and radial acceleration.

Solution Since $I_{\text{sphere}} = \frac{2}{5}mr^2 < mr^2 = I_{\text{hollow cylinder}}$,

> h will decrease. The smaller the rotational inertia, the less gravitational energy will go into rotational energy, and the more will go into translational energy.

Redo the calculation with the solid sphere. At the top of the loop, the sphere's speed must be at least the speed that results in a radial acceleration of g.

$\frac{v^2}{r} = g$, so $v^2 = gr$. Thus, its kinetic energy is $\frac{1}{2}mv^2 + \frac{1}{2}I\omega^2 = \frac{1}{2}mv^2 + \frac{1}{5}mr^2\left(\frac{v}{r}\right)^2 = \frac{7}{10}mv^2 = \frac{7}{10}mgr$, and the

kinetic energy must equal the potential energy difference $mgh - mg(2r)$. Find h.

$\frac{7}{10}mgr = mgh - 2mgr$, so $\frac{7}{10}r = h - 2r$, or $h = 2.7r$.

> Problem 68 had a minimum of $h = 3r$. With a solid sphere, the minimum is $h = 2.7r$, which is a little less than $3r$.

73. Strategy The rotational inertia of a hoop is $I = MR^2$. Use Eq. (8-14).

Solution Find the magnitude of the angular momentum of the flywheel.

$$L = I\omega = MR^2\omega = (5.6\times10^4 \text{ kg})(2.6 \text{ m})^2\left(\frac{350 \text{ rev}}{1 \text{ min}}\right)\left(\frac{2\pi \text{ rad}}{\text{rev}}\right)\left(\frac{1 \text{ min}}{60 \text{ s}}\right) = \boxed{1.4\times10^7 \text{ kg}\cdot\text{m}^2/\text{s}}$$

75. Strategy Since the torque is constant, it is equal to the change in angular momentum divided by the time interval.

Solution Find the time to stop the spinning wheel

$$\tau = \frac{\Delta L}{\Delta t}, \text{ so } \Delta t = \frac{\Delta L}{\tau} = \frac{-6.40 \text{ kg}\cdot\text{m}^2/\text{s}}{-4.00 \text{ N}\cdot\text{m}} = \boxed{1.60 \text{ s}}.$$

77. Strategy Use conservation of angular momentum and Eq. (8-14).

Solution Find the skater's new rate of rotation.

$$L_i = I_i\omega_i = L_f = I_f\omega_f, \text{ so } \omega_f = \frac{I_i}{I_f}\omega_i = \frac{1}{0.67}(1.0 \text{ rev/s}) = \boxed{1.5 \text{ rev/s}}.$$

81. Strategy The rotational inertias of the wheel and guinea pig are $I_w = MR^2$ and $I_g = mR^2$, respectively, where M is the mass of the wheel, m is the mass of the guinea pig, and R is the radius of the wheel. Use conservation of angular momentum and $v = r\omega$.

Solution Find the angular velocity of the wheel.

$$L_w = L_g, \text{ so } I_w\omega_w = MR^2\omega_w = I_g\omega_g = mR^2\omega_g = mRv_g.$$

$$\text{Thus, } \omega_w = \frac{mv_g}{MR} = \frac{(0.500 \text{ kg})(0.200 \text{ m/s})}{(2.00 \text{ kg})(0.400 \text{ m})} = \boxed{0.125 \text{ rad/s}}.$$

85. Strategy The average torque is equal to the magnitude of the change in angular momentum divided by the time interval.

Solution Let $\vec{L}_i = L$ in the $+y$-direction. Then $\Delta\vec{L}$ has components $\Delta L_x = L\sin\theta$ and $\Delta L_y = L\cos\theta - L = L(\cos\theta - 1)$. So,

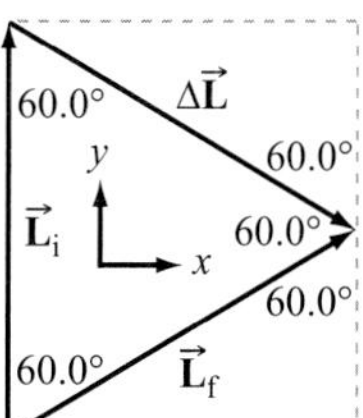

$$\left|\Delta\vec{L}\right| = \sqrt{(L\sin\theta)^2 + [L(\cos\theta - 1)]^2} = L\sqrt{\sin^2 60.0° + (\cos 60.0° - 1)^2} = 1.00L.$$

Compute the magnitude of the required torque.

$$\tau = \left|\frac{\Delta\vec{L}}{\Delta t}\right| = \frac{1.00L}{\Delta t} = \frac{1.00I\omega}{\Delta t} = \frac{\frac{1}{2}mr^2\omega}{\Delta t}$$

$$= \frac{(1.00\times10^5 \text{ kg})(2.00 \text{ m})^2(300.0 \text{ rpm})}{2(3.00 \text{ s})}\left(\frac{2\pi \text{ rad}}{\text{rev}}\right)\left(\frac{1 \text{ min}}{60 \text{ s}}\right) = \boxed{2.10\times10^6 \text{ N}\cdot\text{m}}$$

89. (a) Strategy The rotational inertia of a uniform solid sphere is $I = \frac{2}{5}MR^2$. Use Eq. (8-1).

Solution Find the kinetic energy of the Earth.

$$K_{\text{rot}} = \frac{1}{2}I\omega^2 = \frac{1}{2}\left(\frac{2}{5}MR^2\right)\omega^2 = \frac{1}{5}(5.974\times10^{24} \text{ kg})(6.371\times10^6 \text{ m})^2\left(\frac{2\pi \text{ rad}}{24 \text{ h}}\right)^2\left(\frac{1 \text{ h}}{3600 \text{ s}}\right)^2 = \boxed{2.6\times10^{29} \text{ J}}$$

(b) Strategy and Solution $T = \dfrac{2\pi}{\omega}$ and $K_{\text{rot}} \propto \omega^2$, so $\omega \propto \sqrt{K_{\text{rot}}}$ and $\dfrac{T_f}{T_i} = \dfrac{\omega_i}{\omega_f} = \sqrt{\dfrac{K_i}{K_f}}$. The change in the

period is $T_f - T_i = \left(\sqrt{\dfrac{K_i}{K_f}} - 1\right)T_i = \left(\sqrt{\dfrac{1}{0.990}} - 1\right)(24 \text{ h})\left(\dfrac{60 \text{ min}}{1 \text{ h}}\right) = 7 \text{ min}.$

$\boxed{\text{The length of the day would increase by 7 minutes.}}$

(c) Strategy Divide 1.0% of the Earth's rotational kinetic energy by the world's energy usage.

Solution One percent of the Earth's rotational kinetic energy would supply the world's energy needs (at today's usage) for $\dfrac{0.010(2.6\times10^{29} \text{ J})}{1.0\times10^{21} \text{ J/yr}} = \boxed{2.6 \text{ million years}}$.

93. Strategy The rotational inertia of each blade (uniform rod) is $I = \frac{1}{3}ML^2$, where L is the length of each blade. Find the angular acceleration of the fan using the definition; and use Eq. (8-9) to find the torque applied to the fan by the motor.

Solution The angular acceleration is $\alpha = \Delta\omega/\Delta t$. Find the torque.

$$\Sigma\tau = I\alpha = 4\left(\frac{1}{3}ML^2\right)\frac{\Delta\omega}{\Delta t} = \frac{4ML^2\Delta\omega}{3\Delta t} = \frac{4(0.35 \text{ kg})(0.60 \text{ m})^2(1.8 \text{ rev/s})}{3(4.35 \text{ s})}\left(\frac{2\pi \text{ rad}}{\text{rev}}\right) = \boxed{0.44 \text{ N}\cdot\text{m}}$$

95. Strategy The rotational inertia of the rod is $I = \frac{1}{3}mL^2$. Use conservation of energy.

Solution Find the speed of the lower end of the uniform rod when moving at its lowest point.

$$\Delta K = K_{\text{rot}} = \frac{1}{2}I\omega^2 = \frac{1}{2}\left(\frac{1}{3}mL^2\right)\left(\frac{v}{L}\right)^2 = \frac{1}{6}mv^2 = -\Delta U = mgh = mg\frac{L}{2}, \text{ so } v = \boxed{\sqrt{3gL}}.$$

97. Strategy Use Eqs. (8-8). Choose the axis of rotation at the hinge attaching the crane to the cab (the pivot).

Solution Find T_1.

$\Sigma\tau = 0 = T_2(12.2 \text{ m})\sin 10.0° + T_1(12.2 \text{ m})\sin 5.0° - (18 \text{ kN})(6.1 \text{ m})\sin 40.0° - (67 \text{ kN})(12.2 \text{ m})\sin 40.0°$ and

$\Sigma F_y = T_1 - 67 \text{ kN} = 0$, so $\boxed{T_1 = 67 \text{ kN}}$.

Find T_2.

$$T_2 = \frac{[(18 \text{ kN})(6.1 \text{ m}) + (67 \text{ kN})(12.2 \text{ m})]\sin 40.0° - (67 \text{ kN})(12.2 \text{ m})\sin 5.0°}{(12.2 \text{ m})\sin 10.0°}, \text{ so } \boxed{T_2 = 250 \text{ kN}}.$$

At the pivot:

$\Sigma F_y = F_{py} - 18 \text{ kN} - 67 \text{ kN} - T_1 \cos 45.0° - T_2 \cos 50.0° = 0$, so

$F_{py} = 18 \text{ kN} + 67 \text{ kN} + (247.7 \text{ kN})\cos 50.0° + (67 \text{ kN})\cos 45.0° = 291.6 \text{ kN}$.

$\Sigma F_x = F_{px} - T_1 \sin 45.0° - T_2 \sin 50.0° = 0$, so

$F_{px} = (247.7 \text{ kN})\sin 50.0° + (67 \text{ kN})\sin 45.0° = 237.1 \text{ kN}$.

Find the magnitude.

$$F_p = \sqrt{(237.1 \text{ kN})^2 + (291.6 \text{ kN})^2} = 380 \text{ kN}$$

Find the direction.

$$\theta = \tan^{-1}\frac{291.6}{237.1} = 51°$$

So, $\boxed{\vec{F}_p = 380 \text{ kN at } 51° \text{ with the horizontal}}$.

101. Strategy Let the subscripts be 1 for the painter, 2 for the can, and 3 for the plank.

Solution

(a) Choose the axis of rotation at the point of contact between the plank and the right sawhorse.

$$\Sigma\tau = 0 = m_3 g d_3 - m_1 g d_1 - m_2 g d_2, \text{ so } d_1 = \frac{m_3 d_3 - m_2 d_2}{m_1}.$$

The distance from the right-hand edge is $1.40 \text{ m} - d_1 = d$.

$$d = 1.40 \text{ m} - \frac{m_3 d_3 - m_2 d_2}{m_1} = 1.40 \text{ m} - \frac{(20.0 \text{ kg})(3.00 \text{ m} - 1.40 \text{ m}) - (4.0 \text{ kg})(1.40 \text{ m} - 0.14 \text{ m})}{61 \text{ kg}}$$

$$= \boxed{0.96 \text{ m from the RH edge}}$$

(b) Choose the axis of rotation at the point of contact between the plank and the left sawhorse.

$$\Sigma\tau = 0 = m_1 g d_1 - m_2 g d_2 - m_3 g d_3, \text{ so } d_1 = \frac{m_2 d_2 + m_3 d_3}{m_1}.$$

The distance from the left-hand edge is $1.40 \text{ m} - d_1 = d$.

$$d = 1.40 \text{ m} - \frac{m_2 d_2 + m_3 d_3}{m_1} = 1.40 \text{ m} - \frac{(4.0 \text{ kg})(6.00 \text{ m} - 1.40 \text{ m} - 0.14 \text{ m}) + (20.0 \text{ kg})(1.60 \text{ m})}{61 \text{ kg}}$$

$$= \boxed{0.58 \text{ m from the LH edge}}$$

105. Strategy The system is in equilibrium. Use Eqs. (8-8).

Solution Find h.

$$\frac{h}{(1.26 \text{ m})/2} = \tan 75°, \text{ so } h = [(1.26 \text{ m})/2]\tan 75° = (0.630 \text{ m})\tan 75°.$$

At the top of the ladder, each leg exerts a horizontal force on the other. These forces are equal in magnitude and opposite in direction, since the system is in equilibrium. Let the magnitude of this force be F. The tension T in the rope is directed to the left at the connection point on the right leg, so for the right leg, we have $\Sigma F_x = F - T = 0$ or $T = F$.

Calculate the torque about the contact point of the right leg of the ladder and the ground.

$$\Sigma \tau = (0.630 \text{ m})mg - Fh = 0, \text{ so } T = F = \frac{(0.630 \text{ m})mg}{h} = \frac{(0.630 \text{ m})mg}{(0.630 \text{ m})\tan 75°} = \frac{mg}{\tan 75°}.$$

The tension in the rope is the same along its length, so

$$T_{\text{rope}} = \frac{mg}{\tan 75°} = \frac{(42 \text{ kg})(9.80 \text{ N/kg})}{\tan 75°} = \boxed{110 \text{ N}}.$$

109. Strategy The rotational inertial of a uniform disk is $I = \frac{1}{2}MR^2$. Use Eq. (8-14).

Solution Find the magnitude of the angular momentum of the disk.

$$L = I\omega = \frac{1}{2}MR^2\omega = \frac{1}{2}(2.0 \text{ kg})(0.100 \text{ m})^2(3.0 \text{ rev/s})(2\pi \text{ rad/rev}) = \boxed{0.19 \text{ kg}\cdot\text{m}^2/\text{s}}$$

113. Strategy The system is in equilibrium. Choose the axis of rotation at the ankle.

Solution Find the force that each calf muscle needs to exert while the woman is standing.

$$\Sigma \tau = 0 = 2F(4.4 \text{ cm})\sin 81° - mg(3.0 \text{ cm}), \text{ so}$$

$$F = \frac{mg(3.0 \text{ cm})}{2(4.4 \text{ cm})\sin 81°} = \frac{(68 \text{ kg})(9.80 \text{ N/kg})(3.0 \text{ cm})}{2(4.4 \text{ cm})\sin 81°} = \boxed{230 \text{ N}}.$$

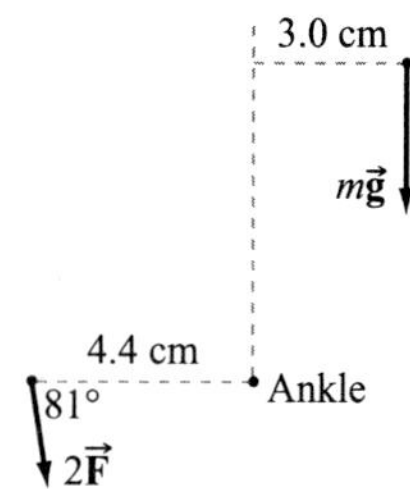

117. Strategy Use conservation of energy.

Solution

(a) Find the speed with which the roustabout reaches the ground.

$$\Delta K = \frac{1}{2}mv^2 = -\Delta U = mgL, \text{ so } v = \boxed{\sqrt{2gL}}.$$

(b) Find the speed with which the roustabout reaches the ground.

$$\Delta K = \frac{1}{2}I\omega^2 = \frac{1}{2}\left(\frac{1}{3}ML^2\right)\left(\frac{v}{L}\right)^2 = \frac{1}{6}Mv^2 = -\Delta U = Mg\frac{L}{2}, \text{ so } v = \boxed{\sqrt{3gL}}.$$

(c) Since $\sqrt{2gL} < \sqrt{3gL}$, $\boxed{\text{the roustabout should jump}}$.

REVIEW AND SYNTHESIS: CHAPTERS 6–8

Review Exercises

1. (a) Strategy Multiply the extension per mass by the mass to find the maximum extension required.

Solution

$$\left(\frac{1.0 \text{ mm}}{25 \text{ g}}\right)(5.0 \text{ kg})\left(\frac{1000 \text{ g}}{1 \text{ kg}}\right)\left(\frac{1 \text{ m}}{1000 \text{ mm}}\right) = \boxed{0.20 \text{ m}}$$

(b) Strategy Set the weight of the mass equal to the magnitude of the force due to the spring scale. Use Hooke's law.

Solution

$$\text{Weight} = mg = kx, \text{ so } k = \frac{mg}{x} = \frac{(5.0 \text{ kg})(9.80 \text{ N/kg})}{0.20 \text{ m}} = \boxed{250 \text{ N/m}}.$$

5. (a) Strategy Use the conservation of energy.

Solution Find the work done by friction.

$$W_{\text{total}} = W_{\text{friction}} + W_{\text{grav}} = W_{\text{friction}} + mgd \sin\theta = \Delta K = 0 - \frac{1}{2}mv_i^2, \text{ so}$$

$$W_{\text{friction}} = -\frac{1}{2}mv_i^2 - mgd \sin\theta = -m\left(\frac{1}{2}v_i^2 + gd\sin\theta\right)$$

$$= -(100 \text{ kg})\left[\frac{1}{2}(2.00 \text{ m/s})^2 + (9.80 \text{ m/s}^2)(1.50 \text{ m})\sin 30.0°\right]$$

$$= -940 \text{ J}.$$

Thus, the energy dissipated by friction was $\boxed{940 \text{ J}}$.

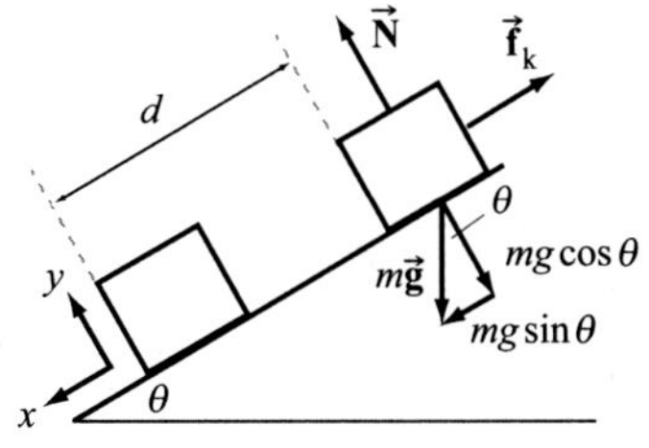

(b) Strategy Use Newton's second law.

Solution Find the normal force on the crate.
$\Sigma F_y = N - mg \cos\theta = 0$, so $N = mg \cos\theta$.

Since $v_{fx}^2 - v_{ix}^2 = 0 - v_i^2 = 2a_x \Delta x = 2a_x d$, the acceleration of the crate is $-v_i^2/(2d)$.
Find the force of sliding friction.

$$\Sigma F_x = -f_k + mg \sin\theta = -\mu_k mg \cos\theta + mg \sin\theta = ma_x = -m\frac{v_i^2}{2d}, \text{ so}$$

$$\mu_k = \tan\theta + \frac{v_i^2}{2dg \cos\theta} = \tan 30.0° + \frac{(2.00 \text{ m/s})^2}{2(1.50 \text{ m})(9.80 \text{ m/s}^2)\cos 30.0°} = \boxed{0.734}.$$

9. **Strategy** The collision is inelastic. Use conservation of momentum and energy.

   Solution Write equations using conservation of momentum and energy.

   momentum: $mv_i = (m + M)v_f$

   energy: $\dfrac{1}{2}(m + M)v_f^2 = (m + M)g\Delta y$

   Find the initial speed of the putty.

   $$\frac{1}{2}(m + M)\left(\frac{mv_i}{m + M}\right)^2 = (m + M)g\Delta y$$

   $$\left(\frac{m}{m + M}\right)^2 v_i^2 = 2g\Delta y$$

   $$v_i = \sqrt{2g\Delta y\left(\frac{m + M}{m}\right)^2} = \sqrt{2(9.8 \text{ m/s}^2)(1.50 \text{ m})\left(\frac{0.50 \text{ kg} + 2.30 \text{ kg}}{0.50 \text{ kg}}\right)^2} = \boxed{30 \text{ m/s}}$$

13. **Strategy** Use conservation of energy. Let $d = 2.05$ m. Then, the ramp rises $h = d\sin 5.00°$. The rotational inertia of a uniform sphere is $\frac{2}{5}mr^2$.

   Solution Find the speed of the ball when it reaches the top of the ramp.

   $$0 = \Delta K + \Delta U = \frac{1}{2}mv_f^2 + \frac{1}{2}I\omega_f^2 - \frac{1}{2}mv_i^2 - \frac{1}{2}I\omega_i^2 + mgh$$

   $$= \frac{1}{2}mv_f^2 + \frac{1}{2}\left(\frac{2}{5}mr^2\right)\left(\frac{v_f}{r}\right)^2 - \frac{1}{2}mv_i^2 - \frac{1}{2}\left(\frac{2}{5}mr^2\right)\left(\frac{v_i}{r}\right)^2 + mgh$$

   $$= \frac{7}{10}mv_f^2 - \frac{7}{10}mv_i^2 + mgh, \text{ so}$$

   $$v_f = \sqrt{v_i^2 - \frac{10}{7}gh} = \sqrt{(2.20 \text{ m/s})^2 - \frac{10}{7}(9.80 \text{ m/s}^2)(2.05 \text{ m})\sin 5.00°} = \boxed{1.53 \text{ m/s}}.$$

17. **Strategy** Use conservation of energy. The energy delivered to the fluid in the beaker plus the kinetic energies of the pulley, spool, axle, paddles, and the block are equal to the work done by gravity on the block, which is negative the change in the block's gravitational potential energy. The rotational inertia of the pulley (uniform solid disk) is $\frac{1}{2}m_p r^2$.

   Solution Let the energy delivered to the fluid be E, the distance the block falls be h, and the rotational inertia of the spool, axle, and paddles be $I_s = 0.00140 \text{ kg}\cdot\text{m}^2$. Since the radii of the pulley and the spool are the same (r), their tangential speeds are the same, so let $v_p = v_s = v$.

   $$m_b gh = \frac{1}{2}m_b v_b^2 + \frac{1}{2}I_p\omega_p^2 + \frac{1}{2}I_s\omega_s^2 + E = \frac{1}{2}m_b v_b^2 + \frac{1}{2}\left(\frac{1}{2}m_p r^2\right)\left(\frac{v}{r}\right)^2 + \frac{1}{2}I_s\left(\frac{v}{r}\right)^2 + E$$

   The tangential speeds of the pulley and spool are equal to the speed of the block.

   $$m_b gh = \frac{1}{2}m_b v_b^2 + \frac{1}{4}m_p v^2 + \frac{1}{2}I_s\frac{v^2}{r^2} + E = \frac{1}{2}m_b v^2 + \frac{1}{4}m_p v^2 + \frac{1}{2}I_s\frac{v^2}{r^2} + E, \text{ so}$$

   $$E = m_b gh - \frac{v^2\left(2m_b + m_p + 2I_s/r^2\right)}{4}$$

   $$= (0.870 \text{ kg})(9.80 \text{ m/s}^2)(2.50 \text{ m}) - \frac{(3.00 \text{ m/s})^2[2(0.870 \text{ kg}) + 0.0600 \text{ kg} + 2(0.00140 \text{ kg}\cdot\text{m}^2)/(0.0300 \text{ m})^2]}{4}$$

   $$= \boxed{10.3 \text{ J}}.$$

21. Strategy Use energy conservation to find the speed of Jones just before he grabs Smith. Then, use momentum conservation to find the speed of both just after. Finally, again use energy conservation to find the final height.

Solution Find Jones's speed, v_J.

$$\frac{1}{2}m_\text{J}v_\text{J}^2 = m_\text{J}gh_\text{J}, \text{ so } v_\text{J} = \sqrt{2gh_\text{J}}.$$

Find the speed of both, v.

$$p_\text{i} = m_\text{J}v_\text{J} = p_\text{f} = (m_\text{J} + m_\text{S})v, \text{ so } v = \frac{m_\text{J}v_\text{J}}{m_\text{J} + m_\text{S}} = \frac{m_\text{J}\sqrt{2gh_\text{J}}}{m_\text{J} + m_\text{S}}.$$

Find the final height, h.

$$(m_\text{J} + m_\text{S})gh = \frac{1}{2}(m_\text{J} + m_\text{S})\left(\frac{m_\text{J}\sqrt{2gh_\text{J}}}{m_\text{J} + m_\text{S}}\right)^2, \text{ so } h = \frac{m_\text{J}^2 h_\text{J}}{(m_\text{J} + m_\text{S})^2} = \frac{(78.0 \text{ kg})^2 (3.70 \text{ m})}{(78.0 \text{ kg} + 55.0 \text{ kg})^2} = \boxed{1.27 \text{ m}}.$$

25. Strategy Use conservation of linear momentum.

Solution

$$p_{\text{i}x} = m_\text{b}v_\text{b} = p_{\text{f}x} = (m_\text{b} + m_\text{c})v_{\text{f}x}, \text{ so } v_{\text{f}x} = \frac{m_\text{b}v_\text{b}}{m_\text{b} + m_\text{c}}.$$

$$p_{\text{i}y} = m_\text{c}v_\text{c} = p_{\text{f}y} = (m_\text{b} + m_\text{c})v_{\text{f}y}, \text{ so } v_{\text{f}y} = \frac{m_\text{c}v_\text{c}}{m_\text{b} + m_\text{c}}.$$

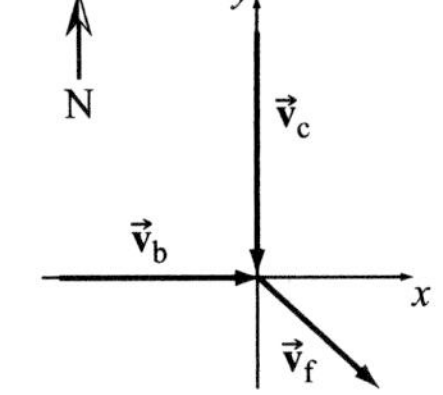

Compute the magnitude of the final velocity.

$$v = \sqrt{v_{\text{f}x}^2 + v_{\text{f}y}^2} = \sqrt{\left(\frac{m_\text{b}v_\text{b}}{m_\text{b} + m_\text{c}}\right)^2 + \left(\frac{m_\text{c}v_\text{c}}{m_\text{b} + m_\text{c}}\right)^2} = \frac{\sqrt{(m_\text{b}v_\text{b})^2 + (m_\text{c}v_\text{c})^2}}{m_\text{b} + m_\text{c}}$$

$$= \frac{\sqrt{[(2.00 \text{ kg})(2.70 \text{ m/s})]^2 + [(1.50 \text{ kg})(-3.20 \text{ m/s})]^2}}{2.00 \text{ kg} + 1.50 \text{ kg}} = 2.06 \text{ m/s}$$

Compute the angle.

$$\theta = \tan^{-1}\frac{v_{\text{f}y}}{v_{\text{f}x}} = \tan^{-1}\frac{\frac{m_\text{c}v_\text{c}}{m_\text{b}+m_\text{c}}}{\frac{m_\text{b}v_\text{b}}{m_\text{b}+m_\text{c}}} = \tan^{-1}\frac{m_\text{c}v_\text{c}}{m_\text{b}v_\text{b}} = \tan^{-1}\frac{(1.50 \text{ kg})(-3.20 \text{ m/s})}{(2.00 \text{ kg})(2.70 \text{ m/s})} = -41.6°$$

The velocity of the block and the clay after the collision is $\boxed{2.06 \text{ m/s at } 41.6° \text{ S of E}}$.

29. (a) Strategy Consider the work-kinetic energy theorem and the impulse momentum theorem.

Solution Since the Romulan ship is twice as massive as the Vulcan ship, the Romulan ship will not travel as far as the Vulcan ship for the same engine force, since $\Delta x = (1/2)a(\Delta t)^2 = (1/2)(F/m)(\Delta t)^2$. Since $W = F\Delta x = \Delta K$, $\boxed{\text{the Vulcan ship will have the greater kinetic energy}}$. Since $\Delta p = F\Delta t$, $\boxed{\text{the ships will have the same momentum}}$.

(b) Strategy Consider the work-kinetic energy theorem and the impulse momentum theorem.

Solution Since the distances and the forces are the same, and since $W = F\Delta x = \Delta K$, $\boxed{\text{the ships will have the same kinetic energy}}$. Since $\Delta x = (1/2)a(\Delta t)^2 = (1/2)(F/m)(\Delta t)^2$, the more massive Romulan ship will have to fire its engines longer than the Vulcan ship to travel the same distance. Since $\Delta p = F\Delta t$ and the forces are the same, $\boxed{\text{the Romulan ship will have the greater momentum}}$.

(c) Strategy Refer to parts (a) and (b).

Solution For part (a), we have the following:
Vulcan:

$$\Delta K = W = F\Delta x = F\left[\frac{F}{2m}(\Delta t)^2\right] = \frac{(9.5\times10^6 \text{ N})^2(100 \text{ s})^2}{2(65,000 \text{ kg})} = 6.9\times10^{12} \text{ J}$$

$$\Delta p = F\Delta t = (9.5\times10^6 \text{ N})(100 \text{ s}) = 9.5\times10^8 \text{ kg}\cdot\text{m/s}$$

Romulan:

$$\Delta K = W = F\Delta x = F\left[\frac{F}{2m}(\Delta t)^2\right] = \frac{(9.5\times10^6 \text{ N})^2(100 \text{ s})^2}{2(130,000 \text{ kg})} = 3.5\times10^{12} \text{ J}$$

$$\Delta p = F\Delta t = (9.5\times10^6 \text{ N})(100 \text{ s}) = 9.5\times10^8 \text{ kg}\cdot\text{m/s}$$

> In part (a), the momenta are the same, 9.5×10^8 kg$\cdot$m/s, but the kinetic energies differ:
> Vulcan at 6.9×10^{12} J and Romulan at 3.5×10^{12} J.

For part (b), we have the following:
Vulcan:

$$\Delta K = W = F\Delta x = (9.5\times10^6 \text{ N})(100 \text{ m}) = 9.5\times10^8 \text{ J}$$

Since $K = \frac{1}{2}mv^2 = \frac{p^2}{2m}$, $p = \sqrt{2mK} = \sqrt{2(65,000 \text{ kg})(9.5\times10^8 \text{ J})} = 1.1\times10^7$ kg$\cdot$m/s.

Romulan:

$$\Delta K = W = F\Delta x = (9.5\times10^6 \text{ N})(100 \text{ m}) = 9.5\times10^8 \text{ J}$$

$$p = \sqrt{2mK} = \sqrt{2(2\times65,000 \text{ kg})(9.5\times10^8 \text{ J})} = 1.6\times10^7 \text{ kg}\cdot\text{m/s}.$$

> In part (b), the kinetic energies are the same, 9.5×10^8 J, but the momenta differ:
> Vulcan at 1.1×10^7 kg$\cdot$m/s and Romulan at 1.6×10^7 kg$\cdot$m/s.

33. Strategy Use conservation of energy. m is the mass of one wheel. M is the total mass of the system. v is the speed of the center of mass of the system (which is the same as the speed of a point on either wheel).

Solution

(a) $K_{\text{rot}} = \frac{1}{2}I\omega^2 = \frac{1}{2}mv^2 = K_{\text{trans}}$ for one wheel. $K_{\text{rot,total}} = 2\cdot\frac{1}{2}mv^2 = mv^2$ and $K_{\text{trans,total}} = \frac{1}{2}Mv^2$.

$$K_{\text{total}} = U_i$$

$$mv^2 + \frac{1}{2}Mv^2 = MgH$$

$$v^2(2m + M) = 2MgH$$

$$v = \sqrt{\frac{2MgH}{2m+M}} = \sqrt{\frac{2(80.0 \text{ kg})(9.80 \text{ m/s}^2)(20.0 \text{ m})}{2(1.5 \text{ kg}) + 80.0 \text{ kg}}} = \boxed{19.4 \text{ m/s}}$$

(b) Since the speed depends upon the combined total mass of the system, the speed at the bottom would not be the same for a less massive rider. The answer is $\boxed{\text{no}}$.

37. Strategy Use conservation of momentum and the equations for motion with a constant acceleration.

Solution

(a) The banana will fall at the same rate as the monkey; therefore, you should throw the banana directly at the monkey.

$$\tan\theta = \frac{3.33\ \text{m} + 1.67\ \text{m}}{3.00\ \text{m}}, \text{ so } \theta = \tan^{-1}\frac{5.00}{3.00} = \boxed{59.0°\ \text{above the horizontal}}.$$

(b) $\boxed{\text{Since the banana will fall at the same rate as the monkey, regardless of the launch speed of the banana, the launch angle is the same for all launch speeds}}$. Relatively high launch speeds will reach the monkey relatively sooner (and higher); relatively low launch speeds will reach the monkey relatively later (and lower).

(c) Find the time it takes the banana to reach the monkey.

$$\Delta y = -\frac{1}{2}g(\Delta t)^2, \text{ so } \Delta t = \sqrt{-\frac{2\Delta y}{g}} = \sqrt{-\frac{2(-1.67\ \text{m})}{9.80\ \text{m/s}^2}} = 0.5838\ \text{s} = t.$$

The banana reaches the monkey when it has traveled 3.00 m.

$$\Delta x = v_x \Delta t = v\cos\theta\,\Delta t, \text{ so } v = \frac{\Delta x}{\Delta t\cos\theta} = \frac{3.00\ \text{m}}{(0.5838\ \text{s})\cos 59.0°} = \boxed{9.98\ \text{m/s}}.$$

(d) The speed of the monkey just before the collision is given by $v_{my} = -gt$ and $v_{mx} = 0$. The speed of the banana at this time is given by $v_{by} = v\sin\theta - gt$ and $v_{bx} = v\cos\theta$. Use conservation of momentum.

$$mv_{bx} = (m+M)v_{fx}, \text{ so}$$

$$v_{fx} = \frac{m}{m+M}v_{bx} = \frac{m}{m+M}v\cos\theta = \frac{0.20\ \text{kg}}{0.20\ \text{kg} + 3.00\ \text{kg}}(9.98\ \text{m/s})\cos 59.0° = 0.321\ \text{m/s}.$$

$$mv_{by} + Mv_{my} = (m+M)v_{fy}, \text{ so}$$

$$v_{fy} = \frac{mv_{by} + Mv_{my}}{m+M} = \frac{m(v\sin\theta - gt) + M(-gt)}{m+M} = \frac{m}{m+M}v\sin\theta - gt$$

$$= \frac{0.20\ \text{kg}}{0.20\ \text{kg} + 3.00\ \text{kg}}(9.98\ \text{m/s})\sin 59.0° - (9.80\ \text{m/s}^2)(0.5835\ \text{s}) = -5.18\ \text{m/s}.$$

The time it takes for the monkey to hit the ground is given by

$$\Delta y = v_{fy}t_2 - \frac{1}{2}gt_2^2, \text{ or } 0 = \frac{1}{2}gt_2^2 - v_{fy}t_2 + \Delta y = (4.90\ \text{m/s}^2)t_2^2 + (5.18\ \text{m/s})t_2 - 5.33\ \text{m}.$$

Using the quadratic formula, we find $t = 0.64$ s. The horizontal distance is

$$d = v_{fx}t = (0.321\ \text{m/s})(0.64\ \text{s}) = \boxed{0.21\ \text{m}}.$$

MCAT Review

1. Strategy Use conservation of momentum.

Solution

$$p_i = mv_i = p_f = mv_f + p_{wall}, \text{ so } p_{wall} = m(v_i - v_f) = (0.2\ \text{kg})[2.0\ \text{m/s} - (-1.0\ \text{m/s})] = 0.6\ \text{kg}\cdot\text{m/s}.$$

The correct answer is $\boxed{\text{D}}$.

2. **Strategy** Use Hooke's law.

   Solution Let up be the positive direction. The gravitational force on the mass is
   $F = mg = (0.10 \text{ kg})(-9.80 \text{ m/s}^2) = -0.98 \text{ N}.$ Solving for the spring constant in Hooke's law, we have
   $$k = -\frac{F}{x} = -\frac{-0.98 \text{ N}}{0.15 \text{ m}} = 6.5 \text{ N/m}. \text{ Thus, the correct answer is } \boxed{D}.$$

3. **Strategy** The net torque is zero.

   Solution
   $\Sigma \tau = 0 = F(0.60 \text{ m}) - (1.0 \times 10^{-7} \text{ kg})(9.80 \text{ m/s}^2)(0.40 \text{ m}),$ so
   $$F = \frac{(1.0 \times 10^{-7} \text{ kg})(9.80 \text{ m/s}^2)(0.40 \text{ m})}{0.60 \text{ m}} = 6.5 \times 10^{-7} \text{ N}.$$
   The correct answer is $\boxed{B}$.

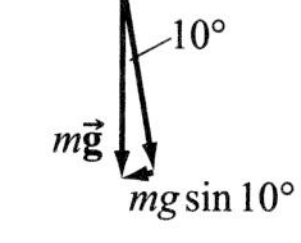

4. **Strategy** Determine the speed of the first ball just before in collides with the second. The collision is completely inelastic; that is, the balls stick together. Use conservation of momentum to find the speed of the balls after the collision.

   Solution Find the speed of the first ball just before the collision.
   $v_{fx} - v_{ix} = v_1 - 0 = a_x \Delta t,$ so $v_1 = (10 \text{ m/s}^2)(2.0 \text{ s}) = 20 \text{ m/s}.$
   Find the speed v of the balls just after the collision.
   $$p_i = m_1 v_1 = p_f = (m_1 + m_2)v, \text{ so } v = \frac{m_1 v_1}{m_1 + m_2} = \frac{(0.50 \text{ kg})(20 \text{ m/s})}{0.50 \text{ kg} + 1.0 \text{ kg}} = 6.7 \text{ m/s}.$$
   The correct answer is $\boxed{B}$.

5. **Strategy** Use Newton's second law and Eq. (6-27).

   Solution The gravitational force working against the motion of the car as it climbs the hill is
   $mg \sin 10°,$ so the additional power required is
   $$P_{car} = -P_{grav} = -Fv \cos 180° = (mg \sin 10°)v = (1000 \text{ kg})(10 \text{ m/s}^2)\sin 10°(15 \text{ m/s})$$
   $$= 1.5 \times 10^5 \times \sin 10° \text{ W}.$$
   The correct answer is $\boxed{D}$.

6. **Strategy** Find the vertical distance the patient would have climbed had the treadmill been stationary (and very long). Then, find the work done by the patient on the treadmill.

   Solution The "distance" walked along the incline is $(2 \text{ m/s})(600 \text{ s}) = 1200 \text{ m}.$
   Thus, the vertical distance climbed is $(1200 \text{ m})\sin 30° = 600 \text{ m}.$ The work done is
   $$W = Fd = mgd = (90 \text{ kg})(10 \text{ m/s}^2)(600 \text{ m}) = 0.54 \text{ MJ}.$$
   The correct answer is $\boxed{C}$.

7. **Strategy** Find the angle between the force exerted by the patient and the patient's velocity. Use Eq. (6-27).

   Solution The force due to gravity is down, so the force exerted by the patient is up. The velocity is directed at the angle of the incline, or 30° above the horizontal, so the angle between the force and the velocity is 60°. Compute the mechanical power output of the patient. 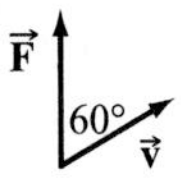

   $$P = Fv\cos\theta = mgv\cos\theta = (100 \text{ kg})(10 \text{ m/s}^2)(3 \text{ m/s})\cos 60° = 1500 \text{ W}$$

   The correct answer is $\boxed{\text{B}}$.

8. **Strategy and Solution** The force pushing each friction pad is normal to the wheel; that is, it is the normal force in $f_k = \mu_k N$. Solve for the normal force.

   $$N = \frac{f_k}{\mu_k} = \frac{20 \text{ N}}{0.4} = 50 \text{ N}$$

   This is the total force. The force pushing each friction pad is half this, or 25 N. The correct answer is $\boxed{\text{B}}$.

9. **Strategy** Find the average tangential speed at the friction pads. Then, use the relationship between tangential speed and radial acceleration.

   Solution

   The average tangential speed is $v = \dfrac{4800 \text{ m}}{20 \text{ min}} \times \dfrac{1 \text{ min}}{60 \text{ s}} = 4.0 \text{ m/s}$. The radial acceleration is

   $$a_r = \frac{v^2}{r} = \frac{(4.0 \text{ m/s})^2}{0.3 \text{ m}} = 50 \text{ m/s}^2. \text{ The correct answer is } \boxed{\text{D}}.$$

10. **Strategy** Use the work-kinetic energy theorem.

    Solution The work done by friction on the wheel is $W = -f_k d$, where d is the linear distance the wheel passes between the pads before it stops. Relate d to the kinetic energy of the wheel.

    $$W_{\text{total}} = -f_k d = \Delta K = 0 - K_i, \text{ so } d = \frac{K_i}{f_k}.$$

    Divide d by the circumference of a circle with radius 0.3 m to find the number of rotations.

    $$\frac{d}{2\pi r} = \frac{K_i}{2\pi r f_k} = \frac{30 \text{ J}}{2\pi(0.3 \text{ m})(20 \text{ N})} = 0.8 \text{ rotations}$$

    Since $0.8 < 1$, the correct answer is $\boxed{\text{A}}$.

11. **Strategy** Compute the average mechanical power output of the cyclist and compare it to the power consumed by the wheel at the friction pads.

    Solution The metabolic power available for work is $535 \text{ W} - 85 \text{ W} = 450 \text{ W}$. Since the efficiency is 20%, the average mechanical power output of the cyclist is $0.20 \times 450 \text{ W} = 90 \text{ W}$. The average tangential speed of the wheel is $v = \dfrac{4800 \text{ m}}{20 \text{ min}} \times \dfrac{1 \text{ min}}{60 \text{ s}} = 4.0 \text{ m/s}$. Therefore, the power consumed by the friction pads is

    $P = f_k v = (20 \text{ N})(4.0 \text{ m/s}) = 80 \text{ W}$. Thus, the difference between the average mechanical power output of the cyclist and the power consumed by the wheel at the friction pads is $90 \text{ W} - 80 \text{ W} = 10 \text{ W}$.

    The correct answer is $\boxed{\text{B}}$.

12. Strategy and Solution Increasing the force on the friction pads would increase the power consumed by the wheel at the friction pads (because $P = Fv$). So, if the cyclist is pedaling at the same rate and the power consumed by the friction pads increases, the difference between the two decreases and the fraction of mechanical power output of the cyclist consumed by the wheel at the friction pad increases. Thus, the correct answer is $\boxed{D}$.

13. Strategy Relate the cyclist's average metabolic rate to the energy released per volume of oxygen consumed, the time on the bike, and volume of oxygen consumed.

Solution The cyclist's average metabolic rate while riding is 535 W. The total energy used during 20 minutes is
$$(535 \text{ W})(20 \text{ min})\frac{60 \text{ s}}{1 \text{ min}} = 642{,}000 \text{ J}.$$ The total energy released by the consumption of oxygen is $(20{,}000 \text{ J/L})V$, where V is the volume of oxygen consumed. Equating these two expressions and solving for V gives the number of liters of oxygen the cyclist consumes.
$$(20{,}000 \text{ J/L})V = 642{,}000 \text{ J, so } V = \frac{642{,}000 \text{ J}}{20{,}000 \text{ J/L}} = 32 \text{ L} \approx 30 \text{ L.}$$ The correct answer is $\boxed{B}$.

14. Strategy and Solution Since the force has been reduced by 50% and the distance has been doubled, the cyclist does the same amount of work $[W = 0.50F(2\Delta x) = F\Delta x]$. So, the energy transmitted in the second workout is equal to the energy transmitted in the first. The correct answer is $\boxed{C}$.

15. Strategy The circumference of a circle is $C = 2\pi r$. A wheel moves a distance equal to its circumference during each rotation. The wheel rotates twice during each rotation of the pedals.

Solution The circumference of a circle with a radius of 0.15 m is $2\pi(0.15 \text{ m})$. The circumference of a circle with a radius of 0.3 m is $2\pi(0.3 \text{ m})$. During each rotation of the pedals, a point on the wheel at a radius of 0.3 m moves a distance $2[2\pi(0.3 \text{ m})]$. The ratio of the distance moved by a pedal to the distance moved by a point on the wheel located at a radius of 0.3 m in the same amount of time is $\dfrac{2\pi(0.15 \text{ m})}{2[2\pi(0.3 \text{ m})]} = 0.25$.

The correct answer is $\boxed{A}$.

16. Strategy Use the definition of power.

Solution
$$P = \frac{\Delta E}{\Delta t}, \text{ so } \Delta t = \frac{\Delta E}{P} = \left(\frac{300 \text{ kcal}}{500 \text{ W}}\right)\left(\frac{4186 \text{ J}}{1 \text{ kcal}}\right)\left(\frac{1 \text{ min}}{60 \text{ s}}\right) = 41.9 \text{ min.}$$ The correct answer is $\boxed{D}$.

17. Strategy Consider the distance a point on the wheel travels for each situation.

Solution The circumference of a circle with a radius of 0.3 m is $2\pi(0.3 \text{ m})$. The circumference of a circle with a radius of 0.4 m is $2\pi(0.4 \text{ m})$. During each rotation, a point on a wheel travels a distance equal to the circumference. The force on the wheel is the same in each case, but the distance traveled by a point on the wheel is greater for a greater radius. In this case, the distance is $0.4 \text{ m}/(0.3 \text{ m}) = 1.33$ times farther or 33%. Since work is equal to the product of force times distance, the work done on the wheel per revolution is 33% more. Thus, the correct answer is $\boxed{C}$.

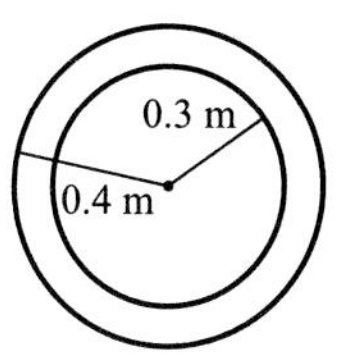

Chapter 9

FLUIDS

Conceptual Questions

1. A manometer (with one side open) measures gauge pressure. A barometer measures absolute pressure. A tire pressure gauge and a sphygmomanometer both measure gauge pressure.

5. There is practically no atmosphere on the Moon and hence practically zero pressure. Drinking from a glass with a straw would be impossible, since a pressure difference is required to push the liquid up the straw. With a sealed juice box, the astronaut could supply the necessary pressure by squeezing the box.

9. The pressure of the atmosphere decreases with altitude. Therefore, the balloon gradually expands as it rises.

13. The rate at which air moves up a chimney is determined by the size of the pressure change along the chimney's length. Bernoulli's equation tells us that on a windless day, the pressure difference between the two ends of the chimney is proportional to the height of the chimney. On a windy day, the velocity difference between the two ends causes an additional pressure difference—resulting in an increased draft.

17. To get optimal use from hydraulic systems, the fluid used in mediating the operation of the system must be very nearly incompressible. Liquids meet this requirement, but gases like air are compressible. Thus, proper operation of a hydraulic device requires that gases be "bled" from the system.

Problems

1. **Strategy** Use the definition of average pressure.

   Solution Compute the average pressure.

   $$P_{av} = \frac{F}{A} = \frac{500 \text{ N}}{1.0 \text{ cm}^2} \left(\frac{100 \text{ cm}}{1 \text{ m}}\right)^2 \left(\frac{1 \text{ atm}}{101.3 \times 10^3 \text{ Pa}}\right) = \boxed{49 \text{ atm}}$$

5. **Strategy** The average pressure is the force applied to the floor divided by the contact area.

   Solution

   The baby applies a pressure of $P_b = \dfrac{F}{A} = \dfrac{m_b g}{3\left(\frac{1}{4}\pi d_s^2\right)} = \dfrac{4m_b g}{3\pi d_s^2}$.

   The adult applies a pressure of $P_a = \dfrac{F}{A} = \dfrac{m_a g}{4\left(\frac{1}{4}\pi d_c^2\right)} = \dfrac{m_a g}{\pi d_c^2}$.

   The ratio of these two pressures is $\dfrac{P_b}{P_a} = \dfrac{4m_b g}{3\pi d_s^2}\left(\dfrac{m_a g}{\pi d_c^2}\right)^{-1} = \dfrac{4m_b d_c^2}{3m_a d_s^2} = \dfrac{4(10 \text{ kg})(0.060 \text{ m})^2}{3(60 \text{ kg})(0.020 \text{ m})^2} = 2.0.$

   $\boxed{\text{The baby applies 2.0 times as much pressure as the adult.}}$

all piston must equal that done on the car.

small piston must be pushed downward to raise the car 1.0 cm.

$$d_c = \frac{A}{a}d_c = 100.0(0.010 \text{ m}) = \boxed{1.0 \text{ m}}.$$

fish.

$$n^3)(9.80 \text{ m/s}^2)(10 \text{ m})\left(\frac{1 \text{ atm}}{1.013 \times 10^5 \text{ Pa}}\right) = \boxed{2.0 \text{ atm}}$$

ume for each of the six barrels. The pressure at the plug of each barrel is
ic pressure. The force on the plug is equal to the pressure at the plug
(and plug).

$$/\text{s}^2)(1.0 \text{ m} - 0.20 \text{ m})]\pi(0.01 \text{ m})^2 = 33.9 \text{ N}$$
$$/\text{s}^2)(1.2 \text{ m} - 0.20 \text{ m})]\pi(0.01 \text{ m})^2 = 34.5 \text{ N}$$
$${}^3)(1.2 \text{ m} - 0.20 \text{ m})]\pi(0.0125 \text{ m})^2 = 52.9 \text{ N}$$
$$^2)(1.0 \text{ m} - 0.20 \text{ m})]\pi(0.0125 \text{ m})^2 = 52.2 \text{ N}$$
$$\text{s}^2)(1.45 \text{ m} - 0.20 \text{ m})]\pi(0.01 \text{ m})^2 = 35.3 \text{ N}$$

the plug due to the liquid in the barrel, from largest to smallest, we

ace is 1025 kg/m^3. Use Eq. (9-4).

's skin.

$$n^3)(9.80 \text{ m/s}^2)(2500 \text{ m}) = \boxed{2.5 \times 10^7 \text{ Pa}}$$

$$\text{m/s}^2)(35.0 \text{ m}) = \boxed{343 \text{ kPa}}$$

$$..._{} \text{ kg/m}^{})(9.80 \text{ m/s}^2)(35 \text{ m}) = -410 \text{ Pa}$$

The pressure decreases by $\boxed{410 \text{ Pa}}$.

25. Strategy Use Eq. (9-5).

Solution Find the change in pressure of the gas.
$$\Delta P = \rho g d = (1.0 \times 10^3 \text{ kg/m}^3)(9.80 \text{ m/s}^2)(0.040 \text{ m}) = \boxed{390 \text{ Pa}}$$

29. Strategy Use the appropriate conversion factors to convert the woman's systolic blood pressure into the various pressure units.

Solution

(a) $(160 \text{ mm Hg})\left(\dfrac{101.3 \text{ kPa}}{760.0 \text{ mm Hg}}\right) = \boxed{21 \text{ kPa}}$

(b) $(160 \text{ mm Hg})\left(\dfrac{14.70 \text{ lb}/\text{in}^2}{760.0 \text{ mm Hg}}\right) = \boxed{3.1 \text{ lb}/\text{in}^2}$

(c) $(160 \text{ mm Hg})\left(\dfrac{1 \text{ atm}}{760.0 \text{ mm Hg}}\right) = \boxed{0.21 \text{ atm}}$

(d) $(160 \text{ mm Hg})\left(\dfrac{760.0 \text{ torr}}{760.0 \text{ mm Hg}}\right) = \boxed{160 \text{ torr}}$

33. Strategy The weight of the barge must equal the weight of the displaced water.

Solution Find the depth of the barge below the waterline.

$$Weight\ water = \rho g V = \rho g d A = Weight\ barge = mg, \text{ so } d = \frac{m}{\rho A} = \frac{3.0 \times 10^5 \text{ kg}}{(1.0 \times 10^3 \text{ kg}/\text{m}^3)(20.0 \text{ m})(10.0 \text{ m})} = \boxed{1.5 \text{ m}}.$$

35. Strategy The relationship between the fraction of a floating object's volume that is submerged to the ratio of the object's density to the fluid in which it floats is $V_{\text{f}}/V_{\text{o}} = \rho_{\text{o}}/\rho_{\text{f}}$.

Solution

(a) Find the density of the object.

$$\frac{V_{\text{f}}}{V_{\text{o}}} = \frac{\rho_{\text{o}}}{\rho_{\text{f}}}, \text{ so } \rho_{\text{o}} = \frac{V_{\text{f}}}{V_{\text{o}}}\rho_{\text{f}} = 0.14(999.87 \text{ kg}/\text{m}^3) = \boxed{140 \text{ kg}/\text{m}^3}.$$

(b) Find the percentage of the object that is submerged if it is placed in ethanol by forming a proportion.

$$\frac{\text{fraction submerged in ethanol}}{\text{fraction submerged in water}} = \frac{\rho_{\text{o}}/\rho_{\text{ethanol}}}{\rho_{\text{o}}/\rho_{\text{water}}} = \frac{\rho_{\text{water}}}{\rho_{\text{ethanol}}} = \frac{999.87}{790}, \text{ so the fraction submerged in ethanol is}$$

$$\frac{999.87}{790}(0.14) = 0.18, \text{ or } \boxed{18\%}.$$

37. Strategy The ratio of the density of the wood to that of the oil is equal to the fraction of the volume of the wood that is submerged.

Solution Find the density of the oil.

$$\frac{\rho_{\text{w}}}{\rho_{\text{o}}} = \frac{V_{\text{o}}}{V_{\text{w}}}, \text{ so } \rho_{\text{o}} = \frac{V_{\text{w}}}{V_{\text{o}}}\rho_{\text{w}} = \frac{V_{\text{w}}}{0.900 V_{\text{w}}}\rho_{\text{w}} = \frac{0.67 \text{ g}/\text{cm}^3}{0.900} = \boxed{0.74 \text{ g}/\text{cm}^3}.$$

41. Strategy Find an expression for the new density of the fish; then set this equal to the density of the water and solve for the volume of the bladder.

Solution The new density of the fish is
$$\rho = \frac{m_f + m_a}{V_f + V_a} = \frac{m_f + \rho_a V_a}{\dfrac{m_f}{\rho_f} + V_a} = \rho_w.$$

Solve for V_a.

$$m_f + \rho_a V_a = \frac{\rho_w}{\rho_f} m_f + \rho_w V_a, \text{ so } V_a = m_f \frac{1 - \dfrac{\rho_w}{\rho_f}}{\rho_w - \rho_a} = (0.0100 \text{ kg}) \frac{1 - \dfrac{1060 \text{ kg}/\text{m}^3}{1080 \text{ kg}/\text{m}^3}}{1060 \text{ kg}/\text{m}^3 - 1.20 \text{ kg}/\text{m}^3} = \boxed{0.17 \text{ cm}^3}.$$

45. Strategy Let the $+y$-direction be upward. Use Newton's second law and Eq. (9-7).

Solution

(a) $\Sigma F_y = F_B - mg = ma$, so
$$a = \frac{F_B}{m} - g = \frac{\rho_w g V}{m} - g = g\left(\frac{\rho_w V}{\rho V} - 1\right) = (9.80 \text{ m}/\text{s}^2)\left(\frac{1.00 \text{ g}/\text{cm}^3}{0.50 \text{ g}/\text{cm}^3} - 1\right) = 9.8 \text{ m}/\text{s}^2.$$

Thus, $\vec{a} = \boxed{9.8 \text{ m}/\text{s}^2 \text{ upward}}$.

(b) $a = (9.80 \text{ m}/\text{s}^2)\left(\dfrac{1.00 \text{ g}/\text{cm}^3}{0.750 \text{ g}/\text{cm}^3} - 1\right) = 3.3 \text{ m}/\text{s}^2$, so $\vec{a} = \boxed{3.3 \text{ m}/\text{s}^2 \text{ upward}}$.

(c) $a = (9.80 \text{ m}/\text{s}^2)\left(\dfrac{1.00 \text{ g}/\text{cm}^3}{0.125 \text{ g}/\text{cm}^3} - 1\right) = 68.6 \text{ m}/\text{s}^2$, so $\vec{a} = \boxed{68.6 \text{ m}/\text{s}^2 \text{ upward}}$.

47. Strategy Use Eq. (9-13).

Solution Find the speed of the water as it passes through the nozzle.
$$A_2 v_2 = A_1 v_1, \text{ so } v_2 = \frac{A_1}{A_2} v_1 = \frac{\pi r_1^2}{\pi r_2^2} v_1 = \left(\frac{1.0 \text{ cm}}{0.20 \text{ cm}}\right)^2 (2.0 \text{ m}/\text{s}) = \boxed{50 \text{ m}/\text{s}}.$$

49. (a) Strategy Use Eq. (9-13).

Solution Find the speed of the water in the hose.
$$v_2 = \frac{A_1}{A_2} v_1 = \frac{\pi r_1^2}{\pi r_2^2} v_1 = \left(\frac{r_1}{r_2}\right)^2 v_1 = \left(\frac{1.00 \text{ mm}}{8.00 \text{ mm}}\right)^2 (25.0 \text{ m}/\text{s}) = \boxed{39.1 \text{ cm}/\text{s}}$$

(b) Strategy Use Eq. (9-12).

Solution Compute the volume flow rate.
$$\frac{\Delta V}{\Delta t} = A_1 v_1 = \pi (1.00 \times 10^{-3} \text{ m})^2 (25.0 \text{ m}/\text{s}) = \boxed{78.5 \text{ cm}^3/\text{s}}$$

(c) Strategy Use Eq. (9-11).

Solution Compute the mass flow rate.
$$\frac{\Delta m}{\Delta t} = \rho A_1 v_1 = (1.00 \text{ g}/\text{cm}^3)(78.5 \text{ cm}^3/\text{s}) = \boxed{78.5 \text{ g}/\text{s}}$$

53. Strategy Use Eq. (9-14).

Solution The potential energy difference is relatively small, so Bernoulli's equation becomes

$$P_1 + \frac{1}{2}\rho v_1^2 = P_2 + \frac{1}{2}\rho v_2^2, \text{ or } P_1 - P_2 = \frac{1}{2}\rho v_2^2 - \frac{1}{2}\rho v_1^2.$$

Estimate the force.

$$F = \Delta PA = (P_1 - P_2)A = \left(\frac{1}{2}\rho v_2^2 - \frac{1}{2}\rho v_1^2\right)A = \frac{1}{2}A\rho(v_2^2 - v_1^2)$$

$$= \frac{1}{2}(28 \text{ m}^2)(1.3 \text{ kg}/\text{m}^3)[(190 \text{ m/s})^2 - (160 \text{ m/s})^2] = \boxed{1.9 \times 10^5 \text{ N}}$$

57. Strategy Use Eq. (9-12) to determine the speed of the water at the faucet. Then, use Eq. (9-14) to find the height difference. Assume that the diameter of the tower is so large compared to that of the faucet that the water at the top of the tower does not move. Also, the pressure at the tower and the faucet is the same.

Solution Let the tower be labeled as 1. Find the speed of the water at the faucet.

$$v_2 = \frac{1}{A}\frac{\Delta V}{\Delta t} = \frac{\frac{1}{4}\pi d_c^2 h_c}{\frac{1}{4}\pi d_f^2 \Delta t} = \frac{d_c^2 h_c}{d_f^2 \Delta t}$$

With the above assumptions, Bernoulli's equation becomes $\rho g y_1 = \rho g y_2 + \frac{1}{2}\rho v_2^2 = \rho g y_2 + \frac{1}{2}\rho\left(\frac{d_c^2 h_c}{d_f^2 \Delta t}\right)^2$, so

the height difference is $y_1 - y_2 = \dfrac{1}{2g}\left(\dfrac{d_c^2 h_c}{d_f^2 \Delta t}\right)^2 = \dfrac{1}{2(9.80 \text{ m}/\text{s}^2)}\left[\dfrac{(44 \text{ cm})^2 (0.52 \text{ m})}{(2.54 \text{ cm})^2 (12 \text{ s})}\right]^2 = \boxed{8.6 \text{ m}}$.

61. (a) Strategy Use Eq. (9-15).

Solution Find the pressure of the fluid in the syringe, P_s.

$$\frac{\Delta V}{\Delta t} = \frac{\pi \Delta P r^4}{8\eta L}, \text{ so } \left(\frac{8\eta L}{\pi r^4}\right)\frac{\Delta V}{\Delta t} = P_s - P_v \text{ and}$$

$$P_s = (16.0 \text{ mm Hg})\left(\frac{1.013 \times 10^5 \text{ Pa}}{760.0 \text{ mm Hg}}\right) + \frac{8(2.00 \times 10^{-3} \text{ Pa} \cdot \text{s})(0.0300 \text{ m})}{\pi(0.000300 \text{ m})^4}(0.250 \text{ cm}^3/\text{s})\left(\frac{\text{m}}{100 \text{ cm}}\right)^3 = \boxed{6850 \text{ Pa}}.$$

(b) Strategy Use the definition of average pressure.

Solution Find the force the must be applied to the plunger.

$$F = P_{av}A = (6850 \text{ Pa})(1.00 \text{ cm}^2)(10^{-2} \text{ m/cm})^2 = \boxed{0.685 \text{ N}}$$

63. Strategy and Solution The volume flow rate for each of the pipes in system C is one quarter that of the pipe in system A, since the total rates are the same and system C has four times as many pipes. So, since the flow speed in each of the pipes in C is 3.0 m/s, the flow speed in A must be four times this, or $\boxed{12 \text{ m/s}}$.

65. Strategy Use Eq. (9-15). Form a ratio of the volume flow rates.

Solution Find the pressure supplied by the pump in system A.

$$\frac{\dfrac{\pi \Delta P_A r^4}{8\eta L}}{2\left[\dfrac{\pi \Delta P_B r^4}{8\eta\left(\frac{L}{2}\right)}\right]} = \frac{\Delta P_A}{4\Delta P_B} = \frac{P_A - P_{atm}}{4(P_B - P_{atm})} = \frac{\frac{\Delta V}{\Delta t}\,A}{\frac{\Delta V}{\Delta t}\,B} = 1, \text{ so } P_A = 4P_B - 3P_{atm} = [4(5.0 \text{ atm}) - 3(1.0 \text{ atm})] = \boxed{17 \text{ atm}}.$$

69. Strategy Use Eq. (9-15).

Solution

(a) Show that Poiseuille's law can be written in the form $\Delta P = IR$.

$$\frac{\pi \Delta P r^4}{8\eta L} = \frac{\Delta V}{\Delta t}, \text{ so } \Delta P = \frac{\Delta V}{\Delta t}\left(\frac{8\eta L}{\pi r^4}\right) = IR. \text{ Therefore, } \Delta P = IR \text{ where } I = \frac{\Delta V}{\Delta t} \text{ and } R = \frac{8\eta L}{\pi r^4}.$$

(b) From part (a), $\boxed{R = \dfrac{8\eta L}{\pi r^4}}$.

71. Strategy Use Eq. (9-16) and Newton's second law.

Solution Find the viscosity of the second liquid.

$$\Sigma F_y = F_D + F_B - m_s g = 6\pi\eta r v + m_l g - m_s g = 6\pi\eta r v + (m_l - m_s)g = 6\pi\eta r v - \frac{4}{3}\pi r^3 (\rho_s - \rho_l)g = 0,$$

so $\eta = \dfrac{\frac{4}{3}\pi r^3 (\rho_s - \rho_l)g}{6\pi r v} = \dfrac{2r^2 (\rho_s - \rho_l)g}{9v}$.

Find the viscosity of the second liquid by forming a proportion.

$$\frac{\eta_2}{\eta_1} = \frac{\dfrac{2r^2(\rho_s - \rho_l)g}{9v_2}}{\dfrac{2r^2(\rho_s - \rho_l)g}{9v_1}} = \frac{v_1}{v_2}, \text{ so } \eta_2 = \frac{v_1}{1.2v_1}\eta_1 = \frac{\eta_1}{1.2} = \frac{0.5 \text{ Pa·s}}{1.2} = \boxed{0.4 \text{ Pa·s}}.$$

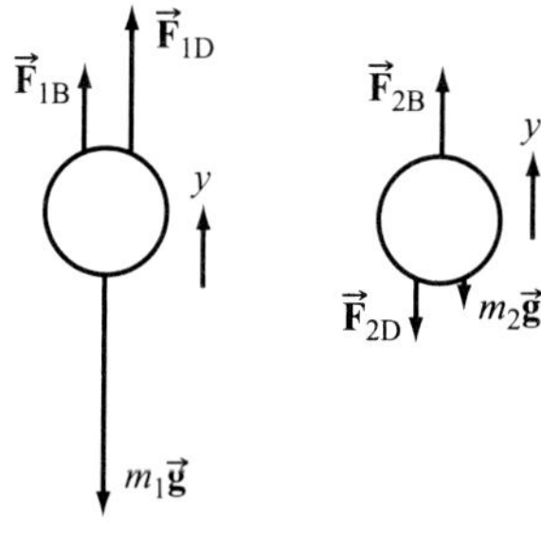

73. (a) Strategy Use Eq. (9-16).

Solution Find the drag force on the dinoflagellate in seawater.

$$F_D = 6\pi\eta r v = 6\pi(0.0010 \text{ Pa·s})(35.0\times 10^{-6} \text{ m})\frac{1.0\times 10^{-3} \text{ m}}{5.0 \text{ s}} = \boxed{1.3\times 10^{-10} \text{ N}}$$

77. Strategy Use Eq. (9-16) and Newton's second law.

Solution Find the terminal speed of the air bubble.

Aluminum sphere: $\Sigma F_y = F_{1D} + F_{1B} - m_1 g = 0$

Air bubble: $\Sigma F_y = F_{2B} - F_{2D} - m_2 g = 0$

Divide F_{2D} by F_{1D}.

$$\frac{F_{2D}}{F_{1D}} = \frac{6\pi\eta r v_2}{6\pi\eta r v_1} = \frac{F_{2B} - m_2 g}{m_1 g - F_{1B}} = \frac{m_w g - m_2 g}{m_1 g - m_w g}, \text{ so}$$

$$v_2 = \frac{m_w - m_2}{m_1 - m_w}v_1 = \frac{1 - \dfrac{m_2}{m_w}}{\dfrac{m_1}{m_w} - 1}v_1 = \frac{1 - \dfrac{\rho_a}{\rho_w}}{\dfrac{2.7\rho_w}{\rho_w} - 1}v_1 = \frac{1 - \dfrac{1.20}{1001.8}}{2.7 - 1}(5.0 \text{ cm/s}) = \boxed{2.9 \text{ cm/s}}.$$

81. (a) Strategy and Solution The net force on the hemisphere is vertical and equal to the force on the flat surface of the hemisphere. Each horizontal component of the force has an equal and opposite horizontal component at the opposite side of the hemisphere that cancels its contribution to the net force. The flat part of the hemisphere ultimately has the net vertical force exerted on it via the body of the hemisphere: $F = AP = \pi r^2 P$.

(b) **Strategy** Use Eq. (9-17) and the definition of average pressure.

Solution Show that the air pressure inside the bubble must exceed the water pressure outside by $\Delta P = 2\gamma/r$.

$$\Delta P = \frac{F}{A} = \frac{2\pi r \gamma}{\pi r^2} = \frac{2\gamma}{r}$$

85. Strategy Use Eqs. (9-13) and (9-14).

Solution

(a) Find the speed of the water as it exits the showerhead (v_1).

$$P_1 + \rho g y_1 + \frac{1}{2}\rho v_1^2 = P_2 + \rho g y_2 + \frac{1}{2}\rho v_2^2$$

$$\frac{1}{2}\rho(v_1^2 - v_2^2) = P_2 - P_1 + \rho g(y_2 - y_1)$$

$$v_1^2 - \frac{A_1^2}{A_2^2}v_1^2 = \frac{2P_{\text{gauge}}}{\rho} - 2gh$$

$$v_1 = \sqrt{\frac{\frac{2P_{\text{gauge}}}{\rho} - 2gh}{1 - \left[N\pi r_1^2/(\pi r_2^2)\right]^2}} = \sqrt{\frac{\frac{2(410\times10^3\ \text{Pa})}{1.00\times10^3\ \text{kg/m}^3} - 2(9.80\ \text{m/s}^2)(6.7\ \text{m})}{1 - 36^2\left(\frac{0.33\ \text{mm}}{6.3\ \text{mm}}\right)^4}} = \boxed{26\ \text{m/s}}$$

(b) Find the speed of the water as it moves through the output pipe of the pump.

$$v_2 = \frac{A_1}{A_2}v_1 = \frac{N\pi r_1^2}{\pi r_2^2}v_1 = 36\left(\frac{0.33\ \text{mm}}{6.3\ \text{mm}}\right)^2(26\ \text{m/s}) = \boxed{2.6\ \text{m/s}}$$

89. (a) Strategy Compute the volume of the block and use its density ($\rho = 2702\ \text{kg/m}^3$) to find its mass. Then, compute its weight.

Solution The volume of the block is $(0.0200\ \text{m})(0.0300\ \text{m})(0.0500\ \text{m}) = 3.00\times10^{-5}\ \text{m}^3$. The weight of the block is $mg = \rho V g = (2702\ \text{kg/m}^3)(3.00\times10^{-5}\ \text{m}^3)(9.80\ \text{m/s}^2) = \boxed{0.794\ \text{N}}$.

(b) Strategy Find the weight of the oil displaced by the block and subtract it from the weight of the block to find the reading of the scale.

Solution The weight of the displaced oil is $m_{\text{oil}}g = \rho_{\text{oil}}V_{\text{block}}g$. Thus, the scale reading is

$$0.794\ \text{N} - \rho_{\text{oil}}V_{\text{block}}g = 0.794\ \text{N} - (850\ \text{kg/m}^3)(3.00\times10^{-5}\ \text{m}^3)(9.80\ \text{m/s}^2) = \boxed{0.544\ \text{N}}.$$

93. Strategy The magnitude of the force on the spring is equal to the difference between the weight of the displaced water and the weight of the cube. Use Hooke's law to find the spring constant.

Solution

$$k = \frac{F}{x} = \frac{(\rho_w - \rho_c)Vg}{x} = \frac{(1.00\times10^3\ \text{kg/m}^3 - 8.00\times10^2\ \text{kg/m}^3)(0.0400\ \text{m})^3(9.80\ \text{m/s}^2)}{0.0100\ \text{m}} = \boxed{12.5\ \text{N/m}}$$

97. Strategy Use Eq. (9-3).

Solution Find the height of a hill you must ascend for the barometer to read a pressure drop of 1.0 cm Hg.

$$\Delta P = \rho g h,\ \text{so}\ h = \frac{\Delta P}{\rho g} = \frac{1.0\ \text{cm Hg}}{(1.20\ \text{kg/m}^3)(9.80\ \text{m/s}^2)}\left(\frac{1.013\times10^5\ \text{Pa}}{76.0\ \text{cm Hg}}\right) = \boxed{110\ \text{m}}.$$

101. (a) Strategy Use Eqs. (9-7), (9-16), and Newton's second law.

Solution Find the terminal velocity of the bubbles.

$$\Sigma F_y = F_B - F_D - m_a g = \rho_w g V - 6\pi\eta r v_t - \rho_a g V = -6\pi\eta r v_t + (\rho_w - \rho_a)\tfrac{4}{3}\pi r^3 g = 0, \text{ so}$$

$$v_t = \frac{(\rho_w - \rho_a)\tfrac{4}{3}\pi r^3 g}{6\pi\eta r} = \frac{2r^2 g(\rho_w - \rho_a)}{9\eta}$$

$$= \frac{2(1.0\times10^{-3}\ \text{m})^2(9.80\ \text{m/s}^2)(1.00\times10^3\ \text{kg/m}^3 - 1.20\ \text{kg/m}^3)}{9(1.0\times10^{-3}\ \text{Pa}\cdot\text{s})} = 2.2\ \text{m/s}.$$

Thus, $\vec{v}_t = \boxed{2.2\ \text{m/s up}}$.

(b) Strategy Divide the change in pressure by the change in time and use the result from part (a).

Solution

$$\Delta P = \rho g \Delta y, \text{ so } \frac{\Delta P}{\Delta t} = \rho g \frac{\Delta y}{\Delta t} = \rho g v_t = (1.00\times10^3\ \text{kg/m}^3)(9.80\ \text{m/s}^2)(2.175\ \text{m/s}) = \boxed{21\ \text{kPa/s}}.$$

105. Strategy Since the net force is zero, the weight of the lead plus the weight of the wood is equal to the weight of the displaced water. In addition, the volume of the lead is equal to the volume of the water displaced less the volume of the wood.

Solution Find the volume of the displaced water in terms of the mass of the lead and wood.

$$m_{\text{lead}}g + m_{\text{wood}}g = m_w g = \rho_w g V_w, \text{ so } V_w = \frac{m_{\text{lead}} + m_{\text{wood}}}{\rho_w}.$$

Find the mass of the lead.

$$V_w - V_{\text{wood}} = V_{\text{lead}}$$

$$\frac{m_{\text{lead}} + m_{\text{wood}}}{\rho_w} - V_{\text{wood}} = \frac{m_{\text{lead}}}{\rho_{\text{lead}}}$$

$$m_{\text{lead}} + m_{\text{wood}} - \rho_w V_{\text{wood}} = \frac{\rho_w}{\rho_{\text{lead}}} m_{\text{lead}}$$

$$m_{\text{lead}}\left(1 - \frac{\rho_w}{\rho_{\text{lead}}}\right) = \rho_w V_{\text{wood}} - m_{\text{wood}} = \rho_w V_{\text{wood}} - \rho_{\text{wood}} V_{\text{wood}}$$

$$m_{\text{lead}} = \frac{(\rho_w - \rho_{\text{wood}})V_{\text{wood}}}{1 - \dfrac{\rho_w}{\rho_{\text{lead}}}} = \frac{(1.00\times10^3\ \text{kg/m}^3 - 0.78\times10^3\ \text{kg/m}^3)(0.330\ \text{m})^3}{1 - \dfrac{1.00\times10^3\ \text{kg/m}^3}{11.3\times10^3\ \text{kg/m}^3}} = \boxed{8.7\ \text{kg}}$$

107. (a) Strategy Let V be the volume of the liquid displaced and V_h be the volume of the hydrometer. Then $V_h - V = Ah$ where A is the cross-sectional area of the stem and h is the height above the liquid. Use Newton's second law.

Solution Find the distance from the top of the cylinder where the mark should be placed.

$$\Sigma F_y = F_B - m_h g = \rho g V - m_h g = (\text{S.G.})\rho_w g V - m_h g = 1.00\rho_w g(V_h - Ah) - m_h g = 0, \text{ so}$$

$$h = \frac{1}{A}\left(V_h - \frac{m_h}{1.00\rho_w}\right) = \frac{1}{0.400\ \text{cm}^2}\left[8.80\ \text{cm}^3 - \frac{4.80\ \text{g}}{1.00(1.00\ \text{g/cm}^3)}\right] = \boxed{10.0\ \text{cm}}.$$

(b) Strategy Use the results of part (a),

Solution Find the specific gravity of the alcohol.

$(\text{S.G.})\rho_w g V - m_h g = (\text{S.G.})\rho_w g (V_h - Ah) - m_h g = 0$, so

$$\text{S.G.} = \frac{m_h}{\rho_w(V_h - Ah)} = \frac{4.80 \text{ g}}{(1.00 \text{ g}/\text{cm}^3)[8.80 \text{ cm}^3 - (0.400 \text{ cm}^2)(7.25 \text{ cm})]} = \boxed{0.814}.$$

(c) Strategy For the minimum S.G., the volume of the displaced liquid is equal to the volume of the hydrometer.

Solution Find the lowest specific gravity that can be measured with this hydrometer.

$(\text{S.G.})\rho_w V_h = m_h$, so $\text{S.G.}_{\text{min}} = \dfrac{m_h}{\rho_w V_h} = \dfrac{4.80 \text{ g}}{(1.00 \text{ g}/\text{cm}^3)(8.80 \text{ cm}^3)} = \boxed{0.545}.$

109. Strategy Use the relationships between pressure, density, force, area, and height.

Solution Find the density of the liquid.

$$\Delta P = \frac{W_1}{A} = \frac{\rho_1 g V_1}{A} = \rho_w g \Delta y_w, \text{ so}$$

$$\rho_1 = \frac{\rho_w \Delta y_w A}{V_1} = \frac{\rho_w \Delta y_w \pi r^2}{\pi r^2 h} = \frac{\rho_w \Delta y_w}{h} = \frac{(1.0 \text{ g}/\text{cm}^3)[0.45 \text{ m} - (0.50 \text{ m} - 0.30 \text{ m})]}{0.30 \text{ m}} = \boxed{0.83 \text{ g}/\text{cm}^3}.$$

Chapter 10

ELASTICITY AND OSCILLATIONS

Conceptual Questions

1. Young's modulus does not tell us which is stronger. Instead, it tells us which is more resistant to deformation for a given stress. The ultimate strength would tell us which is stronger—i.e., which can withstand the greatest stress.

5. The compressive force experienced by the columns is greater at the bottom than at the top, because the bottom must support the weight of the column itself in addition to whatever the column is holding up. By increasing the cross-sectional area of the bottom of the column, the stress it experiences is reduced. Tapering columns so that they are thicker at the base prevents the stress at the bottom from being too large.

9. The tension in the bungee cord at the lowest point would be greater than the person's weight, because there is an upward acceleration. In fact, the tension would have its maximum value at the bottom, because that is where the upward acceleration is the greatest.

13. To produce the same strain, the ratio of the force to the cross-sectional area must remain unchanged. The total cross-sectional area of the two wires together is twice the original area. Thus, the force applied to the two wires must be doubled as well. Modeling a thick wire as a bundle of thin wires, the preceding argument explains why the force to produce a given strain must be proportional to the cross-sectional area—and thus why the strain depends on the stress.

17. In the mass-spring system, the restoring force supplied by the spring is independent of the object's mass. Thus, the larger inertia of a more massive object produces a longer period. The restoring force for small amplitude oscillations of the pendulum is the horizontal component of the tension in the string. In this case, the magnitude of the tension is approximately equal to the weight of the bob. Although a more massive bob has more inertia, it also has a proportionally larger restoring force. Thus, the period of oscillation of the pendulum is independent of the mass.

Problems

1. **Strategy** The stress is proportional to the strain. Use Eq. (10-4).

   Solution Find the vertical compression of the beam.
   $$Y\frac{\Delta L}{L} = \frac{F}{A}, \text{ so } \Delta L = \frac{FL}{YA} = \frac{(5.8\times10^4 \text{ N})(2.5 \text{ m})}{(200\times10^9 \text{ Pa})(7.5\times10^{-3} \text{ m}^2)} = \boxed{0.097 \text{ mm}}.$$

5. **Strategy** The stress is proportional to the strain. Use Eq. (10-4).

   Solution Find Young's modulus for the wire.
   $$Y = \frac{FL}{A\Delta L} = \frac{(1.00\times10^3 \text{ N})(5.00 \text{ m})}{(0.100 \text{ cm}^2)(10^{-2} \text{ m/cm})^2(6.50\times10^{-3} \text{ m})} = \boxed{7.69\times10^{10} \text{ Pa}}$$

9. **(a) Strategy** Use Eq. (10-4).

Solution Find the force the wings must exert to extend the resilin.

$$\frac{F}{A} = Y\frac{\Delta L}{L}, \text{ so } F = \frac{YA\Delta L}{L} = \frac{(1.7\times10^6 \text{ N}/\text{m}^2)(1.0\times10^{-6} \text{ m}^2)(4.0 \text{ cm} - 1.0 \text{ cm})}{1.0 \text{ cm}} = \boxed{5.1 \text{ N}}.$$

(b) Strategy The energy stored in the resilin is elastic potential energy. Use Hooke's law.

Solution Find the energy stored in the resilin.

$$U = \frac{1}{2}kx^2 = \frac{1}{2}\left(\frac{F}{x}\right)x^2 = \frac{1}{2}Fx = \frac{1}{2}(5.1 \text{ N})(0.030 \text{ m}) = \boxed{7.7\times10^{-2} \text{ J}}$$

13. **Strategy** Refer to Fig. 10.4c. The stress is proportional to the strain.

Solution Calculate Young's moduli for tension and compression of bone.
Tension:
For tensile stress and strain, the graph is far from being linear, but for relatively small values of stress and strain, it is approximately linear. So, for small values of tensile stress and strain, Young's Modulus is

$$Y = \frac{\text{stress}}{\text{strain}} = \frac{5.0\times10^7 \text{ N}/\text{m}^2}{0.0033} = \boxed{1.5\times10^{10} \text{ N}/\text{m}^2}.$$

Compression:

Similarly, for small values of compressive stress and strain, $Y = \dfrac{-4.5\times10^7 \text{ N}/\text{m}^2}{-0.0050} = \boxed{9.0\times10^9 \text{ N}/\text{m}^2}$.

17. **Strategy** Set the stresses equal to the compressive strengths to determine the effective cross-sectional areas.

Solution Find the effective cross-sectional areas.

Human: $\dfrac{F}{A} = 1.6\times10^8 \text{ Pa}$, so $A = \dfrac{5\times10^4 \text{ N}}{1.6\times10^8 \text{ Pa}} = \boxed{3 \text{ cm}^2}$.   Horse: $A = \dfrac{10\times10^4 \text{ N}}{1.4\times10^8 \text{ Pa}} = \boxed{7.1 \text{ cm}^2}$

19. **Strategy** The stress on the copper wire must be less than its elastic limit.

Solution Find the maximum load that can be suspended from the copper wire.

$$\frac{F}{A} < \text{elastic limit, so } F < \pi r^2(\text{elastic limit}) = \pi(0.0010 \text{ m})^2(2.0\times10^8 \text{ Pa}) = \boxed{630 \text{ N}}.$$

21. **Strategy** Assume that the stress is proportional to the strain up to the breaking point. Use Eq. (10-4).

Solution Find the stress at the breaking point of the steel wire.

$$\text{stress at breaking point} = Y\frac{\Delta L}{L} = (2.0\times10^{11} \text{ N}/\text{m}^2)\left(\frac{0.20}{100}\right) = \boxed{4.0\times10^8 \text{ Pa}}$$

25. **Strategy** Use $\Delta P = \rho g d$ and the relationship between volume, mass, and density.

Solution Find the percent increase in the density of water at a depth of 1.0 km.

$$\frac{\Delta\rho}{\rho} = \frac{\rho'-\rho}{\rho} = \frac{\rho'}{\rho} - 1 = \frac{\frac{m}{V'}}{\frac{m}{V}} - 1 = \frac{V}{V'} - 1 = \frac{V-V'}{V'} = -\frac{\Delta V}{V'}$$

Since $V' \approx V$, $\dfrac{\Delta\rho}{\rho} \approx -\dfrac{\Delta V}{V}$. Use Hooke's law for volume deformations.

$$100\%\times\frac{\Delta\rho}{\rho} \approx \frac{\Delta P}{B}\times100\% = \frac{\rho_w g d}{B}\times100\% = \frac{(1.0\times10^3 \text{ kg}/\text{m}^3)(9.80 \text{ N}/\text{kg})(1.0\times10^3 \text{ m})}{2.2\times10^9 \text{ Pa}}\times100\% = \boxed{0.45\%}$$

27. Strategy Use Hooke's law for volume deformations. The pressure of the Moon is roughly 10^{-9} Pa.

Solution Find the change in volume of the aluminum.

$$\Delta P = -B\frac{\Delta V}{V}, \text{ so } \Delta V = -\frac{V\Delta P}{B} = -\frac{(1.00 \text{ cm}^3)(10^{-9} \text{ Pa} - 1.013 \times 10^5 \text{ Pa})}{70 \times 10^9 \text{ Pa}} = 1.4 \times 10^{-6} \text{ cm}^3.$$

The volume of the aluminum sphere would increase by 1.4×10^{-6} cm^3.

29. Strategy Set the shear stress equal to the total shear strength to find the maximum shearing force.

Solution Find the maximum shearing force F on the plates that the four bolts can withstand.

$$\frac{F}{4A_{\text{bolt}}} = \text{shear strength, so } F = 4\pi r^2(\text{shear strength}) = 4\pi(0.010 \text{ m})^2(6.0\times10^8 \text{ Pa}) = \boxed{7.5 \times 10^5 \text{ N}}.$$

31. Strategy Use Hooke's law for shear deformations.

Solution Find the magnitude of the tangential force.

$$\frac{F}{A} = \frac{F}{L^2} = S\frac{\Delta x}{L}, \text{ so } F = S\Delta xL = (940 \text{ Pa})(0.64\times10^{-2} \text{ m})(0.050 \text{ m}) = \boxed{0.30 \text{ N}}.$$

33. Strategy At the maximum extension of the spring, $x = A$ and the magnitude of the acceleration is maximum. Use Eq. (10-22).

Solution Find the magnitude of the acceleration at the point of maximum extension of the spring.

$$a_{\text{m}} = \omega^2 A = \frac{4\pi^2 A}{T^2} = \frac{4\pi^2(0.050 \text{ m})}{(0.50 \text{ s})^2} = \boxed{7.9 \text{ m/s}^2}$$

37. Strategy In SHM, the frequency of oscillation is given by $\sqrt{k/m}/2\pi$.

Solution Ranking the spring constant divided by the mass in decreasing order is equivalent to ranking the frequency of oscillations in decreasing order.

(a) $\dfrac{k}{m}$; (b) $\dfrac{k}{2m}$; (c) $\dfrac{k}{m}$; (d) $\dfrac{k/2}{2m} = \dfrac{k}{4m}$; (e) $\dfrac{2k}{2m} = \dfrac{k}{m}$

In decreasing order, we have $\boxed{\text{(a) = (c) = (e), (b), (d)}}$.

41. Strategy Use Eqs. (10-21) and (10-22).

Solution

(a) Find v_{m} and a_{m} in terms of f. Then compare high- and low-frequency sounds.

$$v_{\text{m}} = \omega A = 2\pi fA \propto f \text{ and } a_{\text{m}} = \omega^2 A = 4\pi^2 f^2 A \propto f^2, \text{ so } v_{\text{m}} \text{ and } a_{\text{m}} \text{ are greatest for } \boxed{\text{high frequency}}.$$

(b) $v_{\text{m}} = 2\pi(20.0 \text{ Hz})(1.0\times10^{-8} \text{ m}) = \boxed{1.3\times10^{-6} \text{ m/s}}$

$a_{\text{m}} = 4\pi^2(20.0 \text{ Hz})^2(1.0\times10^{-8} \text{ m}) = \boxed{1.6\times10^{-4} \text{ m/s}^2}$

(c) $v_{\text{m}} = 2\pi(20.0\times10^3 \text{ Hz})(1.0\times10^{-8} \text{ m}) = \boxed{0.0013 \text{ m/s}}$

$a_{\text{m}} = 4\pi^2(20.0\times10^3 \text{ Hz})^2(1.0\times10^{-8} \text{ m}) = \boxed{160 \text{ m/s}^2}$

45. Strategy The angular frequency of oscillation is inversely proportional to the square root of the mass. Form a proportion.

Solution Find the new value of ω.

$$\omega \propto \sqrt{\frac{1}{m}}, \text{ so } \frac{\omega_f}{\omega_i} = \frac{\sqrt{1/m_f}}{\sqrt{1/m_i}} = \sqrt{\frac{m_i}{m_f}} = \sqrt{\frac{1}{4.0}} = \frac{1}{2.0}. \text{ Therefore, } \omega_f = \frac{\omega_i}{2.0} = \frac{10.0 \text{ rad/s}}{2.0} = \boxed{5.0 \text{ rad/s}}.$$

49. Strategy Use Eqs. (10-21) and (10-22) and Newton's second law.

Solution Find the radio's maximum displacement and maximum speed, and the maximum net force exerted on it.

(a) $a_m = \omega^2 A$, so $A = \dfrac{a_m}{\omega^2} = \dfrac{98 \text{ m/s}^2}{4\pi^2(120 \text{ Hz})^2} = \boxed{1.7 \times 10^{-4} \text{ m}}$.

(b) $v_m = \omega A = \omega \dfrac{a_m}{\omega^2} = \dfrac{a_m}{\omega} = \dfrac{98 \text{ m/s}^2}{2\pi(120 \text{ Hz})} = \boxed{0.13 \text{ m/s}}$.

(c) According to Newton's second law, $F_m = m a_m = (5.24 \text{ kg})(98 \text{ m/s}^2) = \boxed{510 \text{ N}}$.

53. (a) Strategy The speed is maximum when the spring and mass system is at its equilibrium point. Use Newton's second law.

Solution Find the extension of the spring.

$$\Sigma F_y = kx - mg = 0, \text{ so } x = \frac{mg}{k} = \frac{(0.60 \text{ kg})(9.80 \text{ N/kg})}{15 \text{ N/m}} = \boxed{0.39 \text{ m}}.$$

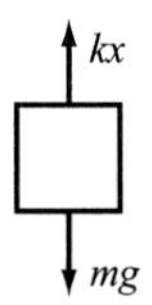

(b) Strategy Use Eqs. (10-20a) and (10-21).

Solution Find the maximum speed of the body.

$$v_m = \omega A = \sqrt{\frac{k}{m}} x = \sqrt{\frac{15 \text{ N/m}}{0.60 \text{ kg}}} (0.39 \text{ m}) = \boxed{2.0 \text{ m/s}}$$

57. Strategy and Solution Since $y(t) = A \sin \omega t$, $f = \dfrac{\omega}{2\pi} = \dfrac{1.57 \text{ rad/s}}{2\pi} = \boxed{0.250 \text{ Hz}}$.

61. Strategy Use the definition of average speed and Eq. (10-21). In (d), graph v_x on the vertical axis and t on the horizontal axis.

Solution

(a) The average speed is the total distance traveled divided by the time of travel.
$$v_{av} = \frac{\Delta x}{\Delta t} = \frac{4A}{T} = \frac{4A}{2\pi/\omega} = \boxed{\frac{2}{\pi}\omega A}$$

(b) The maximum speed for SHM is $v_m = \boxed{\omega A}$.

(c) $\dfrac{v_{av}}{v_m} = \dfrac{\frac{2}{\pi}\omega A}{\omega A} = \boxed{\dfrac{2}{\pi}}$

(d) Graph $v_x(t)$ and a line from the origin to v_m.

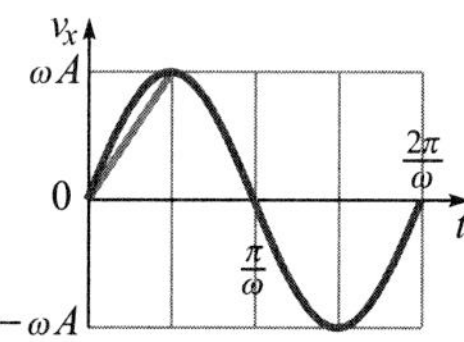

> If the acceleration were constant so that the speed varied linearly, the average speed would be 1/2 of the maximum velocity. Since the actual speed is always larger than what it would be for constant acceleration, the average speed must be larger.

65. Strategy Use Eq. (10-26b).

Solution Compute the period of the pendulum.
$$T = 2\pi\sqrt{\frac{L}{g}} = 2\pi\sqrt{\frac{4.0 \text{ m}}{9.80 \text{ m/s}^2}} = \boxed{4.0 \text{ s}}$$

67. Strategy and Solution According to Eq. (10-26b), $T = 2\pi\sqrt{\dfrac{L}{g}}$, which does not depend upon the mass.

Therefore, $T = \boxed{1.5 \text{ s}}$.

69. Strategy The frequency for small-amplitude oscillations of a pendulum is given by $\sqrt{g/L}$.

Solution Ranking the lengths of the strings from smallest to greatest is equivalent to ranking the frequencies from greatest to smallest, since the frequency is inversely proportional to the square root of the length. Therefore, the answer is $\boxed{\text{(c), (a) = (b), (d) = (e)}}$.

71. Strategy Use Eq. (10-26b).

Solution Find the length of the pendulum.
$$T = 2\pi\sqrt{\frac{L}{g}}, \text{ so } L = \frac{gT^2}{4\pi^2} = \frac{(9.80 \text{ m/s}^2)(1.0 \text{ s})^2}{4\pi^2} = \boxed{0.25 \text{ m}}.$$

73. **(a)** **Strategy and Solution** Since the period of a pendulum is inversely proportional to gravitational field strength, the greater period implies that the gravitational field strength on the other planet is $\boxed{\text{less}}$ than that on Earth.

(b) **Strategy** Use Eq. (10-26b). Form a proportion.

Solution Refer to the mystery planet as X.

$$T_E = 2\pi\sqrt{\frac{L}{g_E}} \text{ and } T_X = 2\pi\sqrt{\frac{L}{g_X}}, \text{ so } \frac{T_E}{T_X} = 2\pi\sqrt{\frac{L}{g_E}}\left(2\pi\sqrt{\frac{L}{g_X}}\right)^{-1} = \sqrt{\frac{g_X}{g_E}}.$$

Thus, $g_X = g_E\left(\dfrac{T_E}{T_X}\right)^2 = (9.80 \ \text{m/s}^2)\left(\dfrac{0.650 \ \text{s}}{0.862 \ \text{s}}\right)^2 = \boxed{5.57 \ \text{m/s}^2}$.

77. **Strategy** The total mechanical energy of a pendulum is $E = \frac{1}{2}m\omega^2 A^2$. Form a proportion.

Solution Find the mechanical energy of the pendulum.

$$\frac{E_2}{E_1} = \frac{A_2^2}{A_1^2}, \text{ so } E_2 = \left(\frac{A_2}{A_1}\right)^2 E_1 = \left(\frac{3.0 \ \text{cm}}{2.0 \ \text{cm}}\right)^2 (5.0 \ \text{mJ}) = \boxed{11 \ \text{mJ}}.$$

79. **Strategy** $E = \frac{1}{2}m\omega^2 A^2 \propto A^2$ for a pendulum. Form a proportion.

Solution Find by what factor the energy has decreased.

$$\frac{E_2}{E_1} = \frac{A_2^2}{A_1^2} = \frac{(A_1/20.0)^2}{A_1^2} = \frac{1}{20.0^2} = \frac{1}{400}$$

$\boxed{\text{The energy has decreased by a factor of } 400}$.

81. **Strategy** Use Eq. (10-4) for each situation.

Solution Initially, for the stretch, we have
$$Y\frac{\Delta L}{L} = \frac{F}{A} = \frac{T}{A} = \frac{W}{A}, \text{ so } \Delta L = \frac{LW}{YA}; \text{ where } T \text{ is the tension in the steel cable.}$$
For the three cables, the tension is one-third the previous value and the lengths of cable are one-third of the previous length. Find the stretch for each of the three cables.
$$Y\frac{\Delta L'}{L'} = Y\frac{\Delta L'}{L/3} = \frac{F'}{A} = \frac{T'}{A} = \frac{W/3}{A}, \text{ so } \Delta L' = \frac{(L/3)(W/3)}{YA} = \frac{LW}{9YA} = \boxed{\frac{\Delta L}{9}}.$$

85. Strategy Use Eqs. (10-26b) and (10-20c).

Solution

(a) The period of a simple pendulum is inversely proportional to the square root of the gravitational acceleration. Since the gravitational field on the Moon is one-sixth that on Earth, the period of the simple pendulum on the Moon is $\boxed{\text{greater than } T}$.

(b) Find the ratio of the pendulum's period on the Moon to its period on Earth.
$$\frac{T_M}{T} = \frac{2\pi\sqrt{L/g_M}}{2\pi\sqrt{L/g}} = \sqrt{\frac{g}{g_M}} = \sqrt{\frac{g}{g/6}} = \boxed{\sqrt{6}}$$

(c) The period of a mass-spring system is directly proportional to the square root of the mass and inversely proportional to the square root of the spring constant. Since the mass and spring constant are the same on the Moon as on Earth, the period of the mass-spring system on the Moon is $\boxed{\text{equal to } T}$.

(d) The mass-spring system's period on the Moon is the same as its period on Earth; therefore $T_M/T = \boxed{1}$.

89. Strategy Assume the weight of the cable is negligible compared to the weight of the aviator. Use Eq. (10-26b).

Solution Find the period for a pendulum assuming SHM.
$$T = 2\pi\sqrt{\frac{L}{g}} = 2\pi\sqrt{\frac{45\text{ m}}{9.80\text{ m/s}^2}} = \boxed{13\text{ s}}$$

93. Strategy Graph x on the vertical axis and t on the horizontal axis. Analyze the slope of the graph (the magnitude of which is the speed) in terms of the distance between the dots to determine the fastest and slowest speeds of the mass.

Solution Graph $x(t) = -(10\text{ cm})\cos[(1.57\text{ s}^{-1})t]$.

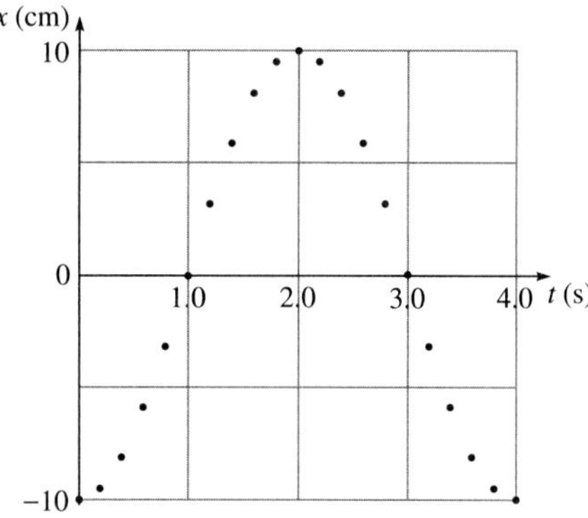

The distance between adjacent dots should be the least at the endpoints and greatest at the center, so its speed is lowest at the endpoints and fastest at its equilibrium position.

97. Strategy Since the body begins with its maximum amplitude at $t = 0$, the body oscillates according to a cosine function $(\cos 0 = 1)$. Use Newton's second law, Hooke's law, and Eq. (10-20a).

Solution Find the amplitude A.

$$\Sigma F_y = kA - mg = 0, \text{ so } A = \frac{mg}{k} = \frac{4.0 \text{ N}}{250 \text{ N/m}} = 1.6 \text{ cm}.$$

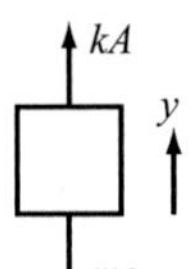

Find the angular frequency ω.

$$\omega = \sqrt{\frac{k}{m}} = \sqrt{\frac{kg}{mg}} = \sqrt{\frac{(250 \text{ N/m})(9.80 \text{ m/s}^2)}{4.0 \text{ N}}} = 25 \text{ rad/s}$$

Thus, the equation describing the motion of the body is $\boxed{y = (1.6 \text{ cm})\cos\left[(25 \text{ rad/s})t\right]}$.

101. Strategy Use Eq. (10-2) to find the tensile stress. Then compare the tensile stress with the elastic limit of steel piano wire.

Solution Find the tensile stress in the piano wire in Problem 90.

$$\text{tensile stress} = \frac{F}{A} = \frac{T}{\frac{1}{4}\pi d^2} = \frac{4T}{\pi d^2} = \frac{4(402 \text{ N})}{\pi(0.80\times10^{-3} \text{ m})^2} = 8.0\times10^8 \text{ Pa} < 8.26\times10^8 \text{ Pa}$$

The tensile stress is $\boxed{8.0\times10^8 \text{ Pa; it is just under the elastic limit.}}$

105. (a) Strategy Use conservation of energy. Do *not* assume SHM.

Solution Find the speed of the pendulum bob at the bottom of its swing.

$$\Delta K = \frac{1}{2}mv^2 = -\Delta U = mgL, \text{ so } v = \boxed{\sqrt{2gL}}.$$

(b) Strategy Assume (incorrectly, for such a large amplitude) that the motion *is* SHM. Use Eqs. (10-21) and (10-26a).

Solution Find the speed of the pendulum bob at the bottom of its swing.

The amplitude A is a quarter of the circumference of a circle with radius L, or $\frac{2\pi L}{4} = \frac{\pi}{2}L$.

Assuming SHM, $v_{\rm m} = \omega A = \boxed{\sqrt{\frac{g}{L}\left(\frac{\pi}{2}L\right)} = \frac{\pi}{2}\sqrt{gL}}$. Since $v \propto \omega \propto \frac{1}{T}$, a smaller speed implies a larger

period. Since $\dfrac{v_{\rm m}}{v} = \dfrac{\frac{\pi}{2}\sqrt{gL}}{\sqrt{2gL}} = \dfrac{\pi}{2\sqrt{2}} > 1$, the period of a pendulum for large amplitudes is $\boxed{\text{larger}}$ than that

given by Eq. (10-26b).

Chapter 11

WAVES

Conceptual Questions

1. Piano, guitar, and violin strings produce transverse waves when they are plucked or bowed—they do not produce longitudinal sound waves. The transverse motion of the strings causes longitudinal sound waves to be produced in the surrounding air.

5. **(a)** The wavelength of the fundamental will decrease.

   (b) The frequency of the fundamental will increase.

   (c) The wave velocity is constant, thus the time for a pulse to travel the length of the string will decrease.

   (d) The maximum velocity of a point on the string is proportional to the frequency and will therefore increase.

   (e) The maximum acceleration of a point on the string is proportional to the square of the frequency and will therefore increase.

9. In the figure below, the shape of a wave traveling to the right is depicted at two times separated by a short interval. Arrows indicate the direction of the string's movement.

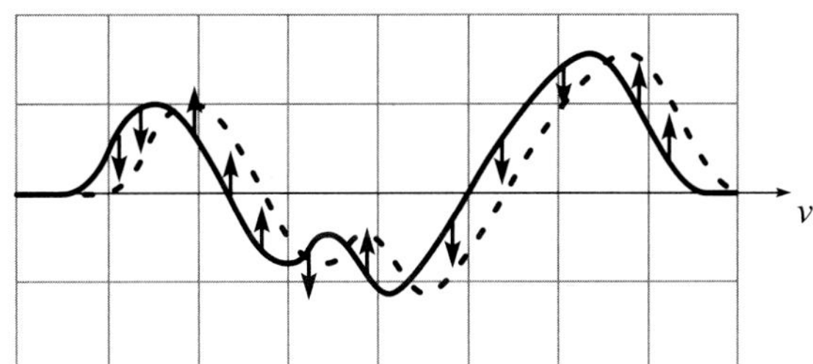

Problems

1. **Strategy** Form a proportion with the intensities, treating the Sun as an isotropic source. Use Eq. (11-1).

   Solution Find the intensity of the sunlight that reaches Jupiter.
   $$\frac{I_J}{I_E} = \frac{\frac{P}{4\pi r_J^2}}{\frac{P}{4\pi r_E^2}} = \frac{r_E^2}{r_J^2}, \text{ so } I_J = \left(\frac{r_E}{r_J}\right)^2 I_E = \left(\frac{1}{5.2}\right)^2 (1400 \text{ W}/\text{m}^2) = \boxed{52 \text{ W}/\text{m}^2}.$$

5. **Strategy** The power equals the intensity times the area.

   Solution Find the power radiated by the jet airplane in the form of sound waves.
   $$P = IA = \left(\frac{5.0 \text{ m}}{120 \text{ m}}\right)^2 (1.0\times10^2 \text{ W}/\text{m}^2)4\pi(120 \text{ m})^2 = \boxed{31 \text{ kW}}$$

9. **Strategy** Refer to the figure. Use the definition of average speed.

   Solution

   (a) Find the speed.
   $$v_x = \frac{\Delta x}{\Delta t} = \frac{1.80 \text{ m} - 1.50 \text{ m}}{0.20 \text{ s}} = 1.5 \text{ m/s}$$
   Find the position.
   $$x_f = x_i + v\Delta t = 1.80 \text{ m} + (1.5 \text{ m/s})(3.00 \text{ s} - 0.20 \text{ s}) = \boxed{6.0 \text{ m}}$$

   (b) $\displaystyle t_f = \frac{x_f - x_i}{v_x} + t_i = \frac{4.00 \text{ m} - 1.80 \text{ m}}{1.5 \text{ m/s}} + 0.20 \text{ s} = \boxed{1.7 \text{ s}}$

11. **Strategy** Use Eq. (11-4).

    Solution Find the speed of the transverse waves on the string.
    $$v = \sqrt{\frac{F}{\mu}} = \sqrt{\frac{90.0 \text{ N}}{3.20 \times 10^{-3} \text{ kg/m}}} = \boxed{168 \text{ m/s}}$$

13. **Strategy** Use Eq. (11-2) for the speed of the transverse waves.

    Solution The weight of the string divided by the load is
    $$\frac{0.25 \text{ N}}{1.00 \times 10^3 \text{ N}} = 2.5 \times 10^{-4} = 0.025\%.$$
    The weight of the string is negligible since the result will be limited to two significant figures (by 0.25 N).
    Find the time it takes the wave pulse to travel to the upper end of the string.
    $$\Delta t = \frac{\Delta y}{v} = \frac{L}{\sqrt{FL/m}} = \sqrt{\frac{mL}{F}} = \sqrt{\frac{mgL}{Fg}} = \sqrt{\frac{(0.25 \text{ N})(10.0 \text{ m})}{(1.00 \times 10^3 \text{ N})(9.80 \text{ N/kg})}} = \boxed{16 \text{ ms}}$$

15. **Strategy** Use Eq. (11-5).

    Solution Find the wavelength.
    $$\lambda = vT = (75.0 \text{ m/s})(5.00 \times 10^{-3} \text{ s}) = \boxed{0.375 \text{ m}}$$

17. **Strategy** Use Eq. (11-6).

    Solution Find the frequencies.

    (a) $\displaystyle f = \frac{v}{\lambda} = \frac{340 \text{ m/s}}{1.0 \text{ m}} = \boxed{340 \text{ Hz}}$

    (b) $\displaystyle f = \frac{v}{\lambda} = \frac{3.0 \times 10^8 \text{ m/s}}{1.0 \text{ m}} = \boxed{3.0 \times 10^8 \text{ Hz}}$

21. **Strategy and Solution** If another swimmer were 9.6 m away from you, your motions would be the same (in phase). For periodic motion, the motion half a wavelength from any point along a line parallel to the motion of the waves is opposite to the motion at that point, so the other swimmer should be half a wavelength away, or
    $$\frac{9.6 \text{ m}}{2} = \boxed{4.8 \text{ m}}.$$

25. (a) Strategy Use Eq. (10-21).

Solution Find the maximum transverse speed of a point on the string.
$$v_{\mathrm{m}} = \omega A = (130 \ \mathrm{rad/s})(0.0220 \ \mathrm{m}) = \boxed{2.9 \ \mathrm{m/s}}$$

(b) Strategy Use Eq. (10-22).

Solution Find the maximum transverse acceleration of a point on the string.
$$a_{\mathrm{m}} = \omega^2 A = (130 \ \mathrm{rad/s})^2 (0.0220 \ \mathrm{m}) = \boxed{370 \ \mathrm{m/s^2}}$$

(c) Strategy Use Eq. (11-7).

Solution Find the wave speed.
$$v = \frac{\omega}{k} = \frac{130 \ \mathrm{rad/s}}{15 \ \mathrm{rad/m}} = \boxed{8.7 \ \mathrm{m/s}}$$

(d) Strategy Consider why transverse speed and wave speed are different.

Solution The answer to part (c) is different from the answer to part (a) because
$\boxed{\text{the motion of the particles on the string is not the same as the motion of the wave along the string}}$.

27. Strategy The equation for a transverse sinusoidal wave moving in the positive *x*-direction and in the negative *y*-direction in the next instant of time according to the situation given in the problem statement can be written in the form $y(x, t) = A\sin(kx - \omega t)$. Use Eqs. (10-21) and (11-7).

Solution Find the wave speed v by finding the maximum speed of a point on the string.
$$v = 5.00 v_{\mathrm{m}} = 5.00 \omega A$$
Find the wave number.
$$k = \frac{\omega}{v} = \frac{\omega}{5.00 \omega A} = \frac{1}{5.00 A} = \frac{1}{5.00(0.0250 \ \mathrm{m})} = 8.00 \ \mathrm{rad/m}$$

The equation for the transverse sinusoidal wave is $\boxed{y(x, t) = (2.50 \ \mathrm{cm})\sin[(8.00 \ \mathrm{rad/m})x - (2.90 \ \mathrm{rad/s})t]}$.

29. Strategy Since the displacement axes use the same scale, the "taller" the wave, the greater the amplitude.

Solution Ranking the amplitudes, largest to smallest, we have $\boxed{\text{(b) = (e), (a), (c) = (d)}}$.

33. Strategy Use Eqs. (10-21) and (10-22). Plot the graphs.

Solution Find the maximum speed and maximum acceleration of a point on the string.
$$v_{\mathrm{m}} = \omega A = (4.0\pi \ \mathrm{rad/s})(0.0050 \ \mathrm{m}) = \boxed{0.063 \ \mathrm{m/s}} \ \text{and} \ a_{\mathrm{m}} = \omega^2 A = (4.0\pi \ \mathrm{rad/s})^2 (0.0050 \ \mathrm{m}) = \boxed{0.79 \ \mathrm{m/s^2}}.$$
Plot the graphs.
$$y(0, t) = (0.0050 \ \mathrm{m})\cos[(4.0\pi \ \mathrm{rad/s})t]$$

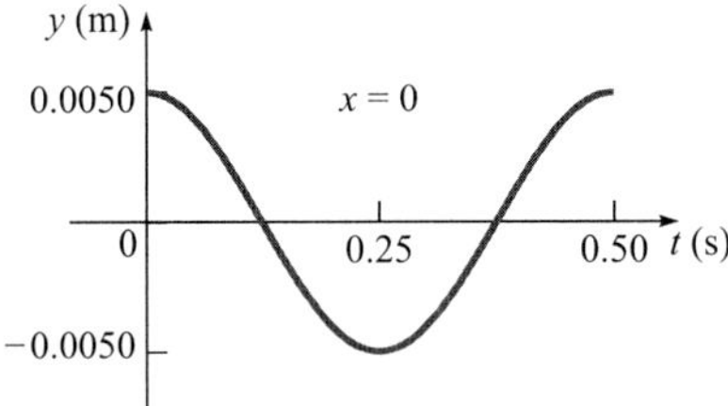

v_y leads y by $\frac{1}{4}$ cycle; $v_y(0, t) = -(0.063 \text{ m/s})\sin[(4.0\pi \text{ rad/s})t]$

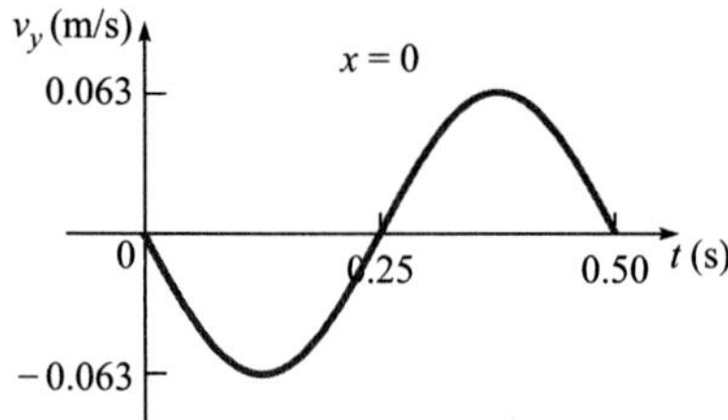

a_y leads v_y by $\frac{1}{4}$ cycle; $a_y(0, t) = -(0.79 \text{ m/s}^2)\cos[(4.0\pi \text{ rad/s})t]$

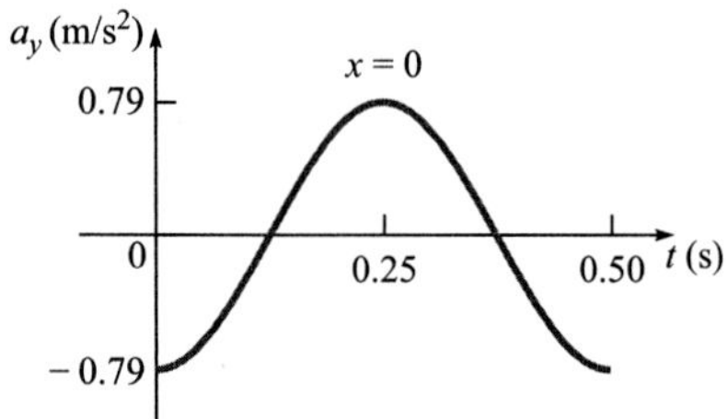

37. **Strategy** Compute the positions of the peaks of each pulse for the given times. Then use the principle of superposition to graph the shape of the cord for each time.

Solution

t (s)	Short Pulse Position	Tall Pulse Position
0.15	$10 \text{ cm} + (40 \text{ cm/s})(0.15 \text{ s}) = 16 \text{ cm}$	$30 \text{ cm} - (40 \text{ cm/s})(0.15 \text{ s}) = 24 \text{ cm}$
0.25	$10 \text{ cm} + (40 \text{ cm/s})(0.25 \text{ s}) = 20 \text{ cm}$	$30 \text{ cm} - (40 \text{ cm/s})(0.25 \text{ s}) = 20 \text{ cm}$
0.30	$10 \text{ cm} + (40 \text{ cm/s})(0.30 \text{ s}) = 22 \text{ cm}$	$30 \text{ cm} - (40 \text{ cm/s})(0.30 \text{ s}) = 18 \text{ cm}$

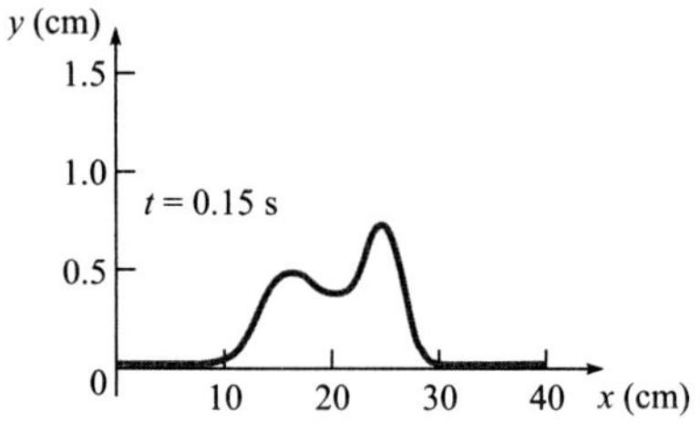 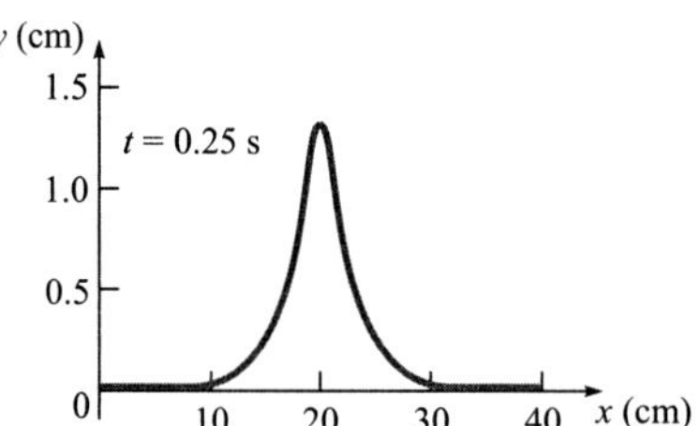

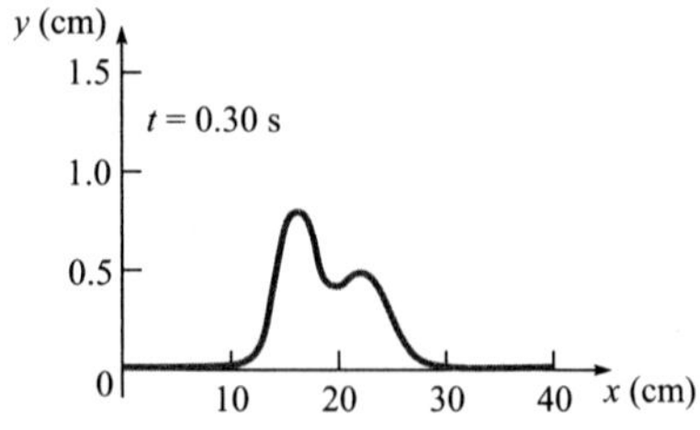

41. Strategy $f = v/\lambda$ and the frequency is the same in both mediums.

Solution Find the wavelength of the light in water.

$$\frac{v_a}{\lambda_a} = \frac{v_w}{\lambda_w}, \text{ so } \lambda_w = \frac{v_w}{v_a}\lambda_a = 0.750(0.500\times10^{-6} \text{ m}) = \boxed{375 \text{ nm}}.$$

45. Strategy Refer to the figure. Use $\Delta x = v_x \Delta t$.

Solution The pulse moves $1.80 \text{ m} - 1.50 \text{ m} = 0.30 \text{ m}$ in 0.20 s. So, the speed of the wave is

$v = \dfrac{0.30 \text{ m}}{0.20 \text{ s}} = 1.5 \text{ m/s}$. The pulse travels to the right until it reaches the endpoint; it is then reflected and inverted.

It then travels to the left until it hits the left endpoint; it is again reflected and inverted. The pulse travels to the right until it reaches $x = 1.5$ m. The total distance traveled is $2.5 \text{ m} + 4.0 \text{ m} + 1.5 \text{ m} = 8.0 \text{ m}$. The elapsed time is

$$t = \frac{x}{v} = \frac{8.0 \text{ m}}{1.5 \text{ m/s}} = \boxed{5.3 \text{ s}}.$$

47. Strategy The waves are coherent. The sound waves interfere destructively. Use the principle of superposition.

Solution Find the wavelength and half the wavelength.

$$\lambda = \frac{v}{f} = \frac{343 \text{ m/s}}{523 \text{ Hz}} = 0.6558 \text{ m and } \frac{\lambda}{2} = \frac{v}{2f} = \frac{343 \text{ m/s}}{2(523 \text{ Hz})} = 0.3279 \text{ m}.$$

If the waves from both speakers travel the same distance plus half a wavelength, they arrive 180° out of phase and interfere destructively. Therefore, the first possible distance between 2 m and 4 m is $2.28 \text{ m} + 0.3279 \text{ m} = 2.61 \text{ m}$. Adding integral multiples of the wavelength to this distance gives the other possibilities.

$2.608 \text{ m} + 0.6558 \text{ m} = 3.26 \text{ m}; 2.608 \text{ m} + 2\times0.6558 \text{ m} = 3.92 \text{ m}$

The possible distances from speaker #2, between 2 m and 4 m, are $\boxed{2.61 \text{ m}, 3.26 \text{ m}, \text{ and } 3.92 \text{ m}}$.

49. Strategy The waves are coherent. Use the principle of superposition. Intensity is proportional to the square of the amplitude.

Solution

(a) The resulting wave will have its largest amplitude if the waves interfere constructively. The amplitude is

$$A = A_1 + A_2 = 6.0 \text{ cm} + 3.0 \text{ cm} = \boxed{9.0 \text{ cm}}.$$

(b) The resulting wave will have its smallest amplitude if the waves interfere destructively. The amplitude is

$$A = |A_1 - A_2| = |6.0 \text{ cm} - 3.0 \text{ cm}| = \boxed{3.0 \text{ cm}}.$$

(c) Form a proportion.

$$\frac{I_1}{I_2} = \left(\frac{A_1}{A_2}\right)^2 = \left(\frac{9.0 \text{ cm}}{3.0 \text{ cm}}\right)^2 = \boxed{9.0}$$

51. Strategy Intensity is proportional to the amplitude squared. For destructive interference, the amplitude of the superposition is the absolute value of the difference of the original amplitudes.

Solution Find A_1/A_2.

$$\frac{A_1}{A_2} = \sqrt{\frac{I_1}{I_2}} = \sqrt{\frac{25}{28}}$$

Find the amplitude of the superposition.

$$A = |A_1 - A_2| = A_2\left(1 - \sqrt{\frac{25}{28}}\right)$$

Find the intensity of the superposition.

$$\sqrt{\frac{I}{I_2}} = \frac{A}{A_2} = 1 - \sqrt{\frac{25}{28}}, \text{ so } I = \left(1 - \sqrt{\frac{25}{28}}\right)^2 I_2 = \left(1 - \sqrt{\frac{25}{28}}\right)^2 (28\times10^{-3} \text{ W/m}^2) = \boxed{80 \text{ }\mu\text{W/m}^2}.$$

53. Strategy The intensity minimums imply that the distance $37.1 \text{ m} - 25.8 \text{ m} = 11.3 \text{ m}$ is equal to a whole number of wavelengths m plus one-half wavelength. Determine the number (or numbers) of wavelengths m that gives a frequency (or frequencies) between 100 Hz and 150 Hz. Use Eq. (11-6).

Solution The number of wavelengths m is related to the distance between intensity minimums by

$$\left(m + \frac{1}{2}\right)\lambda = 11.3 \text{ m}. \text{ In terms of frequency, we have } f = \frac{v}{\lambda} = \left(m + \frac{1}{2}\right)\frac{343 \text{ m/s}}{11.3 \text{ m}} = (30.35 \text{ Hz})m + 15.18 \text{ Hz}.$$

Substitute 100 Hz and 150 Hz for f and solve for m to find the range of possible values.
$(30.35 \text{ Hz})m + 15.18 \text{ Hz} = 100 \text{ Hz}$, so $m > 2.8$ and $(30.35 \text{ Hz})m + 15.18 \text{ Hz} = 150 \text{ Hz}$, so $m < 4.4$.
The two possible values of m are 3 and 4. Try them both.

$$f(3) = \left(3 + \frac{1}{2}\right)\frac{343 \text{ m/s}}{11.3 \text{ m}} = 106 \text{ Hz and } f(4) = \left(4 + \frac{1}{2}\right)\frac{343 \text{ m/s}}{11.3 \text{ m}} = 137 \text{ Hz}.$$

Both values are within the range of allowed frequencies, so the possible frequencies of the sound waves coming from the speakers are $\boxed{106 \text{ Hz and } 137 \text{ Hz}}$.

57. Strategy Nodes are separated by a distance of $\lambda/2$. Use Eq. (11-7).

Solution Find the distance between adjacent nodes.

$$\text{distance} = \frac{\lambda}{2} = \frac{2\pi}{2k} = \frac{\pi}{2.0\times10^2 \text{ rad/m}} = \boxed{0.016 \text{ m}}$$

61. Strategy The frequencies are given by $f_n = nv/(2L)$. The speed of the transverse waves is related to the tension by $v = \sqrt{T/\mu}$.

Solution

(a) Find the speed of the transverse waves.

$$\frac{v}{2L} = f_1, \text{ so } v = 2Lf_1 = 2(1.50 \text{ m})(450.0 \text{ Hz}) = \boxed{1350 \text{ m/s}}.$$

(b) Find the tension.

$$f_1 = \frac{v}{2L} = \frac{1}{2L}\sqrt{\frac{T}{\mu}}, \text{ so } T = 4\mu L^2 f_1^2 = 4(25.0\times10^{-6} \text{ kg/m})(1.50 \text{ m})^2(450.0 \text{ Hz})^2 = \boxed{45.6 \text{ N}}.$$

(c) The frequencies are the same for both mediums, but the wavelength depends upon the wave speed.

$$f = f_1 = 450.0 \text{ Hz, so } \lambda = \frac{v}{f} = \frac{340 \text{ m/s}}{450.0 \text{ Hz}} = 0.76 \text{ m.}$$

The wavelength and frequency are $\boxed{0.76 \text{ m and } 450.0 \text{ Hz}}$, respectively.

65. (a) Strategy and Solution All frequencies higher than the fundamental are integral multiples of the fundamental. Since there are no other frequencies between the two given, the fundamental is the difference between those two. Thus, the fundamental frequency is $1040 \text{ Hz} - 780 \text{ Hz} = \boxed{260 \text{ Hz}}$.

(b) Strategy Use Eqs. (11-2) and (11-13).

Solution Find the total mass of the string.

$$f_1 = \frac{v}{2L} = \frac{1}{2L}\sqrt{\frac{FL}{m}} = \sqrt{\frac{F}{4mL}}, \text{ so } m = \frac{F}{4f_1^2 L} = \frac{1200 \text{ N}}{4(260 \text{ Hz})^2(1.6 \text{ m})} = \boxed{2.8 \text{ g}}.$$

69. Strategy According to Eq. (11-13), the frequency of the string is inversely proportional to the length of the string.

Solution Form a proportion.

$$f \propto \frac{1}{L}, \text{ so } \frac{L_2}{L_1} = \frac{f_1}{f_2}.$$

Relate the distance between frets to the frequencies and the length of the string.

$$\Delta L = L_1 - L_2 = L_1 - \frac{f_1}{f_2}L_1 = L_1\left(1 - \frac{f_1}{f_2}\right)$$

First fret: $\Delta L = (64.8 \text{ cm})\left(1 - \dfrac{1}{1.0595}\right) = \boxed{3.64 \text{ cm}}$

Second fret: $3.64 \text{ cm} + (64.8 \text{ cm} - 3.64 \text{ cm})\left(1 - \dfrac{1}{1.0595}\right) = \boxed{7.07 \text{ cm}}$

Third fret: $7.074 \text{ cm} + (64.8 \text{ cm} - 7.074 \text{ cm})\left(1 - \dfrac{1}{1.0595}\right) = \boxed{10.32 \text{ cm}}$

73. Strategy Destructive interference occurs when the path length difference of the two sound waves is an odd multiple of half of the wavelength.

Solution The wavelength is $\lambda = v/f = (340 \text{ m/s})/(680 \text{ Hz}) = 0.50 \text{ m}$. The largest possible path length difference is equal to the distance between the speakers, 1.5 m. $\lambda/2 = 0.25$ m, so the path length differences that cause destructive interference are 0.25 m, 0.75 m, and 1.25 m. Let the speakers lie along the x-axis at $x = \pm 0.75$ m. Then the path length difference is zero along the y-axis and 1.5 m along the x-axis. As the listener walks along the circle of radius 1 m, the path length difference varies from 0 to 1.5 m. The path length difference equals 0.25 m, 0.75 m, and 1.25 m once for each quadrant of the circle (three occurrences of destructive interference). There are four quadrants, so the listener observes destructive interference at $\boxed{12}$ points along the circle.

77. Strategy The fundamental frequency depends on the wavelength of the fundamental and the wave speed. The wavelength is determined by the length of the wire. The wave speed depends on the tension and the linear mass density. So we need to find out how those three quantities—length, tension, linear mass density—change when the wire is cut in half and then find the new fundamental frequency. The only numerical value we know is the original frequency, so we work by proportions. According to Eq. (11-4), the speed of waves on a wire is directly proportional to the square root of the tension. According to Eq. (11-13), the frequency of the waves on a wire is directly proportional to the speed of the waves. Therefore, the frequency is directly proportional to the square root of the tension in a wire. Use Newton's second law.

Solution For a wire fixed at both ends, the wavelength of the fundamental is $2L$. When the wire is cut in half, the wavelength is cut in half: $\lambda_f = \lambda_i/2$. The linear mass density does not change—half the length and half the mass—but the tension does—two wires supporting the sign instead of one. For the single wire supporting the sign (see the FBD): $\Sigma F_y = T_1 - mg = 0$, so $T_1 = mg$.

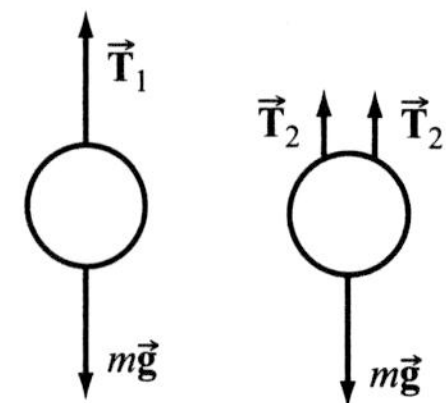

For the two wires: $\Sigma F_y = 2T_2 - mg = 0$, so $T_2 = mg/2 = T_1/2$.

The frequency is $f = v/\lambda$ and the wave speed is $v = \sqrt{F/\mu}$, so $f = \lambda^{-1}\sqrt{F/\mu}$.

Form a proportion.

$$\frac{f_2}{f_1} = \frac{\lambda_1}{\lambda_2}\sqrt{\frac{T_2}{T_1}\frac{\mu_1}{\mu_2}} = 2\sqrt{\frac{1}{2}} \times 1 = \sqrt{2}, \text{ so } f_2 = \sqrt{2}f_1 = \sqrt{2} \times 660 \text{ Hz} = \boxed{930 \text{ Hz}}.$$

81. Strategy Refer to Problem 80. The function is $y = A[\sin(kx - \omega t) + \sin(kx + \omega t)]$. Use the trigonometric identity $\sin\alpha + \sin\beta = 2\sin[(\alpha+\beta)/2]\cos[(\alpha-\beta)/2]$ and the principle of superposition.

Solution Use the identity.
$$y = A[\sin(kx - \omega t) + \sin(kx + \omega t)] = A\sin(\omega t - kx) + A\sin(\omega t + kx)$$
$$= 2A\sin\left(\frac{kx - \omega t + kx + \omega t}{2}\right)\cos\left[\frac{kx - \omega t - (kx + \omega t)}{2}\right] = 2A\sin(kx)\cos(-\omega t) = [2A\cos(\omega t)]\sin(kx) = A'\sin(kx)$$

Therefore, $A' = 2A\cos(\omega t)$. Prove this holds for each of the amplitudes of the graphs from Problem 80.
$$y(x,t) = (5.0 \text{ cm})\{\sin[(\pi/5.0 \text{ rad/cm})x - (\pi/6.0 \text{ rad/s})t] + \sin[(\pi/5.0 \text{ rad/cm})x + (\pi/6.0 \text{ rad/s})t]\}$$
$$= 2(5.0 \text{ cm})\sin[(\pi/5.0 \text{ rad/cm})x]\cos[(\pi/6.0 \text{ rad/s})t]$$
$$= (10 \text{ cm})\sin[(\pi/5.0 \text{ rad/cm})x]\cos[(\pi/6.0 \text{ rad/s})t]$$
So the maximum amplitude is 10 cm for $x = 2.5$ cm.
At $t = 1.0$ s,
$$y(2.5 \text{ cm}, 1.0 \text{ s}) = (10 \text{ cm})\sin[(\pi/5.0 \text{ rad/cm})(2.5 \text{ cm})]\cos[(\pi/6.0 \text{ rad/s})(1.0 \text{ s})] = (10 \text{ cm})(1)\cos(\pi/6.0) = 8.7 \text{ cm}.$$
At $t = 2.0$ s,
$$y(2.5 \text{ cm}, 2.0 \text{ s}) = (10 \text{ cm})\sin[(\pi/5.0 \text{ rad/cm})(2.5 \text{ cm})]\cos[(\pi/6.0 \text{ rad/s})(2.0 \text{ s})] = (10 \text{ cm})(1)\cos(\pi/3.0) = 5.0 \text{ cm}.$$
The values correspond nicely with those shown in the graphs in the solution for Problem 80.
Using the original function, we have
$$y(2.5 \text{ cm}, 1.0 \text{ s}) = (5.0 \text{ cm})\{\sin[(\pi/5.0 \text{ rad/cm})(2.5 \text{ cm}) - (\pi/6.0 \text{ rad/s})(1.0 \text{ s})]$$
$$+ \sin[(\pi/5.0 \text{ rad/cm})(2.5 \text{ cm}) + (\pi/6.0 \text{ rad/s})(1.0 \text{ s})]\} = 8.7 \text{ cm}$$
and
$$y(2.5 \text{ cm}, 2.0 \text{ s}) = (5.0 \text{ cm})\{\sin[(\pi/5.0 \text{ rad/cm})(2.5 \text{ cm}) - (\pi/6.0 \text{ rad/s})(2.0 \text{ s})]$$
$$+ \sin[(\pi/5.0 \text{ rad/cm})(2.5 \text{ cm}) + (\pi/6.0 \text{ rad/s})(2.0 \text{ s})]\} = 5.0 \text{ cm}.$$

Therefore, the amplitudes of the graphs of Problem 80 satisfy the equation $A' = 2A\cos(\omega t)$, where A' is the amplitude of the wave plotted and A is 5.0 cm.

83. Strategy Use dimensional analysis.

Solution λ has units m. g has units m/s^2. $\lambda \cdot g$ has units m^2/s^2. $\sqrt{\lambda g}$ has units m/s. So, $\boxed{v \propto \sqrt{\lambda g}}$.

85. Strategy The position of the particle will follow the shape of the wave. The velocity is a bit more complicated. As the wave passes the point under consideration, the particle moves upward rapidly until it reaches the "top" of the wave. When it reaches the top, its velocity is instantaneously zero. The point then moves downward at an average velocity less than that with which it rose. The velocity is equal to the slope of the position graph for any time t.

Solution The plot of the position as a function of time:

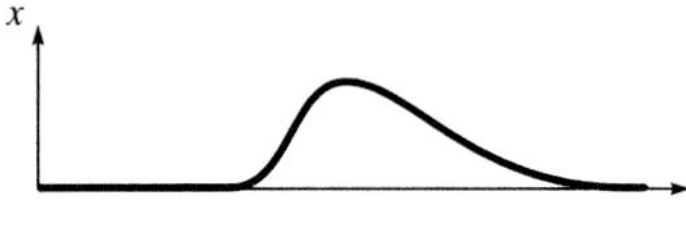

The plot of the velocity as a function of time:

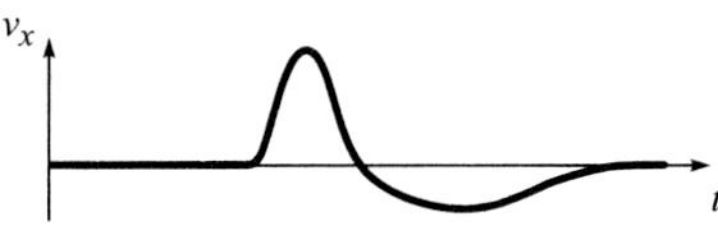

87. Strategy Use the principle of superposition.

Solution $\Delta x = 1.80 \text{ m} - 1.50 \text{ m} = 0.30 \text{ m}$ in $\Delta t = 0.20 \text{ s}$, so $v = 0.30 \text{ m}/(0.20 \text{ s}) = 1.5 \text{ m}/\text{s}$.

Find the position of the peak at $t = 1.6$ s.

$x_{\text{peak}} = x_{\text{i}} + vt = 1.5 \text{ m} + (1.5 \text{ m}/\text{s})(1.6 \text{ s}) = 3.9 \text{ m}$

The peak of the pulse is nearly to the end of the string. The reflected pulse is below the string, so most of the height of the original pulse is cancelled.

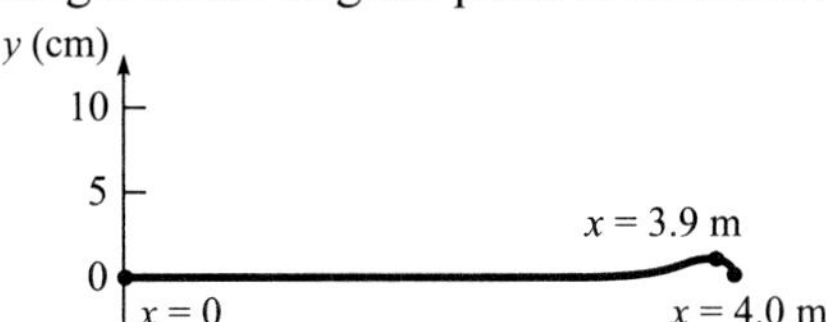

Chapter 12

SOUND

Conceptual Questions

1. The wavelength of the standing waves inside a bassoon is determined by the length of its air chamber. With a fixed fundamental wavelength, the frequency of these waves depends only upon the speed of sound in air, which itself depends significantly upon the air temperature. As a result of thermal contraction and expansion, changes in air temperature also affect the frequency of waves generated on a cello string—albeit much less significantly than for the bassoon.

5. For high-frequency sounds the wavelength is relatively small. The phase difference of the sounds arriving at each ear is then very sensitive to small variations in head position, wind speed, and several other factors. This makes the phase difference method unreliable for high-frequency sounds.

9. The maximum loudness of a stereo is proportional to the maximum intensity level that it can produce. Doubling the power of the stereo's amplifier also doubles its intensity, but it increases the intensity level by only 3 dB.

13. Doubling the pressure amplitude of a sound wave doubles the displacement amplitude, quadruples the intensity, and increases the intensity level by 6 dB.

17. Although the fundamental frequency of the highest note on a piano is approximately 4 kHz, the harmonics of that note occur at higher frequencies. An instrument's timbre is determined by the presence or absence of these harmonics. High quality audio equipment must therefore be able to reproduce frequencies up to the limit of the human audible range in order to faithfully reproduce the sound of an instrument.

Problems

1. **Strategy** The speed of sound in air at 273.15 K (0°C) is 331 m/s. Use Eq. (12-3).

   Solution Find the speed of sound in air at 56.7°C.
   $$v = v_0\sqrt{\frac{T}{T_0}} = (331 \text{ m/s})\sqrt{\frac{273.15 \text{ K} + 56.7 \text{ K}}{273.15}} = \boxed{364 \text{ m/s}}$$

5. **Strategy** Use Eq. (11-6).

   Solution Find the wavelength of the ultrasonic waves.
   $$\lambda = \frac{v}{f} = \frac{1533 \text{ m/s}}{2.5\times10^5 \text{ Hz}} = \boxed{6.1 \text{ mm}}$$

7. **(a) Strategy** Use Eqs. (12-2) and (12-3).

   Solution Compute the speed of sound at $T = 12°C$.

   $$v = v_0\sqrt{\frac{T}{T_0}} = (331 \text{ m/s})\sqrt{\frac{273.15 \text{ K} + 12 \text{ K}}{273.15 \text{ K}}} = \boxed{338 \text{ m/s}}$$

   (b) Strategy The speed of light is so much faster than the speed of sound in air that the time it takes for light to reach an observer is negligible in this case. Use $\Delta x = v\Delta t$.

   Solution Find the distance to the lightning strike.

   $$\Delta x = v\Delta t = (338 \text{ m/s})(8.2 \text{ s}) = \boxed{2.8 \text{ km}}$$

9. **Strategy** Use Eq. (12-5).

   Solution Find the speed of sound in the thin rod.

   $$v = \sqrt{\frac{Y}{\rho}} = \sqrt{\frac{1.1\times10^{11} \text{ Pa}}{8.92\times10^3 \text{ kg/m}^3}} = \boxed{3.5 \text{ km/s}}$$

   The copper alloy has a slightly lower speed of sound than that listed in Table 12.1 for copper, 3560 m/s.

13. **Strategy** Let the distance to the train from Stan and Ollie be d. Then, $d = v_{\text{steel}}\Delta t_{\text{steel}} = v_{\text{air}}\Delta t_{\text{air}}$. Also, $\Delta t_{\text{air}} = \Delta t_{\text{steel}} + 2.1 \text{ s}$.

    Solution Find Δt_{steel}.

    $$v_{\text{steel}}\Delta t_{\text{steel}} = v_{\text{air}}\Delta t_{\text{air}} = v_{\text{air}}(\Delta t_{\text{steel}} + 2.1 \text{ s}) = v_{\text{air}}\Delta t_{\text{steel}} + v_{\text{air}}(2.1 \text{ s}), \text{ so } \Delta t_{\text{steel}} = \frac{v_{\text{air}}(2.1 \text{ s})}{v_{\text{steel}} - v_{\text{air}}} = \frac{2.1 \text{ s}}{\frac{v_{\text{steel}}}{v_{\text{air}}} - 1}.$$

    Compute d.

    $$d = v_{\text{steel}}\Delta t_{\text{steel}} = (5790 \text{ m/s})\frac{2.1 \text{ s}}{\frac{5790}{343} - 1} = \boxed{770 \text{ m}}$$

17. **Strategy** Use Eq. (12-9). Let $\beta_1 = 100.0$ dB.

    Solution Find the reduced intensity level, β_2.

    $$\beta_2 - \beta_1 = (10 \text{ dB})\log_{10}\frac{I_2}{I_1}, \text{ so } \beta_2 = \beta_1 + (10 \text{ dB})\log_{10}\frac{I_2}{I_1} = 100.0 \text{ dB} + (10 \text{ dB})\log_{10}\frac{1}{2.0} = \boxed{97.0 \text{ dB}}.$$

21. **Strategy** Intensity is power per unit area. Power is the rate at which energy is produced. Assume the loudspeaker to be an isotropic source of sound waves. Use Eq. (12-8).

    Solution Solve for the intensity.

    $$\beta = (10 \text{ dB})\log\frac{I}{I_0}$$

    $$10^{\frac{\beta}{10 \text{ dB}}} = \frac{I}{I_0}$$

    $$I = I_0 10^{\frac{\beta}{10 \text{ dB}}}$$

    Find the rate at which sound energy is produced by the loudspeaker.

    $$P = IA = I_0 10^{\frac{\beta}{10 \text{ dB}}}(4\pi r^2) = 4\pi(10^{-12} \text{ W/m}^2)10^{\frac{71 \text{ dB}}{10 \text{ dB}}}(25 \text{ m})^2 = \boxed{0.099 \text{ W}}$$

23. Strategy Solve for the intensity in Eq. (12-8). The sound is incoherent, so add the eight intensities; then solve for the combined intensity level of the eight cars at point P.

Solution Solve for the intensity.

$$\beta = (10\ \mathrm{dB})\log_{10}\frac{I}{I_0}, \quad \text{so } 10^{\frac{\beta}{10\ \mathrm{dB}}} = \frac{I}{I_0} \quad \text{and } I = I_0 10^{\frac{\beta}{10\ \mathrm{dB}}}.$$

Find the combined intensity level.

$$I_{\text{total}} = I_1 + \cdots + I_8 = I_0 10^{\frac{\beta_1}{10\ \mathrm{dB}}} + \cdots + I_0 10^{\frac{\beta_8}{10\ \mathrm{dB}}},\ \text{so } \frac{I_{\text{total}}}{I_0} = 10^{\frac{\beta_1}{10\ \mathrm{dB}}} + \cdots + 10^{\frac{\beta_8}{10\ \mathrm{dB}}}.\ \text{Thus,}$$

$$\beta_{\text{total}} = (10\ \mathrm{dB})\log_{10}\frac{I_{\text{total}}}{I_0} = (10\ \mathrm{dB})\log_{10}\left(10^{\frac{\beta_1}{10\ \mathrm{dB}}} + \cdots + 10^{\frac{\beta_8}{10\ \mathrm{dB}}}\right) = (10\ \mathrm{dB})\log_{10}\left(10^{\frac{98.0\ \mathrm{dB}}{10\ \mathrm{dB}}} + \cdots + 10^{\frac{98.0\ \mathrm{dB}}{10\ \mathrm{dB}}}\right)$$

$$= \boxed{107\ \mathrm{dB}}.$$

25. Strategy Use Eq. (12-8) to find an expression for the intensity. Then use each relationship given for the intensities to obtain the relationships for the intensity levels.

Solution Solve for the intensity.

$$\beta = (10\ \mathrm{dB})\log\frac{I}{I_0}$$

$$10^{\frac{\beta}{10\ \mathrm{dB}}} = \frac{I}{I_0}$$

$$I = I_0 10^{\frac{\beta}{10\ \mathrm{dB}}}$$

(a) Show that if $I_2 = 10.0 I_1$, $\beta_2 = \beta_1 + 10.0\ \mathrm{dB}$.

$$I_2 = 10.0 I_1$$

$$I_0 10^{\frac{\beta_2}{10\ \mathrm{dB}}} = 10.0 I_0 10^{\frac{\beta_1}{10\ \mathrm{dB}}}$$

$$10^{\frac{\beta_2}{10\ \mathrm{dB}}} = (10.0)10^{\frac{\beta_1}{10\ \mathrm{dB}}}$$

$$\log 10^{\frac{\beta_2}{10\ \mathrm{dB}}} = \log\left[(10.0)10^{\frac{\beta_1}{10\ \mathrm{dB}}}\right] = \log 10.0 + \log 10^{\frac{\beta_1}{10\ \mathrm{dB}}}$$

$$\frac{\beta_2}{10\ \mathrm{dB}} = 1.00 + \frac{\beta_1}{10\ \mathrm{dB}}$$

$$\beta_2 = \beta_1 + 10.0\ \mathrm{dB}$$

(b) Show that if $I_2 = 2.0 I_1$, $\beta_2 = \beta_1 + 3.0\ \mathrm{dB}$.

$$I_2 = 2.0 I_1$$

$$I_0 10^{\frac{\beta_2}{10\ \mathrm{dB}}} = 2.0 I_0 10^{\frac{\beta_1}{10\ \mathrm{dB}}}$$

$$10^{\frac{\beta_2}{10\ \mathrm{dB}}} = (2.0)10^{\frac{\beta_1}{10\ \mathrm{dB}}}$$

$$\log 10^{\frac{\beta_2}{10\ \mathrm{dB}}} = \log\left[(2.0)10^{\frac{\beta_1}{10\ \mathrm{dB}}}\right] = \log 2.0 + \log 10^{\frac{\beta_1}{10\ \mathrm{dB}}}$$

$$\frac{\beta_2}{10\ \mathrm{dB}} = 0.30 + \frac{\beta_1}{10\ \mathrm{dB}}$$

$$\beta_2 = \beta_1 + 3.0\ \mathrm{dB}$$

29. (a) Strategy $f_n = nv/(4L)$ for a pipe closed at one end and $v = 343$ m/s for $T = 20.0°C$.

Solution Find the length of the organ pipe.

$$f_1 = \frac{v}{4L}, \text{ so } L = \frac{v}{4f_1} = \frac{343 \text{ m/s}}{4(261.5 \text{ Hz})} = \boxed{32.8 \text{ cm}}$$

(b) Strategy The frequency of the organ pipe is proportional to the speed of the waves and the speed is proportional to the square root of temperature, so $f \propto v \propto \sqrt{T}$.

Solution Find the fundamental frequency after the temperature drop.

$$f_{0.0°} = f_{20°}\sqrt{\frac{T_{0.0°}}{T_{20°}}} = (261.5 \text{ Hz})\sqrt{\frac{273.15 \text{ K} + 0.0 \text{ K}}{273.15 \text{ K} + 20.0 \text{ K}}} = \boxed{252.4 \text{ Hz}}$$

31. Strategy The frequency of the organ pipe is proportional to the speed of the waves and the speed is proportional to the square root of temperature, so $f \propto v \propto \sqrt{T}$.

Solution Find the fundamental frequency after the temperature increase.

$$f_{20.0°} = f_{0.0°}\sqrt{\frac{T_{20.0°}}{T_{0.0°}}} = (382 \text{ Hz})\sqrt{\frac{273.15 \text{ K} + 20.0 \text{ K}}{273.15 \text{ K} + 0.0 \text{ K}}} = \boxed{396 \text{ Hz}}$$

33. Strategy $f_n = nv/(2L)$ for a pipe open at both ends and $v = v_0\sqrt{T/T_0}$.

Solution Find an expression for the temperature in terms of n.

$$f_n = \frac{nv}{2L} = \frac{nv_0}{2L}\sqrt{\frac{T}{T_0}}, \text{ so } T = T_0\left(\frac{2Lf_n}{nv_0}\right)^2 = \frac{273.15 \text{ K}}{n^2}\left[\frac{2(2.0 \text{ m})(702 \text{ Hz})}{331 \text{ m/s}}\right]^2 = \frac{19{,}658 \text{ K}}{n^2}.$$

The assumed temperature range is 293 K (20°C) to 308 K (35°C). We need to find n such that T falls within this range. By trial and error, n is found to be 8. So, $T = 19{,}658 \text{ K}/8^2 = 307 \text{ K} = \boxed{34°C}$.

37. (a) Strategy Use Eq. (12-11).

Solution Find the frequency of Brian's string.

$$f_{\text{beat}} = \Delta f = f_2 - f_1, \text{ so } f_1 = f_2 - \Delta f = 440.0 \text{ Hz} - 2.0 \text{ Hz} = \boxed{438.0 \text{ Hz}}.$$

(b) Strategy and Solution Since the frequency of the string is directly proportional to the square root of its tension. Brian needs to $\boxed{\text{tighten}}$ his A string to increase its frequency.

41. Strategy The third harmonic of the cello string has a frequency of three times the fundamental. Use Eq. (12-11).

Solution Find the beat frequency between the cello and the violin.

$$f_{\text{beat}} = \Delta f = 3(65.40 \text{ Hz}) - 196.0 \text{ Hz} = \boxed{0.2 \text{ Hz}}$$

45. Strategy Since the source is moving and the observer is stationary, use Eq. (12-12).

Solution Compute the frequencies of the sound received by the stationary observer.

(a) The source is moving toward the stationary observer ($v_s > 0$).
$$f_o = \frac{f_s}{1 - \frac{v_s}{v}} = \frac{1.0\ \text{kHz}}{1 - 0.50} = \boxed{2.0\ \text{kHz}}$$

(b) The source is now moving away ($v_s < 0$).
$$f_o = \frac{1.0\ \text{kHz}}{1 + 0.50} = \boxed{670\ \text{Hz}}$$

49. Strategy First treat the cell as the observer moving toward a source; then treat the cell as a source moving toward an observer. Use Eqs. (12-12) and (12-13).

Solution Cell as observer moving toward a source ($v_o < 0$):
$$f_1 = \left(1 - \frac{v_o}{v_{\text{sound}}}\right) f_s = \left(1 + \frac{v}{u}\right) f$$

Cell as source moving toward an observer ($v_s > 0$):
$$f_o = \frac{f_s}{1 - \frac{v_s}{v_{\text{sound}}}} = f_r = \frac{f_1}{1 - \frac{v}{u}} = \frac{\left(1 + \frac{v}{u}\right) f}{1 - \frac{v}{u}} = \frac{u}{u} \times \frac{\left(1 + \frac{v}{u}\right) f}{1 - \frac{v}{u}} = f \frac{u + v}{u - v}$$

53. Strategy The distance traveled (round trip) by the sound wave in time Δt is $v\Delta t$. The depth d of the lake is half this distance.

Solution Find the depth of the lake.
$$d = \frac{1}{2} v\Delta t = \frac{1}{2}(0.540\ \text{s})(1493\ \text{m/s}) = \boxed{403\ \text{m}}$$

57. Strategy Use the result of Problem 49 for the frequency of reflected waves during angiodynography and Eq. (12-11).

Solution Find the beat frequency.
$$f_{\text{beat}} = \Delta f = f_r - f = \frac{1 + \frac{v_{\text{cell}}}{v_{\text{sound}}}}{1 - \frac{v_{\text{cell}}}{v_{\text{sound}}}} f - f = (5.0 \times 10^6\ \text{Hz}) \left(\frac{1 + \frac{0.10\ \text{m/s}}{1570\ \text{m/s}}}{1 - \frac{0.10\ \text{m/s}}{1570\ \text{m/s}}} - 1\right) = \boxed{640\ \text{Hz}}$$

61. (a) Strategy The fundamental frequency produced by the chimney when both of its ends are open is $(261.6\ \text{Hz})(1/2)^3 = 32.70\ \text{Hz}$. Use Eq. (11-13).

Solution Find the height of the chimney.
$$f_1 = \frac{v}{2L}, \text{ so } L = \frac{v}{2f_1} = \frac{330\ \text{m/s}}{2(32.70\ \text{Hz})} = \boxed{5.05\ \text{m}}.$$

(b) Strategy Use Eq. (12-10b).

Solution Find the fundamental frequency produced by the chimney when its bottom is closed.
$$f_1 = \frac{v}{4L} = \frac{330\ \text{m/s}}{4} \left[\frac{330\ \text{m/s}}{2(32.70\ \text{Hz})}\right]^{-1} = \frac{32.70\ \text{Hz}}{2} = \boxed{16.35\ \text{Hz}}$$

65. Strategy Find the time is takes for Kyle's voice to travel the 5.00 m to the surface of the water. Then, use this time, the given time of 0.0210 s, and the speed of sound in water to find Rob's depth below the boat.

Solution Find Δt_{air}.

$d_{\text{air}} = v_{\text{air}} \Delta t_{\text{air}}$, so $\Delta t_{\text{air}} = \dfrac{d_{\text{air}}}{v_{\text{air}}}$. The time of travel of the sound in the water is $\Delta t_{\text{water}} = 0.0210 \text{ s} - \Delta t_{\text{air}}$.

Find Rob's depth below the boat, d_{water}.

$$d_{\text{water}} = v_{\text{water}} \Delta t_{\text{water}} = v_{\text{water}}(0.0210 \text{ s} - \Delta t_{\text{air}}) = v_{\text{water}}\left(0.0210 \text{ s} - \frac{d_{\text{air}}}{v_{\text{air}}}\right) = (1533 \text{ m/s})\left(0.0210 \text{ s} - \frac{5.00 \text{ m}}{343 \text{ m/s}}\right)$$

$$= \boxed{9.8 \text{ m}}$$

69. Strategy Use Eq. (11-6).

Solution Compute the frequency of the sound wave.

$$f = \frac{v}{\lambda} = \frac{343 \text{ m/s}}{0.15 \text{ m}} = \boxed{2.3 \text{ kHz}}$$

73. Strategy The distance traveled (round trip) by the sound pulse in time Δt is $v\Delta t$. The distance to the ocean floor is half this distance.

Solution Find the elapsed time between an emitted pulse and the return of its echo at the correct depth d.

$$2d = v\Delta t, \text{ so } \Delta t = \frac{2d}{v} = \frac{2(40.0 \text{ fathoms})\left(1.83 \; \frac{\text{m}}{\text{fathom}}\right)}{1533 \text{ m/s}} = \boxed{0.0955 \text{ s}}.$$

75. Strategy Find the greatest common factor of the frequencies to determine the fundamental. Use Eq. (11-6).

Solution Factor the numerical value of each frequency into its prime factors.
$36 = 2\times2\times3\times3; \; 60 = 2\times2\times3\times5; \; 84 = 2\times2\times3\times7$
The factors in common are 2, 2, and 3. The greatest common factor is $2\times2\times3 = 12$, so the fundamental frequency is 12 Hz. Compute the wavelength of the wave.

$$\lambda = \frac{v}{f} = \frac{180 \text{ m/s}}{12 \text{ Hz}} = \boxed{15 \text{ m}}$$

77. Strategy Since the combined sound intensity is incoherent, the sound intensity of one of the violins is 1/8 of the combined total ($I_2 = I_1/8$). Use Eq. (12-9).

Solution Find the sound intensity level for one violin.

$$\beta_2 - \beta_1 = (10 \text{ dB})\log\frac{I_2}{I_1} = (10 \text{ dB})\log\frac{1}{8} = -(10 \text{ dB})\log 8, \text{ so}$$

$$\beta_2 = \beta_1 - (10 \text{ dB})\log 8 = 38.0 \text{ dB} - (10 \text{ dB})\log 8 = \boxed{29.0 \text{ dB}}.$$

REVIEW AND SYNTHESIS: CHAPTERS 9–12

Review Exercises

1. **Strategy** The magnitude of the buoyant force on an object in water is equal to the weight of the water displaced by the object.

   Solution

   (a) Lead is much denser than aluminum, so for the same mass, its volume is much less. Therefore, aluminum has the larger buoyant force acting on it; since it is less dense it occupies more volume.

   (b) Steel is denser than wood. Even though the wood is floating, it displaces more water than does the steel. Therefore, wood has the larger buoyant force acting on it; since it displaces more water than the steel.

   (c) Lead: $\rho_w g V_{Pb} = \rho_w g \dfrac{m_{Pb}}{\rho_{Pb}} = (1.00 \times 10^3 \ \text{kg/m}^3)(9.80 \ \text{m/s}^2)\dfrac{1.0 \ \text{kg}}{11{,}300 \ \text{kg/m}^3} = \boxed{0.87 \ \text{N}}$

   Aluminum: $\rho_w g V_{Al} = \rho_w g \dfrac{m_{Al}}{\rho_{Al}} = (1.00 \times 10^3 \ \text{kg/m}^3)(9.80 \ \text{m/s}^2)\dfrac{1.0 \ \text{kg}}{2702 \ \text{kg/m}^3} = \boxed{3.6 \ \text{N}}$

   Steel: $\rho_w g V_{Steel} = \rho_w g \dfrac{m_{Steel}}{\rho_{Steel}} = (1.00 \times 10^3 \ \text{kg/m}^3)(9.80 \ \text{m/s}^2)\dfrac{1.0 \ \text{kg}}{7860 \ \text{kg/m}^3} = \boxed{1.2 \ \text{N}}$

   Wood: $mg = (1.0 \ \text{kg})(9.80 \ \text{m/s}^2) = \boxed{9.8 \ \text{N}}$ (Since the wood is floating, the buoyant force is equal to its weight.)

5. **Strategy** Use Eqs. (10-20a), (10-21), and (10-22), and Newton's second law.

   Solution The normal force on the 1.0-kg block m_1 is $N = m_1 g$. So, the force of friction on m_1 is $f = \mu N = \mu m_1 g$.

   Find the maximum acceleration that the top block can experience before it starts to slip.

   $\Sigma F = f = m_1 a$, so $a = \dfrac{f}{m_1} = \dfrac{\mu m_1 g}{m_1} = \mu g.$

   For SHM, the maximum acceleration is $a_m = \omega^2 A = \dfrac{k}{m} A$, which in this case is equal to

   $\dfrac{kA}{m_1 + m_2}$. Equate the accelerations and solve for A.

   $\dfrac{kA}{m_1 + m_2} = \mu g$, so $A = \dfrac{\mu(m_1 + m_2)g}{k}.$

   The maximum speed is $v_m = \omega A$. Compute the maximum speed that this set of blocks can have without the top block slipping.

   $v_m = \omega A = \sqrt{\dfrac{k}{m_1 + m_2}}\left[\dfrac{\mu(m_1 + m_2)g}{k}\right] = \mu g\sqrt{\dfrac{m_1 + m_2}{k}} = 0.45(9.80 \ \text{m/s}^2)\sqrt{\dfrac{1.0 \ \text{kg} + 5.0 \ \text{kg}}{150 \ \text{N/m}}} = \boxed{0.88 \ \text{m/s}}$

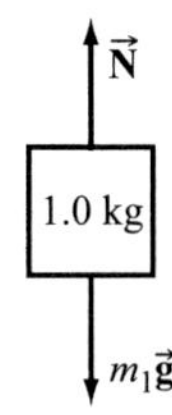

9. (a) Strategy Use Eqs. (11-2) and (11-13).

Solution Find the tension in the guitar string.

$$f_1 = \frac{v}{2L} = \frac{1}{2L}\sqrt{\frac{FL}{m}} = \sqrt{\frac{F}{4Lm}}, \text{ so } F = 4Lmf_1^2 = 4(0.655 \text{ m})(0.00331 \text{ kg})(82 \text{ Hz})^2 = \boxed{58 \text{ N}}.$$

(b) Strategy Use Eqs. (11-4) and (11-13).

Solution Find the length of the lowest frequency string when it is fingered at the fifth fret.

$$f_1 = \frac{v}{2L} = \frac{1}{2L}\sqrt{\frac{F}{\mu}}, \text{ so } L = \frac{1}{2f_1}\sqrt{\frac{F}{\mu}} = \frac{1}{2(110 \text{ Hz})}\sqrt{\frac{58 \text{ N}}{0.00331 \text{ kg}/(0.655 \text{ m})}} = \boxed{49 \text{ cm}}.$$

13. Strategy The frequency of the sound is increased by a factor equal to the number of holes in the disk. Use Eq. (11-6).

Solution The frequency of the sound is

$$f = 25(60.0 \text{ Hz}) = \boxed{1500 \text{ Hz}}.$$

Compute the wavelength that corresponds to this frequency.

$$\lambda = \frac{v}{f} = \frac{343 \text{ m/s}}{1.50 \times 10^3 \text{ Hz}} = \boxed{22.9 \text{ cm}}$$

17. Strategy The source is moving in the direction of propagation of the sound, so $v_s > 0$. The observer is moving in the direction opposite the propagation of the sound, so $v_o < 0$. Use Eq. (12-14).

Solution Find the frequency heard by the passenger in the oncoming boat.

$$f_o = \frac{v - v_o}{v - v_s} f_s = \frac{343 \text{ m/s} - (-15.6 \text{ m/s})}{343 \text{ m/s} - 20.1 \text{ m/s}} (312 \text{ Hz}) = \boxed{346 \text{ Hz}}$$

21. Strategy Use the equations describing waves on a string fixed at both ends.

Solution

(a) The wavelength of the fundamental mode is $\lambda = 2L = 2(0.640 \text{ m}) = \boxed{1.28 \text{ m}}$.

(b) The wave speed on the string is $v = \lambda f = (1.28 \text{ m})(110.0 \text{ Hz}) = \boxed{141 \text{ m/s}}$.

(c) Find the linear mass density of the string.

$$v = \sqrt{\frac{F}{\mu}}, \text{ so } \mu = \frac{F}{v^2} = \frac{133 \text{ N}}{(140.8 \text{ m/s})^2} = \boxed{6.71 \text{ g/m}}.$$

(d) Find the maximum speed of a point on the string.

$$v = \omega A = 2\pi f A = 2\pi(110.0 \text{ Hz})(0.00230 \text{ m}) = \boxed{1.59 \text{ m/s}}$$

(e) The frequency in air is that same as that in the bridge and body of the guitar, which is the same as the frequency of the string: $\boxed{110.0 \text{ Hz}}$.

(f) Find the speed of sound in air.

$$v = 331 \text{ m/s} + (0.60 \text{ m/s})(20) = 343 \text{ m/s}$$

Find the wavelength of the sound wave in air.

$$\lambda = \frac{v}{f} = \frac{343 \text{ m/s}}{110.0 \text{ Hz}} = \boxed{3.12 \text{ m}}$$

25. Strategy Use Newton's second law and Hooke's law.

Solution

(a) Find the tension in the string when hanging straight down.

$\Sigma F = T_1 - mg = 0$, so $T_1 = mg$.

Find the tension in the string while it is swinging.

$\Sigma F_y = T_2 \cos\theta - mg = 0$, so $T_2 = \dfrac{mg}{\cos\theta}$.

Find the stretch of the string.

$Y = \dfrac{\Delta T/A}{\Delta L/L}$, so

$$\Delta L = \frac{L\Delta T}{AY} = \frac{Lmg}{AY}\left(\frac{1}{\cos\theta} - 1\right) = \frac{(2.200 \text{ m})(0.411 \text{ kg})(9.8 \text{ m/s}^2)}{\pi(0.00125 \text{ m})^2(4.00\times10^9 \text{ Pa})}\left(\frac{1}{\cos 65.0°} - 1\right) = \boxed{6.17\times10^{-4} \text{ m}}.$$

(b) Find the kinetic energy.

$$\Sigma F_x = T_2 \sin\theta = ma_x = \frac{mv^2}{r} = \frac{2K}{r}, \text{ so}$$

$$K = \frac{rT_2 \sin\theta}{2} = \frac{L\sin\theta \frac{mg}{\cos\theta}\sin\theta}{2} = \frac{mgL\tan\theta\sin\theta}{2} = \frac{(0.411 \text{ kg})(9.80 \text{ m/s}^2)(2.200 \text{ m})\tan 65.0°\sin 65.0°}{2}$$

$$= \boxed{8.61 \text{ J}}$$

(c) Find the time it takes a transverse wave pulse to travel the length of the string. Neglect the small change due to the stretch when finding the linear mass density.

$$v = \sqrt{\frac{T_2}{\mu}} = \sqrt{\frac{T_2}{\rho A}} \text{ and } L = vt, \text{ so}$$

$$t = \frac{L+\Delta L}{v} = (L+\Delta L)\sqrt{\frac{\rho A}{T_2}} = (L+\Delta L)\sqrt{\frac{\rho A\cos\theta}{mg}} = (2.2006 \text{ m})\sqrt{\frac{(1150 \text{ kg/m}^3)\pi(0.00125 \text{ m})^2\cos 65.0°}{(0.411 \text{ kg})(9.80 \text{ m/s}^2)}}$$

$$= \boxed{0.0536 \text{ s}}$$

MCAT Review

1. Strategy The buoyant force on the brick is equal in magnitude to the weight of the volume of water it displaces.

Solution The brick is completely submerged, so its volume is equal to that of the displaced water. The weight of the displaced water is $30 \text{ N} - 20 \text{ N} = 10 \text{ N}$. Find the volume of the brick.

$$\rho_w g V_w = \rho_w g V_{\text{brick}} = 10 \text{ N}, \text{ so } V_{\text{brick}} = \frac{10 \text{ N}}{\rho_w g} = \frac{10 \text{ N}}{(1000 \text{ kg/m}^3)(10 \text{ m/s}^2)} = 1\times10^{-3} \text{ m}^3$$

The correct answer is $\boxed{A}$.

2. Strategy The expansion of the cable obeys $F = k\Delta L$.

Solution Compute the expansion of the cable.

$$\Delta L = \frac{F}{k} = \frac{5000 \text{ N}}{5.0\times10^6 \text{ N/m}} = 10^{-3} \text{ m}$$

The correct answer is $\boxed{A}$.

3. **Strategy** Solve for the intensity in the definition of sound level.

   Solution Find the intensity of the fire siren.

   $$\text{SL} = 10\log_{10}\frac{I}{I_0}, \text{ so } I = I_0 10^{\text{SL}/10} = (1.0\times10^{-12} \text{ W/m}^2)10^{100/10} = 1.0\times10^{-2} \text{ W/m}^2.$$

   The correct answer is $\boxed{\text{D}}$.

4. **Strategy and Solution** The two centimeters of liquid with a specific gravity of 0.5 are equivalent to one centimeter of water; that is, the 6-cm column of liquid is equivalent to a 5-cm column of water. Therefore, the new gauge pressure at the base of the column is five-fourths the original. The correct answer is $\boxed{\text{C}}$.

5. **Strategy** The wavelength is inversely proportional to the index n. Let $\lambda_n = 8$ m and $\lambda_{n+2} = 4.8$ m.

   Solution Find n.

   $$L = \frac{n\lambda_n}{4} = \frac{(n+2)\lambda_{n+2}}{4}, \text{ so } \frac{\lambda_n}{\lambda_{n+2}} = \frac{n+2}{n} = 1 + \frac{2}{n}, \text{ or } n = 2\left(\frac{\lambda_n}{\lambda_{n+2}} - 1\right)^{-1}.$$

   Compute L.

   $$L = \frac{n\lambda_n}{4} = 2\left(\frac{\lambda_n}{\lambda_{n+2}} - 1\right)^{-1}\frac{\lambda_n}{4} = \frac{1}{2}\left(\frac{8 \text{ m}}{4.8 \text{ m}} - 1\right)^{-1}(8 \text{ m}) = 6 \text{ m}$$

   The correct answer is $\boxed{\text{C}}$.

6. **Strategy** The wave may interfere within the range of possibility of totally constructive or totally destructive interference.

   Solution For totally constructive interference, the amplitude of the combined waves is $5+3 = 8$ units. For totally destructive interference, the amplitude of the combined waves is $5-3 = 2$ units. The correct answer is $\boxed{\text{B}}$.

7. **Strategy and Solution** As the bob repeatedly swings to and fro, it speeds up and slows down, as well as changes direction. Therefore, its linear acceleration must change in both magnitude and direction. The correct answer is $\boxed{\text{D}}$.

8. **Strategy** Since $K \propto v^2$ and $v \propto r^{-4}$, $K \propto r^{-8}$.

   Solution Compute the ratio of kinetic energies.

   $$\frac{K_2}{K_1} = \left(\frac{2}{1}\right)^{-8} = \frac{1}{256} = \frac{1}{4^4} \text{ or } K_2 : K_1 = 1 : 4^4.$$

   The correct answer is $\boxed{\text{B}}$.

9. **Strategy** The buoyant force on a ball is equal to the weight of the volume of water displaced by that ball.

   Solution Since B_1 is not fully submerged and B_2 and B_3 are, the buoyant force on B_1 is less than the buoyant forces on the other two. Since B_2 and B_3 are fully submerged, the buoyant forces on each are the same. The correct answer is $\boxed{\text{B}}$.

10. **Strategy and Solution** Ball 1 is floating, so its density is less than that of the water. Since Ball 2 is submerged within the water, its density is the same as that of the water. Ball 3 is sitting on the bottom of the tank, so its density is greater than that of the water. The correct answer is $\boxed{\text{A}}$.

11. **Strategy** The supporting force of the bottom of the tank is equal to the weight of Ball 3 less the buoyant force of the water.

    Solution Compute the supporting force on Ball 3.
    $$\rho_{\text{Ball 3}}gV_{\text{Ball 3}} - \rho_{\text{water}}gV_{\text{Ball 3}} = gV_{\text{Ball 3}}(\rho_{\text{Ball 3}} - \rho_{\text{water}})$$
    $$= (9.80 \text{ m/s}^2)(1.0\times10^{-6} \text{ m}^3)(7.8\times10^3 \text{ kg/m}^3 - 1.0\times10^3 \text{ kg/m}^3) = 6.7\times10^{-2} \text{ N}$$
    The correct answer is $\boxed{\text{B}}$.

12. **Strategy** The relationship between the fraction of a floating object's volume that is submerged to the ratio of the object's density to the fluid in which it floats is $V_f/V_o = \rho_o/\rho_f$. So, the fraction of the object that is not submerged is $V_{ns} = 1 - V_f/V_o = 1 - \rho_o/\rho_f$.

    Solution Find the fraction of the volume of Ball 1 that is above the surface of the water.
    $$1 - \frac{\rho_{\text{Ball 1}}}{\rho_{\text{water}}} = 1 - \frac{8.0\times10^2 \text{ kg/m}^3}{1.0\times10^3 \text{ kg/m}^3} = \frac{1}{5}$$
    The correct answer is $\boxed{\text{D}}$.

13. **Strategy** The pressure difference at a depth d in water is given by $\rho_{\text{water}}gd$.

    Solution Compute the approximate difference in pressure between the two balls.
    $$\rho_{\text{water}}gd = (1.0\times10^3 \text{ kg/m}^3)(9.80 \text{ m/s}^2)(0.20 \text{ m}) = 2.0\times10^3 \text{ N/m}^2$$
    The correct answer is $\boxed{\text{C}}$.

14. **Strategy** For the force exerted on Ball 3 by the bottom of the tank to be zero, Ball 3 must have the same density as the water.

    Solution Find the volume of the hollow portion of Ball 3, V_H.
    $$\rho_{\text{Ball 3}} = \frac{m_{\text{Ball 3}}}{V_{\text{Ball 3}}} = \frac{\rho_{\text{Fe}}V_{\text{Fe}}}{V_{\text{Ball 3}}} = \rho_{\text{water}} \text{ and } V_H = V_{\text{Ball 3}} - V_{\text{Fe}}, \text{ so}$$
    $$V_H = V_{\text{Ball 3}} - \frac{\rho_{\text{water}}}{\rho_{\text{Fe}}}V_{\text{Ball 3}} = V_{\text{Ball 3}}\left(1 - \frac{\rho_{\text{water}}}{\rho_{\text{Fe}}}\right) = (1.0\times10^{-6} \text{ m}^3)\left(1 - \frac{1.0\times10^3 \text{ kg/m}^3}{7.8\times10^3 \text{ kg/m}^3}\right) = 0.87\times10^{-6} \text{ m}^3.$$
    The correct answer is $\boxed{\text{C}}$.

Chapter 13

TEMPERATURE AND THE IDEAL GAS

Conceptual Questions

1. The development of standard temperature scales requires use of the zeroth law of thermodynamics. This law tells us that if a thermometer is in thermodynamic equilibrium with both a test object and an object used to define a standard, then the test and standard objects must be in thermodynamic equilibrium with each other. If this law did not hold, the temperature scale on a thermometer would have no relation to the actual temperature of the test object, and therefore, no standard could be defined.

5. The thermal expansion coefficients of silver and brass only differ by about 5%. Thus, a bimetallic strip made from these materials would bend only slightly during expansion, contrary to its intended purpose.

9. The SI units of mass density and number density are kg/m^3 and m^{-3}, respectively. An equal number density does not imply an equal mass density because the mass of an individual atom may be different in each gas.

13. The pressure of the air outside the balloon decreases as its distance above the Earth increases. The balloon expands until the pressures inside and outside are equal.

17. We could use any two of the quantities to decide whether the gas is dilute. The gas is dilute if the number density is low enough, or equivalently if the average intermolecular distance is large enough compared to the molecular diameter. For example, if the mean free path is much larger than the diameter, then we know that the intermolecular distance is large enough and the number density small enough so that the gas is dilute.

Problems

1. **Strategy** Use Eqs. (13-2b) and (13-3).

   Solution Convert the temperature.

   (a) $T_C = \dfrac{T_F - 32°F}{1.8°F/°C} = \dfrac{84°F - 32°F}{1.8°F/°C} = \boxed{29°C}$

   (b) $T = 29\text{ K} + 273.15\text{ K} = \boxed{302\text{ K}}$

5. **Strategy and Solution** There are $78 + 114 = 192$ degrees C and 144 degrees J between the freezing and boiling points of ethyl alcohol. Thus, the conversion factor is $144/192 = 0.750$. So, $T_J = (0.750°J/°C)T_C + A,$ where A is the offset to be determined. Find A by setting both temperatures equal to their respective boiling temperatures. $144°J = (0.750°J/°C)(78°C) + A,$ so $A = 144°J - (0.750°J/°C)(78°C) = 85.5°J.$

   Thus, the conversion from °J to °C is given by $\boxed{T_J = (0.750°J/°C)T_C + 85.5°J}$.

9. **Strategy** Since each section of track expands along its entire length, only half of the expansion is considered for a particular gap. Two sections meet at a gap, so the gap should be as wide as the expansion of one section of track. Use Eq. (13-4).

   Solution Find the amount of space that should be left between the track sections.
   $$\frac{\Delta L}{L_0} = \alpha \Delta T, \text{ so } \Delta L = L_0 \alpha \Delta T = (18.30\text{ m})(12 \times 10^{-6}\text{ K}^{-1})(50.0°C - 10.0°C) = \boxed{8.8\text{ mm}}.$$

13. Strategy According to Eq. (13-4), the change in diameter is given by $\Delta d = d_0 \alpha \Delta T = d - d_0$.

Solution Find the diameter of the rivets.
$d = d_0 + \Delta d = d_0 + d_0 \alpha \Delta T = d_0(1 + \alpha \Delta T)$, so
$$d_0 = \frac{d}{1+\alpha \Delta T} = \frac{0.6350 \text{ cm}}{1+(22.5\times10^{-6} \text{ K}^{-1})(-78.5°C - 20.5°C)} = \boxed{0.6364 \text{ cm}}.$$

17. Strategy Use Eq. (13-7) and the fact that $\beta = 3\alpha$.

Solution

(a) Find the change in volume for the dental filling.
$$\frac{\Delta V}{V_0} = \beta \Delta T, \text{ so } \Delta V = \beta V_0 \Delta T = 3\alpha V_0 \Delta T = 3(42\times10^{-6} \text{ K}^{-1})(30 \text{ mm}^3)(-5°C - 37°C) = \boxed{-0.16 \text{ mm}^3}.$$

(b) Find the change in volume for the cavity.
$$\Delta V = 3\alpha V_0 \Delta T = 3(17\times10^{-6} \text{ K}^{-1})(30 \text{ mm}^3)(-5°C - 37°C) = \boxed{-0.064 \text{ mm}^3}$$

21. Strategy Use Eq. (13-4). The internal radius of the ring expands as if it were a solid piece of brass.

Solution Find the temperature at which the internal radius of the ring is 1.0010 cm.
$$\frac{\Delta L}{L_0} = \alpha \Delta T, \text{ so } \frac{\Delta L}{\alpha L_0} = T - T_0. \text{ Compute the temperature.}$$
$$T = \frac{\Delta L}{\alpha L_0} + T_0 = \frac{1.0010 \text{ cm} - 1.0000 \text{ cm}}{(19\times10^{-6} \text{ K}^{-1})(1.0000 \text{ cm})} + 22.0°C = \boxed{75°C}$$

25. Strategy Use Eqs. (13-4) and (13-5).

Solution Find the new scale of the rule.
$$\frac{\Delta L_r}{L_{r0}} = \frac{L_r - L_{r0}}{L_{r0}} = \frac{L_r}{L_{r0}} - 1 = \alpha_r \Delta T, \text{ so } \frac{L_r}{L_{r0}} = 1 + \alpha_r \Delta T.$$
Divide the new length of the brick by the new scale.
$$\frac{L_b}{1+\alpha_r \Delta T} = \frac{L_{b0}(1+\alpha_b \Delta T)}{1+\alpha_r \Delta T} = \frac{(25.00 \text{ cm})[1+(0.75\times10^{-6} \text{ K}^{-1})(60.00 \text{ K})]}{1+(12\times10^{-6} \text{ K}^{-1})(60.00 \text{ K})} = \boxed{24.98 \text{ cm}}$$

27. Strategy Use Eqs. (13-4) and (13-6) and the given initial and final areas.

Solution Find the fractional change.
$$\frac{\Delta A}{A_0} = \frac{A - A_0}{A_0} = \frac{(s_0 + \Delta s)^2 - s_0^2}{s_0^2} = \frac{s_0^2 + 2s_0\Delta s + (\Delta s)^2 - s_0^2}{s_0^2} = \frac{2s_0\Delta s + (\Delta s)^2}{s_0^2} = \frac{\Delta s(2s_0 + \Delta s)}{s_0^2}$$
Now, since $s_0 \gg \Delta s$, we have
$$\frac{\Delta A}{A_0} \approx \frac{\Delta s(2s_0 + 0)}{s_0^2} = \frac{2s_0\Delta s}{s_0^2} = \frac{2\Delta s}{s_0} = 2\alpha \Delta T \text{ since } \frac{\Delta s}{s_0} = \alpha \Delta T.$$

29. Strategy Use the definition of one mol and Avogadro's number.

Solution Find the conversion between atomic mass units u and kg.
$$\frac{\frac{12 \text{ g}}{12 \text{ mol}}}{N_A} = \frac{0.001 \text{ kg}}{6.022\times10^{23}}, \text{ so } 1 \text{ u} = 1.66\times10^{-27} \text{ kg}.$$

31. Strategy Add the molecular masses of each element in carbon dioxide.

Solution Find the mass of carbon dioxide in kg.
$$\text{mass of } CO_2 \text{ in kg} = m_C + 2m_O = [12.011 \text{ u} + 2(15.9994 \text{ u})](1.6605\times10^{-27} \text{ kg/u}) = \boxed{7.31\times10^{-26} \text{ kg}}$$

33. Strategy Divide the mass of the water in the human by the mass of a water molecule.

Solution Estimate the number of water molecules.
$$N = \frac{\text{mass of water}}{\text{molecular mass}} = \frac{0.62(80.2 \text{ kg})}{[2(1.0 \text{ u}) + 16.0 \text{ u}](1.66\times10^{-27} \text{ kg/u})} = \boxed{1.7\times10^{27}}$$

35. Strategy Divide the total mass by the molar mass of sucrose to find the number of moles. Then use Eq. (13-11) to find the number of hydrogen atoms.

Solution $m_{C_{12}H_{22}O_{11}} = 12(12.011 \text{ g/mol}) + 22(1.00794 \text{ g/mol}) + 11(15.9994 \text{ g/mol}) = 342.30 \text{ g/mol}$

There are 342.30 grams of sucrose per mole, so there are $\dfrac{684.6 \text{ g}}{342.30 \text{ g/mol}} = 2.000$ mol of sucrose.

There are 2.000(22) = 44.00 moles of hydrogen. Find the number of hydrogen atoms.
$$N = nN_A = (44.00 \text{ mol})(6.022\times10^{23} \text{ mol}^{-1}) = \boxed{2.650\times10^{25} \text{ atoms}}$$

37. Strategy Divide the total mass of methane by its molar mass.

Solution Find the number of moles.
$$n_{CH_4} = \frac{\text{mass of } CH_4}{\text{molar mass of } CH_4} = \frac{144.36 \text{ g}}{12.011 \text{ g/mol} + 4(1.00794 \text{ g/mol})} = \boxed{8.9985 \text{ mol}}$$

41. Strategy The number of SiO_2 molecules (and the number of Si atoms) N is roughly equal to the volume of a sand grain V_g divided by the volume of a SiO_2 molecule V_m.

Solution Find the order of magnitude of the number of silicon atoms in a grain of sand.
$$N = \frac{V_g}{V_m} = \frac{\frac{4}{3}\pi r_g^3}{\frac{4}{3}\pi r_m^3} = \left(\frac{d_g}{d_m}\right)^3 = \left(\frac{0.5\times10^{-3} \text{ m}}{0.5\times10^{-9} \text{ m}}\right)^3 = \boxed{10^{18} \text{ atoms}}$$

45. Strategy The microscopic form of the ideal gas law is $PV = NkT$. Solving for the temperature gives $T = PV/(kN)$. Since k is a constant, ranking PV/N is equivalent to ranking the temperatures.

Solution Compute PV/N for each case. Ignore units and "$\times 10^{23}$" for simplicity, since this will not affect the ranking.

(a) $\dfrac{PV}{N} = \dfrac{100\times4}{6} = 66.\overline{6}$; (b) $\dfrac{200\times4}{6} = 133.\overline{3}$; (c) $\dfrac{50\times8}{6} = 66.\overline{6}$; (d) $\dfrac{100\times4}{3} = 133.\overline{3}$; (e) $\dfrac{100\times2}{3} = 66.\overline{6}$;

(f) $\dfrac{50\times4}{3} = 66.\overline{6}$

Ranking the cylinders in order of temperature, highest to lowest, we have $\boxed{(b) = (d),\ (a) = (c) = (e) = (f)}$.

49. (a) Strategy The temperature and the number of moles of the gas are constant. Use Boyle's law.

Solution Find the volume of the oxygen at atmospheric pressure.
$$P \propto \frac{1}{V}, \text{ so } V_f = \frac{P_iV_i}{P_f} = \frac{(15.2\times10^6 \text{ Pa})(0.0170 \text{ m}^3)}{101,325 \text{ Pa}} = \boxed{2.55 \text{ m}^3}.$$

(b) Strategy Divide the volume of the oxygen at atmospheric pressure by the volume flow rate.

Solution Find how long the cylinder of oxygen will last.

$$\Delta t = \frac{V}{\frac{\Delta V}{\Delta t}} = \frac{2.55 \text{ m}^3}{8.0 \text{ L/min}}\left(\frac{1 \text{ L}}{10^{-3} \text{ m}^3}\right)\left(\frac{1 \text{ h}}{60 \text{ min}}\right) = \boxed{5.3 \text{ h}}$$

53. Strategy The number of moles of the gas is constant. Use the ideal gas law.

Solution Find the volume of the hydrogen.

$$\frac{P_f V_f}{T_f} = \frac{P_i V_i}{T_i}, \text{ so } V_f = \frac{P_i V_i T_f}{P_f T_i} = \frac{(1.00\times10^5 \text{ N/m}^2)(5.0 \text{ m}^3)(273.15 \text{ K} - 13 \text{ K})}{(0.33\times10^3 \text{ N/m}^2)(273.15 \text{ K} + 27 \text{ K})} = \boxed{1.3\times10^3 \text{ m}^3}.$$

57. Strategy Find the number of moles of molecular oxygen by dividing the total mass by the molar mass. Use the macroscopic form of the ideal gas law, Eq. (13-16).

Solution Compute the volume occupied by the gas.

$$V = \frac{nRT}{P} = \frac{(0.532 \text{ kg})(10^3 \text{ g/kg})}{2(15.9994 \text{ g/mol})} \times \frac{[8.314 \text{ J/(mol} \cdot \text{K)}](0.0 \text{ K} + 273.15 \text{ K})}{1.0\times10^5 \text{ Pa}} = \boxed{0.38 \text{ m}^3}$$

61. Strategy The number of moles of the gas is constant. Use the ideal gas law and Eq. (9-3).

Solution Find the volume of the bubble just before it breaks the surface of the water.

$$\frac{P_f V_f}{T_f} = \frac{P_i V_i}{T_i}, \text{ so}$$

$$V_f = \frac{(P_f + \rho g d)V_i T_f}{P_f T_i} = \frac{\left(1 + \frac{\rho g d}{P_f}\right)V_i T_f}{T_i} = \frac{\left[1 + \frac{(1.00\times10^3 \text{ kg/m}^3)(9.80 \text{ m/s}^2)(20.0 \text{ m})}{1.013\times10^5 \text{ Pa}}\right](1.00 \text{ cm}^3)(273.15 \text{ K} + 25.0 \text{ K})}{273.15 \text{ K} + 10.0 \text{ K}}$$

$$= \boxed{3.09 \text{ cm}^3}.$$

65. Strategy The total translational kinetic energy of the gas molecules is equal to the number of molecules times the average translational kinetic energy per molecule. Use Eqs. (13-13) and (13-20).

Solution Divide the total translational kinetic energy by the volume for each pressure.

(a) $\dfrac{K_{total}}{V} = \dfrac{N\langle K_{tr}\rangle}{V} = \dfrac{N\langle \frac{3}{2}kT\rangle}{V} = \dfrac{3}{2}P = \dfrac{3}{2}(1.00 \text{ atm})(1.013\times10^5 \text{ Pa/atm}) = \boxed{1.52\times10^5 \text{ J/m}^3}$

(b) $\dfrac{K_{total}}{V} = \dfrac{3}{2}(300.0 \text{ atm})(1.013\times10^5 \text{ Pa/atm}) = \boxed{4.559\times10^7 \text{ J/m}^3}$

69. Strategy Use the ideal gas law and Eq. (13-22).

Solution Find the temperature of the nitrogen gas.

$$PV = nRT, \text{ so } T = \frac{PV}{nR}.$$

Compute the rms speed of the nitrogen molecules.

$$v_{rms} = \sqrt{\frac{3kT}{m}} = \sqrt{\frac{3kPV}{mnR}} = \sqrt{\frac{3(1.381\times10^{-23} \text{ J/K})(1.6 \text{ atm})(1.013\times10^5 \text{ Pa/atm})(0.25 \text{ m})^3}{2(14.0 \text{ u})(1.6605\times10^{-27} \text{ kg/u})(2.0 \text{ mol})[8.314 \text{ J/(mol} \cdot \text{K)}]}} = \boxed{370 \text{ m/s}}$$

73. Strategy Use Eq. (13-22).

Solution Find the rms speeds of the atom and molecules.

$$\text{He: } v_{\text{rms}} = \sqrt{\frac{3kT}{m}} = \sqrt{\frac{3(1.38\times10^{-23} \text{ J/K})(273.15 \text{ K}+25 \text{ K})}{(4.00260 \text{ u})(1.66\times10^{-27} \text{ kg/u})}} = \boxed{1360 \text{ m/s}}$$

$$N_2\text{: } v_{\text{rms}} = \sqrt{\frac{3(1.38\times10^{-23} \text{ J/K})(273.15 \text{ K}+25 \text{ K})}{2(14.00674 \text{ u})(1.66\times10^{-27} \text{ kg/u})}} = \boxed{515 \text{ m/s}}$$

$$H_2\text{: } v_{\text{rms}} = \sqrt{\frac{3(1.38\times10^{-23} \text{ J/K})(273.15 \text{ K}+25 \text{ K})}{2(1.00794 \text{ u})(1.66\times10^{-27} \text{ kg/u})}} = \boxed{1920 \text{ m/s}}$$

$$O_2\text{: } v_{\text{rms}} = \sqrt{\frac{3(1.38\times10^{-23} \text{ J/K})(273.15 \text{ K}+25 \text{ K})}{2(15.9994 \text{ u})(1.66\times10^{-27} \text{ kg/u})}} = \boxed{482 \text{ m/s}}$$

77. Strategy Multiplying the average translational kinetic energy, Eq. (13-20), by the number of molecules N gives the total translational kinetic energy. Use Eq. (13-13), the microscopic form of the ideal gas law, to find the number of molecules.

Solution Find the total change in translational kinetic energy of the inhaled air.

$$PV = NkT, \text{ so } N = \frac{P_iV_i}{kT_i}. \ K_{\text{tr,total}} = N\langle K_{\text{tr}}\rangle = N\left(\frac{3}{2}kT\right) = \frac{3}{2}NkT, \text{ so}$$

$$\Delta K_{\text{tr,total}} = \frac{3}{2}Nk\Delta T = \frac{3}{2}\frac{P_iV_i}{kT_i}k\Delta T = \frac{3P_iV_i\Delta T}{2T_i} = \frac{3(101\times10^3 \text{ Pa})(0.50\times10^{-3} \text{ m}^3)[37°\text{C}-(-10°\text{C})]}{2(273.15 \text{ K}-10 \text{ K})} = \boxed{14 \text{ J}}.$$

79. Strategy and Solution The average translational energy of a molecule in an ideal gas is $\langle K_{\text{tr}}\rangle = \frac{1}{2}m\langle v^2\rangle$. The rms speed is $v_{\text{rms}} = \sqrt{\langle v^2\rangle}$, so $\frac{1}{2}mv_{\text{rms}}^2 = \langle K_{\text{tr}}\rangle$. From Eq. (13-19), we know that $\langle K_{\text{tr}}\rangle = \frac{3PV}{2N}$, and using the ideal gas law, $PV = nRT$, we have $\frac{1}{2}mv_{\text{rms}}^2 = \frac{3}{2}\left(\frac{PV}{N}\right) = \frac{3}{2}\left(\frac{nRT}{N}\right)$, so $v_{\text{rms}} = \sqrt{3\left(\frac{n}{mN}\right)RT} = \sqrt{\frac{3RT}{M}}$.

81. Strategy Form a proportion with the two reaction rates and solve for the temperature increase. Use Eq. (13-24).

Solution Find the temperature increase.

$$\frac{1.035}{1} = \frac{e^{-\frac{E_a}{kT_2}}}{e^{-\frac{E_a}{kT_1}}} = e^{\frac{E_a}{k}\left(\frac{1}{T_1}-\frac{1}{T_2}\right)}$$

$$\ln 1.035 = \frac{E_a}{k}\left(\frac{1}{T_1}-\frac{1}{T_2}\right)$$

$$\frac{k\ln 1.035}{E_a} = \frac{1}{T_1}-\frac{1}{T_2}$$

$$\frac{1}{T_2} = \frac{1}{T_1}-\frac{k\ln 1.035}{E_a}$$

$$T_2 = \left(\frac{1}{T_1}-\frac{k\ln 1.035}{E_a}\right)^{-1}$$

$$\Delta T = \left(\frac{1}{T_1} - \frac{k \ln 1.035}{E_a} \right)^{-1} - T_1$$

$$= \left[\frac{1}{273.15 \text{ K} + 10.00 \text{ K}} - \frac{(1.38 \times 10^{-23} \text{ J/K}) \ln 1.035}{2.81 \times 10^{-19} \text{ J}} \right]^{-1} - (273.15 \text{ K} + 10.00 \text{ K}) = \boxed{0.14°\text{C}}$$

85. Strategy Use Eq. (13-26).

Solution Find the time for the perfume molecule to diffuse 5.00 m in one direction.

$$x_{\text{rms}} = \sqrt{2Dt}, \text{ so } t = \frac{x_{\text{rms}}^2}{2D} = \frac{(5.00 \text{ m})^2}{2(1.00 \times 10^{-5} \text{ m}^2/\text{s})} = \boxed{1.25 \times 10^6 \text{ s}}.$$

89. Strategy Use Eq. (13-26).

Solution Find the time it takes for a water molecule to diffuse out through the pore.

$$x_{\text{rms}} = \sqrt{2Dt}, \text{ so } t = \frac{x_{\text{rms}}^2}{2D} = \frac{(2.5 \times 10^{-5} \text{ m})^2}{2(2.4 \times 10^{-5} \text{ m}^2/\text{s})} = \boxed{1.3 \times 10^{-5} \text{ s}}.$$

93. (a) Strategy Find the distance between N_2 molecules and multiply by the scale factor $\frac{0.0375 \text{ m}}{0.30 \times 10^{-9} \text{ m}} = 1.25 \times 10^8$ to get the distance between ping pong balls. Let each N_2 molecule be at the center of a sphere with diameter d. Then the average distance between N_2 molecules is approximately d. The volume of each sphere is V/N with $V = 0.0224 \text{ m}^3$ for $N = 6.02 \times 10^{23}$ molecules, $P = 1.00$ atm, and $T = 0.0°\text{C}$.

Solution Find the average distance between the ping-pong balls.

$$\frac{V}{N} = \frac{1}{6}\pi d^3, \text{ so } d = \sqrt[3]{\frac{6V}{\pi N}}.$$

Therefore, the distance between ping-pong balls is $(1.25 \times 10^8)\sqrt[3]{\frac{6(0.0224 \text{ m}^3)}{\pi(6.02 \times 10^{23})}} = \boxed{52 \text{ cm}}.$

(b) Strategy Find the mean free path for an N_2 molecule and multiply by the scale factor to find the average distance between ping pong ball collisions. Use Eq. (13-25) and the results of part (a).

Solution Find the mean free path.

$$\Lambda = \frac{1}{\sqrt{2}\pi d^2 (N/V)} = \frac{kT}{\sqrt{2}\pi d^2 P}$$

So, the average distance between ping pong ball collisions is

$$\frac{(1.25 \times 10^8)(1.38 \times 10^{-23} \text{ J/K})(273.15 \text{ K} + 0.0 \text{ K})}{\sqrt{2}\pi(0.30 \times 10^{-9} \text{ m})^2(1.00 \text{ atm})(1.013 \times 10^5 \text{ Pa/atm})} = \boxed{12 \text{ m}}.$$

97. Strategy Use Eq. (13-4).

Solution Compute the length change of the titanium rod.

$$\Delta L = L_0 \alpha \Delta T = (0.0500 \text{ m})(8.6 \times 10^{-6} \text{ K}^{-1})(20°\text{C} - 37°\text{C}) = \boxed{-7.3 \text{ μm (probably not enough to notice)}}$$

101. Strategy Use the ideal gas law with the number of moles constant to find V_a, the volume of air at standard temperature and pressure.

Solution Find the number of breaths taken per day.

$$\frac{P_a V_a}{T_a} = \frac{P_L V_L}{T_L}, \text{ so } V_a = \frac{P_L T_a V_L}{P_a T_L}.$$

$$\frac{\text{volume per day}}{\text{volume per breath}} = \frac{\text{volume per day}}{V_a} = \frac{(\text{volume per day})P_a T_L}{P_L T_a V_L}$$

$$= \frac{\left(210 \ \frac{\text{L}}{\text{day}}\right)\left(\frac{10^3 \ \text{cm}^3}{\text{L}}\right)(1 \ \text{atm})(1.013 \times 10^5 \ \text{Pa/atm})(39 \ \text{K} + 273.15 \ \text{K})}{(450 \ \text{mm Hg})\left(133.3 \ \frac{\text{Pa}}{\text{mm Hg}}\right)(0 \ \text{K} + 273.15 \ \text{K})(100 \ \text{cm}^3)} = \boxed{4000 \ \text{breaths}}$$

105. (a) Strategy The slope of a graph is rise over run.

Solution Find the slope of pressure versus temperature.

$$\frac{\Delta P}{\Delta T} = \frac{8.00 \ \text{mm}}{20.0°\text{C}} = \boxed{0.400 \ \text{mm Hg/°C}}$$

(b) Strategy Use the ideal gas law and the result of part (a).

Solution Find the number of moles of gas present.
$(\Delta P)V = nR(\Delta T)$, so

$$n = \frac{V}{R}\frac{\Delta P}{\Delta T} = \frac{0.500 \ \text{L}}{8.314 \ \text{J/(mol} \cdot \text{K)}}(0.400 \ \text{mm Hg/°C})\frac{10^{-3} \ \text{m}^3}{1 \ \text{L}}(1.333 \times 10^2 \ \text{Pa/mm Hg}) = \boxed{3.21 \times 10^{-3} \ \text{mol}}.$$

109. Strategy Use the microscopic form of the ideal gas law, Eq. (13-13).

Solution Find the number of air molecules N.

$$NkT = PV, \text{ so } N = \frac{PV}{kT} = \frac{(1.00 \times 10^5 \ \text{Pa})\frac{4}{3}\pi(0.125 \times 10^{-3} \ \text{m})^3}{(1.38 \times 10^{-23} \ \text{J/K})(310 \ \text{K})} = \boxed{1.9 \times 10^{14} \ \text{molecules}}.$$

113. Strategy The increase in volume of the mercury minus the increase in volume of the glass bulb equals the volume of mercury that moves up the tube. Use Eqs. (13-6) and (13-7).

Solution Find the how far the thread of mercury moves h.

$$\Delta V_{\text{Hg}} - \Delta V_g = A_{\text{tube}} h$$

$$V_0 \beta_{\text{Hg}} \Delta T - V_0 \beta_g \Delta T = (A_0 + 2\alpha_g A_0 \Delta T)h$$

$$h = \frac{V_0(\beta_{\text{Hg}} - \beta_g)\Delta T}{A_0(1 + 2\alpha_g \Delta T)} = \frac{(0.200 \times 10^{-6} \ \text{m}^3)[(182 - 9.75) \times 10^{-6} \ \text{K}^{-1}](1.00 \ \text{K})}{\frac{1}{4}\pi(0.120 \times 10^{-3} \ \text{m})^2[1 + 2(3.25 \times 10^{-6} \ \text{K}^{-1})(1.00 \ \text{K})]} = \boxed{3.05 \ \text{mm}}$$

<h1 style="text-align:center">Chapter 14</h1>

<h1 style="text-align:center">HEAT</h1>

Conceptual Questions

1. Heat flows from the hotter to the colder object.

5. The air between the layers acts as very good thermal insulation.

9. The fins increase the area of the surface responsible for radiating heat from the engine to the air so that the engine cools more efficiently.

13. Water contained in the sauce and cheese of a pizza has a higher specific heat and thermal conductivity than the crust. Thus, for a given time interval, the amount of heat energy transferred to the roof of the mouth by the sauce and cheese is greater than that which is transferred to the hand by the crust.

17. The rate at which heat flows via conduction, convection, and radiation depends upon the temperature difference between the hot and cold bodies. By adding the milk to the coffee immediately, the temperature difference between the coffee and the surrounding air is minimized, so the rate of heat flow is smaller. Additionally, when the color of the coffee is lighter, the radiative heat loss will be further reduced.

21. In winter, the walls of a room are at a lower temperature than the air within the room. In summer, the walls are warmer than the inside air. The amount of heat lost from a person standing within the room via conduction and convection is constant throughout the year since the temperature of the air is held fixed. The temperature difference between the body and the walls is greatest in winter. As a result, more heat is lost via radiative transfer in winter and the room therefore feels cooler.

Problems

1. (a) Strategy The gravitational potential energy of the 1.4 kg of water is converted to internal energy in the 6.4-kg system.

Solution Compute the increase in internal energy.

$$U = mgh = (1.4 \text{ kg})(9.80 \text{ m/s}^2)(2.5 \text{ m}) = \boxed{34 \text{ J}}$$

(b) Strategy and Solution $\boxed{\text{Yes; the increase in internal energy increases the average kinetic energy of the water molecules, thus the temperature is slightly increased}}$.

3. Strategy The amount of internal energy generated is equal to the decrease in kinetic energy of the bullet.

Solution Compute the amount of internal energy generated.

$$\left| \Delta K \right| = \frac{1}{2} mv_i^2 = \frac{1}{2}(0.0200 \text{ kg})(7.00 \times 10^2 \text{ m/s})^2 = \boxed{4.90 \text{ kJ}}$$

5. **(a) Strategy** The decrease in gravitational potential energy of the child is equal to the amount of internal energy generated.

   Solution Compute the amount of internal energy generated.
   $$U = mgh = (15 \text{ kg})(9.80 \text{ m/s}^2)(1.7 \text{ m}) = \boxed{250 \text{ J}}$$

   (b) Strategy and Solution Friction warms the slide and the child, and the air molecules are deflected by the child's body. The energy goes into $\boxed{\text{all three}}$.

9. **Strategy** The conversion factor is $1 \text{ kW} \cdot \text{h} = 3.600 \text{ MJ}$.

   Solution Convert 1.00 kJ to kilowatt-hours.
   $$(1.00 \times 10^3 \text{ J}) \frac{1 \text{ kW} \cdot \text{h}}{3.600 \times 10^6 \text{ J}} = \boxed{2.78 \times 10^{-4} \text{ kW} \cdot \text{h}}$$

11. **Strategy** The heat capacity of an object is equal to its mass times its specific heat.

    Solution Find the heat capacity of the 5.00-g gold ring.
    $$C = mc = (0.00500 \text{ kg})[0.128 \text{ kJ/(kg} \cdot \text{K)}] = \boxed{6.40 \times 10^{-4} \text{ kJ/K}}$$

13. **Strategy** Use Eq. (14-4). Let $Q = 1$ kJ, $m = 400$ g, and $c = 0.45$ kJ/(kg · K).

    Solution $Q = mc\Delta T$, so $\Delta T = Q/(mc)$.
    $$\text{(a) } \frac{Q}{mc}; \text{ (b) } \frac{2Q}{mc}; \text{ (c) } \frac{2Q}{(2m)c} = \frac{Q}{mc}; \text{ (d) } \frac{Q}{m(2c)} = \frac{Q}{2mc}; \text{ (e) } \frac{2Q}{m(2c)} = \frac{Q}{mc}; \text{ (f) } \frac{2Q}{(2m)(2c)} = \frac{Q}{2mc}$$
    Ranking these six situations in order of the temperature increase, largest to smallest, we have $\boxed{\text{(b), (a) = (c) = (e), (d) = (f)}}$.

15. **Strategy** The final kinetic energy equals the energy content of the banana.

    Solution Find the man's speed.
    $$\frac{1}{2}mv_f^2 = U, \text{ so } v_f = \sqrt{\frac{2U}{m}} = \sqrt{\frac{2(418 \times 10^3 \text{ J})}{83 \text{ kg}}} = \boxed{100 \text{ m/s}}.$$

17. **Strategy** The heat capacity of an object is equal to its mass times its specific heat.

    Solution Find the heat capacity of the 30.0-kg block of ice.
    $$C = mc = (30.0 \text{ kg})[2.1 \text{ kJ/(kg} \cdot \text{K)}] = \boxed{63 \text{ kJ/K}}$$

21. **Strategy** Use Eq. (14-4).

    Solution Find the heat required to raise the woman's body temperature.
    $$Q = mc\Delta T = (50.0 \text{ kg})[3.5 \text{ kJ/(kg} \cdot \text{K)}](38.4°\text{C} - 37.0°\text{C}) = \boxed{250 \text{ kJ}}$$

25. **Strategy** Use the definition of average power and Eq. (14-4).

    Solution Solve for Δt using $P = \Delta E/\Delta t = \Delta Q/\Delta t$.
    $$\Delta t = \frac{\Delta Q}{P} = \frac{mc\Delta T}{P} = \frac{(0.50 \text{ kg})[4186 \text{ J/(kg} \cdot \text{K)}](100.0°\text{C} - 20.0°\text{C})}{2.1 \times 10^3 \text{ W}} = \boxed{80 \text{ s}}.$$

27. **Strategy** Use the ideal gas law to find the number of moles of nitrogen gas in the container. Then, solve for the final temperature of the gas in Eq. (14-6).

    Solution Find the number of moles of gas in the container.

    $$PV = nRT, \text{ so } n = \frac{PV}{RT}.$$

    Find the new temperature of the gas after the heat is added.

    $$Q = nC_V\Delta T = \frac{PV}{RT_i}C_V(T_f - T_i), \text{ so}$$

    $$T_f = T_i + \frac{RT_iQ}{PVC_V} = 23°\text{C} + \frac{[8.314 \text{ J/(mol·K)}](23 \text{ K} + 273.15 \text{ K})(26.6\times10^3 \text{ J})}{(3.5 \text{ atm})(1.013\times10^5 \text{ Pa/atm})(425 \text{ L})(10^{-3} \text{ m}^3/\text{L})[20.8 \text{ J/(mol·K)}]} = \boxed{44°\text{C}}.$$

29. **Strategy** Find the number of moles of air using the macroscopic from of the ideal gas law. Then use the molar specific heat at constant pressure to find the heat loss, where $C_p = 7R/2$ (for an ideal diatomic gas).

    Solution

    (a) Find the number of moles of air breathed per day.

    $$PV = nRT_i, \text{ so } n = \frac{PV}{RT_i}.$$

    Find the heat loss per day due to breathing alone.

    $$Q = nC_p\Delta T = \frac{PV}{RT_i}\frac{7R}{2}\Delta T = \frac{7PV\Delta T}{2T_i} = \frac{7(101\times10^3 \text{ Pa})(24 \text{ h}\times5 \text{ m}^3/\text{h})(35°\text{C} - 20°\text{C})}{2(273.15 \text{ K} + 20 \text{ K})} = \boxed{2 \text{ MJ}}$$

    (b) Compare the heat loss from breathing to the total heat loss.

    $$\frac{2 \text{ MJ}}{9 \text{ MJ}}\times100\% = \boxed{20\%, \text{ which is significant}}$$

    (c) Compute the average power loss due to breathing alone.

    $$\mathcal{P} = \frac{Q}{\Delta t} = \frac{2.17\times10^6 \text{ J}}{86,400 \text{ s}} = \boxed{30 \text{ W}}$$

33. **Strategy** The sum of the heat flows is zero. Use Eqs. (14-4) and (14-9).

    Solution Compute the heat of fusion of water.

    $$0 = Q_{ice} + Q_w + Q_c = m_{ice}L_f + m_{ice}c_w\Delta T_{ice} + m_wc_w\Delta T_w + m_cc_c\Delta T_c, \text{ so}$$

    $$L_f = -\frac{c_w(m_{ice}\Delta T_{ice} + m_w\Delta T_w) + m_cc_c\Delta T_c}{m_{ice}}$$

    $$= -\frac{[4.186 \text{ J/(g·K)}][(30.0 \text{ g})(8.5 \text{ K}) + (2.00\times10^2 \text{ g})(-11.5 \text{ K})] + (3.00\times10^2 \text{ g})[0.380 \text{ J/(g·K)}](-11.5 \text{ K})}{30.0 \text{ g}}$$

    $$= \boxed{330 \text{ J/g}}$$

37. Strategy Find the sum of the heats required to raise the temperature of the ice to 0.0°C, melt the ice, raise the resulting water to 100.0°C, evaporate the water, and raise the temperature of the resulting steam to 110.0°C. Use Eqs. (14-4) and (14-9).

Solution Find the required heat.

$$Q = Q_{ice} + Q_w + Q_s$$
$$= mL_f + mc_{ice}\Delta T_{ice} + mc_w\Delta T_w + mL_v + mc_s\Delta T_s$$
$$= (1.0 \text{ kg})\{333.7 \text{ kJ/kg} + [2.1 \text{ kJ/(kg} \cdot \text{K)}](20.0 \text{ K}) + [4.186 \text{ kJ/(kg} \cdot \text{K)}](100.0 \text{ K}) + 2256 \text{ kJ/kg}$$
$$+ [2.01 \text{ kJ/(kg} \cdot \text{K)}](10.0 \text{ K})\}$$
$$= \boxed{3100 \text{ kJ}}$$

41. Strategy Use Eq. (14-9) for fusion.

Solution The heat required to melt the ice is 12.0 kJ − 4.0 kJ = 8.0 kJ. Find the mass of the ice.

$$m = \frac{Q}{L_f} = \frac{8.0 \text{ kJ}}{333.7 \text{ kJ/kg}} = \boxed{24 \text{ g}}$$

43. Strategy Heat flows from the aluminum into the ice. Use Eqs. (14-4) and (14-9).

Solution Find the temperature of the aluminum at which it will melt a mass of the ice equal to its own.

$$m_{Al}c_{Al}\Delta T_{Al} + m_{ice}L_f = m_{Al}c_{Al}(T_f - T_i) + m_{ice}L_f = 0, \text{ so } T_i = \frac{L_f}{c_{Al}} + T_f = \frac{333.7 \text{ J/g}}{0.900 \text{ J/(g} \cdot \text{K)}} + 0.0°\text{C} = \boxed{371°\text{C}}.$$

Since 371°C < 660°C (the melting point of aluminum), the answer is $\boxed{\text{yes}}$.

45. Strategy The rate of heat loss is $\Delta Q/\Delta t = L_v \Delta m/\Delta t$ since $Q = mL_v$ for evaporation and where Δm represents the mass of water evaporated.

Solution Compute the rate of heat lost through transpiration.

$$\frac{\Delta Q}{\Delta t} = (0.618 \text{ g/min})\left(\frac{1 \text{ min}}{60 \text{ s}}\right)(2256 \text{ J/g}) = \boxed{23.2 \text{ W}}$$

49. Strategy The heat supplied heats the substance to its melting point, melts it, then raises the temperature of the resulting liquid to 327°C. Use Eqs. (14-4) and (14-9).

Solution Compute the heat of fusion.

$$Q = mL_f + mc\Delta T, \text{ so } L_f = -c\Delta T + \frac{Q}{m} = -[0.129 \text{ kJ/(kg} \cdot \text{K)}](327 - 21) \text{ K} + \frac{31.15 \text{ kJ}}{0.500 \text{ kg}} = \boxed{22.8 \text{ kJ/kg}}.$$

53. Strategy Add the thermal resistances. Use Eqs. (14-12) and (14-13).

Solution Find the equivalent thermal resistance of the rods.

$$\Sigma R_n = \frac{d_{Cu}}{\kappa_{Cu}A} + \frac{d_{Fe}}{\kappa_{Fe}A} = \frac{d}{A}\left(\frac{2}{\kappa_{Cu}} + \frac{1}{\kappa_{Fe}}\right) \text{ where } d = d_{Fe}.$$

Find the rate of heat transfer.

$$\mathcal{P} = \frac{\Delta T}{\Sigma R_n} = \frac{A\Delta T}{d\left(\frac{2}{\kappa_{Cu}} + \frac{1}{\kappa_{Fe}}\right)} = \frac{(6.0 \times 10^{-6} \text{ m}^2)(100.0 \text{ K})}{(0.25 \text{ m})\left[\frac{2}{401 \text{ W/(m} \cdot \text{K)}} + \frac{1}{80.2 \text{ W/(m} \cdot \text{K)}}\right]} = \boxed{0.14 \text{ W}}$$

55. Strategy Use Eq. (14-11).

Solution Find the required heat output.

$$\mathcal{P} = \frac{\Delta T}{R} = \frac{37°C}{0.33 \ \text{K/W}} = \boxed{110 \ \text{W}}$$

57. Strategy Use Fourier's law of heat conduction, Eq. (14-10).

Solution Find the temperature drop across the epidermis.

$$\mathcal{P} = \kappa A \frac{\Delta T}{d}, \ \text{so} \ \Delta T = \frac{d\mathcal{P}}{\kappa A} = \frac{(0.00200 \ \text{m})(0.050 \ \text{W})}{[0.45 \ \text{W/(m·K)}](10.0 \times 10^{-4} \ \text{m}^2)} = \boxed{0.22°C}.$$

61. Strategy Use Fourier's law of heat conduction, Eq. (14-10).

Solution

(a) Find the temperature at the interface between the wood and cork if the cork is on the inside and the wood on the outside. Set $\mathcal{P}_w = \mathcal{P}_c$.

$$\kappa_w A \frac{\Delta T_w}{d} = \kappa_c A \frac{\Delta T_c}{d}$$
$$\kappa_w(0.0°C - T) = \kappa_c(T - 20.0°C)$$
$$(0.0°C)\kappa_w + (20.0°C)\kappa_c = T(\kappa_c + \kappa_w)$$
$$T = \frac{(20.0°C)(0.046)}{0.046 + 0.13} = \boxed{5.2°C}$$

(b) Find the temperature at the interface between the wood and cork if the wood is inside and the cork is outside.

$$\kappa_w(20.0°C - T) = \kappa_c(T - 0.0°C)$$
$$(20.0°C)\kappa_w = T(\kappa_c + \kappa_w)$$
$$T = \frac{(20.0°C)(0.13)}{0.046 + 0.13} = \boxed{15°C}$$

(c)   The temperature at the interface differs for the two cases, but the total thermal resistance is the same either way, so it doesn't matter whether the cork is placed on the inside or the outside of the wooden wall.

65. Strategy Use Stefan's law of thermal radiation, Eq. (14-16). Note that the surface temperature T is absolute temperature (in kelvins).

Solution $\mathcal{P} = e\sigma A T^4$. $e\sigma$ is the same for each case. Compute AT^4, ignoring units for simplicity.

(a) $1.00(227 + 273.15)^4 = 6.26 \times 10^{10}$; (b) $1.01(227 + 273.15)^4 = 6.32 \times 10^{10}$;
(c) $1.05(227 + 273.15)^4 = 6.57 \times 10^{10}$; (d) $1.00(232 + 273.15)^4 = 6.51 \times 10^{10}$;
(e) $0.99(232 + 273.15)^4 = 6.45 \times 10^{10}$; (f) $0.98(232 + 273.15)^4 = 6.38 \times 10^{10}$

Ranking the wood stoves in order of the power radiated, from greatest to least, we have $\boxed{\text{(c), (d), (e), (f), (b), (a)}}$.

67. Strategy Use Stefan's law of thermal radiation, Eq. (14-16).

Solution Find the surface area of the filament.

$$\mathcal{P} = e\sigma A T^4, \ \text{so} \ A = \frac{\mathcal{P}}{e\sigma T^4} = \frac{40.0 \ \text{W}}{0.32[5.670 \times 10^{-8} \ \text{W/(m}^2 \cdot \text{K}^4)](2.6 \times 10^3 \ \text{K})^4} = \boxed{4.8 \times 10^{-5} \ \text{m}^2}.$$

69. Strategy Use Eq. (14-18).

Solution Find the rate at which the student "burns" calories.

$$\mathscr{P}_{\text{net}} = e\sigma A(T^4 - T_{\text{s}}^{\ 4})$$

$$= 1.0[5.670\times10^{-8}\ \text{W}/(\text{m}^2\cdot\text{K}^4)](1.7\ \text{m}^2)[(35\ \text{K}+273.15\ \text{K})^4 - (16\ \text{K}+273.15\ \text{K})^4]\left(\frac{1\ \text{kcal}}{4186\ \text{J}}\right)\left(\frac{3600\ \text{s}}{1\ \text{h}}\right)$$

$$= \boxed{170\ \text{kcal/h}}$$

71. Strategy Use Eq. (14-18). Assume the man's head is spherical, with a radius of 10 cm.

Solution Estimate the surface area of the man's head.

$$A = 4\pi r^2 = 4\pi(0.10\ \text{m})^2 = 0.1257\ \text{m}^2$$

The total energy loss is given by the power radiated multiplied by the time.

$$\mathscr{P}_{\text{net}} = e\sigma A(T_{\text{skin}}^4 - T_{\text{air}}^4),\ \text{so}\ E = \mathscr{P}\Delta t = e\sigma A(T_{\text{skin}}^4 - T_{\text{air}}^4)\Delta t$$

$$= 0.97[5.670\times10^{-8}\ \text{W}/(\text{m}^2\cdot\text{K}^4)](0.1257\ \text{m}^2)[(35\ \text{K}+273.15\ \text{K})^4 - (-15\ \text{K}+273.15\ \text{K})^4](15\times60\ \text{s}) = \boxed{28\ \text{kJ}}.$$

73. (a) Strategy Power is equal to intensity times area and the heat absorbed is given by Eq. (14-4).

Solution Find the rate of increase of the lizard's temperature $\Delta T/\Delta t$.

$$\frac{Q}{\Delta t} = \frac{mc\Delta T}{\Delta t} = IA,\ \text{so}\ \frac{\Delta T}{\Delta t} = \frac{IA}{mc} = \frac{\frac{1}{2}(1.4\times10^3\ \text{W}/\text{m}^2)(1.6\times10^{-4}\ \text{m}^2)}{(3.0\ \text{g})[4.2\ \text{J}/(\text{g}\cdot{}^\circ\text{C})]} = \boxed{8.9\times10^{-3}\ {}^\circ\text{C/s}}.$$

(b) Strategy The time required to raise the temperature of the lizard is equal to the temperature difference divided by the rate of temperature increase.

Solution

$$\Delta t_1 = \frac{\Delta T_1}{\Delta T/\Delta t} = \frac{5.0^\circ\text{C}}{8.9\times10^{-3}\ {}^\circ\text{C/s}} = \boxed{9.4\ \text{min}}$$

75. (a) Strategy The power absorbed by the leaf must equal that radiated away. Power is equal to intensity times area. Use Stefan's law of thermal radiation, Eq. (14-16).

Solution Absorbed:

$$I_{\text{top}}A + e\sigma A T_{\text{s}}^4 = 0.700(9.00\times10^2\ \text{W}/\text{m}^2)(5.00\times10^{-3}\ \text{m}^2) +$$

$$(1)[5.670\times10^{-8}\ \text{W}/(\text{m}^2\cdot\text{K}^4)](5.00\times10^{-3}\ \text{m}^2)(273.15\ \text{K}+25.0\ \text{K})^4$$
$$= 5.39\ \text{W}$$

Find the temperature of the leaf. The area is now $2(5.00\times10^{-3}\ \text{m}^2) = 10.0\times10^{-3}\ \text{m}^2$ (both sides of the leaf).

$$\mathscr{P} = e\sigma A T^4,\ \text{so}$$

$$T = \left(\frac{\mathscr{P}}{e\sigma A}\right)^{1/4} = \left[\frac{5.39\ \text{W}}{(1)[5.670\times10^{-8}\ \text{W}/(\text{m}^2\cdot\text{K}^4)](10.0\times10^{-3}\ \text{m}^2)}\right]^{1/4} = 312\ \text{K} - 273\ \text{K} = \boxed{39^\circ\text{C}}.$$

(b) Strategy Since the bottom of the leaf absorbs and emits at the same rate, it can be ignored.

Solution Find the power per unit area that must be lost by other methods.

$$\frac{\mathscr{P}_{\text{abs,Sun}}}{A} = \frac{\mathscr{P}_{\text{rad}}}{A} + \frac{\mathscr{P}_{\text{other}}}{A} = e\sigma T^4 + \frac{\mathscr{P}_{\text{other}}}{A},\ \text{so}$$

$$\frac{\mathscr{P}_{\text{other}}}{A} = \frac{\mathscr{P}_{\text{abs,Sun}}}{A} - e\sigma T^4 = 0.700(9.00\times10^2 \text{ W}/\text{m}^2) - (1)[5.670\times10^{-8} \text{ W}/(\text{m}^2 \cdot \text{K}^4)](273.15 \text{ K} + 25.0 \text{ K})^4$$
$$= 182 \text{ W}/\text{m}^2$$

Thus, the power per unit area that must be lost by other methods is $\boxed{182 \text{ W}/\text{m}^2}$.

77. Strategy Use Wien's law, Eq. (14-17).

Solution Compute the temperature of the blackbody.

$$\lambda_{\text{max}}T = 2.898\times10^{-3} \text{ m}\cdot\text{K}, \text{ so } T = \frac{2.898\times10^{-3} \text{ m}\cdot\text{K}}{2.65\times10^{-6} \text{ m}} = \boxed{1090 \text{ K}}.$$

81. Strategy The basal metabolic rate is the minimal energy intake necessary to sustain life in a state of complete inactivity.

Solution Calculate the BMR/kg of body mass and BMR/m^2 of surface area for each animal.

Animal	(a) BMR/kg	(b) BMR/m^2
Mouse	210	1200
Dog	51	1000
Human	32	1000
Pig	18	1000
Horse	11	960

(a) According to the table, since BMR/kg is larger for smaller animals, $\boxed{\text{it is true}}$ that smaller animals must consume more food per kilogram of body mass.

(c) When an animal is resting, the food energy metabolized must be shed as heat (no work). Since $\boxed{\text{radiative loss depends upon surface area}}$, BMR/m^2 must be approximately the same for different-sized animals.

85. Strategy Use Eq. (14-4) and the relationship between power and intensity.

Solution The energy provided by the sunlight is converted to heat in the water. The energy provided is $\mathscr{P}\Delta t = IA\Delta t$. Compute the time to heat the water.

$$Q = mc\Delta T = IA\Delta t, \text{ so } \Delta t = \frac{mc\Delta T}{IA} = \frac{(1.0 \text{ L})(1000 \text{ g/L})[4.186 \text{ J}/(\text{g}\cdot\text{K})](100.0-15.0) \text{ K}}{(750 \text{ W}/\text{m}^2)(1.5 \text{ m}^2)} = \boxed{320 \text{ s}}.$$

89. Strategy The heat loss is given by $\mathscr{P}\Delta t$ and it is equal to $Q = mL_{\text{v}}$.

Solution Find the mass of water required to replenish the fluid loss.

$$m = \frac{\mathscr{P}\Delta t}{L_{\text{v}}} = \frac{(650 \text{ W})(30.0 \text{ min})(60 \text{ s}/\text{min})}{2430 \text{ J/g}} = \boxed{480 \text{ g}}$$

91. (a) Strategy Use Eqs. (14-4) and (14-9).

Solution Find the heat given up by the steam.

$$Q = -mc_w\Delta T + mL_v = (4.0\text{ g})\{-[4.186\text{ J/(g}\cdot\text{K)}](45.0-100.0)\text{ K} + 2256\text{ J/g}\} = \boxed{9.9\text{ kJ}}$$

(b) Strategy Use Eq. (14-4).

Solution Compute the mass of the tissue.

$$m = \frac{Q}{c\Delta T} = \frac{9945\text{ J}}{[3.5\text{ J/(g}\cdot\text{K)}](45.0-37.0)\text{ K}} = \boxed{360\text{ g}}$$

93. Strategy Use Eq. (14-4), substituting for Q the expression given in the problem statement.

Solution Compute the temperature rise.

$$Q = mc\Delta T,\text{ so }\Delta T = \frac{Q}{mc} = \frac{0.544\times10^{-3}\text{ J} + (1.46\times10^{-3}\text{ J/cm)}(1.5\text{ cm})}{(0.10\text{ g})[4.186\text{ J/(g}\cdot\text{K)}]} = \boxed{6.5\times10^{-3}{}^\circ\text{C}}.$$

97. Strategy The potential energy of the spring is equal to $\frac{1}{2}kx^2$. Use Eq. (14-4).

Solution Find the temperature change of the water.

$$Q = mc\Delta T = \Delta U = \frac{1}{2}kx^2,\text{ so }\Delta T = \frac{kx^2}{2mc} = \frac{(8.4\times10^3\text{ N/m)}(0.10\text{ m})^2}{2(1.0\text{ kg})[4186\text{ J/(kg}\cdot\text{K)}]} = \boxed{0.010{}^\circ\text{C}}.$$

101. Strategy Heat flows from the copper block to the water and iron pot. Use Eq. (14-4).

Solution Find the final temperature of the system.

$$0 = Q_w + Q_{Cu} + Q_{Fe}$$
$$Q_w = -Q_{Cu} - Q_{Fe}$$
$$m_w c_w (T_f - T_i) = -m_{Cu} c_{Cu}(T_f - T_{Cu}) - m_{Fe} c_{Fe}(T_f - T_i)$$
$$T_f(m_w c_w + m_{Cu} c_{Cu} + m_{Fe} c_{Fe}) = m_{Cu} c_{Cu} T_{Cu} + (m_{Fe} c_{Fe} + m_w c_w)T_i$$

Solve for T_f.

$$\begin{aligned}
T_f &= \frac{m_{Cu} c_{Cu} T_{Cu} + (m_{Fe} c_{Fe} + m_w c_w)T_i}{m_w c_w + m_{Cu} c_{Cu} + m_{Fe} c_{Fe}} \\
&= \frac{(2.0\text{ kg})[385\text{ J/(kg}\cdot\text{K)}](100.0{}^\circ\text{C}) + \{(2.0\text{ kg})[440\text{ J/(kg}\cdot\text{K)}] + (1.0\text{ kg})[4186\text{ J/(kg}\cdot\text{K)}]\}(25.0{}^\circ\text{C})}{(1.0\text{ kg})[4186\text{ J/(kg}\cdot\text{K)}] + (2.0\text{ kg})[385\text{ J/(kg}\cdot\text{K)}] + (2.0\text{ kg})[440\text{ J/(kg}\cdot\text{K)}]} = \boxed{35{}^\circ\text{C}}
\end{aligned}$$

105. Strategy Heat flows from the drink to the ice. Use Eqs. (14-4) and (14-9).

Solution Find the mass of the drink required to just melt the ice.

$$0 = Q_{drink} + Q_{ice} = m_{drink}c\Delta T + m_{ice}L_f,\text{ so}$$
$$m_{drink} = -\frac{m_{ice}L_f}{c\Delta T} = -\frac{(0.10\text{ kg})(333.7\text{ kJ/kg)}}{[4.186\text{ kJ/(kg}\cdot\text{K)}](0{}^\circ\text{C} - 25{}^\circ\text{C})} = \boxed{0.32\text{ kg}}.$$

109. Strategy Gravitational potential energy is converted into internal energy. Use Eq. (14-9) and $U = mgh$.

Solution Find the mass of the ice melted by friction.

$$Q = m_m L_f = 0.75U = 0.75mgh,\text{ so }m_m = \frac{0.75mgh}{L_f} = \frac{0.75(75\text{ kg})(9.80\text{ m/s}^2)(2.43\text{ m})}{333{,}700\text{ J/kg}} = \boxed{4.0\text{ g}}.$$

Chapter 15

THERMODYNAMICS

Conceptual Questions

1. Yes, but it wouldn't be a very good heat pump. Like an electric heater, the heat output would be equal to the work input, with no heat being taken from the cold reservoir.

5. Energy is always conserved, so there would be just as much energy as before when the fossil fuels are exhausted. There wouldn't be as much high quality or useful energy though, so it would be better to call it something like a "high quality energy crisis."

9. No, entropy changes don't require a flow of heat. For example, when a gas expands freely into a vacuum, its entropy increases but there is no heat flow. Beating an egg is another example of a process that increases entropy with no flow of heat.

13. This is not a violation. Although the salt crystals are in a more ordered state, the surroundings of the bucket are in a more disordered state. The gaseous state of the evaporated seawater is less ordered than the water.

Problems

1. **Strategy** The work done by Ming is equal to the magnitude of the force of friction $f = \mu N$ times the total "rubbing" distance. Use the first law of thermodynamics.

   Solution Find the change in internal energy.
   $$\Delta U = Q + W = 0 + \mu N d = 0.45(5.0 \text{ N})[8(0.16 \text{ m})] = \boxed{2.9 \text{ J}}$$

3. **Strategy** Use the first law of thermodynamics. $\Delta U > 0$ and $W > 0$.

   Solution Find the heat flow.
   $$Q = \Delta U - W = 400 \text{ J} - 500 \text{ J} = -100 \text{ J}$$
   $$\boxed{100 \text{ J of heat flows out of the system.}}$$

5. **Strategy** From conservation of energy, the change in the internal energy of the paint is equal to the heat flow into the paint plus the work done on the paint.

   Solution

   (a) In 5.00 min, the work done by the paddle on the paint is
   $$W = 0.448 \text{ kJ/s} \times 5.00 \text{ min} \times 60 \text{ s/min} = 134.4 \text{ kJ}$$
   Since we assume no heat flow ($Q = 0$), the internal energy of the paint changes by $\Delta U = Q + W = +134.4$ kJ. The temperature increases 1.00 K for every 12.5 kJ of increased internal energy, so
   $$\Delta T = 134.4 \text{ kJ} \times \frac{1.00 \text{ K}}{12.5 \text{ kJ}} = \boxed{10.8 \text{ K}}$$

(b) To apply the first law, we first find the internal energy change:
$$\Delta U = \frac{12.5 \text{ kJ}}{1.00 \text{ K}} \times 6.3 \text{ K} = 78.75 \text{ kJ}$$
Now we apply the first law:
$$\Delta U = Q + W$$
$$Q = \Delta U - W = 78.75 \text{ kJ} - 134.4 \text{ kJ} = -56 \text{ kJ}$$
Q is negative because $\boxed{56 \text{ kJ}}$ of heat flows *out of* the paint.

Discussion How did we know the work done by the paddle on the paint was *positive*? Think of the force the paddle exerts on the paint as it pushes paint out of its way; the force and the displacement are in the same direction. The quantity 12.5 kJ/K is the heat capacity of the paint—it tells us how many kJ the internal energy of the paint must increase for its temperature to increase 1 K, *regardless of whether the internal energy increase is caused by heat, work, or a combination of the two.*

7. **Strategy** No work is done during the constant volume process, but work is done during the constant pressure process. Use Eq. (15-3).

   Solution Compute the total work done by the gas.
   $$W = P_{\text{i}}\Delta V = (2.000 \text{ atm})(1.013 \times 10^5 \text{ Pa/atm})(2.000 \text{ L} - 1.000 \text{ L})(10^{-3} \text{ m}^3/\text{L}) = \boxed{202.6 \text{ J}}$$

9. **(a) Strategy** Use the ideal gas law. Refer to the figure.

   Solution According to the graph, the pressure at point C is $\boxed{98.0 \text{ kPa}}$. Find the temperature at point C.
   $$PV = nRT, \text{ so } T = \frac{PV}{nR} = \frac{(98.0 \times 10^3 \text{ Pa})(2.00 \text{ L})(10^{-3} \text{ m}^3/\text{L})}{(0.0200 \text{ mol})[8.314 \text{ J}/(\text{mol} \cdot \text{K})]} = \boxed{1180 \text{ K}}.$$

   (b) Strategy This is a constant volume process. Use Eq. (15-6) and the ideal gas law.

   Solution Find the change in internal energy of the gas.
   $$\Delta U = nC_{\text{v}}\Delta T = n\left(\frac{3}{2}R\right)\left(\frac{P_{\text{f}}V}{nR} - \frac{P_{\text{i}}V}{nR}\right) = \frac{3V}{2}(P_{\text{f}} - P_{\text{i}})$$
   $$= \frac{3(1.00 \text{ L})(10^{-3} \text{ m}^3/\text{L})}{2}(98.0 \times 10^3 \text{ Pa} - 230 \times 10^3 \text{ Pa}) = \boxed{-200 \text{ J}}$$

   (c) Strategy The work done per cycle is equal to the area contained within the curve.

   Solution Compute the work done per cycle.
   $$W = \frac{1}{2}(230 \times 10^3 \text{ Pa} - 98.0 \times 10^3 \text{ Pa})(2.00 \text{ L} - 1.00 \text{ L})(10^{-3} \text{ m}^3/\text{L}) = \boxed{66 \text{ J}}$$

   (d) Strategy and Solution At the beginning and end of a complete cycle, regardless of the starting point, the temperature is the same, so $\boxed{\Delta U = 0 \text{ because } \Delta T = 0 \text{ in a cycle.}}$

13. (a) Strategy No work is done during the constant volume process, but work is done during the constant pressure process. Use Eq. (15-3).

Solution Compute the total work done on the gas.

$$W = -P_i \Delta V = -(1.000 \text{ atm})(1.013 \times 10^5 \text{ Pa/atm})(16.00 \text{ L} - 4.00 \text{ L})(10^{-3} \text{ m}^3/\text{L}) = \boxed{-1216 \text{ J}}$$

(b) Strategy Use the first law of thermodynamics and the ideal gas law, as well as Eqs. (15-3), (15-6), and (15-7).

Solution Calculate the total change in internal energy of the gas and the total heat flow into the gas during the entire process.

A to B (constant volume):

$$\Delta U_1 = nC_V \Delta T_1 = \frac{3}{2} nR\Delta T_1$$

Using the ideal gas law, $\Delta T_1 = \dfrac{\Delta PV}{nR}$, so $\Delta U_1 = \dfrac{3}{2}\Delta PV$.

B to C to D (constant pressure):

$$\Delta U_2 = Q + W = nC_p \Delta T_2 - P_i \Delta V = \frac{5}{2} nR\Delta T_2 - P_i \Delta V$$

Using the ideal gas law, $\Delta T_2 = \dfrac{P_i \Delta V}{nR}$, so $\Delta U_2 = \dfrac{5}{2} P_i \Delta V - P_i \Delta V = \dfrac{3}{2} P_i \Delta V$.

The total change in internal energy is

$$\Delta U = \Delta U_1 + \Delta U_2 = \frac{3}{2}\Delta PV + \frac{3}{2} P_i \Delta V$$

$$= \frac{3}{2}[(1.000 \text{ atm} - 2.000 \text{ atm})(4.000 \text{ L}) + (1.000 \text{ atm})(16.00 \text{ L} - 4.000 \text{ L})](1.013 \times 10^5 \text{ Pa/atm})(10^{-3} \text{ m}^3/\text{L})$$

$$= \boxed{1216 \text{ J}}$$

The total heat flow is $Q = \Delta U - W = 1215.6 \text{ J} + 1215.6 \text{ J} = \boxed{2431 \text{ J}}$.

15. (a) Strategy For A–E (constant pressure), $W = -P_i \Delta V$, and for E–D (constant temperature), $W = nRT \ln V_i/V_f$. Use the ideal gas law to find T.

Solution

$$W_{\text{total}} = -P_i \Delta V + nRT \ln \frac{V_i}{V_f} = -P_i \Delta V + nR\left(\frac{P_E V_E}{nR}\right) \ln \frac{V_E}{V_D}$$

$$= \left[-(2.000 \text{ atm})(8.000 \text{ L} - 4.000 \text{ L}) + (2.000 \text{ atm})(8.000 \text{ L}) \ln \frac{8.000 \text{ L}}{16.00 \text{ L}} \right](1.013 \times 10^5 \text{ Pa/atm})(10^{-3} \text{ m}^3/\text{L})$$

$$= \boxed{-1934 \text{ J}}$$

(b) Strategy For constant temperature, $\Delta U = 0$. For constant pressure,

$$\Delta U = Q + W = nC_p \Delta T - P_i \Delta V = \frac{5}{2} nR\left(\frac{P_i \Delta V}{nR}\right) - P_i \Delta V = \frac{3}{2} P_i \Delta V.$$

Solution

$$\Delta U = \frac{3}{2}(2.000 \text{ atm})(8.000 \text{ L} - 4.000 \text{ L})(1.013 \times 10^5 \text{ Pa/atm})(10^{-3} \text{ m}^3/\text{L}) = \boxed{1216 \text{ J}}.$$

The total heat flow is $Q = \Delta U - W = 1215.6 \text{ J} + 1933.85 \text{ J} = \boxed{3149 \text{ J}}$.

17. Strategy The work done by the system is positive for a heat engine and negative for a heat pump or refrigerator. The net work is positive for cycles moving clockwise and negative for cycles moving counterclockwise.

Solution

(a) Cycles $\boxed{B \text{ and } D}$ might describe a heat engine, since the $\boxed{\text{cycles move clockwise}}$.

(b) Cycles $\boxed{A,\ C,\ \text{and}\ E}$ might describe a heat pump, since the $\boxed{\text{cycles move counterclockwise}}$.

(c) Cycles $\boxed{A,\ C,\ \text{and}\ E}$ might describe a refrigerator, since $\boxed{\text{heat pumps and refrigerators work the same way}}$.

21. (a) Strategy Use Eq. (15-12).

Solution Find the heat absorbed by the engine.
$$Q_{\text{H}} = \frac{W_{\text{net}}}{e} = \frac{1.00 \times 10^3 \text{ J}}{0.333} = \boxed{3.00 \text{ kJ}}$$

(b) Strategy The net work done by an engine during one cycle is equal to the net heat flow into the engine during the cycle.

Solution Find the heat exhausted by the engine.
$$W_{\text{net}} = Q_{\text{H}} - Q_{\text{C}}, \text{ so } Q_{\text{C}} = Q_{\text{H}} - W_{\text{net}} = 3.00 \text{ kJ} - 1.00 \text{ kJ} = \boxed{2.00 \text{ kJ}}.$$

25. Strategy The power of the solar power plant is equal to the intensity of the sunlight times the area of the collectors times the efficiency of the collectors.

Solution Find the area required.
$$eIA = P, \text{ so } A = \frac{P}{eI} = \frac{1.0 \times 10^9 \text{ W}}{0.200(0.20 \times 10^3 \text{ W/m}^2)} = \boxed{2.5 \times 10^7 \text{ m}^2 \text{ or } 25 \text{ km}^2}.$$

27. Strategy The work done by the engine is equal to the increase in gravitational potential energy of the crate plus the increase in kinetic energy. Use Eq. (15-12).

Solution Find the required heat input.
$$Q_{\text{in}} = \frac{W_{\text{net}}}{e} = \frac{mgh + \frac{1}{2}mv^2}{e} = \frac{m}{e}\left(gh + \frac{v^2}{2}\right) = \frac{5.00 \text{ kg}}{0.300}\left[(9.80 \text{ m/s}^2)(10.0 \text{ m}) + \frac{(4.00 \text{ m/s})^2}{2}\right] = \boxed{1770 \text{ J}}$$

29. Strategy Use Eq. (15-16).

Solution Find the daily cost to run the air conditioning unit.
$$W_{\text{net}} = \frac{Q_{\text{C}}}{K_{\text{r}}}$$
$$\left(\frac{\$0.10}{\text{kW} \cdot \text{h}}\right)W_{\text{net}} = \left(\frac{\$0.10}{\text{kW} \cdot \text{h}}\right)\frac{Q_{\text{C}}}{K_{\text{r}}} = \left(\frac{\$0.10}{\text{kW} \cdot \text{h}}\right)\frac{1.73 \times 10^8 \text{ J}}{2.00}\left(\frac{1 \text{ kW} \cdot \text{h}}{3.6 \times 10^6 \text{ J}}\right) = \boxed{\$2.40}$$

33. Strategy The maximum efficiency is that of a reversible heat engine. Use Eq. (15-17).

Solution Calculate the maximum possible efficiency.
$$e_{\text{r}} = 1 - \frac{T_{\text{C}}}{T_{\text{H}}} = 1 - \frac{273.15 \text{ K} + 4.0 \text{ K}}{273.15 \text{ K} + 18.0 \text{ K}} = \boxed{0.0481}$$

35. Strategy The pump requires the minimum possible work if it is reversible. Use Eqs. (15-15) and (15-19).

Solution

$$W_{net} = \frac{Q_H}{K_{p,rev}} = Q_H\left(1 - \frac{T_C}{T_H}\right) = (1.0\times10^3 \text{ J})\left(1 - \frac{273.15 \text{ K} - 10.0 \text{ K}}{273.15 \text{ K} + 20.0 \text{ K}}\right) = \boxed{100 \text{ J}}$$

37. Strategy The rate at which the engine stores energy in the battery is 5.0 nJ/day. The rate at which the body supplies energy is equal to the rate that the engine stores energy in the battery divided by the efficiency. Convert days into seconds.

Solution Find the rate at which the body supplies energy.

$$\frac{\Delta W}{\Delta t}\bigg|_{body} = \frac{1}{e}\frac{\Delta W}{\Delta t}\bigg|_{batt} = \frac{1}{e_r/2}\frac{\Delta W}{\Delta t}\bigg|_{batt} = 2\left(1 - \frac{T_C}{T_H}\right)^{-1}\frac{\Delta W}{\Delta t}\bigg|_{batt}$$

$$= 2\left(1 - \frac{273.15 \text{ K} + 20 \text{ K}}{273.15 \text{ K} + 37 \text{ K}}\right)^{-1}(5.0\times10^{-9} \text{ J/day})\frac{1 \text{ day}}{86,400 \text{ s}} = \boxed{2.11 \text{ pW}}$$

39. (a) Strategy Use Eq. (15-17).

Solution Find the efficiency of the reversible engine.

$$e_r = 1 - \frac{T_C}{T_H} = 1 - \frac{273.15 \text{ K} + 300.0 \text{ K}}{273.15 \text{ K} + 600.0 \text{ K}} = \boxed{0.3436}$$

(b) Strategy Use Eq. (15-18).

Solution Find the amount of heat exhausted to the cold reservoir.

$$\frac{Q_C}{Q_H} = \frac{T_C}{T_H}, \text{ so } Q_C = \frac{T_C}{T_H}Q_H = \frac{273.15 \text{ K} + 300.0 \text{ K}}{273.15 \text{ K} + 600.0 \text{ K}}(420.0 \text{ kJ}) = \boxed{275.7 \text{ kJ}}.$$

41. Strategy For maximum efficiency, assume reversibility. Use Eq. (15-17).

Solution Find the percent decrease in theoretical maximum efficiency.

$$\frac{\Delta e_r}{e_r}\times100\% = \frac{1 - \frac{T_{Cf}}{T_H} - \left(1 - \frac{T_{Ci}}{T_H}\right)}{1 - \frac{T_{Ci}}{T_H}}\times100\% = \frac{-\frac{T_{Cf}}{T_H} + \frac{T_{Ci}}{T_H}}{1 - \frac{T_{Ci}}{T_H}}\times100\% = \frac{T_{Ci} - T_{Cf}}{T_H - T_{Ci}}\times100\% = \frac{27°C - 47°C}{500.0°C - 27°C}\times100\%$$

$$= -4.2\%$$

The theoretical maximum efficiency would decrease by $\boxed{4.2\%}$.

45. Strategy Use Eq. (15-16). Since the water is initially at 0°C, $Q_C = mL_f$ is the amount of heat that must be removed from the water to freeze it.

Solution Find the work required to freeze the water.

$$K_r = \frac{Q_C}{W_{net}}, \text{ so } W_{net} = \frac{Q_C}{K_r} = \frac{mL_f}{K_r} = \frac{(1.0 \text{ kg})(333.7 \text{ kJ/kg})}{3.0} = \boxed{110 \text{ kJ}}.$$

49. Strategy Use $W_{\text{net}} = Q_{\text{H}} - Q_{\text{C}}$, Eq. (15-17), and the definition of efficiency of an engine.

Solution

$$Q_{\text{C}} = Q_{\text{H}} - W_{\text{net}} = \frac{W_{\text{net}}}{e_{\text{r}}} - W_{\text{net}} = W_{\text{net}}\left(\frac{1}{e_{\text{r}}} - 1\right) = W_{\text{net}}\left(\frac{1}{1 - T_{\text{C}}/T_{\text{H}}} - 1\right) = W_{\text{net}}\left(\frac{T_{\text{H}}}{T_{\text{H}} - T_{\text{C}}} - \frac{T_{\text{H}} - T_{\text{C}}}{T_{\text{H}} - T_{\text{C}}}\right) = \frac{T_{\text{C}}}{T_{\text{H}} - T_{\text{C}}}W_{\text{net}}$$

51. Strategy and Solution The mass is the same for each case. For equal masses, water has more entropy than ice, and warmer water has more entropy than cooler water, so the order is $\boxed{\text{(b), (a), (c), (d)}}$.

53. Strategy The temperature is constant and the heat entering the system is $Q = mL_{\text{f}}$. Use Eq. (15-20).

Solution Find the change in the ice cube's entropy.
$$\Delta S = \frac{Q}{T} = \frac{mL_{\text{f}}}{T} = \frac{(1.00 \text{ g})(333.7 \text{ J/g})}{273.15 \text{ K} + 0.0 \text{ K}} = \boxed{1.22 \text{ J/K}}$$

57. Strategy A small amount of heat is transferred from the water to the iron, but the temperatures change little during the 10.0 s. Use Eq. (15-20).

Solution Calculate the change in entropy of the system.
$$\Delta S = -\frac{Q}{T_{\text{H}}} + \frac{Q}{T_{\text{C}}} = Q\left(\frac{1}{T_{\text{C}}} - \frac{1}{T_{\text{H}}}\right) = (41.86\times10^3 \text{ J})\left(\frac{1}{273.15 \text{ K} + 0.0 \text{ K}} - \frac{1}{273.15 \text{ K} + 100.0 \text{ K}}\right) = \boxed{+41.1 \text{ J/K}}$$

61. Strategy Use Eq. (15-20).

Solution Compute the entropy change of the universe.
$$\Delta S_{\text{tot}} = \Delta S_{\text{H}} + \Delta S_{\text{C}} = -\frac{Q}{T_{\text{H}}} + \frac{Q}{T_{\text{C}}} = Q\left(\frac{1}{T_{\text{C}}} - \frac{1}{T_{\text{H}}}\right) = mL_{\text{V}}\left(\frac{1}{T_{\text{C}}} - \frac{1}{T_{\text{H}}}\right) = (\rho V)L_{\text{V}}\left(\frac{1}{T_{\text{C}}} - \frac{1}{T_{\text{H}}}\right)$$
$$= (1.00 \text{ g/mL})(150 \text{ mL})(2256 \text{ J/g})\left(\frac{1}{273.15 \text{ K} + 28.0 \text{ K}} - \frac{1}{273.15 \text{ K} + 35.0 \text{ K}}\right) = \boxed{+0.026 \text{ kJ/K}}$$

63. Strategy Use Eq. (15-20) to find the change in entropy. Determine the number of moles of the sample and use Avogadro's number to find the number of molecules.

Solution Find the change in entropy.
$$\Delta S = \frac{Q}{T} = \frac{2.20 \text{ J}}{273.15 \text{ K} + 60.0 \text{ K}} = 0.00660 \text{ J/K}$$
Find the number of molecules.
$$\frac{35.0\times10^{-3} \text{ kg}}{29.5 \text{ kg/mol}}(6.022\times10^{23} \text{ molecules/mol}) = 7.14\times10^{20} \text{ molecules}$$
Find the entropy change per protein molecule.
$$\frac{0.00660 \text{ J/K}}{7.14\times10^{20} \text{ molecules}} = \boxed{9.24\times10^{-24} \text{ J/K}}$$

65. Strategy and Solution $\boxed{\text{The engine will not work.}}$ There is no energy available to do the work necessary to extract the water's internal energy.

69. Strategy Use Eq. (15-17) and $W_{net} = Q_H - Q_C$.

Solution Find the waste heat exhausted.
Coal:
$$e_r = 1 - \frac{T_C}{T_H} = 1 - \frac{273.15 \text{ K} + 27 \text{ K}}{273.15 \text{ K} + 727 \text{ K}} = 0.700 \text{ and}$$

$$Q_C = Q_H - W_{net} = \frac{W_{net}}{e} - W_{net} = W_{net}\left(\frac{1}{e} - 1\right) = (1.00 \times 10^{14} \text{ J})\left(\frac{1}{0.700} - 1\right) = 4.3 \times 10^{13} \text{ J}.$$

Nuclear:
$$e_r = 1 - \frac{T_C}{T_H} = 1 - \frac{273.15 \text{ K} + 27 \text{ K}}{273.15 \text{ K} + 527 \text{ K}} = 0.625 \text{ and } Q_C = (1.00 \times 10^{14} \text{ J})\left(\frac{1}{0.625} - 1\right) = 6.0 \times 10^{13} \text{ J}.$$

The coal-fired plant and the nuclear plant exhaust 4.3×10^{13} J and 6.0×10^{13} J of heat per day, respectively.

73. (a) Strategy The net work done per cycle is equal to the area contained within the curve.

Solution Compute the work done per cycle.
$$W = \frac{1}{2}(5.00 \text{ atm} - 1.00 \text{ atm})(1.013 \times 10^5 \text{ Pa/atm})(2.00 \text{ m}^3 - 0.500 \text{ m}^3) = \boxed{304 \text{ kJ}}$$

(b) Strategy Use the ideal gas law to compute the temperatures at the upper left and lower right points on the curve.

Solution Compute the temperatures.
$$PV = nRT, \text{ so } \frac{P_2 V_2}{P_1 V_1} = \frac{T_2}{T_1}.$$

$$T_{ul} = \frac{P_2 V_2 T_1}{P_1 V_1} = \frac{(5.00)(0.500)(470.0 \text{ K})}{(1.00)(0.500)} = 2350 \text{ K}$$

$$T_{lr} = \frac{P_2 V_2 T_1}{P_1 V_1} = \frac{(1.00)(2.00)(470.0 \text{ K})}{(1.00)(0.500)} = 1880 \text{ K}$$

The maximum temperature is $\boxed{2350 \text{ K}}$.

(c) Strategy Use the ideal gas law at the lower-left corner of the diagram.

Solution Find the number of moles of gas used in the engine.
$$n = \frac{PV}{RT} = \frac{(1.00 \text{ atm})(1.013 \times 10^5 \text{ Pa/atm})(0.500 \text{ m}^3)}{[8.314 \text{ J/(mol} \cdot \text{K)}](470.0 \text{ K})} = \boxed{13.0 \text{ mol}}$$

77. Strategy The energy of the mixed state, U, must equal the sum of the original (unmixed) states, U_1 and U_2.
The energy for a monatomic ideal gas is related to the temperature by $U = \frac{3}{2}NkT$.

Solution Find the final temperature T of the mixture.
$$U = U_1 + U_2$$
$$\frac{3}{2}(N_1 + N_2)kT = \frac{3}{2}N_1 k T_1 + \frac{3}{2}N_2 k T_2$$
$$\frac{3}{2}(n_1 + n_2)RT = \frac{3}{2}n_1 R T_1 + \frac{3}{2}n_2 R T_2$$
$$T = \frac{n_1 T_1 + n_2 T_2}{n_1 + n_2} = \frac{(4.0 \text{ mol})(20.0°\text{C}) + (3.0 \text{ mol})(30.0°\text{C})}{4.0 \text{ mol} + 3.0 \text{ mol}} = \boxed{24°\text{C}}$$

81. **Strategy and Solution** The refrigerator removes heat from the room at a rate of 450 W. The electric motor exhausts heat to the room equal to the work done by the motor plus the heat removed from the room at a rate of 250 W + 450 W = 700 W. By conservation of energy, the net rate of heat added to the room must be $+700\text{ W} - 450\text{ W} = \boxed{+250\text{ W}}$.

85. **Strategy** Assume the freezer is reversible. Use Eqs. (14-4), (14-9), (15-16), and (15-19).

Solution Find the minimum work input required to freeze the ice.

$$K_r = \frac{\text{heat removed}}{\text{net work input}} = \frac{Q_C}{W_{\text{net}}} = \frac{1}{T_H/T_C - 1}, \text{ so}$$

$$W_{\text{net}} = Q_C\left(\frac{T_H}{T_C} - 1\right) = (mL_f - mc_w\Delta T_w - mc_{\text{ice}}\Delta T_{\text{ice}})\left(\frac{T_H}{T_C} - 1\right)$$

$$= (1.20\text{ kg})\{333{,}700\text{ J/kg} - [4186\text{ J/(kg}\cdot\text{K})](-20.0\text{ K}) - [2100\text{ J/(kg}\cdot\text{K})](-20.0\text{ K})\}\left(\frac{273.15\text{ K} + 20.0\text{ K}}{273.15\text{ K} - 20.0\text{ K}} - 1\right)$$

$$= \boxed{87.1\text{ kJ}}$$

87. (a) **Strategy and Solution** Find the work done by the engine, the heat input, and the change in internal energy for each step in the cycle.

Stage A:

$$W = nRT\ln\frac{V_f}{V_i} = (1.000\text{ mol})[8.314\text{ J/(mol}\cdot\text{K})](373\text{ K})\ln\frac{2.50}{2.00} = 692\text{ J}$$

$\Delta U = 0$ for an isothermal process.
To perform the work, $Q = W = 692$ J of heat was required.

State B:
$W = 0$ for a constant volume process.

$\Delta U = Q$ for a constant volume process, and $C_V = \frac{5}{2}R$ for a diatomic ideal gas.

$$\Delta U = Q = nC_V\Delta T = \frac{5}{2}nR\Delta T = \frac{5}{2}(1.000\text{ mol})[8.314\text{ J/(mol}\cdot\text{K})](273\text{ K} - 373\text{ K}) = -2080\text{ J}$$

Stage C:

$$W = nRT\ln\frac{V_f}{V_i} = (1.000\text{ mol})[8.314\text{ J/(mol}\cdot\text{K})](273\text{ K})\ln\frac{2.00}{2.50} = -506\text{ J}$$

$\Delta U = 0$ and $Q = W = -506$ J.

Stage D:

$$W = 0 \text{ and } \Delta U = Q = \frac{5}{2}nR\Delta T = \frac{5}{2}(1.000\text{ mol})[8.314\text{ J/(mol}\cdot\text{K})](373\text{ K} - 273\text{ K}) = 2080\text{ J}.$$

Stage	W (J)	Q (J)	ΔU (J)
A	692	692 into the gas	0
B	0	2080 out of the gas	−2080
C	−506	506 out of the gas	0
D	0	2080 into the gas	2080
ABCD	186	186 into the gas	0

(b) Strategy Use Eq. (15-12).

Solution Find the efficiency of the engine.

$$e = \frac{W_{net}}{Q_{in}} = \frac{185.5 \text{ J}}{692 \text{ J} + 2078.5 \text{ J}} = \boxed{0.0670}$$

(c) Strategy Use Eq. (15-17).

Solution Compute the efficiency of a reversible engine and compare it to the efficiency of this engine.

$$e_r = 1 - \frac{T_C}{T_H} = 1 - \frac{273 \text{ K}}{373 \text{ K}} = \boxed{0.268} \text{ and } \frac{e_r}{e} = \frac{0.268}{0.0670} = 4.00, \text{ or } \boxed{e_r = 4.00e}.$$

89. (a) Strategy For an isobaric process, $W = P\Delta V$ is the work done by the gas, and $P\Delta V = nR\Delta T$ according to the ideal gas law.

Solution Find the work done by the air in the swim bladder.

$$W = nR\Delta T = \frac{PV_i}{T_i R} R\Delta T = \frac{(1.1 \text{ atm})(1.013 \times 10^5 \text{ Pa/atm})(8.16 \times 10^{-3} \text{ L})(10^{-3} \text{ m}^3/\text{L})}{273.15 \text{ K} + 20.0 \text{ K}} (2.0 \text{ K}) = \boxed{6.2 \text{ mJ}}$$

(b) Strategy $C_P = \frac{7}{2}R$ for a diatomic ideal gas. Use Eq. (15-7).

Solution Find the heat gained by the air in the swim bladder.

$$Q = nC_P\Delta T = \frac{7}{2}nR\Delta T = \frac{7}{2}W = \frac{7}{2}(6.2 \text{ mJ}) = \boxed{22 \text{ mJ}}$$

(c) Strategy Use Eq. (14-4).

Solution Find the temperature change.

$$\Delta T = \frac{Q}{mc} = \frac{-21.7 \times 10^{-3} \text{ J}}{(5.00 \text{ g})[3.5 \text{ J}/(\text{g} \cdot {}^\circ\text{C})]} = -1.2 \text{ mK}$$

The temperature will decrease $\boxed{1.2 \text{ mK}}$.

Review Exercises

1. **Strategy** Assume no heat is lost to the air. The potential energy of the water is converted into heating of the water. The internal energy of the water increases by an amount equal to the initial potential energy.

   Solution Find the change in internal energy.
   $$\Delta U = mgh = (1.00 \text{ m}^3)(1.00 \times 10^3 \text{ kg/m}^3)(9.80 \text{ m/s}^2)(11.0 \text{ m}) = \boxed{108 \text{ kJ}}$$

5. **Strategy** Set the sum of the heat flows equal to zero. Use Eqs. (14-4) and (14-9).

   Solution

   (a) Find the mass of ice required.
   $$0 = Q_w + Q_{ice} = m_w c_w \Delta T_w + m_{ice} L_f + m_{ice} c_{ice} \Delta T_{ice} = m_w c_w \Delta T_w + m_{ice}(L_f + c_{ice}\Delta T_{ice}), \text{ so}$$
   $$m_{ice} = -\frac{m_w c_w \Delta T_w}{L_f + c_{ice}\Delta T_{ice}} = -\frac{(0.250 \text{ kg})[4.186 \text{ kJ/(kg} \cdot \text{K)}](-25.0 \text{ K})}{333.7 \text{ kJ/kg} + [2.1 \text{ k J/(kg} \cdot \text{K)}](10.0 \text{ K})} = \boxed{74 \text{ g}}.$$

   (b) Find the final temperature of the water, T, which includes the melted ice.
   $$0 = Q_w + Q_{ice}$$
   $$0 = m_w c_w \Delta T_w + m_{ice} L_f + m_{ice} c_{ice} \Delta T_1 + m_{ice} c_w \Delta T_2$$
   $$0 = m_w c_w (T - T_w) + m_{ice} L_f + m_{ice} c_{ice} \Delta T_{ice} + m_{ice} c_w (T - 273.15 \text{ K})$$
   $$0 = (m_w + m_{ice}) c_w T + m_{ice}(L_f + c_{ice}\Delta T_{ice}) - c_w[m_w T_w + m_{ice}(273.15 \text{ K})]$$
   $$T = \frac{c_w[m_w T_w + m_{ice}(273.15 \text{ K})] - m_{ice}(L_f + c_{ice}\Delta T_{ice})}{(m_w + m_{ice})c_w}$$
   $$T = \frac{\begin{matrix}[4.186 \text{ kJ/(kg} \cdot \text{K)}][(0.250 \text{ kg})(273.15 \text{ K} + 25.0 \text{ K}) + (0.037 \text{ kg})(273.15 \text{ K})] \\ - (0.037 \text{ kg})\{333.7 \text{ kJ/kg} + [2.1 \text{ kJ/(kg} \cdot \text{K)}](10.0 \text{ K})\}\end{matrix}}{(0.250 \text{ kg} + 0.037 \text{ kg})[4.186 \text{ kJ/(kg} \cdot \text{K)}]} - 273.15 \text{ K} = \boxed{11°\text{C}}$$

9. (a) **Strategy** The maximum power emission is inversely proportional to the absolute temperature. Use Wien's law.

   Solution Compute the surface temperature of the star.
   $$T = \frac{2.898 \times 10^{-3} \text{ m} \cdot \text{K}}{\lambda_{max}} = \frac{2.898 \times 10^{-3} \text{ m} \cdot \text{K}}{700.0 \times 10^{-9} \text{ m}} = \boxed{4140 \text{ K}}$$

   (b) **Strategy** Use Stefan's law of blackbody radiation.

   Solution Compute the power radiated.
   $$\mathscr{P} = \sigma A T^4 = [5.670 \times 10^{-8} \text{ W/(m}^2 \cdot \text{K}^4)][4\pi(7.20 \times 10^8 \text{ m})^2](4140 \text{ K})^4 = \boxed{1.09 \times 10^{26} \text{ W}}$$

   (c) **Strategy** Intensity is power radiated per unit area.

   Solution Compute the intensity measured by the Earth-based observer.
   $$I = \frac{\mathscr{P}}{A} = \frac{1.085 \times 10^{26} \text{ W}}{4\pi(9.78 \text{ ly})^2(9.461 \times 10^{15} \text{ m/ly})^2} = \boxed{1.01 \times 10^{-9} \text{ W/m}^2}$$

13. Strategy The temperature is constant and the heat entering the system is $Q = mL_f$.

Solution Find the change in entropy of the ice.
$$\Delta S = \frac{Q}{T} = \frac{mL_f}{T} = \frac{(2.00 \text{ kg})(333.7 \text{ kJ/kg})}{273.15 \text{ K} + 0.0 \text{ K}} = \boxed{2.44 \text{ kJ/K}}$$

17. Strategy The heat loss is proportional to the temperature difference.

Solution Compute the reduction in heat loss.
$$\mathcal{P}_2 = \frac{\Delta T_2}{\Delta T_1} \mathcal{P}_1 = \frac{81°C - 36°C}{81°C - 21°C} \mathcal{P}_1 = 0.75 \mathcal{P}_1$$

The heat loss was $\boxed{\text{reduced to 75\% of the original}}$.

21. (a) Strategy and Solution

> The boiling temperature of water varies with pressure. If the pressure is high, the water molecules are pushed close together, making it harder for them to form a gas. (Gas molecules are farther apart from each other than are liquid molecules.) A higher pressure raises the temperature at which the coolant fluid will boil.

(b) Strategy and Solution

> If you were to remove the cap on your radiator without first bringing the radiator pressure down to atmospheric pressure, the fluid would suddenly boil, sending out a jet of hot steam that could burn you.

25. (a) Strategy Use the ideal gas law, $PV = nRT$. Draw a qualitative diagram.

Solution First, the temperature is constant, so $P \propto V^{-1}$. Since the volume is reduced to one-eighth of its initial size, the pressure increases by a factor of eight. Next, the volume is constant, while the temperature and pressure increases. Then, the temperature is again constant. Finally, the volume is constant as the temperature and pressure decreases. The P-V diagram is shown.

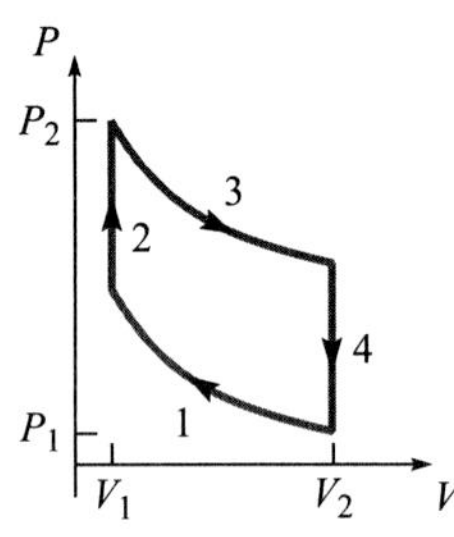

(b) Strategy Refer to the diagram in part (a). Calculate the quantities for each step in the cycle. Note that the gas is diatomic.

Solution Step 1, isothermal process:

The work done on the gas is $W = nRT \ln \dfrac{V_i}{V_f} = (2.00 \text{ mol})[8.314 \text{ J/(mol} \cdot \text{K)}](325 \text{ K}) \ln 8 = 11.2 \text{ kJ}$.

The change in the internal energy of the gas is 0 for an isothermal process.
The heat transferred is $Q = -W = -11.2 \text{ kJ}$.

Step 2, isochoric process:
Without a displacement, work cannot be done, so $W = 0$.
The change in the internal energy of the gas is equal to the heat that enters the system, so the change in internal energy and the heat transferred are
$$\Delta U = Q = nC_v \Delta T = n\left(\frac{5}{2}R\right)\Delta T = \frac{5}{2}(2.00 \text{ mol})[8.314 \text{ J/(mol} \cdot \text{K)}](985 \text{ K} - 325 \text{ K}) = 27.4 \text{ kJ}.$$

Step 3, isothermal process:

The work done on the gas is $W = nRT \ln \dfrac{V_i}{V_f} = (2.00 \text{ mol})[8.314 \text{ J/(mol} \cdot \text{K)}](985 \text{ K}) \ln \dfrac{1}{8} = -34.1 \text{ kJ}.$

The change in the internal energy of the gas is 0 for an isothermal process.
The heat transferred is $Q = -W = 34.1 \text{ kJ}.$

Step 4, isochoric process:
Without a displacement, work cannot be done, so $W = 0.$
The change in the internal energy of the gas is equal to the heat that enters the system, so the change in internal energy and the heat transferred are

$$\Delta U = Q = nC_v\Delta T = n\left(\frac{5}{2}R\right)\Delta T = \frac{5}{2}(2.00 \text{ mol})[8.314 \text{ J/(mol} \cdot \text{K)}](325 \text{ K} - 985 \text{ K}) = -27.4 \text{ kJ}.$$

The results of the processes and the totals are shown in the table. (Note that the totals for work and heat differ slightly from the sums of the values for each step due to round-off error.)

Process	W (kJ)	ΔU (kJ)	Q (kJ)
Step 1	11.2	0	−11.2
Step 2	0	27.4	27.4
Step 3	−34.1	0	34.1
Step 4	0	−27.4	−27.4
Total	−22.8	0	22.8

(c) **Strategy** The efficiency is equal to the ratio of the net work done by the gas to the heat transferred into the gas.

Solution The work done by the gas is negative the work done on the gas.
$W_{\text{net}} = -(-22.8 \text{ kJ}) = 22.8 \text{ kJ}$ and the heat transferred into the gas is $Q_{\text{in}} = 27.4 \text{ kJ} + 34.1 \text{ kJ} = 61.5 \text{ kJ}.$

The efficiency of the engine is $e = \dfrac{W_{\text{net}}}{Q_{\text{in}}} = \dfrac{22.8 \text{ kJ}}{61.5 \text{ kJ}} = \boxed{0.371 \text{ or } 37.1\%}.$

(d) **Strategy** Use Eq. (15-17).

Solution Compute the efficiency of a Carnot engine operating at the same extreme temperatures.
$$e_r = 1 - \frac{T_C}{T_H} = 1 - \frac{325 \text{ K}}{985 \text{ K}} = \boxed{0.670 \text{ or } 67.0\%}$$

29. (a) **Strategy** Use conservation of energy. Ignore air resistance.

Solution Find the escape speed.
$$K_i + U_i = K_f + U_f$$
$$\frac{1}{2}mv^2 - \frac{GMm}{R_E} = 0 + 0$$
$$v = \sqrt{\frac{2GM}{R_E}} = \sqrt{\frac{2(6.674 \times 10^{-11} \text{ N} \cdot \text{m}^2/\text{kg}^2)(5.974 \times 10^{24} \text{ kg})}{6.37 \times 10^6 \text{ m}}} = \boxed{11.2 \text{ km/s}}$$

(b) Strategy Use Eq. (13-22).

Solution Calculate the average speed.
$$v = \sqrt{\frac{3kT}{m}} = \sqrt{\frac{3(1.381\times10^{-23}\ \text{J/K})(273.15\ \text{K})}{(2.00\ \text{u})(1.6605\times10^{-27}\ \text{kg/u})]}} = \boxed{1850\ \text{m/s}}$$

(b) Strategy Use Eq. (13-22).

Solution Calculate the average speed.
$$v = \sqrt{\frac{3kT}{m}} = \sqrt{\frac{3(1.381\times10^{-23}\ \text{J/K})(273.15\ \text{K})}{(32.0\ \text{u})(1.6605\times10^{-27}\ \text{kg/u})]}} = \boxed{461\ \text{m/s}}$$

(d) Strategy and Solution The escape speed is about 6 times the rms speed for hydrogen and more than 24 times the rms speed for oxygen. A small but significant fraction of the hydrogen molecules have speeds greater than the escape speed and can escape from the atmosphere. The fraction of oxygen molecules with speeds greater than the escape speed is negligibly small (see Fig. 13.14).

MCAT Review

1. **Strategy and Solution** According to the second law of thermodynamics, heat never flows spontaneously from a colder body to a hotter body, therefore, heat will not flow from bar A to bar B. The correct answer is $\boxed{\text{C}}$.

2. **Strategy** Assume that the specific heat capacity of seawater is approximately the same at $0°\text{C}$ and $5°\text{C}$.

   Solution Find the approximate temperature T.
   $0 = Q_0 + Q_5 = mc(T - 0°\text{C}) + mc(T - 5°\text{C})$, so $2T = 5°\text{C}$ or $T = 2.50°\text{C}$.
   The correct answer is $\boxed{\text{B}}$.

3. **Strategy** Use the latent heat of fusion for water.

   Solution The heat gained by the ice when melting is $Q = mL_\text{f} = (0.0180\ \text{kg})(333.7\ \text{kJ/kg}) = 6.01\ \text{kJ}$.
   The correct answer is $\boxed{\text{C}}$.

4. **Strategy and Solution** Since $e = 1 - Q_\text{C}/Q_\text{H} = 1 - T_\text{C}/T_\text{H}$, decreasing the exhaust temperature will increase the steam engine's efficiency. The correct answer is $\boxed{\text{B}}$.

5. **Strategy and Solution** Since refrigerators remove heat by transferring it to a liquid that vaporizes, refrigerators are primarily dependent upon the heat of vaporization of the refrigerant liquid. The correct answer is $\boxed{\text{A}}$.

6. **Strategy and Solution** Steam is generally at a higher temperature than water and the specific heat of steam is lower than that of water, so water would be more effective than steam for changing steam to water. Circulating water brings more mass of water in contact with the condenser than stationary water, so it can carry away heat at a faster rate, therefore, it would be more effective for changing steam to water. The correct answer is $\boxed{\text{D}}$.

7. **Strategy and Solution** Since it is not possible to convert all of the input heat into output work, the amount of useful work that can be generated from a source of heat can only be less than the amount of heat. The correct answer is $\boxed{\text{A}}$.

8. **Strategy and Solution** The internal energy of the steam is converted into mechanical energy as it expands and moves the piston of the steam engine to the right, therefore, the correct answer is $\boxed{\text{C}}$.

9. **Strategy and Solution** The refrigerant must be able to vaporize (boil) at temperatures lower than the freezing point of water so that it can carry away heat (as a gas) from the contents of the refrigerator (which contain water) to cool and possibly freeze the contents. The correct answer is $\boxed{\text{B}}$.

10. **Strategy** The heat transferred to the water by the heaters was $Q_w = m_w c_w \Delta T_w$. The heat required for the oil is $Q_o = m_o c_o \Delta T_o$.

    Solution Form a proportion and use the temperature changes of the oil and water and the specific heat and the specific gravity of the oil to obtain a ratio of heat required for the oil to that transferred to the water.
    $$\frac{Q_o}{Q_w} = \frac{m_o c_o \Delta T_o}{m_w c_w \Delta T_w} = \frac{(0.7 m_w)(0.60 c_w)(60 - 20)}{m_w c_w (100 - 20)} = 0.21$$
    So, 21% of the amount of heat transferred to the water is required to heat the oil to 60°C. Assuming the heaters work at the same rate for both the water and the oil, the time required to raise the temperature of the oil from 20°C to 60°C is $0.21(15\text{ h}) = 3.2\text{ h}$. The correct answer is $\boxed{\text{A}}$.

11. **Strategy and Solution** The high pressure would increase the pressure on the plug, making it more difficult to lift. The pressure difference between the air in the tank and the air outside of the tank would increase the fluid velocity when the tank is drained, thus, decreasing the time required to drain the tank. The time required to heat the oil would be the least likely affected, since the oil is fairly incompressible. The correct answer is $\boxed{\text{A}}$.

Chapter 16

ELECTRIC FORCES AND FIELDS

Conceptual Questions

1. This proposition would not work because even with small net charges some objects would be observed to repel each other, which does not happen with gravity. To account for the weight of an object one may say, for example, that the Earth is slightly positively charged and the object negatively charged. A slightly positively charged object should then be repelled from the Earth and fall upward. Furthermore, increasing the charge on an object would increase the force from the Earth, but the weight is not observed to change by increasing an object's charge.

5. **(a)** Since the sphere is positively charged, it must have lost electrons to the charged rod, so its mass will be smaller. We know that electrons were transferred from the sphere to the rod, and that positive ions were not transferred from the rod to the sphere, because electrons in a metal are much more mobile than ions in any solid.

   (b) The rod must have been positively charged since it attracted electrons off of the sphere. Also, after the two objects are touched they must be at the same potential. Since the sphere ends up positively charged, the rod must end up positively charged as well.

9. No, the net charge—sum of all the charges with signs—is the same before and after.

13. **(a)** The foils are positively charged and will repel each other even after the rod is removed.

    (b) As another positively charged rod is brought close to the conducting sphere, the foils become further positively charged by induction. They will move farther apart, but will return to their previous position if the rod is removed.

    (c) If a negatively charged rod is now brought near the sphere, the foils will become less positively charged and will move closer together.

17. Electric flux is the total "amount" of electric field "flowing" through a surface. It is analogous to the volume flow rate of water through a pipe. For water, the product of the perpendicular velocity and the surface area gives the volume of water flowing through the surface per unit time. Electric flux originates from positive charges just as the flow of water originates from a faucet. Both are therefore given the name "sources". Electric flux flows toward negative charges just as water flows toward drains. Both are therefore given the name "sinks".

Problems

1. **Strategy** There are 10 protons in each water molecule. Multiply the elementary charge by Avogadro's number and the number of protons per molecule.

   Solution Find the total positive charge.

   $$10(1.0 \text{ mol})(6.022 \times 10^{23} \text{ mol}^{-1})(1.602 \times 10^{-19} \text{ C}) = \boxed{9.6 \times 10^5 \text{ C}}$$

3. **(a) Strategy and Solution** Since electrons have negative charge, and since the balloon acquired a negative net charge, electrons were $\boxed{\text{added}}$ to the balloon.

   (b) Strategy Divide the net charge by the charge of an electron.

   Solution Compute the number of electrons transferred.

   $$\frac{-0.60 \times 10^{-9} \text{ C}}{-1.602 \times 10^{-19} \text{ C}} = \boxed{3.7 \times 10^9}$$

5. Strategy and Solution

(a) When the rod is brought near sphere A, negative charge flows from sphere B to sphere A. The spheres are then moved apart and the rod is removed, so A is left with a net $\boxed{\text{negative charge}}$.

(b) Sphere B has $\boxed{\text{an equal magnitude of positive charge}}$, since the two spheres were initially uncharged.

7. Strategy Each time a pair of spheres makes contact, their net charge is shared equally, with the exception of the time when C is grounded.

Solution After A and B make contact and are separated, each sphere has a charge of $Q/2$. After B and C make contact and are separated, each sphere has zero charge because sphere C was grounded. After A and C make contact and are separated, each sphere has a charge of $(Q/2+0)/2 = Q/4$. The charges on spheres A and C are $\boxed{Q/4}$. The charge on sphere B is $\boxed{0}$.

9. Strategy Use Coulomb's law, Eq. (16-2). Ignore the constant k and units for simplicity.

Solution Coulomb's law written with the given variables is
$$F = \frac{k|q_1||q_2|}{r^2} = \frac{k|Q_1||Q_2|}{d^2}.\text{Compute the magnitudes.}$$
(a) $\dfrac{1\times2}{1^2} = 2$; (b) $\dfrac{2\times1}{1^2} = 2$; (c) $\dfrac{2\times4}{4^2} = \dfrac{1}{2}$; (d) $\dfrac{2\times2}{2^2} = 1$; (e) $\dfrac{4\times2}{4^2} = \dfrac{1}{2}$

Ranking the situations in order of the magnitude of the electric force on Q_1, from largest to smallest, we have $\boxed{\text{(a) = (b), (d), (c) = (e)}}$.

11. Strategy Use Coulomb's law, Eq. (16-2).

Solution Find the charge on each sphere.
$$\frac{kq^2}{r^2} = F, \text{ so } q = \sqrt{\frac{Fr^2}{k}} = \sqrt{\frac{(0.036 \text{ N})(0.250 \text{ m})^2}{8.988\times10^9 \text{ N}\cdot\text{m}^2/\text{C}^2}} = 5.0\times10^{-7} \text{ C.}$$

The charge is negative, so each sphere has $\boxed{-5.0\times10^{-7} \text{ C}}$ of charge on it.

13. Strategy Set the magnitudes of the Coulomb force and the gravitational force equal and solve for the charge; then divide the charge by the elementary charge to find the number of electrons.

Solution Set the magnitudes of the forces equal.
$$F_{\text{q}} = \frac{kq^2}{r^2} = F_{\text{g}} = \frac{Gm^2}{r^2}, \text{ so } q = \sqrt{\frac{Gm^2}{k}}.$$

$$\text{number of electrons} = N = \frac{q}{e} = \frac{1}{1.602\times10^{-19} \text{ C}}\sqrt{\frac{(6.674\times10^{-11} \text{ N}\cdot\text{m}^2/\text{kg}^2)(5.0 \text{ kg})^2}{8.988\times10^9 \text{ N}\cdot\text{m}^2/\text{C}^2}} = \boxed{2.7\times10^9}$$

17. Strategy Use Coulomb's law, Eq. (16-2).

Solution The original force magnitude is
$$F = \frac{kq^2}{r^2}, \text{ so the new magnitude is } F_{0.25} = \frac{kq^2}{(0.25r)^2} = 16\left(\frac{kq^2}{r^2}\right) = \boxed{16F}.$$

21. Strategy Use Coulomb's law, Eq. (16-2). The force on the 1.0-μC charge due to the -0.60-μC charge is to the left and that due to the 0.80-μC charge is along the line between the charges and away from the 0.80-μC charge.

Solution Calculate the components of the force.

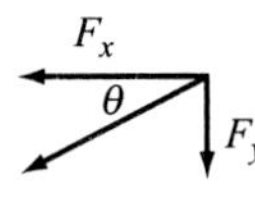

$$F_x = -\frac{(8.988\times10^9 \ \text{N}\cdot\text{m}^2/\text{C}^2)(0.60\times10^{-6}\ \text{C})(1.0\times10^{-6}\ \text{C})}{\left(\sqrt{(0.100\ \text{m})^2 - (0.080\ \text{m})^2}\right)^2}$$

$$+\frac{(8.988\times10^9 \ \text{N}\cdot\text{m}^2/\text{C}^2)(0.80\times10^{-6}\ \text{C})(1.0\times10^{-6}\ \text{C})}{(0.100\ \text{m})^2}\left(\frac{\sqrt{(0.100\ \text{m})^2 - (0.080\ \text{m})^2}}{0.100\ \text{m}}\right) = -1.1\ \text{N}$$

$$F_y = -\frac{(8.988\times10^9 \ \text{N}\cdot\text{m}^2/\text{C}^2)(0.80\times10^{-6}\ \text{C})(1.0\times10^{-6}\ \text{C})}{(0.100\ \text{m})^2}\left(\frac{0.080\ \text{m}}{0.100\ \text{m}}\right) = -0.58\ \text{N}$$

Calculate the magnitude and direction of the force.

$$F = \sqrt{F_x^2 + F_y^2} = \sqrt{(-1.067\ \text{m})^2 + (-0.575\ \text{m})^2} = 1.2\ \text{N and } \theta = \tan^{-1}\frac{F_y}{F_x} = \tan^{-1}\frac{-0.575}{-1.067} = 28°.$$

So, $\vec{\mathbf{F}} = \boxed{1.2\ \text{N at } 28° \text{ below the negative } x\text{-axis}}$.

25. Strategy Use Coulomb's law, Eq. (16-2).

Solution The net force is $\vec{\mathbf{F}} = \vec{\mathbf{F}}_{12} + \vec{\mathbf{F}}_{13}$. Separate $\vec{\mathbf{F}}$ into components.

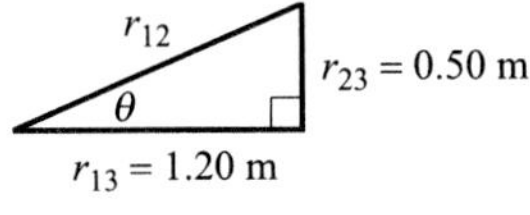

$$F_x = -\frac{k|q_1||q_2|}{r_{12}^2}\cos\theta\left(\frac{r_{13}}{r_{12}}\right) + \frac{k|q_1||q_3|}{r_{13}^2} = -\frac{k|q_1||q_2|}{r_{12}^2}\left(\frac{r_{13}}{r_{12}}\right) + \frac{k|q_1||q_3|}{r_{13}^2} \text{ and}$$

$$F_y = -\frac{k|q_1||q_2|}{r_{12}^2}\sin\theta = -\frac{k|q_1||q_2|}{r_{12}^2}\left(\frac{r_{23}}{r_{12}}\right).$$

Calculate the magnitude of $\vec{\mathbf{F}}$.

$$F = \sqrt{F_x^2 + F_y^2} = \sqrt{\left(-\frac{k|q_1||q_2|}{r_{12}^2}\left(\frac{r_{13}}{r_{12}}\right) + \frac{k|q_1||q_3|}{r_{13}^2}\right)^2 + \left(-\frac{k|q_1||q_2|}{r_{12}^2}\left(\frac{r_{23}}{r_{12}}\right)\right)^2} = \frac{k|q_1||q_2|}{r_{12}^3}\sqrt{\left(-r_{13} + \frac{|q_3|r_{12}^3}{|q_2|r_{13}^2}\right)^2 + r_{23}^2}$$

$$= \frac{(8.988\times10^9 \ \text{N}\cdot\text{m}^2/\text{C}^2)(1.2\times10^{-6}\ \text{C})(0.60\times10^{-6}\ \text{C})}{\left(\sqrt{(1.20\ \text{m})^2 + (0.50\ \text{m})^2}\right)^3}$$

$$\times\sqrt{\left[-1.20\ \text{m} + \frac{(0.20\times10^{-6}\ \text{C})\left(\sqrt{(1.20\ \text{m})^2 + (0.50\ \text{m})^2}\right)^3}{(0.60\times10^{-6}\ \text{C})(1.20\ \text{m})^2}\right]^2 + (0.50\ \text{m})^2} = \boxed{2.5\ \text{mN}}$$

27. Strategy Use Eq. (16-4b).

Solution Compute the force on the sphere.

$$\vec{\mathbf{F}} = q\vec{\mathbf{E}} = (-6.0\times10^{-7}\ \text{C})(1.2\times10^6 \ \text{N/C west}) = \boxed{0.72\ \text{N to the east}}$$

29. Strategy Use Newton's second law and Eq. (16-4b).

Solution $\vec{F} = q\vec{E} = e\vec{E}$ for a proton. Find the acceleration.

$$m\vec{a} = e\vec{E}, \text{ so } \vec{a} = \frac{e\vec{E}}{m} = \frac{(1.602\times10^{-19} \text{ C})(33\times10^{3} \text{ N/C up})}{1.673\times10^{-27} \text{ kg}} = \boxed{3.2\times10^{12} \text{ m/s}^2 \text{ up}}.$$

33. Strategy The magnitude of the electric field is greater where the field lines are close together and weaker where they are far apart. Use this fact to determine the relative magnitudes of the electric field at each point.

Solution Ranking the points in order of the distance between field lines, from closest apart to farthest apart, is equivalent to ranking them in order of the magnitude of the electric field, from largest to smallest. Ranking the points, we have $\boxed{A, B, C, D, E}$.

35. Strategy The electric field at $x = d$ is the vector sum of the electric fields due to each positive charge. Use Eq. (16-5).

Solution Find the electric field.

$$E_x = \frac{k|q|}{d^2} - \frac{k|2q|}{(2d)^2} = \frac{k|q|}{d^2} - \frac{k|q|}{2d^2} = \frac{k|q|}{2d^2}$$

The electric field is $\boxed{\dfrac{k|q|}{2d^2} \text{ to the right}}$.

37. Strategy and Solution Since both charges are positive, the electric field is nonzero at all locations not on the x-axis. The answer is $\boxed{\text{no}}$.

39. Strategy The electric field at any point x is the vector sum of the electric fields due to each positive charge. Use Eq. (16-5) and the quadratic formula.

Solution Find where the electric field is equal to zero.

$$E_x = 0 = \frac{k|q|}{x^2} - \frac{k|2q|}{(3d-x)^2}, \text{ so } (x-3d)^2 = 2x^2. \text{ Solve for } x.$$

$$x^2 - 6dx + 9d^2 = 2x^2$$

$$0 = x^2 + 6dx - 9d^2$$

$$x = \frac{-6d \pm \sqrt{(6d)^2 - 4(1)(-9d^2)}}{2(1)}$$

$$x = \frac{-6d \pm \sqrt{72d^2}}{2}$$

$$x = -3d \pm 3d\sqrt{2} = 3d(-1 \pm \sqrt{2}) \approx 1.24d \text{ or } -7.24d$$

The value $-7.24d$ is extraneous, since the electric field is nonzero at $x = -7.24d$.

The electric field is zero at $\boxed{x = 3d(-1+\sqrt{2}) \approx 1.24d}$.

41. **Strategy** Electric field lines begin on positive charges and end on negative charges. In each situation, the magnitudes of the charges are equal, so the number of field lines beginning (for positive) or ending (for negative) on each is the same. Field lines never cross. Use the principles of superposition and symmetry.

Solution

(a) Electric field lines due to two positive point charges:

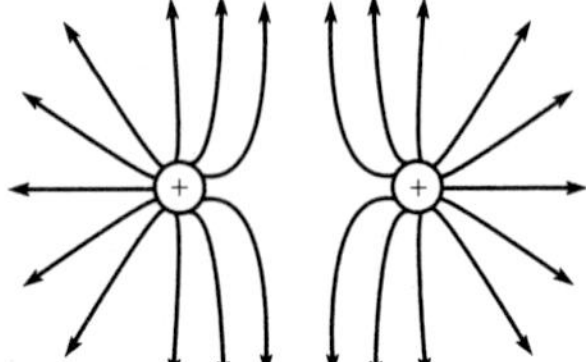

(b) Electric field lines due to two negative point charges:

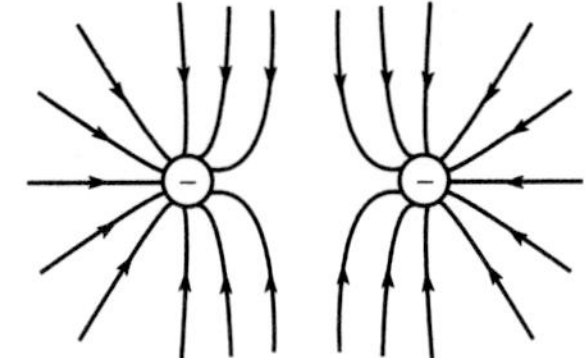

43. **Strategy** Let the x-direction be to the right and the y-direction be up. Due to symmetry, the x-components of the fields due to the two charges add to zero at point C; the vector sum of the y-components of the two fields is equal to twice that due to either one.

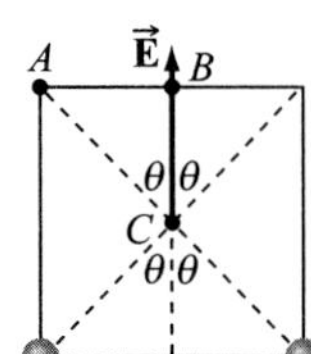

Solution Find the electric field at point C. The angle θ is $45°$.

$$E = 2E_y = 2\frac{k|q|}{r^2}\cos\theta = \frac{2(8.988\times10^9 \text{ N}\cdot\text{m}^2/\text{C}^2)(7.00\times10^{-6} \text{ C})}{(0.300/2 \text{ m})^2 + (0.300/2 \text{ m})^2}\times\frac{1}{\sqrt{2}} = 1.98\times10^6 \text{ N/C}$$

The electric field is $\boxed{1.98\times10^6 \text{ N/C up}}$.

45. **Strategy** Let the origin of the coordinate system be at A. The charges are positive, so the electric field due to the left-hand charge (1) points upward at A and that due to the right-hand charge (2) is directed at an angle of $135°$ as measured from the positive x-axis. Find the net electric field at A. Then, locate the third object such that its electric field is equal in magnitude and opposite in direction to the field generated by the first two objects.

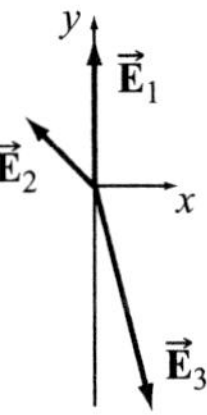

Solution Let $d = 0.300$ m. Find the net electric field due to the two objects.

$$E_x = E_{1x} + E_{2x} = 0 + \frac{k|q|}{2d^2}\cos135° = \frac{k|q|\cos135°}{2d^2}$$

$$E_y = E_{1y} + E_{2y} = \frac{k|q|}{d^2} + \frac{k|q|}{2d^2}\sin135° = \frac{k|q|(2+\sin135°)}{2d^2}$$

Compute the magnitude.

$$E = \sqrt{\left(\frac{k|q|\cos135°}{2d^2}\right)^2 + \left[\frac{k|q|(2+\sin135°)}{2d^2}\right]^2} = \frac{k|q|}{2d^2}\sqrt{(\cos135°)^2 + (2+\sin135°)^2}$$

Set the magnitude of the electric field due to the third object equal to E and solve for the distance r that the third object should be from A.

$$\frac{k|q|}{r^2} = \frac{k|q|}{2d^2}\sqrt{(\cos135°)^2 + (2+\sin135°)^2}, \text{ so}$$

$$r = d\sqrt{\frac{2}{\sqrt{(\cos135°)^2 + (2+\sin135°)^2}}} = (0.300 \text{ m})\sqrt{\frac{2}{\sqrt{(\cos135°)^2 + (2+\sin135°)^2}}} = 0.254 \text{ m}.$$

Compute the direction.

$$\theta = \tan^{-1}\frac{E_y}{E_x} = \tan^{-1}\frac{2+\sin 135°}{\cos 135°} = 75.4° \text{ above the } -x\text{-axis}$$

The charge should be placed on a line that makes an angle of $75.4°$ above the negative x-axis at a distance of 0.254 m along that line from point A. Thus, the charge is above and to the left of A.

49. Strategy Use Eq. (16-5) and the principles of superposition and symmetry. Set the sum of the electric fields equal to 0. $x = d$ is the point at which the sum is zero.

Solution Find the point where the electric field is zero.

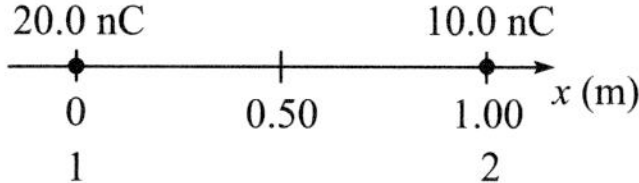

$$0 = \bar{E}_1 + \bar{E}_2,\text{ so } 0 = \frac{kq_1}{d^2} - \frac{kq_2}{(x_2-d)^2}. \text{ Solve for } d.$$

$$\frac{(x_2-d)^2}{d^2} = \left(\frac{x_2}{d}-1\right)^2 = \frac{q_2}{q_1},\text{ so } d = \frac{x_2}{1\pm\sqrt{q_2/q_1}} = \frac{1.00\text{ m}}{1\pm\sqrt{10.0/20.0}} = 0.586\text{ m or }3.41\text{ m}.$$

3.41 m is extraneous (the field lines both point in the same direction at that point), so $d = \boxed{0.586\text{ m}}$.

53. (a) Strategy Use Eq. (16-4b).

Solution Find the force on the electron.

$$\bar{F} = -e\bar{E} = -(1.602\times10^{-19}\text{ C})(500.0\text{ N/C up}) = \boxed{8.010\times10^{-17}\text{ N down}}$$

(b) Strategy Use the work-kinetic energy theorem. The work done on the electron is equal to the force on the electron times the deflection.

Solution Find the increase in the kinetic energy of the electron.

$$\Delta K = W = Fd = (8.010\times10^{-17}\text{ N})(0.00300\text{ m}) = \boxed{2.40\times10^{-19}\text{ J}}$$

55. (a) Strategy Compare the electrical and gravitational forces.

Solution The gravitational force is $mg = (0.00230\text{ kg})(9.80\text{ m/s}^2) = 2.25\times10^{-2}$ N. The electrical force is $qE = (10.0\times10^{-6}\text{ C})(6.50\times10^3\text{ N/C}) = 6.50\times10^{-2}$ N.

The gravitational force is about $1/3$ of the electrical force, so the gravitational force can't be neglected.

(b) Strategy Add the forces and find the total acceleration using Newton's second law. Then, use the formula for the range of a projectile.

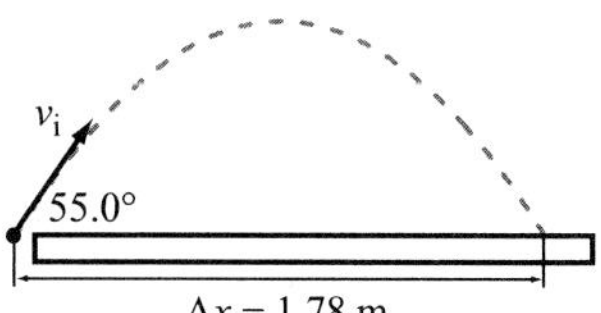

Solution Find the downward acceleration.

$$a = \frac{F_g + F_e}{m}. \text{ Find } \Delta x.$$

$$\Delta x = R = \frac{v_i^2 \sin 2\theta}{a} = \frac{mv_i^2 \sin 2\theta}{F_g + F_e} = \frac{(0.00230\text{ kg})(8.50\text{ m/s})^2 \sin[2(55.0°)]}{2.25\times10^{-2}\text{ N} + 6.50\times10^{-2}\text{ N}} = \boxed{1.78\text{ m}}$$

57. Strategy Find the necessary acceleration in terms of the average electric field. Use Eq. (4-5).

Solution The acceleration is related to the average electric field by $a = qE/m$.

$$v_{fx}^2 - v_{ix}^2 = v_{fx}^2 - 0 = 2a_x \Delta x = \frac{2qE\Delta x}{m}, \text{ so } E = \frac{mv_{fx}^2}{2q\Delta x} = \frac{(1.673\times10^{-27} \text{ kg})(1.0\times10^7 \text{ m/s})^2}{2(1.602\times10^{-19} \text{ C})(4.0 \text{ m})} = \boxed{1.3\times10^5 \text{ N/C}}.$$

61. Strategy and Solution The $-6 \ \mu C$ of charge on the conducting sphere induces a positive charge of $6 \ \mu C$ on the inner surface of the conducting shell. This, in turn, induces a negative charge on the outer surface of the shell. The conducting shell has a net 1-μC charge, so $-6 \ \mu C + 1 \ \mu C = \boxed{-5 \ \mu C}$ is the charge on the outer surface.

65. Strategy Since the spheres have charge of equal magnitude and opposite sign, an equal number of field lines begin on the positively-charged sphere and end on the negatively-charged sphere.

Solution The sketch is shown.

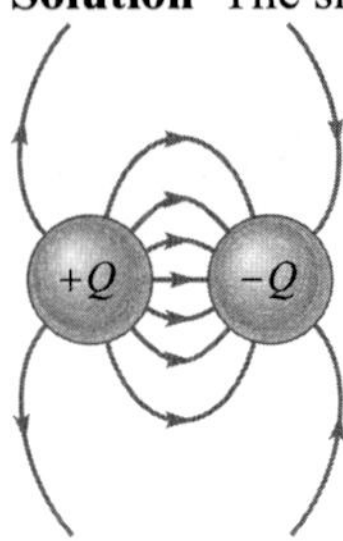

69. Strategy Use the definition of electric flux, Eq. (16-8).

Solution

(a) $\Phi_E = E_\perp A = E_\perp a^2$, so $\boxed{\Phi_{E\parallel} = 0, \ \Phi_{E\perp \text{out}} = Ea^2, \text{ and } \Phi_{E\perp \text{in}} = -Ea^2.}$

(b) $\Phi_E = Ea^2 - Ea^2 = \boxed{0}$

73. Strategy Use Gauss's law, Eq. (16-9), and symmetry.

Solution

(a) The electric field outside a spherically symmetrical charge distribution (with total charge q) is directed radially away from its center, and is parallel at any point to the normal of a spherical Gaussian surface outside the distribution and concentric with it. So, $E_\perp = E$.

$\Phi_E = E_\perp A = EA = E(4\pi r^2) = 4\pi kq$, so $E = kq/r^2$.
The result is the same as the electric field due to a point charge q.

(b) If a spherical Gaussian surface is placed within the charge distribution, $q = 0$ (no enclosed charge).
$\Phi_E = E_\perp A = EA = 4\pi kq$, so $E = 4\pi kq/A = 4\pi k(0)/A = 0$.

77. (a) Strategy and Solution The electric field is zero inside a conductor in static equilibrium, so the electric field inside the sheet is $\boxed{0}$.

(b) Strategy Use Gauss's law, Eq. (16-9).

Solution The Gaussian surface is a "pill box." It is a cylinder with its top and bottom circular surfaces parallel to the surface of the sheet. The electric field lines are approximately parallel to the side of the cylinder, so $\Phi_{E\,\text{side}} = E_{\perp} A_{\text{side}} = 0$, or $E_{\perp} = 0$.

$$\Phi_{E\,\text{net}} = E_{\text{top}} A_{\text{top}} + E_{\text{bottom}} A_{\text{bottom}} = \frac{q}{\epsilon_0},$$ where the top is outside the sheet and the bottom is inside the sheet.

$E_{\text{bottom}} = 0$, and let $A_{\text{top}} = A_{\text{bottom}} = A$ and $E_{\text{top}} = E$, the field just outside the sheet. Find E.

$$EA + (0)A = EA = \frac{q}{\epsilon_0}, \text{ so } E = \frac{1}{\epsilon_0}\left(\frac{q}{A}\right) = \frac{\sigma}{\epsilon_0}.$$

(c) Strategy and Solution $\boxed{\text{No}}$, the results do not contradict each other. Applying the superposition principle to two parallel sheets of charge gives the same result.

81. Strategy Use Coulomb's law and the principle of superposition.

Solution

(a) Find the electric field at point P.

$$\vec{E}(P) = \frac{k|q_t|}{r_t^2} \text{ down} + \frac{k|q_b|}{r_b^2} \text{ up} = k\left(-\frac{|q_t|}{r_t^2} + \frac{|q_b|}{r_b^2}\right) \text{ up}$$

$$= (8.988 \times 10^9 \text{ N} \cdot \text{m}^2/\text{C}^2)\left[-\frac{50 \text{ C}}{(10,000 \text{ m})^2} + \frac{20 \text{ C}}{(2000 \text{ m})^2}\right] \text{ up} = \boxed{4.0 \times 10^4 \text{ N/C up}}$$

(b) Since the electric field is directed upward at the surface, $\boxed{\text{positive}}$ charge would accumulate there.

85. (a) Strategy Set the magnitude of the electric force equal to the magnitude of the gravitational force.

Solution Find the required magnitudes of the net charges on the Earth and the Sun.

$$\frac{k|q_S||q_E|}{r^2} = \frac{Gm_S m_E}{r^2}, \text{ so } |q_S||q_E| = \frac{G}{k} m_S m_E.$$

Let $\dfrac{|q_E|}{m_E} = \dfrac{|q_S|}{m_S}$. Then, $|q_E| = \dfrac{m_E}{m_S}|q_S|$. Find $|q_S|$.

$$|q_S| = \frac{G}{k} m_S m_E \left(\frac{1}{|q_E|}\right) = \frac{G}{k} m_S m_E \left(\frac{m_S}{m_E |q_S|}\right)$$

$$q_S^2 = \frac{Gm_S^2}{k}$$

$$|q_S| = m_S \sqrt{\frac{G}{k}} = (1.987 \times 10^{30} \text{ kg}) \sqrt{\frac{6.674 \times 10^{-11} \text{ N} \cdot \text{m}^2/\text{kg}^2}{8.988 \times 10^9 \text{ N} \cdot \text{m}^2/\text{C}^2}} = \boxed{1.712 \times 10^{20} \text{ C}}$$

Find $|q_E|$.

$$|q_E| = \frac{G}{k} m_S m_E \left(\frac{1}{|q_S|} \right) = \frac{G}{k} m_S m_E \left(\frac{m_E}{m_S |q_E|} \right)$$

$$q_E^2 = \frac{G m_E^2}{k}$$

$$|q_E| = m_E \sqrt{\frac{G}{k}} = (5.974 \times 10^{24} \text{ kg}) \sqrt{\frac{6.674 \times 10^{-11} \text{ N} \cdot \text{m}^2 / \text{kg}^2}{8.988 \times 10^9 \text{ N} \cdot \text{m}^2 / \text{C}^2}} = \boxed{5.148 \times 10^{14} \text{ C}}$$

(b) **Strategy and Solution** If the magnitude of the charges of the proton and electron were not exactly equal, astronomical bodies would have net charges with the same sign, so the force between them would be repulsive. The force responsible for the Earth's orbit is attractive, so this charge imbalance could not possibly be the explanation for Earth's orbit. The answer is $\boxed{\text{no}}$.

89. **Strategy** Since the charge is negative and the electric field is directed upward, the force on the charge is directed downward. Use Eq. (16-4b).

Solution Compute the magnitude of the force on the raindrop.

$$F = 8eE = 8(1.602 \times 10^{-19} \text{ C})(2.0 \times 10^6 \text{ N/C}) = 2.6 \text{ pN}$$

The force on the raindrop is $\boxed{2.6 \text{ pN down}}$.

93. **Strategy** Use Coulomb's law and the principle of superposition.

Solution Find q.

$$0 = E_x(1.0 \text{ m}, 0) = \frac{kq_0}{d^2} + \frac{kq}{(d/2)^2} = q_0 + 4q, \text{ so}$$

$$q = -\frac{1}{4} q_0 = -\frac{1}{4}(6.0 \text{ nC}) = \boxed{-1.5 \text{ nC}}.$$

97. (a) **Strategy** Use the acceleration of a charged particle in an electric field and Eq. (4-5).

Solution Find the speed of the electrons.

$$v_f^2 - v_i^2 = v_f^2 - 0 = 2a\Delta x = 2\left(\frac{eE}{m}\right)\Delta x, \text{ so}$$

$$v_f = \sqrt{2\left(\frac{eE}{m}\right)\Delta x} = \sqrt{\frac{2(1.602 \times 10^{-19} \text{ C})(4.0 \times 10^5 \text{ N/C})(0.050 \text{ m})}{9.109 \times 10^{-31} \text{ kg}}} = \boxed{8.4 \times 10^7 \text{ m/s}}.$$

(b) **Strategy** Use Eq. (4-4) and $\Delta x = v \Delta t$.

Solution Find the time intervals during the acceleration of a the electrons and their motion at constant speed.

$$\Delta x_1 = \frac{1}{2} a (\Delta t_1)^2, \text{ so } \Delta t_1 = \sqrt{\frac{2\Delta x_1}{a}} = \sqrt{\frac{2m\Delta x_1}{eE}}. \quad \Delta x_2 = v\Delta t_2, \text{ so } \Delta t_2 = \frac{\Delta x_2}{v} = \frac{\Delta x_2}{\sqrt{\frac{2eE\Delta x_1}{m}}}.$$

Find the total time to travel the length of the tube.

$$\Delta t = \Delta t_1 + \Delta t_2 = \sqrt{\frac{2m\Delta x_1}{eE}} + \Delta x_2 \sqrt{\frac{m}{2eE\Delta x_1}} = \left(1 + \frac{\Delta x_2}{2\Delta x_1}\right)\sqrt{\frac{2m\Delta x_1}{eE}}$$

$$= \left[1 + \frac{45 \text{ cm}}{2(5.0 \text{ cm})}\right]\sqrt{\frac{2(9.109 \times 10^{-31} \text{ kg})(5.0 \times 10^{-2} \text{ m})}{(1.602 \times 10^{-19} \text{ C})(4.0 \times 10^5 \text{ N/C})}} = \boxed{6.6 \text{ ns}}$$

101. (a) Strategy Use Coulomb's law and the principle of superposition. The direction of the electric field is up if $E > 0$ and down if $E < 0$. Let up be the $+y$-direction.

Solution Write the expression for the electric field and specify the direction.

$$E\left(0,\ y > \frac{1}{2}d\right) = \frac{kq}{\left(y - \frac{d}{2}\right)^2} - \frac{kq}{\left(y + \frac{d}{2}\right)^2};\ +y\text{-direction}$$

(b) Strategy Use the binomial approximation $(1 \pm x)^n \approx 1 \pm nx$ for $x \ll 1$.

Solution Approximate the expression found in part (a) at distant points from the dipole $(y \gg d)$.

$$E\left(0,\ y > \frac{1}{2}d\right) = \frac{kq}{y^2\left(1 - \frac{d}{2y}\right)^2} - \frac{kq}{y^2\left(1 + \frac{d}{2y}\right)^2} \approx \frac{kq}{y^2}\left[1 + \frac{d}{y} - \left(1 - \frac{d}{y}\right)\right] = \frac{2kqd}{y^3}$$

So, $\boxed{\ \vec{\mathbf{E}}(y \gg d) \approx \dfrac{2kqd}{y^3};\ +y\text{-direction}\ }$.

The field is proportional to $\boxed{1/y^3}$. $\boxed{\text{No}}$, the result does not conflict with Coulomb's law, which applies to individual point charges. E is due to the superposition of the fields due to two point charges. The field due to each charge obeys Coulomb's law.

Chapter 17

ELECTRIC POTENTIAL

Conceptual Questions

1. **(a)** The electric field does positive work on $-q$ as it moves closer to $+Q$.

   (b) The potential increases as $-q$ moves closer to $+Q$.

   (c) The potential energy of $-q$ decreases.

   (d) If the fixed charge instead has a value $-Q$, the electric field does negative work, the potential decreases, and the potential energy increases.

5. Zero work is required to move a charge between two points at the same potential. An external force may need to be applied to move the charge but the work done to start the charge in motion will be negated by the work done to stop it.

9. If the electric field is zero throughout a region of space, the electric potential must be constant throughout that region.

13. It doesn't matter which points we choose because the potential on each plate is constant over the whole plate.

17. We can't say anything about the electric field if all we know is the potential at a single point. The electric field tells us how the potential *changes* if we move from one point to another.

21. The plates are isolated so the charge remains constant. The capacitance is given by $C = \kappa \epsilon_0 A / d$. As the plates are moved closer, all these quantities remain constant except d, which decreases. Therefore, the capacitance increases. Since $Q = C\Delta V$, and Q is constant, ΔV must decrease. The electric field remains constant. The energy stored in the capacitor decreases as well.

Problems

1. **Strategy** Use Eq. (17-1). Ignore units and the constant k for simplicity.

   Solution The electric potential energy is given by $U_E = kq_1q_2/r$.

   (a) $\dfrac{1\times2}{1} = 2$; (b) $\dfrac{2\times-1}{1} = -2$; (c) $\dfrac{2\times-4}{2} = -4$; (d) $\dfrac{-2\times-2}{2} = 2$; (e) $\dfrac{4\times-2}{4} = -2$

   Ranking these situations in order of electric potential energy, from highest to lowest, we have $\boxed{(a) = (d), (b) = (e), (c)}$.

3. **(a) Strategy** Use Eq. (17-1).

   Solution Compute the electric potential energy.
   $$U_E = k\frac{q_1q_2}{r} = \frac{(8.988\times10^9 \ \text{N}\cdot\text{m}^2/\text{C}^2)(1.602\times10^{-19} \ \text{C})(-1.602\times10^{-19} \ \text{C})}{0.0529\times10^{-9} \ \text{m}} = \boxed{-4.36\times10^{-18} \ \text{J}}$$

   (b) Strategy and Solution The negative sign signifies that $\boxed{\text{the force between the two charges is attractive; the potential energy is lower than if the two were separated by a larger distance}}$.

5. **Strategy** The work done by the external agent is positive since the potential energy increases. Use Eq. (17-1).

Solution Find the work done by the external agent.

$$W = \Delta U = U_f - U_i = \frac{ke^2}{r_f} - U_\infty = \frac{(8.988\times10^9 \ \text{N}\cdot\text{m}^2/\text{C}^2)(1.602\times10^{-19} \ \text{C})^2}{1.0\times10^{-15} \ \text{m}} - 0 = \boxed{2.3\times10^{-13} \ \text{J}}$$

9. **Strategy** Let $q_1 = -q_2 = q = 10.0$ nC, $d = 4.00$ cm, and $q_3 = -4.2$ nC. Use Eq. (17-2).

Solution Find the total electric potential energy of the three charges at point a.

$$U_E = k\left(\frac{q_1q_2}{r_{12}} + \frac{q_1q_3}{r_{13}} + \frac{q_2q_3}{r_{23}}\right) = k\left[-\frac{q^2}{2d} + \frac{qq_3}{d} - \frac{qq_3}{3d}\right] = \frac{kq}{d}\left[-\frac{q}{2} + q_3\left(1-\frac{1}{3}\right)\right]$$

$$= \frac{(8.988\times10^9 \ \text{N}\cdot\text{m}^2/\text{C}^2)(10.0\times10^{-9} \ \text{C})}{0.0400 \ \text{m}}\left[-\frac{10.0\times10^{-9} \ \text{C}}{2} + \frac{2(-4.2\times10^{-9} \ \text{C})}{3}\right] = \boxed{-17.5 \ \mu\text{J}}$$

11. **Strategy** Let $q_1 = -q_2 = q = 10.0$ nC, $d = 4.00$ cm, and $q_3 = -4.2$ nC. Use Eq. (17-2).

Solution Find the total electric potential energy of the three charges at point c.

$$U_E = k\left(\frac{q_1q_2}{r_{12}} + \frac{q_1q_3}{r_{13}} + \frac{q_2q_3}{r_{23}}\right) = k\left(-\frac{q^2}{2d} + \frac{qq_3}{2d} - \frac{qq_3}{2d}\right) = -\frac{kq^2}{2d}$$

$$= -\frac{(8.988\times10^9 \ \text{N}\cdot\text{m}^2/\text{C}^2)(10.0\times10^{-9} \ \text{C})^2}{2(0.0400 \ \text{m})} = \boxed{-11.2 \ \mu\text{J}}$$

13. **Strategy** Use Eqs. (6-8) and (17-1).

Solution Compute the work done by the electric field.

$$W_{\text{field}} = -\Delta U = U_i - U_f = U_{12} - (U_{12} + U_{13} + U_{23}) = -U_{13} - U_{23} = -\frac{kq_1q_3}{r_{13}} - \frac{kq_2q_3}{r_{23}}$$

$$= -(8.988\times10^9 \ \text{N}\cdot\text{m}^2/\text{C}^2)(2.00\times10^{-9} \ \text{C})\left(\frac{8.00\times10^{-9} \ \text{C}}{0.0400 \ \text{m}} + \frac{-8.00\times10^{-9} \ \text{C}}{0.0400 \ \text{m} + 0.1200 \ \text{m}}\right) = \boxed{-2.70 \ \mu\text{J}}$$

15. **Strategy** Use Eqs. (6-8) and (17-1).

Solution Compute the work done by the electric field.

$$W_{\text{field}} = -\Delta U = U_i - U_f = U_{12} + U_{13i} + U_{23i} - (U_{12} + U_{13f} + U_{23f}) = U_{13i} - U_{13f} + U_{23i} - U_{23f}$$

$$= k\left[q_1q_3\left(\frac{1}{r_{13i}} - \frac{1}{r_{13f}}\right) + q_2q_3\left(\frac{1}{r_{23i}} - \frac{1}{r_{23f}}\right)\right]$$

$$= (8.988\times10^9 \ \text{N}\cdot\text{m}^2/\text{C}^2)\left[(8.00\times10^{-9} \ \text{C})(2.00\times10^{-9} \ \text{C})\left(\frac{1}{0.0400 \ \text{m}} - \frac{1}{0.0800 \ \text{m}}\right)\right.$$

$$\left. + (-8.00\times10^{-9} \ \text{C})(2.00\times10^{-9} \ \text{C})\left(\frac{1}{0.0400 \ \text{m} + 0.1200 \ \text{m}} - \frac{1}{0.0400 \ \text{m}}\right)\right] = \boxed{4.49 \ \mu\text{J}}$$

17. Strategy The potential difference is the change in electric potential energy per unit charge. Use $\Delta U_E = q\Delta V$.

Solution Find the change in the electric potential energy.

$$\Delta U_E = q\Delta V = (3.0 \text{ nC})(25 \text{ V}) = \boxed{75 \text{ nJ}}$$

19. Strategy Use the principle of superposition and Eq. (17-9).

Solution Sum the electric fields at the center due to each charge.

$$\vec{E} = \vec{E}_a + \vec{E}_b + \vec{E}_c + \vec{E}_d = \vec{E}_a + \vec{E}_b - \vec{E}_a - \vec{E}_b = \boxed{0}$$

Do the same for the potential at the center.

$$V = \Sigma \frac{kQ_i}{r_i} = \frac{4kQ}{r} = \frac{4(8.988\times10^9 \text{ N}\cdot\text{m}^2/\text{C}^2)(9.0\times10^{-6} \text{ C})}{\frac{\sqrt{(0.020 \text{ m})^2 + (0.020 \text{ m})^2}}{2}} = \boxed{2.3\times10^7 \text{ V}}$$

21. Strategy Use Eqs. (17-7), (17-8), (17-9), and (6-8).

Solution Find the potentials, potential difference, change in electric potential energy, and work done by the electric field.

(a) $V = \dfrac{kQ}{r} = \dfrac{(8.988\times10^9 \text{ N}\cdot\text{m}^2/\text{C}^2)(-50.0\times10^{-9} \text{ C})}{0.30 \text{ m}} = \boxed{-1.5 \text{ kV}}$

(b) $V = \dfrac{(8.988\times10^9 \text{ N}\cdot\text{m}^2/\text{C}^2)(-50.0\times10^{-9} \text{ C})}{0.50 \text{ m}} = \boxed{-900 \text{ V}}$

(c) $\Delta V = kQ\left(\dfrac{1}{r_B} - \dfrac{1}{r_A}\right) = (8.988\times10^9 \text{ N}\cdot\text{m}^2/\text{C}^2)(-50.0\times10^{-9} \text{ C})\left(\dfrac{1}{0.50 \text{ m}} - \dfrac{1}{0.30 \text{ m}}\right) = \boxed{600 \text{ V}}$

$\Delta V > 0$, so the potential $\boxed{\text{increases}}$.

(d) $\Delta U_E = q\Delta V = (-1.0\times10^{-9} \text{ C})(6.0\times10^2 \text{ V}) = \boxed{-6.0\times10^{-7} \text{ J}}$

$\Delta U_E < 0$, so the potential energy $\boxed{\text{decreases}}$.

(e) $W_{\text{field}} = -\Delta U_E = \boxed{6.0\times10^{-7} \text{ J}}$

25. (a) Strategy and Solution Since $V = \Sigma\dfrac{kq_i}{r_i}$, the minimum or most negative value of the potential is the case where the two negative charges are closer to $x = 0$ than the two positive charges.

(b) Strategy Let $d = 1.0$ m and $q = 1.0$ μC. Use Eq. (17-9).

Solution Find the potential at the origin.

$$V = \Sigma\frac{kQ_i}{r_i} = k\left(\frac{q}{\frac{3}{2}d} + \frac{q}{\frac{d}{2}} + \frac{q}{\frac{d}{2}} - \frac{q}{\frac{3}{2}d}\right) = \frac{kq}{d}\left(\frac{2}{3} + 2 + 2 - \frac{2}{3}\right) = \frac{4(8.988\times10^9 \text{ N}\cdot\text{m}^2/\text{C}^2)(1.0\times10^{-6} \text{ C})}{1.0 \text{ m}} = \boxed{36 \text{ kV}}$$

29. (a) Strategy Use Eq. (17-9).

Solution Find the potential at the points.

$$V_a = \frac{kQ_1}{r_1} + \frac{kQ_2}{r_2} = (8.988 \times 10^9 \ \mathrm{N \cdot m^2/C^2}) \left(\frac{2.50 \times 10^{-9} \ \mathrm{C}}{0.050 \ \mathrm{m}} + \frac{-2.50 \times 10^{-9} \ \mathrm{C}}{0.150 \ \mathrm{m}} \right) = \boxed{300 \ \mathrm{V}}$$

If $Q = 2.50 \times 10^{-9}$ C, $V_b = \dfrac{kQ_1}{r_1} + \dfrac{kQ_2}{r_2} = \dfrac{kQ}{r} - \dfrac{kQ}{r} = \boxed{0}$.

(b) Strategy The work done by an external agent is equal to the change in electric potential energy of the point charge when moved from infinity to b. Use Eq. (17-7).

Solution Compute the work done.

$$W = q\Delta V = q(V_b - V_\infty) = q(0 - 0) = \boxed{0}$$

33. Strategy Since the electric field points in the direction of maximum potential decrease, points farther down the electric field lines are lower in potential. The ratio of the field lines is $12/8 = 1.5$, which means that the magnitude of the positive charge is 1.5 times that of the negative charge; use this to find where the potential is zero—changes from positive to negative, moving from left to right.

Solution Find where the potential is zero.

$$V = \frac{kq_+}{r_+} + \frac{kq_-}{r_-} = \frac{k(1.5q)}{x} + \frac{k(-q)}{d-x} = 0, \ \text{so} \ \frac{1.5}{x} = \frac{1}{d-x} \ \text{and} \ x = 0.6d. \ \text{This is 0.6 times the distance } d \text{ from the}$$

positive charge to the negative charge. For points to the left of this vertical line, the potential is positive. For points to the right of this line, the potential is negative. Only A and B are to the left of the line, so the potential is positive. Since A is closer to the positive charge, the potential is higher at A than at B. The potential is negative at the other three points. The magnitude of the potential is higher the closer a point is to the negative charge; thus, the potential is most negative at C and least negative at E. Ranking these points in order of the potential, from highest to lowest, we have $\boxed{A, B, E, D, C}$.

37. Strategy Equipotential surfaces are perpendicular to electric field lines at all points. For equipotential surfaces drawn such that the potential difference between adjacent surfaces is constant, the surfaces are closer together where the field is stronger. The electric field always points in the direction of maximum potential decrease.

Solution Outside the sphere, $\vec{E}$ is radially directed (toward the sphere), and $V \propto r^{-1}$. The equipotential surfaces are perpendicular to $\vec{E}$ at any point, so they are $\boxed{\text{spheres}}$. Inside the sphere, $\vec{E} = 0$ and V is constant.

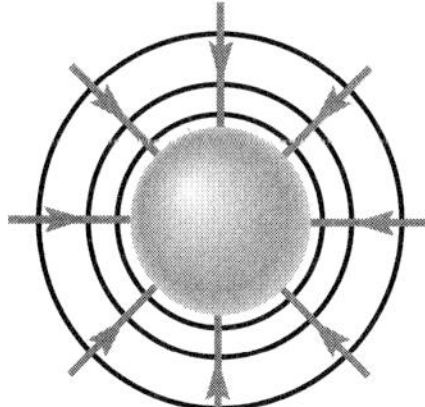

41. Strategy Since the electric field is uniform, we can use Eq. (17-10).

Solution Find the magnitude of the charge on the drop in terms of the elementary charge e.

$$F = qE \text{ and } \Delta V = -Ed, \text{ so } |q| = \left| \frac{F}{E} \right| = \left| \frac{Fe}{Ee} \right| = \left| \frac{Fe}{(-\Delta V/d)e} \right| = \frac{dFe}{e\Delta V} = \frac{(0.16 \ \mathrm{m})(9.6 \times 10^{-16} \ \mathrm{N})}{(1.602 \times 10^{-19} \ \mathrm{C})(480 \ \mathrm{V})} e = \boxed{2e}.$$

45. **(a) Strategy** The electric field always points in the direction of maximum potential decrease. Electrons, being negatively charged, move in the direction opposite the direction of the electric field; that is, in the direction of potential increase.

    Solution Since the speed of the electron decreased, it must have traveled in the direction of the electric field, so it moved in the direction of potential decrease; that is, $\boxed{\text{to a lower potential}}$.

    (b) Strategy The kinetic energy of the electron decreased, so its potential energy increased. Use conservation of energy and Eq. (17-7).

    Solution Compute the potential difference the electron moved through.

$$\Delta V = \frac{\Delta U}{q} = \frac{-\Delta K}{-e} = \frac{m(v_f^2 - v_i^2)}{2e} = \frac{(9.109\times10^{-31}\ \text{kg})[(2.50\times10^6\ \text{m/s})^2 - (8.50\times10^6\ \text{m/s})^2]}{2(1.602\times10^{-19}\ \text{C})} = \boxed{-188\ \text{V}}$$

47. **Strategy** According to Example 17.8, the speed of the electrons at the anode is proportional to the square root of the potential difference. Use a proportion.

    Solution Find the potential difference.

$$v \propto \sqrt{\Delta V},\ \text{so}\ \frac{\Delta V_2}{\Delta V_1} = \left(\frac{v_2}{v_1}\right)^2\ \text{and}\ \Delta V_2 = \Delta V_1 \left(\frac{v_2}{v_1}\right)^2 = (12\ \text{kV})\left(\frac{3.0\times10^7\ \text{m/s}}{6.5\times10^7\ \text{m/s}}\right)^2 = \boxed{2.6\ \text{kV}}.$$

49. **Strategy** Use conservation of energy and Eq. (17-7).

    Solution Find the final kinetic energy.
$$\Delta K = K_f - K_i = -\Delta U = -q\Delta V = -2e\Delta V,\ \text{so}$$
$$K_f = K_i - 2e\Delta V = 1.20\times10^{-16}\ \text{J} - 2(1.602\times10^{-19}\ \text{C})(-0.50\times10^3\ \text{V}) = \boxed{2.8\times10^{-16}\ \text{J}}.$$

51. **Strategy** The electron must have enough kinetic energy at point A to overcome the potential decrease between A and C. Use conservation of energy and Eq. (17-7).

    Solution Find the required kinetic energy.
$$K_A = \Delta U = -e\Delta V = -(1.602\times10^{-19}\ \text{C})(-60.0\ \text{V} - 100.0\ \text{V}) = \boxed{2.56\times10^{-17}\ \text{J}}$$

53. **Strategy** Use conservation of energy and Eq. (17-7). Ignore units for simplicity.

    Solution The change in kinetic energy is given by $\Delta K = -\Delta U = -q\Delta V$.
    (a) $-(-5)\times(-50-100) = -750$; (b) $-(-5)\times[50-(-50)] = 500$; (c) $-25\times(20-50) = 750$;
    (d) $-(-1)\times(-100-400) = -500$; (e) $-1\times[-250-(-100)] = 150$; (f) $-5\times(250-100) = -750$
    Ranking the changes in kinetic energy, from largest to smallest, we have $\boxed{\text{(c), (b), (e), (d), (a) = (f)}}$.

55. **Strategy** Use the definition of capacitance, Eq. (17-14).

    Solution Find the magnitude of the charge on each plate.
$$Q = C\Delta V = (2.0\ \mu\text{F})(9.0\ \text{V}) = \boxed{18\ \mu\text{C}}$$

57. Strategy Use the definition of capacitance, Eq. (17-14).

Solution

$$Q = C\Delta V = (10.2\times10^{-6}\text{ F})(-60.0\text{ V}) = -6.12\times10^{-4}\text{ C}$$

$\boxed{612\ \mu\text{C}}$ of charge must be removed from each plate.

59. Strategy and Solution

(a) Since $\bar{E}$ does not depend upon the separation of the plates $(E = \sigma/\epsilon_0)$, it $\boxed{\text{stays the same}}$.

(b) Since $\Delta V \propto d$, ΔV $\boxed{\text{increases}}$ if d increases.

61. Strategy and Solution

(a) The battery maintains a constant potential difference between the plates, so ΔV $\boxed{\text{stays the same}}$.

(b) The electric field magnitude $\boxed{\text{increases}}$ because the same potential difference occurs over a shorter distance. $(E = \Delta V/d$ for a uniform field.)

(c) To maintain a constant potential difference while the plate spacing changes, the battery must change the charge on the plates. A larger electric field means that the charge $\boxed{\text{increases}}$. Check: $Q = C\Delta V$; C increases and ΔV doesn't change, so Q increases.

65. Strategy Use the definition of capacitance, Eq. (17-14).

Solution Find the capacitance of the spheres.

$$Q = C\Delta V,\text{ so } C = \frac{Q}{\Delta V} = \frac{3.2\times10^{-14}\text{ C}}{0.0040\text{ V}} = \boxed{8.0\text{ pF}}.$$

69. (a) Strategy Use Eq. (16-6).

Solution The electric field between the plates is

$$E = \frac{Q}{\epsilon_0 A} = \frac{4.0\times10^{-11}\text{ C}}{[8.854\times10^{-12}\text{ C}^2/(\text{N}\cdot\text{m}^2)](0.062\text{ m})(0.022\text{ m})} = \boxed{3.3\times10^3\text{ V/m}}.$$

(b) **Strategy** Use the definition of the dielectric constant, Eq. (17-17).

Solution Find the electric field between the plates of the capacitor with the dielectric.

$$\kappa = \frac{E_0}{E},\text{ so } E = \frac{E_0}{\kappa} = \frac{3.3\times10^3\text{ V/m}}{5.5} = \boxed{6.0\times10^2\text{ V/m}}.$$

73. Strategy The spark flies between the spheres when the electric field between them exceeds the dielectric strength. The magnitude of the electric field is given by $\Delta V/d$, where d is the distance between the spheres.

Solution Find d.

$$E = \frac{\Delta V}{d},\text{ so } d = \frac{\Delta V}{E} = \frac{900\text{ V}}{3.0\times10^6\text{ V/m}} = \boxed{0.30\text{ mm}}.$$

75. Strategy Use Eq. (17-16).

Solution Compute the capacitance of the capacitor.

$$C = \kappa \frac{\epsilon_0 A}{d} = \frac{2.5[8.854 \times 10^{-12} \text{ C}^2/(\text{N} \cdot \text{m}^2)](0.30 \text{ m})(0.40 \text{ m})}{0.030 \times 10^{-3} \text{ m}} = \boxed{89 \text{ nF}}$$

77. (a) Strategy Use Eq. (17-18c).

Solution Compute the capacitance.

$$U = \frac{Q^2}{2C}, \text{ so } C = \frac{Q^2}{2U} = \frac{(8.0 \times 10^{-2} \text{ C})^2}{2(450 \text{ J})} = \boxed{7.1 \text{ μF}}.$$

(b) Strategy Use Eq. (17-18a).

Solution Compute the potential difference.

$$U = \frac{1}{2} Q \Delta V, \text{ so } \Delta V = \frac{2U}{Q} = \frac{2(450 \text{ J})}{8.0 \times 10^{-2} \text{ C}} = \boxed{1.1 \times 10^4 \text{ V}}.$$

81. (a) Strategy Use Eq. (17-15).

Solution Find the capacitance for the thundercloud.

$$C = \frac{\epsilon_0 A}{d} = \frac{[8.854 \times 10^{-12} \text{ C}^2/(\text{N} \cdot \text{m}^2)](4500 \text{ m})(2500 \text{ m})}{550 \text{ m}} = \boxed{0.18 \text{ μF}}$$

(b) Strategy Use Eq. (17-18c).

Solution Find the energy stored in the capacitor.

$$U = \frac{Q^2}{2C} = \frac{(18 \text{ C})^2}{2(0.1811 \times 10^{-6} \text{ F})} = \boxed{8.9 \times 10^8 \text{ J}}$$

83. (a) Strategy Use the definition of capacitance, Eq. (17-14), and Eq. (17-15).

Solution Find the charge on the capacitor.

$$Q = C \Delta V = \frac{\epsilon_0 A}{d} \Delta V = \frac{[8.854 \times 10^{-12} \text{ C}^2/(\text{N} \cdot \text{m}^2)](0.100 \text{ m})^2 (150 \text{ V})}{0.75 \times 10^{-3} \text{ m}} = \boxed{18 \text{ nC}}$$

(b) Strategy Use Eq. (17-18a).

Solution Compute the energy stored in the capacitor.

$$U = \frac{1}{2} Q \Delta V = \frac{1}{2} (17.7 \times 10^{-9} \text{ C})(150 \text{ V}) = \boxed{1.3 \text{ μJ}}$$

85. (a) Strategy $U = P \Delta t$ where $P = 10.0$ kW and $\Delta t = 2.0$ ms. Use Eq. (17-18b).

Solution Find the initial potential difference.

$$U = \frac{1}{2} C (\Delta V)^2, \text{ so } \Delta V = \sqrt{\frac{2U}{C}} = \sqrt{\frac{2(10.0 \text{ kW})(2.0 \text{ ms})}{100.0 \times 10^{-6} \text{ F}}} = \boxed{630 \text{ V}}.$$

(b) Strategy Use Eq. (17-18c).

Solution Find the initial charge.

$$U = \frac{Q^2}{2C}, \text{ so } Q = \sqrt{2CU} = \sqrt{2(100.0 \times 10^{-6} \text{ F})(10.0 \text{ kW})(2.0 \text{ ms})} = \boxed{0.063 \text{ C}}.$$

89. (a) Strategy Use the definition of capacitance, Eq. (17-14).

Solution Compute the charge that passes through the body tissues.
$$Q = C\Delta V = (15\times10^{-6}\ \text{F})(9.0\times10^{3}\ \text{V}) = \boxed{0.14\ \text{C}}$$

(b) Strategy Use Eq. (17-18b) and the definition of average power.

Solution Find the average power delivered to the tissues.
$$P_{\text{av}} = \frac{\Delta E}{\Delta t} = \frac{U}{\Delta t} = \frac{C(\Delta V)^2}{2\Delta t} = \frac{(15\times10^{-6}\ \text{F})(9.0\times10^{3}\ \text{V})^2}{2(2.0\times10^{-3}\ \text{s})} = \boxed{0.30\ \text{MW}}$$

93. (a) Strategy Let $q_{\text{L}} = -q_{\text{R}} = 10.0$ nC. Use Eqs. (17-5) and (17-9).

Solution Find the electric potential energy of the point charge at each location.
$$U_b = qV_b = q\left(\frac{kq_{\text{L}}}{r_{\text{L}}} + \frac{kq_{\text{R}}}{r_{\text{R}}}\right) = kqq_{\text{L}}\left(\frac{1}{r_{\text{L}}} - \frac{1}{r_{\text{R}}}\right)$$
$$= (8.988\times10^{9}\ \text{N}\cdot\text{m}^2/\text{C}^2)(-4.2\times10^{-9}\ \text{C})(10.0\times10^{-9}\ \text{C})\left(\frac{1}{0.0400\ \text{m}} - \frac{1}{0.0400\ \text{m}}\right) = 0$$
$$U_c = qV_c = (8.988\times10^{9}\ \text{N}\cdot\text{m}^2/\text{C}^2)(-4.2\times10^{-9}\ \text{C})(10.0\times10^{-9}\ \text{C})\left(\frac{1}{0.0800\ \text{m}} - \frac{1}{0.0800\ \text{m}}\right) = 0$$
The change in electric potential energy is $\Delta U = 0 - 0 = \boxed{0}$.

(b) Strategy Compute the electric potential energy at point a as in part (a). The work done by the external force is negative the work done by the field. Use Eq. (6-8).

Solution Find the electric potential energy of the point charge at point a.
$$U_a = qV_a = q\left(\frac{kq_{\text{L}}}{r_{\text{L}}} + \frac{kq_{\text{R}}}{r_{\text{R}}}\right) = kqq_{\text{L}}\left(\frac{1}{r_{\text{L}}} - \frac{1}{r_{\text{R}}}\right)$$
$$= (8.988\times10^{9}\ \text{N}\cdot\text{m}^2/\text{C}^2)(-4.2\times10^{-9}\ \text{C})(10.0\times10^{-9}\ \text{C})\left(\frac{1}{0.0400\ \text{m}} - \frac{1}{0.1200\ \text{m}}\right) = -6.3\ \mu\text{J}$$
Find the work required to move the point charge.
$$W = -W_{\text{field}} = \Delta U = U_a - U_b = -6.3\ \mu\text{J} - 0 = \boxed{-6.3\ \mu\text{J}}$$

97. Strategy Use Eqs. (6-8) and (17-7).

Solution Compute the work done by the electric field.
$$W_{\text{field}} = -\Delta U = -(-e)\Delta V = (1.602\times10^{-19}\ \text{C})[100.0\ \text{V} - (-100.0\ \text{V})] = \boxed{3.204\times10^{-17}\ \text{J}}$$

101. Strategy Use Newton's second law and Eq. (4-9).

Solution Find the acceleration.
$$\Sigma F_y = eE = ma_y,\ \text{so}\ a_y = \frac{eE}{m}.$$
Find the time to reach the lower plate.
$$\Delta y = \frac{1}{2}a_y(\Delta t)^2,\ \text{so}\ \Delta t = \sqrt{\frac{2m\Delta y}{eE}} = \sqrt{\frac{2(9.109\times10^{-31}\ \text{kg})(0.040\ \text{m})}{(1.602\times10^{-19}\ \text{C})(5.0\times10^{4}\ \text{N/C})}} = \boxed{3.0\ \text{ns}}.$$

105. Strategy Use Eqs. (6-8) and (17-7).

Solution Find the work done by the electric field.

$$W_{\text{field}} = -\Delta U = -q\Delta V = -e\Delta V = -(1.602\times10^{-19}\ \text{C})(-90.0\times10^{-3}\ \text{V}) = \boxed{1.44\times10^{-20}\ \text{J}}$$

109. Strategy Use Newton's second law, $\Delta x = v_x \Delta t$, and Eqs. (4-7) and (4-9).

Solution

(a) Since $\bar{E}$ points downward, the negatively charged electron's change in velocity is directed upward. Find the acceleration.

$$\Sigma F_y = eE = ma_y,\ \text{so}\ a_y = \frac{eE}{m}.$$

Find the time interval.

$$\Delta t = \frac{\Delta x}{v_x}$$

Find the change in velocity.

$$\Delta v_y = a_y \Delta t = \left(\frac{eE}{m}\right)\left(\frac{\Delta x}{v_x}\right) = \frac{(1.602\times10^{-19}\ \text{C})(2.0\times10^4\ \text{N/C})(0.060\ \text{m})}{(9.109\times10^{-31}\ \text{kg})(3.0\times10^7\ \text{m/s})} = 7.0\times10^6\ \text{m/s}$$

So, $\Delta \bar{v} = \boxed{7.0\times10^6\ \text{m/s upward}}$.

(b) Find the deflection of the electrons.

$$\Delta y = \frac{1}{2}a_y(\Delta t)^2 = \frac{1}{2}\left(\frac{eE}{m}\right)\left(\frac{\Delta x}{v_x}\right)^2 = \frac{(1.602\times10^{-19}\ \text{C})(2.0\times10^4\ \text{N/C})(0.060\ \text{m})^2}{2(9.109\times10^{-31}\ \text{kg})(3.0\times10^7\ \text{m/s})^2} = \boxed{7.0\ \text{mm}}$$

113. (a) Strategy Use the definition of capacitance, Eq. (17-14).

Solution Compute the capacitance.

$$C = \frac{Q}{\Delta V} = \frac{0.020\times10^{-6}\ \text{C}}{240\ \text{V}} = \boxed{83\ \text{pF}}$$

(b) Strategy Use Eq. (17-15).

Solution Find the area of a single plate.

$$C = \frac{\epsilon_0 A}{d},\ \text{so}\ A = \frac{dC}{\epsilon_0} = \frac{(0.40\times10^{-3}\ \text{m})(8.33\times10^{-11}\ \text{F})}{8.854\times10^{-12}\ \text{C}^2/(\text{N}\cdot\text{m}^2)} = \boxed{3.8\times10^{-3}\ \text{m}^2}.$$

(c) Strategy Use Eq. (17-10).

Solution Compute the voltage required to ionize the air between the plates.

$$\Delta V = Ed = (3.0\times10^3\ \text{V/mm})(0.40\ \text{mm}) = \boxed{1.2\ \text{kV}}$$

117. Strategy Use conservation of energy, $W_{\text{field}} = -\Delta U$, and the fact that the field is uniform.

Solution Find the final kinetic energy of the alpha particle.

$$\Delta K = K_f - K_i = -\Delta U = W_{\text{field}} = 2eEd,\ \text{so}$$

$$K_f = K_i + 2eEd = 0 + 2(1.602\times10^{-19}\ \text{C})(10.0\times10^3\ \text{V/m})(0.010\ \text{m}) = \boxed{3.2\times10^{-17}\ \text{J}}.$$

121. Strategy Let $E_0 = 20.0$ V/m, E_1 be the field outside of the dielectric after it is inserted, and E_2 be the field inside the dielectric. Use the principle of superposition for the potential after the dielectric is inserted and the fact that $E_2 = E_1/\kappa$.

Solution Find the electric field inside the dielectric.

Initially: $\Delta V = E_0 d$

Finally: $\Delta V = E_1 \dfrac{d}{2} + E_2 \dfrac{d}{2} = E_1 \dfrac{d}{2} + \left(\dfrac{E_1}{\kappa} \right) \dfrac{d}{2} = \dfrac{E_1 d}{2} \left(1 + \dfrac{1}{\kappa} \right)$

Solve for E_1 in terms of E_0.

$\dfrac{E_1 d}{2} \left(1 + \dfrac{1}{\kappa} \right) = E_0 d$, so $E_1 = \dfrac{2E_0}{1 + \dfrac{1}{\kappa}}$.

Calculate E_2.

$E_2 = \dfrac{E_1}{\kappa} = \dfrac{2E_0}{\kappa + 1} = \dfrac{2(20.0 \text{ V/m})}{4.0 + 1} = \boxed{8.0 \text{ V/m}}$

Chapter 18

ELECTRIC CURRENT AND CIRCUITS

Conceptual Questions

1. An electric field is required to start a current flowing within a conductor. The electric field inside a conductor is therefore not equal to zero if a current is flowing through it.

5. If he connects three 300 Ω resistors in parallel, the equivalent resistance will be the desired 100 Ω.

9. Electric stoves and clothes dryers require relatively large amounts of power to operate. Supplying them with 240 V instead of 120 V decreases the magnitude of required current to supply the power. This reduces the rate $(P = I^2R)$ at which energy is dissipated in the wiring.

13. The bird perched on the power line is at the same electric potential as the line but it is isolated from the ground so that no current flows through its body. When a person standing on the ground touches a power line with a metal pole, a potential difference exists between the line and the person's grounded feet—a current will therefore flow through their body.

17. The inverse of the total resistance of two resistors connected in parallel is equal to the sum of the inverse resistances of each resistor. If the resistance of a wire is inversely proportional to its cross-sectional area, then the above statement tells us that the total resistance of two resistors of cross-sectional area A connected in parallel is proportional to $0.5A$. If resistance had any other relationship to cross-sectional area, the total resistance would not agree with that given by the first statement.

21. **(a)** Bulbs C and D are equally bright and they are brighter than bulbs A or B. Bulbs A and B are also equally bright, but less so than bulbs C and D.

    (b) The brightness of bulb B increases.

    (c) The brightness of bulb C remains the same.

Problems

1. **Strategy** Use the definition of electric current.

    Solution Compute the total charge.

    $$I = \frac{\Delta q}{\Delta t}, \text{ so } \Delta q = I\Delta t = (3.0 \text{ A})(4.0 \text{ h})(3600 \text{ s/h}) = \boxed{4.3 \times 10^4 \text{ C}}.$$

3. **(a) Strategy and Solution** The electrons flow from the filament to the anode; since they are negatively charged, the current flows $\boxed{\text{from the anode to the filament}}$.

    (b) Strategy Use the definition of electric current.

    Solution Compute the current in the tube.

    $$I = \frac{\Delta q}{\Delta t} = \Delta q \times f = (1.602 \times 10^{-19} \text{ C})(6.0 \times 10^{12} \text{ s}^{-1}) = \boxed{0.96 \text{ μA}}$$

5. **Strategy** Use the definition of electric current and the elementary charge of an electron.

   Solution Find the number of electrons per second that hit the screen.

   $$I = \frac{\Delta q}{\Delta t} = \frac{Ne}{\Delta t}, \text{ so } \frac{N}{\Delta t} = \frac{I}{e} = \frac{320 \times 10^{-6} \text{ A}}{1.602 \times 10^{-19} \frac{\text{C}}{\text{electron}}} = \boxed{2.0 \times 10^{15} \text{ electrons/s}}.$$

7. **Strategy** Since the oppositely charged ions move in opposite directions, they both contribute to the current in the same direction. Use the definition of electric current.

   Solution Compute the current in the solution.

   $$I = \frac{\Delta q}{\Delta t} = \frac{Ne}{\Delta t} = \frac{[2(3.8 \times 10^{16}) + 6.2 \times 10^{16}](1.602 \times 10^{-19} \text{ C})}{1.0 \text{ s}} = \boxed{22.1 \text{ mA}}$$

9. **Strategy** The total energy stored in a battery is equal to the total work the battery is able to do. Use Eq. (18-2).

   Solution Compute the energy stored in the battery.

   $$W = \mathscr{E}q = (1.20 \text{ V})(675 \text{ C}) = \boxed{810 \text{ J}}$$

11. **(a) Strategy** Use the definition of electric current.

    Solution Compute the amount of charge pumped by the battery.

    $$\Delta q = I\Delta t = (220.0 \text{ A})(1.20 \text{ s}) = \boxed{264 \text{ C}}$$

    (b) Strategy The electrical energy supplied is equal to the work done by the battery. Use Eq. (18-2).

    Solution Compute the amount of electrical energy supplied by the battery.

    $$W = \mathscr{E}q = (12.0 \text{ V})(264 \text{ C}) = \boxed{3.17 \text{ kJ}}$$

13. **Strategy** Use Eq. (18-3).

    Solution Solve for the drift velocity in terms of diameter and current. The lengths are irrelevant.

    $$I = neAv_{\text{D}}, \text{ so } v_{\text{D}} = \frac{I}{neA} = \frac{I}{ne(\pi d^2 / 4)} = \frac{4I}{\pi ned^2}.$$

    Compute I/d^2 for each wire. The other quantities are constants. Ignore units for simplicity.

    (a) $\frac{80}{2^2} = 20$; (b) $\frac{80}{1^2} = 80$; (c) $\frac{40}{4^2} = 2.5$; (d) $\frac{160}{2^2} = 40$; (e) $\frac{20}{1^2} = 20$; (f) $\frac{40}{2^2} = 10$

    Ranking the wires in order of decreasing drift velocity, we have $\boxed{\text{(b), (d), (a) = (e), (f), (c)}}$.

15. **Strategy** Use Eq. (18-3).

    Solution Find the drift speed of the conduction electrons in the wire.

    $$I = neAv_{\text{D}}, \text{ so } v_{\text{D}} = \frac{I}{neA} = \frac{I}{ne(\pi r^2)} = \frac{2.50 \text{ A}}{(8.47 \times 10^{28} \text{ m}^{-3})(1.602 \times 10^{-19} \text{ C})\pi(0.00100 \text{ m})^2} = \boxed{5.86 \times 10^{-5} \text{ m/s}}.$$

17. **Strategy** Find the time of travel Δt using $\Delta x = v_{\text{D}}\Delta t$ and Eq. (18-3).

    Solution Find the time of travel.

    $$\Delta t = \frac{\Delta x}{v_{\text{D}}} = \frac{neAd}{I} = \frac{(5.8 \times 10^{28} \text{ m}^{-3})(1.602 \times 10^{-19} \text{ C})\frac{1}{4}\pi(0.0010 \text{ m})^2(0.010 \text{ m})}{0.15 \text{ A}}\left(\frac{1 \text{ min}}{60 \text{ s}}\right) = \boxed{8.1 \text{ min}}$$

21. Strategy Find the average time Δt it takes for an electron to move 12 m along the wire using $\Delta x = v_{\mathrm{D}} \Delta t$, Eq. (18-3), and $n = 3.5\rho N_{\mathrm{A}}/M$, the number of electrons per unit volume. The cross-sectional area of the wire is $\frac{1}{4}\pi d^2$.

Solution Find Δt.

$$\Delta t = \frac{\Delta x}{v_{\mathrm{D}}} = \frac{neA\Delta x}{I} = \frac{3.5\rho N_{\mathrm{A}}eA\Delta x}{MI}$$

$$= \frac{3.5(2.7\times10^6 \ \mathrm{g/m^3})(6.022\times10^{23} \ \mathrm{mol^{-1}})(1.602\times10^{-19} \ \mathrm{C})\frac{1}{4}\pi(0.0026 \ \mathrm{m})^2(12 \ \mathrm{m})}{(27 \ \mathrm{g/mol})(12 \ \mathrm{A})(3600 \ \mathrm{s/h})} = \boxed{50 \ \mathrm{h}}$$

25. Strategy Use Eq. (18-8). The cross-sectional areas of the wires are given by $\frac{1}{4}\pi d^2$.

Solution Form a proportion to find the ratio of diameters.

$$\frac{R_{\mathrm{Al}}}{R_{\mathrm{Cu}}} = 1 = \frac{\rho_{\mathrm{Al}}\frac{L}{A_{\mathrm{Al}}}}{\rho_{\mathrm{Cu}}\frac{L}{A_{\mathrm{Cu}}}} = \frac{\rho_{\mathrm{Al}}A_{\mathrm{Cu}}}{\rho_{\mathrm{Cu}}A_{\mathrm{Al}}} = \frac{\rho_{\mathrm{Al}}d_{\mathrm{Cu}}^2}{\rho_{\mathrm{Cu}}d_{\mathrm{Al}}^2}, \ \mathrm{so} \ \frac{d_{\mathrm{Cu}}}{d_{\mathrm{Al}}} = \sqrt{\frac{\rho_{\mathrm{Cu}}}{\rho_{\mathrm{Al}}}} = \sqrt{\frac{1.67}{2.65}} = \boxed{0.794}.$$

27. (a) Strategy Use the definition of resistance.

Solution Compute the required potential difference between the electrician's hands.
$$\Delta V = IR = (50 \ \mathrm{mA})(1 \ \mathrm{k\Omega}) = \boxed{50 \ \mathrm{V}}$$

(b) Strategy and Solution An electrician working on a "live" circuit keeps one hand behind his or her back $\boxed{\text{to avoid becoming part of the circuit}}$.

29. Strategy Assume that resistivity is proportional to ion concentration.

Solution Form a proportion to find the concentration of ions in blood plasma.
$$\frac{\rho_{\mathrm{b}}}{\rho_{\mathrm{w}}} = \frac{1\times10^5 \ \Omega\cdot\mathrm{m}}{0.6 \ \Omega\cdot\mathrm{m}} = \frac{[ions]}{1.2\times10^{14} \ \mathrm{ions/cm^3}}, \ \mathrm{so} \ [ions] = 1.2\times10^{14} \ \mathrm{ions/cm^3} \times \frac{1\times10^5 \ \Omega\cdot\mathrm{m}}{0.6 \ \Omega\cdot\mathrm{m}} = \boxed{2\times10^{19} \ \mathrm{ions/cm^3}}.$$

33. Strategy As found in Example 18.4, $R/R_0 = 1 + \alpha\Delta T$. Find T using this and the definition of resistance.

Solution Estimate the temperature of the tungsten filament.
$$1 + \alpha\Delta T = \frac{R}{R_0}, \ \mathrm{so} \ T = \frac{1}{\alpha}\left(\frac{R}{R_0} - 1\right) + T_0 = \frac{1}{4.50\times10^{-3} \ {}^\circ\mathrm{C}^{-1}}\left(\frac{\frac{2.90 \ \mathrm{V}}{0.300 \ \mathrm{A}}}{1.10 \ \Omega} - 1\right) + 20.0^\circ\mathrm{C} = \boxed{1750^\circ\mathrm{C}}.$$

37. Strategy Use the definition of resistance, the relationship between voltage and uniform electric field, and Eq. (18-8).

Solution $V = IR = EL$ and $R = \rho L/A$. Find E.

$$V = EL = IR = I\rho\frac{L}{A}, \ \mathrm{so} \ \boxed{E = \rho\frac{I}{A}, \ \text{where} \ \rho \ \text{is the resistivity}}.$$

41. (a) Strategy Sum the individual emfs with those with their left terminal at the higher potential being positive.

Solution Compute the equivalent emf.

$$\mathscr{E}_{eq} = 3.0\ \text{V} + 3.0\ \text{V} + 2.5\ \text{V} - 1.5\ \text{V} = \boxed{7.0\ \text{V}}$$

(b) Strategy Use the definition of resistance.

Solution Find the value of the resistor.

$$R = \frac{\Delta V}{I} = \frac{\mathscr{E}_{eq}}{I} = \frac{7.0\ \text{V}}{0.40\ \text{A}} = \boxed{18\ \Omega}$$

43. (a) Strategy $C_{eq} = \Sigma C_i$ for the capacitors, which are in parallel.

Solution Compute the equivalent capacitance.

$$C_{eq} = 4.0\ \mu\text{F} + 2.0\ \mu\text{F} + 3.0\ \mu\text{F} + 9.0\ \mu\text{F} + 5.0\ \mu\text{F} = \boxed{23.0\ \mu\text{F}}$$

(b) Strategy Use Eq. (17-14).

Solution Compute the charge on the equivalent capacitor.

$$Q = C\Delta V = C_{eq}\mathscr{E} = (23.0 \times 10^{-6}\ \text{F})(16.0\ \text{V}) = \boxed{368\ \mu\text{C}}.$$

(c) Strategy Use Eq. (17-14).

Solution Compute the charge on the capacitor.

$$Q = C\Delta V = C\mathscr{E} = (3.0 \times 10^{-6}\ \text{F})(16.0\ \text{V}) = \boxed{48\ \mu\text{C}}.$$

45. (a) Strategy Use Eqs. (18-13) and (18-17).

Solution Compute the resistance between points A and B.

$$R_{eq} = \left(\frac{1}{2.0\ \Omega} + \frac{1}{1.0\ \Omega + 1.0\ \Omega} \right)^{-1} + 4.0\ \Omega = \boxed{5.0\ \Omega}$$

(b) Strategy Label the currents on a diagram. Use Kirchhoff's rules.

Solution The current through the emf is $I = \mathscr{E}/R_{eq} = I_1 + I_2$, where the currents labeled 1 and 2 are shown in the diagram. Applying the loop rule, we have $I_2(2R_2) - I_1 R_1 = 0$, so $I_2 = \dfrac{R_1}{2R_2} I_1 = \dfrac{2.0\ \Omega}{2(1.0\ \Omega)} I_1 = 1.0 I_1.$

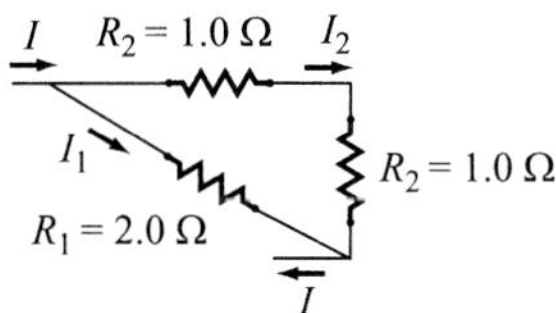

Solve for the current through the 2.0-Ω resistor, I_1.

$$I = I_1 + I_2 = I_1 + 1.0 I_1 = 2.0 I_1, \text{ so } I_1 = \frac{I}{2.0} = \frac{\mathscr{E}}{2.0 R_{eq}} = \frac{20\ \text{V}}{2.0(5.0\ \Omega)} = \boxed{2.0\ \text{A}}.$$

49. Strategy Use the concept of equivalent resistance. The equivalent resistance of two identical resistances R in parallel is half or $R/2$.

Solution

(a) The two 2.0-Ω resistors are in series, so their equivalent resistance is $2.0\ \Omega + 2.0\ \Omega = 4.0\ \Omega$. These two resistors are in parallel with the rightmost 4.0-Ω resistor. Because the resistances of each branch of this parallel circuit are equal, the current is split evenly. Let the current through each branch be called I_3. We must determine I_3. Now, the equivalent resistance of this parallel circuit is 2.0 Ω, and this is in series with the rightmost 3.0-Ω resistor and the rightmost 1.0-Ω, so the equivalent series resistance is 6.0 Ω. This resistance is in parallel with the 6.0-Ω resistor, so the current is again split evenly. Let it be called I_2; then, $I_3 = I_2/2$. The equivalent resistance of this parallel circuit is 3.0-Ω, and this is in series with the middle 1.0-Ω resistor, so the equivalent series resistance is 4.0 Ω. This equivalent resistance is in parallel with the leftmost 4.0-Ω resistor, so the current is again split evenly. Let it be called I_1; then, $I_3 = I_2/2 = I_1/4$. The equivalent resistance of this parallel circuit is 2.0-Ω, and this is in series with the leftmost 1.0-Ω and 3.0-Ω resistors, so the equivalent resistance of the entire circuit is 6.0 Ω. If the current through the emf is I; then, $I_3 = I_2/2 = I_1/4 = I/8$. The current though the emf is given by $I = \mathcal{E}/R_{eq}$. Compute the current through one of the 2.0-Ω resistors.

$$I_3 = \frac{I}{8} = \frac{\mathcal{E}}{8R_{eq}} = \frac{24\ \text{V}}{8(6.0\ \Omega)} = \boxed{0.50\ \text{A}}$$

(b) The current through the 6.0-Ω resistor is I_2, which is one-fourth of the current through the emf.

$$I_2 = \frac{I}{4} = \frac{\mathcal{E}}{4R_{eq}} = \frac{24\ \text{V}}{4(6.0\ \Omega)} = \boxed{1.0\ \text{A}}$$

(c) The current through the leftmost 4.0-Ω resistor is I_1, which is half of the current through the emf.

$$I_1 = \frac{I}{2} = \frac{\mathcal{E}}{2R_{eq}} = \frac{24\ \text{V}}{2(6.0\ \Omega)} = \boxed{2.0\ \text{A}}$$

53. (a) Strategy Use Eqs. (18-15) and (18-18).

Solution Find the equivalent capacitance.

$$C_{eq} = \left(\frac{1}{12\ \mu\text{F}} + \frac{1}{12\ \mu\text{F} + 12\ \mu\text{F}} \right)^{-1} = \boxed{8.0\ \mu\text{F}}$$

(b) Strategy Since the capacitor at the left side of the diagram (1) is in series with the parallel combination of the other two capacitors (2), the charge Q on the capacitor 1 is the same as that on capacitor 2. (Think of the parallel combination as one capacitor with capacitance $C_2 = 12\ \mu\text{F} + 12\ \mu\text{F} = 24\ \mu\text{F}$.) Use the definition of capacitance.

Solution Find the potential difference across C_1. Let this potential difference be V_1 and the potential difference across C_2 be V_2. Then, $\mathcal{E} = V_1 + V_2$. Form a proportion.

$$\frac{V_2}{V_1} = \frac{\mathcal{E} - V_1}{V_1} = \frac{\mathcal{E}}{V_1} - 1 = \frac{Q/C_2}{Q/C_1} = \frac{C_1}{C_2}, \text{ so } V_1 = \frac{\mathcal{E}}{1 + \frac{C_1}{C_2}} = \frac{25\ \text{V}}{1 + \frac{12\ \mu\text{F}}{24\ \mu\text{F}}} = \boxed{17\ \text{V}}.$$

(c) Strategy The charge on the capacitor at the far right of the circuit (1) is half of the charge on the capacitor at the left of the circuit (2).

Solution Find the charge on the capacitor.

$$Q_2 = C_2 V_2 = 2Q_1, \text{ so } Q_1 = \frac{1}{2}C_2 V_2 = \frac{1}{2}(12\times10^{-6} \text{ F})(17 \text{ V}) = \boxed{1.0\times10^{-4} \text{ C}}.$$

57. Strategy Use Kirchhoff's rules. Let I_1 be the top branch, I_2 be the middle branch, and I_3 be the bottom branch. Assume that each current flows right to left.

Solution Find the current in each branch of the circuit.

(1) $I_1 = -I_2 - I_3$ (2) $0 = 25.00 \text{ V} + (5.6 \text{ }\Omega)I_2 - (122 \text{ }\Omega)I_1$

(3) $0 = 25.00 \text{ V} + 5.00 \text{ V} + (75 \text{ }\Omega)I_3 - (122 \text{ }\Omega)I_1$ (4) $0 = 5.00 \text{ V} + (75 \text{ }\Omega)I_3 - (5.6 \text{ }\Omega)I_2$

Substitute (1) into (2).

(5) $0 = 25.00 \text{ V} + (122 \text{ }\Omega + 5.6 \text{ }\Omega)I_2 + (122 \text{ }\Omega)I_3$

Multiply (4) by 5 and subtract from (5).

$$0 = [122 \text{ }\Omega + 5.6 \text{ }\Omega + 5(5.6 \text{ }\Omega)]I_2 + [122 \text{ }\Omega - 5(75 \text{ }\Omega)]I_3, \text{ so } I_2 = \frac{5(75 \text{ }\Omega) - 122 \text{ }\Omega}{122 \text{ }\Omega + 5.6 \text{ }\Omega + 5(5.6 \text{ }\Omega)}I_3 = 1.6I_3 \ (1.626I_3).$$

Substitute the result above into (4).

$$0 = 5.00 \text{ V} + (75 \text{ }\Omega)I_3 - (5.6 \text{ }\Omega)(1.626I_3), \text{ so } I_3 = \frac{5.00 \text{ V}}{1.626(5.6 \text{ }\Omega) - 75 \text{ }\Omega} = -0.076 \text{ A}.$$

So, $I_2 = 1.626(-0.076 \text{ A}) = -0.12 \text{ A}$ and $I_1 = -(-0.12 \text{ A}) - (-0.076 \text{ A}) = 0.20 \text{ A}$.

Branch	I (A)	Direction
AB	0.20	right to left
FC	0.12	left to right
ED	0.076	left to right

59. Strategy Use Kirchhoff's rules. Let the current on the left be I, the one in the middle be I_1, and the one on the right be I_2. I_1 flows downward.

Solution Find the unknown emf and the unknown resistor.

$I_1 = I + I_2 = 1.00 \text{ A} + 10.00 \text{ A} = 11.00 \text{ A}$

Loop *ABCFA*:

$0 = -\mathcal{E} - (6.00 \text{ }\Omega)(1.00 \text{ A}) - (4.00 \text{ }\Omega)(11.00 \text{ A}) + 125 \text{ V}, \text{ so } \mathcal{E} = \boxed{75 \text{ V}}.$

Loop *ABCDEFA*:

$0 = -\mathcal{E} - (6.00 \text{ }\Omega)(1.00 \text{ A}) + (10.00 \text{ A})R = -75 \text{ V} - 6.00 \text{ V} + (10.00 \text{ A})R, \text{ so } R = \dfrac{81 \text{ V}}{10.00 \text{ A}} = \boxed{8.1 \text{ }\Omega}.$

61. Strategy Use Eq. (18-20).

Solution Compute the power dissipated by the resistor.

$P = \mathcal{E}I = (2.00 \text{ V})(2.0 \text{ A}) = \boxed{4.0 \text{ W}}$

65. Strategy and Solution $\boxed{\text{Yes}}$; the power rating can be determined by $P = IV = (5.0\text{ A})(120\text{ V}) = \boxed{600\text{ W}}$.

69. (a) Strategy Use Eqs. (8-13) and (8-17) to find the equivalent resistance. Then draw the diagram.

Solution Compute the equivalent resistance.
$$R_{eq} = 20.0\ \Omega + 50.0\ \Omega + \left(\frac{1}{70.0\ \Omega + 20.0\ \Omega} + \frac{1}{40.0\ \Omega + 20.0\ \Omega}\right)^{-1} = 106.0\ \Omega$$
The simplest equivalent circuit contains the emf and one 106.0-Ω resistor.

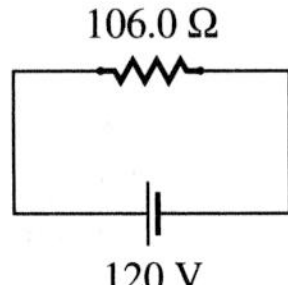

(b) Strategy Use the definition of resistance.

Solution Compute the current that flows from the battery.
$$I = \frac{\mathscr{E}}{R_{eq}} = \frac{120\text{ V}}{106.0\ \Omega} = \boxed{1.1\text{ A}}$$

(c) Strategy Compute the resistance between A and B. Then use the definition of resistance to find the potential difference.

Solution Compute the resistance. It is in series with the two resistors not between A and B.
$106.0\ \Omega - 20.0\ \Omega - 50.0\ \Omega = 36.0\ \Omega$
Compute the potential difference.
$$V = IR = (1.13\text{ A})(36.0\ \Omega) = \boxed{41\text{ V}}$$

(d) Strategy The current that flows through the battery is shared by each branch between points A and B. Draw a diagram with equivalent resistances. Use Kirchhoff's rules.

Solution The diagram is shown.
$$I = I_1 + I_2 \text{ and } I_1(60.0\ \Omega) - I_2(90.0\ \Omega) = 0, \text{ so } I_2 = \frac{60.0}{90.0}I_1 \text{ and}$$
$$I = I_1 + \frac{60.0}{90.0}I_1 = \left(1 + \frac{60.0}{90.0}\right)I_1.$$

So, the current through the upper branch is $I_1 = \left(1 + \frac{60.0}{90.0}\right)^{-1}I = \left(1 + \frac{60.0}{90.0}\right)^{-1}(1.13\text{ A}) = \boxed{0.68\text{ A}}$, and the

current through the lower branch is $I_2 = \frac{60.0}{90.0}I_1 = \frac{60.0}{90.0}(0.68\text{ A}) = \boxed{0.45\text{ A}}$.

(e) Strategy Use $P = I^2 R$.

Solution Determine the power dissipated in the resistors.
$$P_{50} = (1.13\text{ A})^2(50.0\ \Omega) = \boxed{64\text{ W}}, \quad P_{70} = (0.45\text{ A})^2(70.0\ \Omega) = \boxed{14\text{ W}}, \text{ and}$$
$$P_{40} = (0.68\text{ A})^2(40.0\ \Omega) = \boxed{18\text{ W}}.$$

73. (a) Strategy Use Eq. (18-21b). Form a proportion and assume R doesn't change.

Solution Find the power consumed by the lightbulb.

$$\frac{P_{108}}{P_{120}} = \frac{V_{108}^2/R}{V_{120}^2/R}, \text{ so } P_{108} = \left(\frac{V_{108}}{V_{120}}\right)^2 P_{120} = \left(\frac{108}{120}\right)^2 (100.0 \text{ W}) = \boxed{81 \text{ W}}.$$

(b) Strategy and Solution If the filament is at a lower temperature, its resistance will be lower, so its power output will be higher for a particular voltage. Thus, the power drop will be $\boxed{\text{less}}$ than that found in part (a).

75. Strategy Ammeters are connected in series.

Solution Redraw the circuit to include the ammeters.

(a)

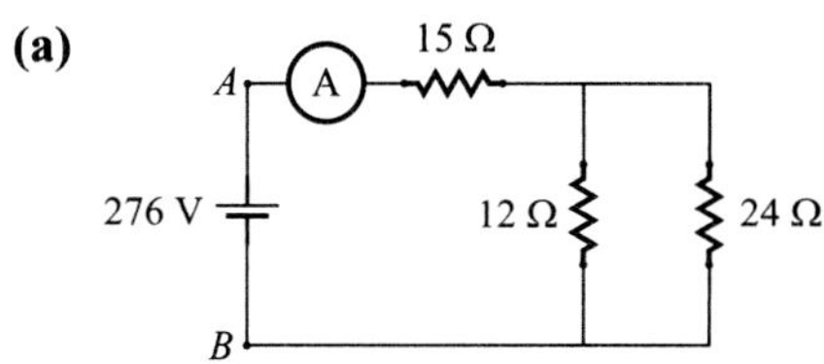

(b)

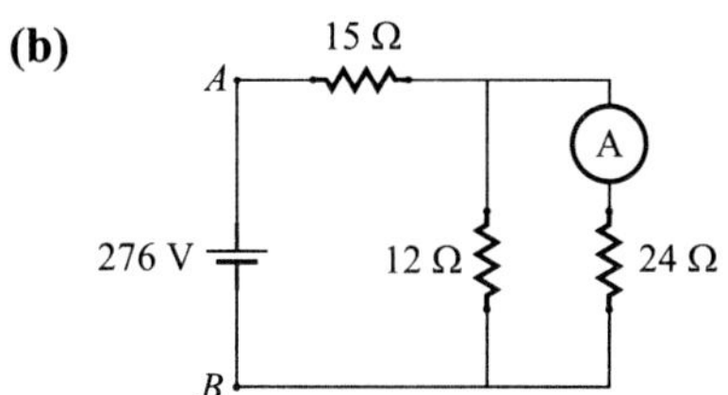

77. Strategy Ammeters are connected in series.

Solution

(a) Redraw the circuit to include the ammeter.

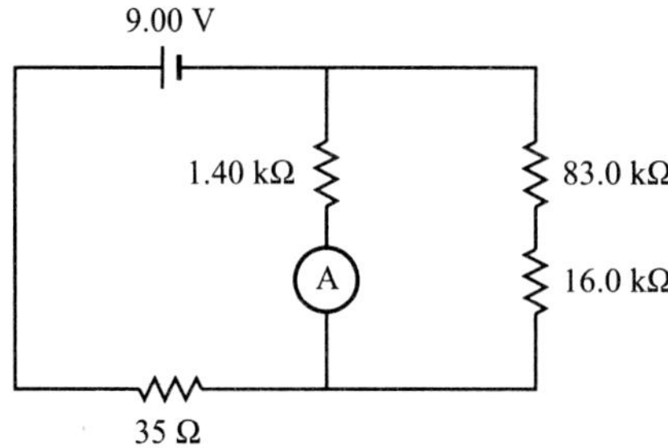

(b) The ammeter is connected in series with the 1.40-kΩ resistor. Find the new current through the resistor, assuming the ammeter to be ideal ($R = 0$).

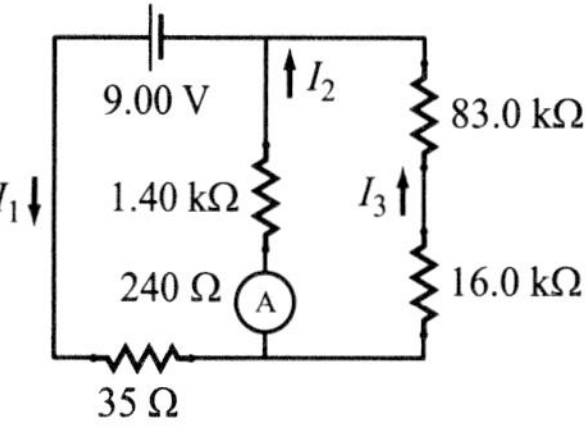

(1) $I_1 = I_2 + I_3$

(2) $0 = 9.00 \text{ V} - I_1(35 \text{ Ω}) - I_2(1.40 \times 10^3 \text{ Ω} + 240 \text{ Ω})$

$\quad 0 = I_2(1.40 \times 10^3 \text{ Ω} + 240 \text{ Ω}) - I_3(99.0 \times 10^3 \text{ Ω})$

$\quad I_3 = \dfrac{1.40 \times 10^3 \text{ Ω} + 240 \text{ Ω}}{99.0 \times 10^3 \text{ Ω}} I_2$

(3) $I_3 = 0.0166 I_2$

Substitute (3) into (1).
$$I_1 = I_2 + 0.0166 I_2$$
$$(4)\ I_1 = 1.0166 I_2$$
Substitute (4) into (2).
$$0 = 9.00\ \text{V} - 1.0166 I_2(35\ \Omega) - I_2(1.40\times10^3\ \Omega + 240\ \Omega),\ \text{so}$$
$$I_2 = \frac{9.00\ \text{V}}{1.0166(35\ \Omega) + 1.40\times10^3\ \Omega + 240\ \Omega} = \boxed{5.37\ \text{mA}}.$$

(c) The ammeter is connected in series with the 1.40-kΩ resistor. Find the new current through the resistor.

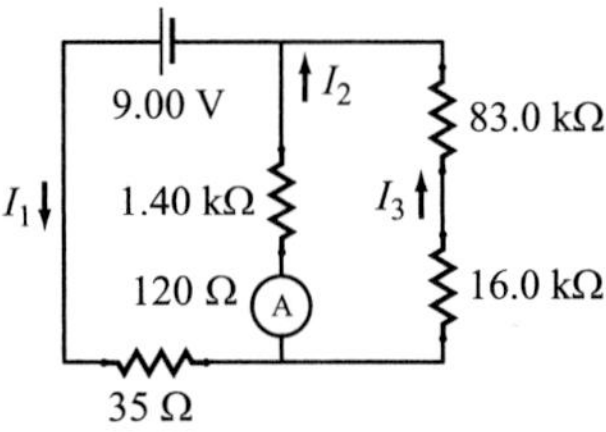

$$(1)\ I_1 = I_2 + I_3$$
$$(2)\ 0 = 9.00\ \text{V} - I_1(35\ \Omega) - I_2(1.40\times10^3\ \Omega + 120\ \Omega)$$
$$0 = I_2(1.40\times10^3\ \Omega + 120\ \Omega) - I_3(99.0\times10^3\ \Omega)$$
$$I_3 = \frac{1.40\times10^3\ \Omega + 120\ \Omega}{99.0\times10^3\ \Omega} I_2$$
$$(3)\ I_3 = 0.01535 I_2$$
Substitute (3) into (1).
$$I_1 = I_2 + 0.01535 I_2$$
$$(4)\ I_1 = 1.01535 I_2$$
Substitute (4) into (2).
$$0 = 9.00\ \text{V} - 1.01535 I_2(35\ \Omega) - I_2(1.40\times10^3\ \Omega + 120\ \Omega),\ \text{so}$$
$$I_2 = \frac{9.00\ \text{V}}{1.01535(35\ \Omega) + 1.40\times10^3\ \Omega + 120\ \Omega} = \boxed{5.79\ \text{mA}}.$$

79. Strategy The resistances are in parallel, so the voltages across the galvanometer and shunt resistor are the same.

Solution Find the required resistance of the shunt resistor.

$$V = (10.0\ \text{A} - 0.250\times10^{-3}\ \text{A})R_S = (0.250\times10^{-3}\ \text{A})(50.0\ \Omega),\ \text{so}\ R_S = \frac{(0.250\times10^{-3}\ \text{A})(50.0\ \Omega)}{10.0\ \text{A} - 0.250\times10^{-3}\ \text{A}} = \boxed{1.25\ \text{m}\Omega}.$$

81. Strategy The resistances are in series, so $V = IR_{eq}$.

Solution Find the required resistance of the series resistor.

$$(2.0\times10^{-3}\ \text{A})(R_S + 75\ \Omega) = 100.0\ \text{V},\ \text{so}\ R_S = \frac{100.0\ \text{V}}{2.0\times10^{-3}\ \text{A}} - 75\ \Omega = \boxed{50\ \text{k}\Omega}.$$

85. Strategy When fully charged, the voltage across the capacitor is equal to that of the emf. According to Kirchhoff's loop rule, when the capacitor is discharging, the voltage across the capacitor is equal to the voltage across the resistor. Use Eqs. (18-24) and (18-26).

Solution Compute the voltage across the resistor.

$$V_R = V_C = \mathcal{E}e^{-t/(RC)} = (90.0\ \text{V})e^{-8.4\times10^{-3}\ \text{s}/[(30.0\times10^3\ \Omega)(0.10\times10^{-6}\ \text{F})]} = \boxed{5.5\ \text{V}}$$

87. Strategy Solve for R using $V_C = \mathcal{E}(1 - e^{-t/\tau})$ where $\tau = RC$.

Solution Find the required resistance.

$$V_C = \mathcal{E}(1 - e^{-t/\tau})$$

$$1 - \frac{V_C}{\mathcal{E}} = e^{-t/\tau}$$

$$\ln\left(1 - \frac{V_C}{\mathcal{E}}\right) = -\frac{t}{RC}$$

$$R = -\frac{t}{C\ln\left(1 - \frac{V_C}{\mathcal{E}}\right)} = -\frac{1.80\ \text{s}}{(125\times10^{-6}\ \text{F})\ln\left(1 - \frac{10.0\ \text{V}}{12.0\ \text{V}}\right)} = \boxed{8.04\ \text{k}\Omega}$$

89. (a) Strategy Use Eq. (18-6).

Solution Compute the required voltage.

$$\Delta V = IR = (40.0\ \text{A})(52.0\ \Omega) = \boxed{2.08\ \text{kV}}$$

(b) Strategy Use Eqs. (18-24) and (18-25).

Solution Find the capacitance.

$$I = I_0 e^{-t/(RC)},\ \text{so}\ e^{t/(RC)} = \frac{I_0}{I}\ \text{and}\ \frac{t}{RC} = \ln\frac{I_0}{I}.\ \text{Thus, } C = \frac{t}{R\ln\frac{I_0}{I}} = \frac{0.00100\ \text{s}}{(52.0\ \Omega)\ln\frac{40.0\ \text{A}}{10.0\ \text{A}}} = \boxed{13.9\ \mu\text{F}}.$$

(c) Strategy and Solution $\boxed{\text{The paramedic shouts “Clear!” to warn others to keep clear of the patient so that they are not shocked as well. The same current that can restart a heart can also stop a heart.}}$

91. (a) Strategy Use Eq. (17-18b).

Solution Find the required initial potential difference.

$$U = \frac{1}{2}C(\Delta V)^2,\ \text{so}\ \Delta V = \sqrt{\frac{2U}{C}} = \sqrt{\frac{2(20.0\ \text{J})}{100.0\times10^{-6}\ \text{F}}} = \boxed{632\ \text{V}}.$$

(b) Strategy Use the definition of capacitance.

Solution Find the initial charge.

$$Q = C(\Delta V) = (100.0\times10^{-6}\ \text{F})(632\ \text{V}) = \boxed{63.2\ \text{mC}}$$

(c) Strategy Solve for R using $I = I_0 e^{-t/\tau}$ where $\tau = RC$.

Solution Find the resistance of the lamp.

$$0.050 I_0 = I_0 e^{-t/\tau}$$

$$\ln 0.050 = -\frac{t}{RC}$$

$$R = -\frac{t}{C\ln 0.050} = -\frac{0.0020\ \text{s}}{(100.0\times10^{-6}\ \text{F})\ln 0.050} = \boxed{6.7\ \Omega}$$

93. Strategy Use Eqs. (18-24) and (18-25) and the definition of resistance.

Solution

(a) Initially ($t = 0$), the capacitor has nearly zero resistance. Find the currents and the voltages.

$$I_1 = I_2 = \frac{V}{R} = \frac{12 \text{ V}}{40.0\times10^3 \ \Omega} = \boxed{0.30 \text{ mA}} \text{ and } V_1 = V_2 = \boxed{12 \text{ V}}.$$

(b) Calculate the time constant.

$$\tau = (40.0\times10^3 \ \Omega)(5.0\times10^{-8} \text{ F}) = 2.0 \text{ ms}$$

Find the currents.

$$I_1 = I_2 = I_0 e^{-t/\tau} = (3.0\times10^{-4} \text{ A})e^{-(1.0 \text{ ms})/(2.0 \text{ ms})} = \boxed{0.18 \text{ mA}}$$

Find V_1 and V_2.

$$V_1 = \boxed{12 \text{ V}} \text{ and } V_2 = I_2 R = (1.82\times10^{-4} \text{ A})(40.0\times10^3 \ \Omega) = \boxed{7.3 \text{ V}}.$$

(c) Find the currents and the voltages.

$$I_1 = I_2 = (3.0\times10^{-4} \text{ A})e^{-(5.0 \text{ ms})/(2.0 \text{ ms})} = \boxed{25 \ \mu\text{A}}, \ V_1 = \boxed{12 \text{ V}}, \text{ and}$$

$$V_2 = I_2 R = (2.463\times10^{-5} \text{ A})(40.0\times10^3 \ \Omega) = \boxed{0.99 \text{ V}}.$$

97. (a) Strategy According to the figure, $I_0 \approx 0.070$ A. Use Eq. (18-25).

Solution Compute the current at $t = \tau$.

$$I(t = \tau) = I_0 e^{-1} = (0.070 \text{ A})e^{-1} = 0.026 \text{ A}$$

So, according to the figure, $\tau \approx 0.060$ s. The final charge is $Q = I_0 \Delta t = I_0 \tau \approx (0.070 \text{ A})(0.060 \text{ s}) = \boxed{4.2 \text{ mC}}$.

(b) Strategy Use the definition of capacitance.

Solution Find the capacitance.

$$C = \frac{Q}{V} \approx \frac{0.0042 \text{ C}}{9.0 \text{ V}} = \boxed{470 \ \mu\text{F}}$$

(c) Strategy Use Eq. (18-24).

Solution Find the total resistance in the circuit.

$$R = \frac{\tau}{C} \approx \frac{0.060 \text{ s}}{470\times10^{-6} \text{ F}} = \boxed{130 \ \Omega}$$

(d) Strategy Use Eqs. (17-18b) and (18-23) and the fact that $U = U_0/2$.

Solution Solve for t.

$$U = \frac{1}{2}CV^2 = \frac{1}{2}CV_0^2(1-e^{-t/\tau})^2 = U_0(1-e^{-t/\tau})^2$$

$$\pm\sqrt{\frac{U}{U_0}} = 1-e^{-t/\tau}$$

$$e^{-t/\tau} = 1\pm\sqrt{\frac{U}{U_0}}$$

$$-\frac{t}{\tau} = \ln\left(1 \pm \sqrt{\frac{U}{U_0}}\right)$$

$$t = -\tau \ln\left(1 \pm \sqrt{\frac{U}{U_0}}\right) \approx -(0.060 \text{ s})\ln\left(1 \pm \sqrt{\frac{1}{2}}\right) = -32 \text{ ms or } 74 \text{ ms}$$

t cannot be negative, so the answer is $\boxed{74 \text{ ms}}$.

101. Strategy Use Eq. (18-19).

Solution The maximum current that can be supplied by the batteries is

$$I_{max} = \frac{P_{max}}{V} = \frac{5.0 \text{ W}}{100.0 \text{ V}} = 0.050 \text{ A}.$$

$$I = \frac{V}{R} = \frac{100.0 \text{ V}}{1.0 \times 10^3 \ \Omega} = 0.10 \text{ A} > I_{max}, \text{ so the current that passes through her is } \boxed{50 \text{ mA}}.$$

105. Strategy Use Eqs. (18-17), (18-21a), and (18-21b).

Solution

(a) $P = V^2/R_{eq}$, so we need to design the circuit such that V is maximized and R_{eq} is minimized. If the batteries are placed in series, $V = 2\mathcal{E}$. If the light bulbs are connected in parallel,

$$R_{eq} = \left(\frac{1}{R} + \frac{1}{R}\right)^{-1} = \frac{R}{2},$$

which is the smallest possible value. The potential across each is $V = 2\mathcal{E}$, which is the largest possible value.

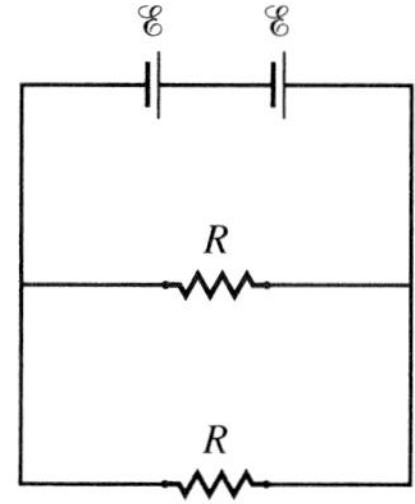

(b) The power dissipated by each bulb is the same.

$$P = \frac{V^2}{R} = \frac{(2\mathcal{E})^2}{R} = \boxed{\frac{4\mathcal{E}^2}{R}}$$

(c) The power through each bulb is given by $P = I^2 R$. The circuit must be designed so that the current through the brighter bulb is larger than that through the dimmer bulb. In the circuit below, the maximum current passes through the bulb on the right, whereas only a fraction of that current passes through the bulb on the left, so $\boxed{\text{the bulb on the right is brighter.}}$

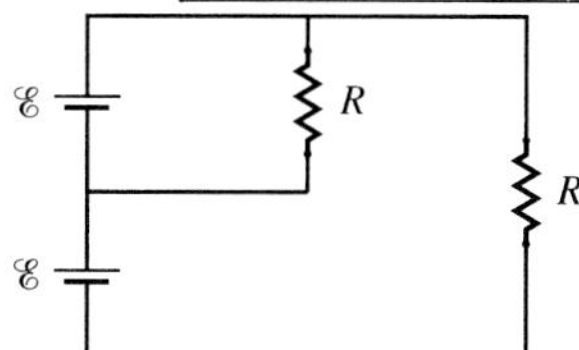

109. Strategy Use the definition of average power and Eq. (18-19).

Solution Find the amount of chemical energy that is converted to electrical energy.

$$P = \frac{\Delta E}{\Delta t} = IV, \text{ so } \Delta E = IV \Delta t = (0.30 \text{ A})(1.5 \text{ V})(4.0 \text{ h})(3600 \text{ s/h}) = \boxed{6.5 \text{ kJ}}.$$

111. Strategy Use the definition of resistance and Eqs. (18-13) and (18-17).

Solution

(a) The current through A_1 is the same as that through the emf.

$$I = \frac{V}{R_{eq}} = \frac{10.0 \text{ V}}{2.00 \text{ } \Omega + \left(\frac{1}{2.00 \text{ } \Omega} + \frac{1}{3.00 \text{ } \Omega} + \frac{1}{6.00 \text{ } \Omega}\right)^{-1} + 2.00 \text{ } \Omega} = \boxed{2.00 \text{ A}}$$

(b) Since $\left(\frac{1}{3.00 \text{ } \Omega} + \frac{1}{6.00 \text{ } \Omega}\right)^{-1} = 2.00 \text{ } \Omega$, the current is split evenly at the first junction to the right of A_1.

So, $I = \dfrac{2.00 \text{ A}}{2} = \boxed{1.00 \text{ A}}$.

113. Strategy The average discharge current is the charge on the capacitor divided by the time to discharge. Use the definition of capacitance.

Solution Find the average discharge current.

$$I_{av} = \frac{Q}{\Delta t} = \frac{CV}{\Delta t} = \frac{(25 \times 10^{-6} \text{ F})(1.0 \text{ V})}{0.80 \text{ s}} = \boxed{31 \text{ } \mu\text{A}}$$

117. Strategy Use Eq. (18-21b).

Solution The equivalent emf for identical emfs in parallel is the same as one emf, so the total power dissipated in

the circuit is $P = \dfrac{V^2}{R_{eq}} = \boxed{\dfrac{\mathscr{E}^2}{2R}}$.

121. Strategy Use the relationships between power, resistance, voltage, charge, and time. Use Eqs. (14-4) and (14-9).

Solution Find an expression for the resistance.

$$P = \frac{Q}{\Delta t} = \frac{mc_w \Delta T + m_{boil} L_v}{\Delta t} = \frac{V^2}{R}, \text{ so}$$

$$R = \frac{V^2 \Delta t}{mc_w \Delta T + m_{boil} L_v} = \frac{(120 \text{ V})^2 (480 \text{ s})}{(1.0 \text{ kg})[4186 \text{ J/(kg} \cdot \text{K})](90 \text{ K}) + 0.050(1.0 \text{ kg})(2,256,000 \text{ J/kg})} = \boxed{14 \text{ } \Omega}.$$

125. (a) Strategy The voltage across each bulb is 120 V. Use Eq. (18-21b).

Solution Find the resistance of each light bulb.

$$P = \frac{V^2}{R}, \text{ so } R = \frac{V^2}{P} = \frac{(120 \text{ V})^2}{9.0 \text{ W}} = \boxed{1600 \ \Omega}.$$

(b) Strategy Use the definition of resistance.

Solution Find the current through each bulb.

$$I = \frac{V}{R} = \frac{120 \text{ V}}{1600 \ \Omega} = \boxed{0.075 \text{ A}}$$

(c) Strategy The bulbs are connected in parallel, so the total current is the sum of individual currents.

Solution The total current is $25(0.075 \text{ A}) = \boxed{1.9 \text{ A}}$.

(d) Strategy Let the number of 10.4-W bulbs be n. Then, the number of 9.0-W bulbs is $25 - n$. Use the definition of resistance and Eq. (18-19).

Solution The current through a single bulb is given by $I = P/V$. The total current for each kind of bulb is $(25 - n)P_1/V$ and nP_2/V, where $P_1 = 9.0$ W and $P_2 = 10.4$ W. The total current must be less than or equal to 2.0 A. Find n.

$$\frac{(25-n)P_1}{V} + \frac{nP_2}{V} \le 2.0 \text{ A}$$
$$25P_1 - nP_1 + nP_2 \le (2.0 \text{ A})V$$
$$n(P_2 - P_1) \le (2.0 \text{ A})V - 25P_1$$
$$n \le \frac{(2.0 \text{ A})V - 25P_1}{P_2 - P_1} = \frac{(2.0 \text{ A})(120 \text{ V}) - 25(9.0 \text{ W})}{10.4 \text{ W} - 9.0 \text{ W}} \approx 10.7$$

Thus, up to $\boxed{10}$ 9.0-W bulbs can be replaced with 10.4-W bulbs without blowing a fuse.

129. Strategy Use Eq. (18-8).

Solution Initially, $R_0 = \rho L_0/A_0$, and finally, $R = \rho L/A$. The volume is constant, so $V = A_0 L_0 = AL$, or $A_0 = AL/L_0$. Divide R by R_0.

$$\frac{R}{R_0} = \frac{\rho \frac{L}{A}}{\rho \frac{L_0}{A_0}}, \text{ so } R = \frac{A_0 L}{AL_0} R_0 = \left(\frac{AL}{L_0}\right)\left(\frac{L}{AL_0}\right) R_0 = \left(\frac{L}{L_0}\right)^2 R_0 = \left(\frac{3L_0}{L_0}\right)^2 R_0 = \boxed{9R_0}.$$

133. (a) Strategy Use the definition of capacitance and Eq. (17-15).

Solution Find the charge on the upper plate.

$$Q = CV = \frac{\epsilon_0 A}{d} V = \frac{\epsilon_0 L^2}{d} V = \frac{[8.854 \times 10^{-12}\ \mathrm{C^2/(N \cdot m^2)}](0.10\ \mathrm{m})^2 (10.0\ \mathrm{V})}{89 \times 10^{-6}\ \mathrm{m}} = \boxed{9.9\ \mathrm{nC}}$$

(b) Strategy Use the definition of capacitance and Eqs. (17-15), (18-24), and (18-25).

Solution Find the time constant.

$$\tau = RC = R\frac{\epsilon_0 A}{d} = \frac{(0.100 \times 10^6\ \Omega)[8.854 \times 10^{-12}\ \mathrm{C^2/(N \cdot m^2)}](0.10\ \mathrm{m})^2}{89 \times 10^{-6}\ \mathrm{m}} = 9.9 \times 10^{-5}\ \mathrm{s}$$

Compute the initial current.

$$\frac{\mathscr{E}}{R} = \frac{10.0\ \mathrm{V}}{0.100 \times 10^6\ \Omega} = 100\ \mu\mathrm{A}$$

The current is given by $I = (100\ \mu\mathrm{A})e^{-t/(9.9 \times 10^{-5}\ \mathrm{s})}$.

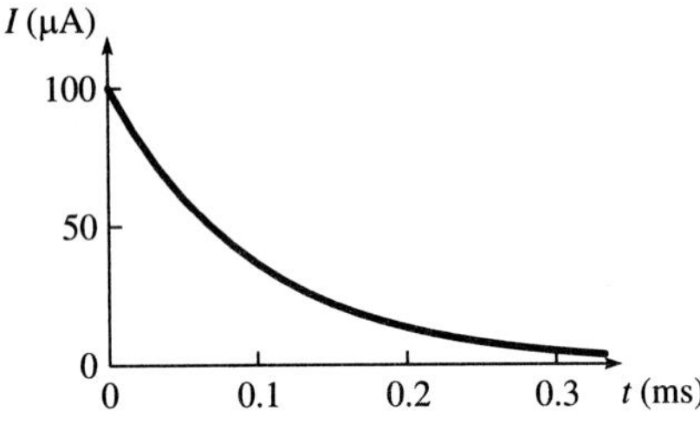

(c) Strategy The energy dissipated in R is that stored in the capacitor, U. Use Eqs. (17-15) and (17-18b).

Solution Compute the energy dissipated over the whole discharging process.

$$U = \frac{1}{2}CV^2 = \frac{1}{2}\left(\frac{\epsilon_0 A}{d}\right)\mathscr{E}^2 = \frac{[8.854 \times 10^{-12}\ \mathrm{C^2/(N \cdot m^2)}](0.10\ \mathrm{m})^2 (10.0\ \mathrm{V})^2}{2(89 \times 10^{-6}\ \mathrm{m})} = \boxed{50\ \mathrm{nJ}}$$

REVIEW AND SYNTHESIS: CHAPTERS 16–18

Review Exercises

1. **Strategy and Solution** Since the spheres are identical, the charge will be shared evenly by the two spheres, so the spheres will have $(18.0\ \mu C + 6.0\ \mu C)/2 = \boxed{12.0\ \mu C}$ of charge each.

5. **Strategy** The force on charge A due to charge B is equal and opposite to the horizontal component of the tension. The sign of charge A is negative, since the sign of charge B is positive and the force is attractive. Use Newton's second law and Eq. (16-2).

   Solution

   (a) Find the tension.

   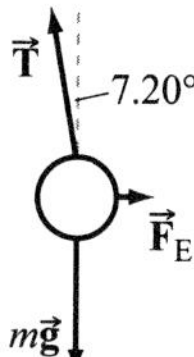

   $$\Sigma F_y = T \cos 7.20° - mg = 0, \text{ so } T = \frac{mg}{\cos 7.20°}.$$

   The horizontal component of the tension is $T_x = \dfrac{mg}{\cos 7.20°} \sin 7.20° = mg \tan 7.20°.$

   Solve for the second charge.

   $$\frac{k|q_A||q_B|}{r^2} = mg \tan 7.20°, \text{ so}$$

   $$|q_A| = \frac{r^2 mg \tan 7.20°}{k|q_B|} = \frac{(0.0500\ \text{m})^2 (0.0900\ \text{kg})(9.80\ \text{m/s}^2) \tan 7.20°}{(8.988 \times 10^9\ \text{N} \cdot \text{m}^2/\text{C}^2)(130 \times 10^{-9}\ \text{C})} = 238\ \text{nC}.$$

   Thus, the charge on A is $\boxed{-238\ \text{nC}}$.

   (b) The tension in the thread is $T = \dfrac{mg}{\cos 7.20°} = \dfrac{(0.0900\ \text{kg})(9.80\ \text{m/s}^2)}{\cos 7.20°} = \boxed{0.889\ \text{N}}.$

9. **Strategy** Use Newton's second law to find the horizontal acceleration of the electron.

   Solution The y-component of the particle's displacement is
   $$\Delta y = v_y \Delta t = (10.0\ \text{m/s})(2.40 \times 10^{-6}\ \text{s}) = \boxed{24\ \mu\text{m}}.$$

   According to Newton's second law, $\Sigma F_x = -eE = ma_x$, so $a_x = -\dfrac{eE}{m}.$

   Thus, the x-component of the particle's displacement is
   $$\Delta x = \frac{1}{2} a_x (\Delta t)^2 = \frac{1}{2}\left(-\frac{eE}{m}\right)(\Delta t)^2 = -\frac{eE(\Delta t)^2}{2m} = -\frac{(1.602 \times 10^{-19}\ \text{C})(200\ \text{V/m})(2.40 \times 10^{-6}\ \text{m})^2}{2(9.109 \times 10^{-31}\ \text{kg})} = \boxed{-100\ \text{m}}.$$

13. Strategy Refer to the figure. Use Kirchhoff's rules.

Solution Use the junction and loop rules.

(1) $I_1 = I_2 + I_3$

(2) $V_1 - I_1 R_1 - I_2 R_2 = 0$

(3) $V_2 + I_2 R_2 - I_3 R_3 = 0$

(a) Solve (2) for I_2.
$$I_2 = \frac{V_1 - I_1 R_1}{R_2} = \frac{30.0 \text{ V} - (2.50 \text{ A})(8.00 \ \Omega)}{5.00 \ \Omega} = \boxed{2.00 \text{ A}}.$$

(b) Solve (1) for I_3.
$$I_3 = I_1 - I_2 = 2.50 \text{ A} - 2.00 \text{ A} = \boxed{0.50 \text{ A}}$$

(c) Solve (3) for R_3.
$$R_3 = \frac{V_2 + I_2 R_2}{I_3} = \frac{9.00 \text{ V} + (2.00 \text{ A})(5.00 \ \Omega)}{0.50 \text{ A}} = \boxed{38 \ \Omega}$$

17. Strategy The rate of heat flow into the water is $P = Q/\Delta t$. Use Eqs. (14-4) and (18-21b).

Solution Find an expression for the resistance.
$$P = \frac{Q}{\Delta t} = \frac{mc\Delta T}{\Delta t} = \frac{V^2}{R}, \text{ so } R = \frac{V^2 \Delta t}{mc\Delta T}.$$
Find the ratio of the resistances.
$$\frac{R_A}{R_B} = \frac{\dfrac{V^2 \Delta t_A}{m_A c \Delta T}}{\dfrac{V^2 \Delta t_B}{m_B c \Delta T}} = \frac{\Delta t_A m_B}{\Delta t_B m_A} = \frac{\Delta t_A \rho V_B}{\Delta t_B \rho V_A} \ \ (V \text{ is now volume.}) = \frac{\Delta t_A V_B}{\Delta t_B V_A} = \frac{(2.0)(5.0)}{(5.0)(1.0)} = \boxed{2.0}$$

21. (a) Strategy Use Eq. (18-26) and the definition of capacitance.

Solution Find the charge as a function of time.

$$Q = C\Delta V = CV_C(t) = Q(t), \text{ so } V_C(t) = \frac{Q(t)}{C} = \mathscr{E}e^{-t/\tau} \text{ or } Q(t) = C\mathscr{E}e^{-t/\tau} = Q_0 e^{-t/\tau}.$$

Find the time constant.

$$\frac{Q(t)}{Q_0} = 0.010 = e^{-t/\tau}, \text{ so } \tau = -\frac{t}{\ln 0.010} = -\frac{4.0\times10^{-3} \text{ s}}{\ln 0.010} = \boxed{8.7\times10^{-4} \text{ s}}.$$

(b) Strategy Use Eq. (17-18b) and (18-24).

Solution Find the capacitance of the flash bulb.

$$U = \frac{1}{2}C(\Delta V)^2, \text{ so } C = \frac{2U}{(\Delta V)^2}.$$

Find the resistance of the flash bulb.

$$\tau = RC, \text{ so } R = \frac{\tau}{C} = \frac{-\frac{t}{\ln 0.010}}{2U/(\Delta V)^2} = -\frac{t(\Delta V)^2}{2U\ln 0.010} = -\frac{(4.0\times10^{-3} \text{ s})(300 \text{ V})^2}{2(32 \text{ J})\ln 0.010} = \boxed{1.2 \ \Omega}.$$

(c) Strategy When the capacitor begins to discharge, its voltage is 300 V. Use Eq. (18-21b).

Solution Compute the maximum power.

$$P_{\text{max}} = \frac{V^2}{R} = \frac{(300 \text{ V})^2}{1.22 \ \Omega} = \boxed{74 \text{ kW}}$$

25. (a) Strategy Use Newton's second law and the equations for motion with constant acceleration.

Solution Find the acceleration.

$$\Delta t = \frac{\Delta x}{v_x} \text{ and } \Delta y = \frac{1}{2}a_y(\Delta t)^2, \text{ where } a_y \text{ is the minimum acceleration. So, } a_y = \frac{2\Delta y}{(\Delta t)^2} = \frac{2v_x^2\Delta y}{(\Delta x)^2} \text{ and according}$$

to Newton's second law, $a_y = \frac{F}{m} = \frac{qE}{m} = \frac{q\Delta V}{m\Delta y}$. Thus, $a_y = \frac{q\Delta V}{m\Delta y} = \frac{2v_x^2\Delta y}{(\Delta x)^2}$. Solve for the minimum ΔV.

$$\frac{q\Delta V}{m\Delta y} = \frac{2v_x^2\Delta y}{(\Delta x)^2}, \text{ so minimum } \Delta V = \frac{2mv_x^2(\Delta y)^2}{q(\Delta x)^2} = \frac{2(2.0\times10^{-13} \text{ kg})(3.0 \text{ m/s})^2(0.010 \text{ m})^2}{1000(1.602\times10^{-19} \text{ C})(0.10 \text{ m})^2} = \boxed{220 \text{ V}}.$$

(b) Strategy Use the results from part (a) and the relationship between speed, acceleration, and time.

Solution Find the speed of the particle.

$$v_f = a\Delta t = \frac{2\Delta y}{(\Delta t)^2}\cdot\Delta t = \frac{2\Delta y}{\Delta t} = \frac{2\Delta y}{\Delta x}\cdot v_x = \frac{2(0.010 \text{ m})}{0.10 \text{ m}}(3.0 \text{ m/s}) = \boxed{0.60 \text{ m/s}}$$

(c) Strategy Use Eq. (9-16).

Solution Calculate the viscous drag force.

$$F_D = 6\pi\eta rv = 6\pi(1.8\times10^{-5} \text{ Pa}\cdot\text{s})(6.0\times10^{-6} \text{ m})(0.60 \text{ m/s}) = \boxed{1.2 \text{ nN}}$$

(d) Strategy and Solution

$$\boxed{\text{It is not realistic to ignore drag, since } F_E \ll F_D. \text{ The potential difference should be larger.}}$$

MCAT Review

1. **Strategy and Solution** At a given temperature, the resistance R of a wire to direct current is given by $R = \rho L / A$, where ρ is the resistivity, L is the length, and A is the cross-sectional area. Therefore, the correct answer is $\boxed{D}$.

2. **Strategy** There are ten electric immersion heaters that each use 5 kW of power. Use Eq. (18-19).

   Solution The total power requirement to run all ten heaters is 50 kW. Find the current.
   $$P = I\Delta V, \text{ so } I = \frac{P}{\Delta V} = \frac{50 \times 10^3 \text{ W}}{600 \text{ V}} = 83 \text{ A}.$$
   The correct answer is $\boxed{C}$.

3. **Strategy** Each heater draws 20 A, so five heaters draw 100 A. Use Eq. (18-19).

   Solution Find the total power usage of the heaters.
   $$P = I\Delta V = (100 \text{ A})(800 \text{ V}) = 80 \text{ kW}$$
   The correct answer is $\boxed{C}$.

4. **Strategy** Use Kirchhoff's rules and the definition of resistance.

   Solution Find the current flowing through R_L. According to the loop rule,
   $$I_L R_L - I_S R_S = 0, \text{ so } I_S = \frac{R_L}{R_S} I_L = \frac{1.0 \ \Omega}{2.0 \ \Omega} I_L = 0.50 I_L. \text{ According to the junction rule,}$$
   $$I = I_L + I_S = I_L + 0.50 I_L = 1.50 I_L, \text{ so } I_L = \frac{I}{1.50}. \text{ Thus, the voltage drop across } R_L \text{ is}$$
   $$V_L = I_L R_L = \frac{I R_L}{1.50} = \frac{(0.5 \text{ A})(1.0 \ \Omega)}{1.50} = 0.33 \text{ V. The correct answer is } \boxed{B}.$$

5. **Strategy** Use Kirchhoff's rules and Eq. (18-21a).

   Solution Find the current flowing through R_S. According to the loop rule,
   $$I_L R_L - I_S R_S = 0, \text{ so } I_L = \frac{R_S}{R_L} I_S = \frac{3.0 \ \Omega}{1.0 \ \Omega} I_S = 3.0 I_S. \text{ According to the junction rule,}$$
   $$I = I_L + I_S = 3.0 I_S + I_S = 4.0 I_S, \text{ so } I_S = 0.25 I. \text{ Thus, the power dissipated in } R_S \text{ is}$$
   $$P_S = I_S{}^2 R_S = (0.25 I)^2 R_S = [0.25(1.2 \text{ A})]^2 (3.0 \ \Omega) = 0.27 \text{ W. The correct answer is } \boxed{A}.$$

6. **Strategy and Solution** As current flows through R_L, power is dissipated at a constant rate by R_L as heat that enters the water, increasing its energy and raising its temperature (and the temperature of the system). Thus, the entropy of the system increases, as well. The correct answer is $\boxed{D}$.

7. **Strategy and Solution** Energy is stored in the battery as chemical energy. This energy is converted into electrical energy when the current flows. The electrical energy is dissipated as heat by the resistor. The correct answer is $\boxed{A}$.

8. **Strategy and Solution** As R_L increases with time, the amount of the current I passing through it decreases and the amount passing through R_S increases. The correct answer is $\boxed{C}$.

9. **Strategy** Use Eq. (14-4).

   Solution The water is heated at a rate of $Q/\Delta t = mc\Delta T/\Delta t = 1.0$ W. So, the time it takes for the temperature of the water to increase $1.0°C$ is $\Delta t = \dfrac{mc\Delta T}{1.0\text{ W}} = \dfrac{(1.0\text{ kg})[4.2\times10^3\text{ J/(kg}\cdot°C)](1.0°C)}{1.0\text{ W}} = 4200$ s.

   The correct answer is $\boxed{D}$.

10. **Strategy** Use Eq. (18-19).

    Solution Compute the current required.

    $P = I\Delta V$, so $I = \dfrac{P}{\Delta V} = \dfrac{1.2\times10^4\text{ W}}{120\text{ V}} = 100$ A. The correct answer is $\boxed{C}$.

11. **Strategy** Refer to the table to compute the initial and final resistances.

    Solution Compute the resistances.

    $R_i = (10^5\text{ m})\dfrac{3.4\times10^{-1}\text{ }\Omega}{10^3\text{ m}} = 34\text{ }\Omega$ and $R_f = (10^5\text{ m})\dfrac{3.8\times10^{-1}\text{ }\Omega}{10^3\text{ m}} = 38\text{ }\Omega$.

    The change in resistance is $R_f - R_i = 38\text{ }\Omega - 34\text{ }\Omega = 4\text{ }\Omega$. The correct answer is $\boxed{C}$.

12. **Strategy** Use Eq. (18-21a).

    Solution Compute the power lost as heat.

    $P = I^2 R = (2\text{ A})^2(3\text{ }\Omega) = 12$ W, so the correct answer is $\boxed{C}$.

13. **Strategy and Solution** The ten residences require 10×10^4 W $= 10^5$ W of power and 5×10^3 W of power is lost as heat, so the total power requirement is 10^5 W $+ 5\times10^3$ W $= 1.05\times10^5$ W. The correct answer is $\boxed{C}$.

Chapter 19

MAGNETIC FORCES AND FIELDS

Conceptual Questions

1. The magnetic field cannot be described as the magnetic force per unit charge because unlike the electric force, the magnetic force depends upon the velocity of the charge.

5. A constant magnetic field does zero work on a moving charge and therefore cannot change its kinetic energy—the speed of a particle with constant kinetic energy must also be constant.

9. It is impossible to completely eliminate the magnetic fields generated by power lines. Current in the lines produces magnetic fields that encircle them—the magnitude of these fields falls fairly rapidly but, in principle, has infinite extent.

13. The magnetic field around the speaker would deflect the electron beams, and thus, distort the image on the computer monitor.

17. **(a)** The metal bar was placed in an external magnetic field and the domains, on average, lined up with the field.

    (b) The metal is ferromagnetic as evidenced by the fact that it contains small magnetic domains in which the magnetic moments of the atoms in the metal are all lined up.

Problems

1. **Strategy** The magnetic field is strong where field lines are close together and weak where they are far apart.

   Solution

   (a) The field lines are farthest apart (lowest density) at point $\boxed{F}$, so the magnetic field strength is smallest there.

   (b) The field lines are closest together (highest density) at point $\boxed{A}$, so the magnetic field strength is largest there.

3. **Strategy** A bar magnet is a magnetic dipole. Field lines emerge from a bar magnet at its north pole and enter at its south pole. Magnetic field lines are closed loops. Use symmetry.

   Solution

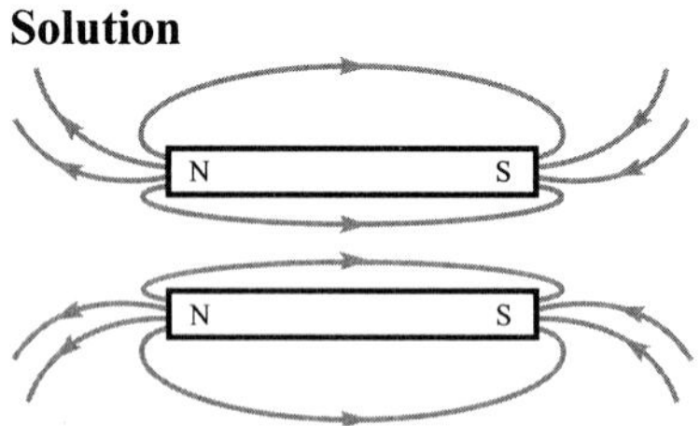

5. **Strategy** A bar magnet is a magnetic dipole. Field lines emerge from a bar magnet at its north pole and enter at its south pole. Magnetic field lines are closed loops. Use symmetry.

   Solution

   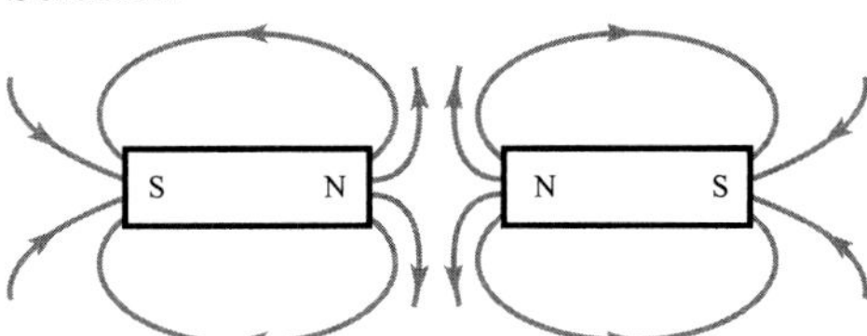

9. **Strategy** The magnetic force on a moving point charge is given by Eq. (19-5). The direction of the force is determined by the right-hand rule and the sign of the charge.

   Solution Find the magnetic force exerted on the proton.

   $$\vec{F} = q\vec{v} \times \vec{B} = e\vec{v} \times \vec{B} = (1.602 \times 10^{-19}\ \text{C})[(6.0 \times 10^{6}\ \text{m/s east}) \times (2.50\ \text{T north})] = \boxed{2.4 \times 10^{-12}\ \text{N up}}$$

   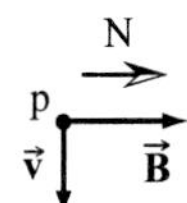

11. **Strategy** Determine the speed of the electron using its kinetic energy. Then determine the magnetic force on it using Eq. (19-5).

    Solution Find the speed.

    $$K = \frac{1}{2}mv^2,\ \text{so}\ v = \sqrt{\frac{2K}{m}}.$$

    Calculate the force.

    $$\vec{F} = -e\vec{v} \times \vec{B} = -(1.602 \times 10^{-19}\ \text{C})\left\{\left[\sqrt{\frac{2(7.2 \times 10^{-18}\ \text{J})}{9.109 \times 10^{-31}\ \text{kg}}}\ \text{east}\right] \times (0.800\ \text{T up})\right\} = \boxed{5.1 \times 10^{-13}\ \text{N north}}$$

13. **Strategy** The magnetic force on a moving point charge is given by Eq. (19-5). The direction of the force is determined by the right-hand rule and the sign of the charge.

    Solution Find the magnetic force on the electron at point a.

    $$\vec{F} = -e\vec{v} \times \vec{B} = -(1.602 \times 10^{-19}\ \text{C})[(8.0 \times 10^{5}\ \text{m/s right}) \times (0.40\ \text{T down})] = \boxed{5.1 \times 10^{-14}\ \text{N out of the page}}$$

15. **Strategy** The magnetic force on a moving point charge is given by Eq. (19-5). The direction of the force is determined by the right-hand rule and the sign of the charge.

    Solution Find the magnetic force on the electron for each of the directions shown in the figure.

    $$\vec{F} = -e[(v\ \text{at}\ 30.0°\ \text{above right}) \times (B\ \text{down})]$$

    $$= (1.602 \times 10^{-19}\ \text{C})(8.0 \times 10^{5}\ \text{m/s})(0.40\ \text{T})\sin 120.0°\ \text{out of the page} = \boxed{4.4 \times 10^{-14}\ \text{N out of the page}}$$

17. **Strategy** The magnetic force on a moving point charge is given by Eq. (19-5). The direction of the force is determined by the right-hand rule and the sign of the charge.

    Solution Find the magnetic force on the dust particle.

    $$\vec{F} = q\vec{v} \times \vec{B} = (-8.0 \times 10^{-18}\ \text{C})[(0.0030\ \text{m/s down}) \times (0.30\ \text{T east})] = \boxed{7.2 \times 10^{-21}\ \text{N north}}$$

    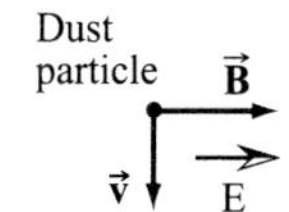

21. **Strategy** Since the (negatively charged) electron is moving due west and the magnetic force is upward, the component of the magnetic field perpendicular to the motion of the electron points north. Use Eq. (19-1).

    Solution Find the angle.

    $$evB\sin\theta = F, \text{ so } \theta = \sin^{-1}\frac{F}{evB} = \sin^{-1}\frac{3.2\times10^{-14}\text{ N}}{(1.602\times10^{-19}\text{ C})(2.0\times10^{5}\text{ m/s})(1.2\text{ T})} = 56°.$$

    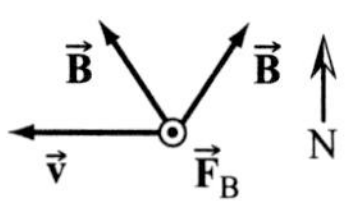

    $\boxed{\text{There are two possibilities: 56° N of W and 56° N of E.}}$

25. **Strategy** $\vec{B}$ and $\vec{v}$ are perpendicular. Use Eq. (19-6).

    Solution Find the magnitude of the magnetic field.

    $$F = evB, \text{ so } B = \frac{F}{ev} = \frac{1.0\times10^{-13}\text{ N}}{(1.602\times10^{-19}\text{ C})(8.0\times10^{5}\text{ m/s})} = \boxed{0.78\text{ T}}.$$

27. **Strategy** Solve Eq. (19-7) for the speed of the proton.

    Solution Find the speed.

    $$\frac{v^2}{r} = \frac{evB}{m}, \text{ so } v = \frac{reB}{m} = \frac{(0.820\text{ m})(1.602\times10^{-19}\text{ C})(0.360\text{ T})}{1.673\times10^{-27}\text{ kg}} = \boxed{2.83\times10^{7}\text{ m/s}}.$$

29. **Strategy** Solve Eq. (19-7) for the magnitude of the magnetic field.

    Solution Find the magnitude of the magnetic field.

    $$\frac{v^2}{r} = \frac{|q|vB}{m} = \frac{2evB}{m_\alpha}, \text{ so } B = \frac{m_\alpha v}{2er} = \frac{(4.003\text{ u})(1.6605\times10^{-27}\text{ kg/u})(0.458)(3.00\times10^{8}\text{ m/s})}{2(1.602\times10^{-19}\text{ C})(1.00\text{ m})} = \boxed{2.85\text{ T}}.$$

31. **Strategy** Use conservation of energy and Equation (19-7).

    Solution Find the speed of the accelerated isotopes.

    $$\Delta K = \frac{1}{2}mv^2 = -\Delta U = |q|V, \text{ so } v = \sqrt{\frac{2|q|V}{m}}.$$

    Relate m to r.

    $$\frac{v^2}{r} = \frac{|q|vB}{m}$$

    $$\frac{1}{r}\sqrt{\frac{2|q|V}{m}} = \frac{|q|B}{m}$$

    $$\frac{1}{r^2}\left(\frac{2|q|V}{m}\right) = \frac{q^2B^2}{m^2}$$

    $$\frac{m}{r^2} = \frac{|q|B^2}{2V} = \text{constant}$$

    m/r^2 is the same for both isotopes. Find m_r, the mass of the rare isotope.

    $$\frac{m_r}{r_r^2} = \frac{m_a}{r_a^2}, \text{ so } m_r = \left(\frac{r_r}{r_a}\right)^2 m_a = \left(\frac{15.6\text{ cm}}{15.0\text{ cm}}\right)^2(12.00\text{ u}) = \boxed{13.0\text{ u}}.$$

33. Strategy Since the ions are accelerated through the same potential difference and they have the same charge, they have the same kinetic energy. Let m_u be the mass of the unknown element. Use Eq. (19-7) and refer to the periodic table.

Solution

(a) Relate the speeds to the masses.

$$\frac{1}{2}m_C v_C^2 = \frac{1}{2}m_u v_u^2, \text{ so } \frac{v_C}{v_u} = \sqrt{\frac{m_u}{m_C}}.$$

$$\frac{v^2}{r} = \frac{|q|vB}{m}, \text{ so } r = \frac{mv}{|q|B}. \text{ Form a proportion.}$$

$$\frac{r_C}{r_u} = \frac{m_C v_C}{m_u v_u} = \frac{m_C}{m_u}\sqrt{\frac{m_u}{m_C}} = \sqrt{\frac{m_C}{m_u}}, \text{ or } m_u = \left(\frac{r_u}{r_C}\right)^2 m_C = \left(\frac{r_C + 1.160 \text{ cm}}{r_C}\right)^2 m_C = \left(1 + \frac{1.160 \text{ cm}}{r_C}\right)^2 m_C.$$

Similarly, we have $\dfrac{r_O}{r_C} = \sqrt{\dfrac{m_O}{m_C}}$, so $r_O = r_C\sqrt{\dfrac{m_O}{m_C}}$. Find r_C.

$$r_O - r_C = r_C\left(\sqrt{\frac{m_O}{m_C}} - 1\right), \text{ so } r_C = \frac{r_O - r_C}{\sqrt{\dfrac{m_O}{m_C}} - 1} = \frac{2.250 \text{ cm}}{\sqrt{\dfrac{16 \text{ u}}{12 \text{ u}}} - 1} = 15 \text{ cm}.$$

Therefore, the mass of the unknown element is $m_u = \left(1 + \dfrac{1.160 \text{ cm}}{15 \text{ cm}}\right)^2 (12 \text{ u}) = \boxed{14 \text{ u}}$.

(b) According to the periodic table, the unknown element is $\boxed{\text{nitrogen}}$.

37. Strategy Since the electron experiences zero net force from the electric and magnetic forces, the forces are equal in magnitude and opposite in direction. Use Eq. (19-10).

Solution Compute the electron's speed.

$$v = \frac{E}{B} = \frac{2.68 \times 10^6 \text{ V/m}}{0.635 \text{ T}} = \boxed{4.22 \times 10^6 \text{ m/s}}$$

39. Strategy Use Eq. (19-10) and the potential due to a uniform electric field.

Solution

$$v_D = \frac{E_H}{B} \text{ where } E_H = \frac{V_H}{w}, \text{ so } v_D = \frac{V_H}{wB} = \frac{7.2 \times 10^{-6} \text{ V}}{(0.035 \text{ m})(0.43 \text{ T})} = \boxed{0.48 \text{ mm/s}}.$$

41. Strategy and Solution

(a) $V_H \propto t^{-1}$, and t changes as the strip is rotated with respect to the field; the probe measures the perpendicular component of the field. Therefore, the answer is $\boxed{\text{no}}$; the Hall probe will not read the correct field strength.

(b) If the field is in the plane of the strip, $\vec{B}$ is parallel to the motion of the electrons and there is no magnetic force. Thus, $\boxed{V_H = 0}$.

45. Strategy Use Eq. (19-10) and conservation of energy.

Solution Derive the charge to mass ratio.

$$q\Delta V = \frac{1}{2}mv^2 = \frac{1}{2}m\left(\frac{E}{B}\right)^2, \text{ so } \frac{q}{m} = \boxed{\frac{E^2}{2B^2\Delta V}}.$$

47. Strategy The magnitude of the force is given by $F = ILB\sin\theta$.

Solution

(a) Solve for B.
$$B = \frac{F}{IL\sin\theta}$$
B is a minimum when $\sin\theta = 1$ $(\theta = 90°)$.
$$B_{min} = \frac{F}{IL(1)} = \frac{4.12 \text{ N}}{(33.0 \text{ A})(0.25 \text{ m})} = \boxed{0.50 \text{ T}}$$

(b) $\boxed{\text{We do not know the directions of the current and the field; therefore, we set } \sin\theta = 1 \text{ and get the minimum field strength.}}$

49. (a) Strategy Use Eq. (19-12a). Refer to the figure.

Solution The current flows from east to west. According to $\vec{F} = I\vec{L}\times\vec{B}$ and the RHR, the force on the rod is directed to the $\boxed{\text{north}}$.

(b) Strategy The rod will accelerate northward. Use Newton's second law, Eq. (19-12b), and Eq. (4-5).

Solution Find the acceleration of the rod.
$\Sigma F = ILB = ma$, so $a = ILB/m$.
Find the speed of the rod.
$$v_f^2 - v_i^2 = v^2 - 0 = 2a\Delta r = 2\frac{ILB}{m}\Delta r, \text{ so}$$
$$v = \sqrt{\frac{2ILB\Delta r}{m}} = \sqrt{\frac{2(2.00 \text{ A})(0.500 \text{ m})(0.750 \text{ T})(8.00 \text{ m})}{0.0500 \text{ kg}}} = \boxed{15.5 \text{ m/s}}.$$

51. Strategy Use Eq. (19-12a).

Solution

(a) Calculate the force on each wire segment.
$$\vec{F}_{top} = I\vec{L}_{top}\times\vec{B} = (1.0 \text{ A})(0.300 \text{ m right})\times(2.5 \text{ T out of the page}) = \boxed{0.75 \text{ N in the } -y\text{-direction}}$$
$$\vec{F}_{bottom} = I\vec{L}_{bottom}\times\vec{B} = -I\vec{L}_{top}\times\vec{B} = \boxed{0.75 \text{ N in the } +y\text{-direction}}$$

$$\vec{F}_{\text{left}} = I\vec{L}_{\text{left}} \times \vec{B} = (1.0 \text{ A})(0.200 \text{ m up}) \times (2.5 \text{ T out of the page}) = \boxed{0.50 \text{ N in the } +x\text{-direction}}$$

$$\vec{F}_{\text{right}} = I\vec{L}_{\text{right}} \times \vec{B} = -I\vec{L}_{\text{left}} \times \vec{B} = \boxed{0.50 \text{ N in the } -x\text{-direction}}$$

(b) Compute the components of the net force.

$$F_{\text{net}, x} = 0.50 \text{ N} - 0.50 \text{ N} = 0 \text{ and } F_{\text{net}, y} = 0.75 \text{ N} - 0.75 \text{ N} = 0, \text{ so } \vec{F}_{\text{net}} = \boxed{0}.$$

53. Strategy Use Eq. (19-12a).

Solution

(a) $\vec{F} = I\vec{L} \times \vec{B}$, where $\vec{L} = L$ west and $\vec{B} = 0.48$ mT at $72°$ below the horizontal with the horizontal component due north. The diagram shows the orientation of $\vec{L}, \vec{B},$ and $\vec{F}$ where $\vec{F}$ was determined by the right hand rule. $\vec{F}$ is directed $\boxed{18° \text{ below the horizontal with the horizontal component due south}}$.

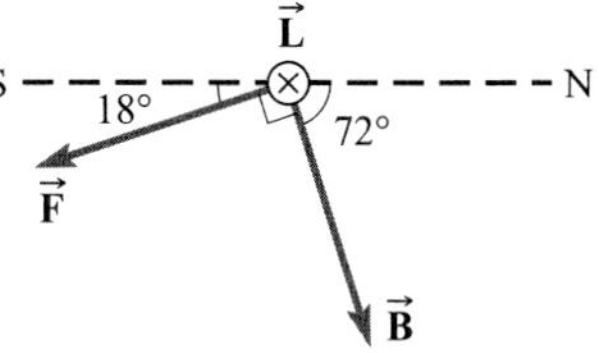

(b) $F = ILB$, so $I = \dfrac{F/L}{B} = \dfrac{0.020 \text{ N/m}}{0.48 \times 10^{-3} \text{ T}} = \boxed{42 \text{ A}}$.

57. Strategy The maximum torque occurs when the angle between the normal to the coil and the magnetic field is $90°$. Use Eq. (19-13a).

Solution Compute the maximum torque.

$$\tau = NIAB = 100(0.0500 \text{ A})\pi(0.020 \text{ cm})^2(0.20 \text{ T}) = \boxed{0.0013 \text{ N} \cdot \text{m}}$$

61. Strategy There are four loops. Label them from 1 to 4 starting with the top as 1, the one below it as 2, etc. The torque of a single loop is given by $\tau = IAB$.

Solution Find the torque for each loop and add them.

$$\Sigma \tau = \tau_1 + \tau_2 + \tau_3 + \tau_4 = I(4a^2)B + I(10a^2)B + I(7a^2)B + I(4a^2)B = I(25a^2)B = IAB$$

where A is the area of the entire irregular loop. The figure shows an irregular loop that is formed by tiny, straight, perpendicular segments. As the segments get smaller and smaller (and more numerous), they approach the smooth curve of the (unmagnified) irregular loop. So, in general, $\tau = IAB$ where A is the area of any shape, planar loop.

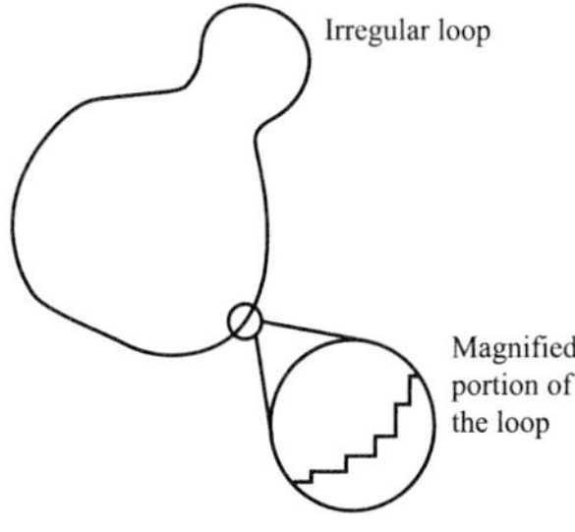

65. Strategy Use Eq. (19-14).

Solution

(a) Compute the distance from the wire.

$$B = \frac{\mu_0 I}{2\pi r}, \text{ so } r = \frac{\mu_0 I}{2\pi B} = \frac{(4\pi \times 10^{-7} \text{ T} \cdot \text{m/A})(5.0 \times 10^3 \text{ A})}{2\pi(45 \times 10^{-6} \text{ T})} = \boxed{22 \text{ m}}.$$

(b) Compute the magnetic field at 700 m.

$$B = \frac{\mu_0 I}{2\pi r} = \frac{(4\pi \times 10^{-7} \text{ T} \cdot \text{m/A})(5.0 \times 10^3 \text{ A})}{2\pi(700 \text{ m})} = \boxed{1.4 \text{ μT}}$$

$$\boxed{\text{This magnetic field is about 3\% of the Earth's field. There would probably not be much effect on the pigeon's navigation because the pigeon would only be over the power lines for a short time.}}$$

67. Strategy Use the principle of superposition and the field due to a long straight current-carrying wire, Eq. (19-14).

Solution According to RHR 2, the field due to each wire is out of the page.

$$B = \frac{\mu_0 I}{2\pi}\left(\frac{1}{r_{\text{bottom}}} + \frac{1}{r_{\text{top}}}\right) = \frac{(4\pi\times10^{-7}\ \text{T}\cdot\text{m/A})(10.0\ \text{A})}{2\pi}\left(\frac{1}{0.25\ \text{m}} + \frac{1}{0.25\ \text{m}+0.0030\ \text{m}}\right) = 1.6\times10^{-5}\ \text{T}$$

So, $\vec{\mathbf{B}}(P) = \boxed{1.6\times10^{-5}\ \text{T out of the page}}$.

69. Strategy Use the principle of superposition and the field due to a long straight current-carrying wire, Eq. (19-14).

Solution Find $\vec{\mathbf{B}}$.

$$\vec{\mathbf{B}} = \frac{\mu_0 I}{2\pi r}\ \text{down} + \frac{\mu_0 I}{2\pi r}\ \text{down} = \frac{(4\pi\times10^{-7}\ \text{T}\cdot\text{m/A})(1.0\ \text{A})}{\pi(0.0050\ \text{m})}\ \text{down} = \boxed{8.0\times10^{-5}\ \text{T down}}$$

71. Strategy Use Eqs. (19-5) and (19-14).

Solution According to RHR 2, the magnetic field at the electron is into the page. Then, according to $\vec{\mathbf{F}} = q\vec{\mathbf{v}}\times\vec{\mathbf{B}}$ and RHR 1, the force on the electron is in the negative y-direction.
Compute the magnitude of the force on the electron.

$$F = |q|vB\sin\theta = ev\frac{\mu_0 I}{2\pi r}\sin 90°$$

$$= \frac{ev\mu_0 I}{2\pi r} = \frac{(1.602\times10^{-19}\ \text{C})(6.8\times10^{6}\ \text{m/s})(4\pi\times10^{-7}\ \text{T}\cdot\text{m/A})(3.2\ \text{A})}{2\pi(0.046\ \text{m})} = 1.5\times10^{-17}\ \text{N}$$

Thus, the force on the electron is $\boxed{1.5\times10^{-17}\ \text{N in the } -y\text{-direction}}$.

73. Strategy The magnetic field at each point in question is equal to the vector sum of the two fields generated by the currents. Use Eqs. (19-12a) and (19-14).

Solution Due to symmetry, the magnitudes of the fields due to each wire are the same at each location, only the directions differ. According to the RHR, the field at C and D due to the vertical wire (1) is directed out of the page and that due to the horizontal wire (2) is directed into the page at C and out of the page at D, so the direction of the total magnetic field is either into or out of the page at C and out of the page at D. Find the magnitudes of the fields due to each wire.

$$B_1 = \frac{\mu_0 I}{2\pi(2d)} = \frac{\mu_0 I}{4\pi d}\ \text{and } B_2 = \frac{\mu_0 I}{2\pi d},\ \text{so } B_2 = 2B_1.$$

Find the field at C.

$$\vec{\mathbf{B}}_1 + \vec{\mathbf{B}}_2 = B_1\ \text{out of the page} + 2B_1\ \text{into the page} = -B_1\ \text{into the page} + 2B_1\ \text{into the page}$$

$$= B_1\ \text{into the page} = \frac{\mu_0 I}{4\pi d}\ \text{into the page} = \frac{(4\pi\times10^{-7}\ \text{T}\cdot\text{m/A})(6.50\ \text{A})}{4\pi(0.033\ \text{m})}\ \text{into the page}$$

$$= \boxed{2.0\times10^{-5}\ \text{T into the page}}$$

Find the field at D.

$$\vec{\mathbf{B}}_1 + \vec{\mathbf{B}}_2 = B_1\ \text{out of the page} + 2B_1\ \text{out of the page} = 3B_1\ \text{out of the page}$$

$$= \frac{3\mu_0 I}{4\pi d}\ \text{out of the page} = \frac{3(4\pi\times10^{-7}\ \text{T}\cdot\text{m/A})(6.50\ \text{A})}{4\pi(0.033\ \text{m})}\ \text{out of the page} = \boxed{5.9\times10^{-5}\ \text{T out of the page}}$$

77. Strategy Use the field due to a long straight current-carrying wire, Eq. (19-14), and Eq. (19-12a).

Solution

(a) According to RHR 2, the direction of the field is perpendicular to the plane containing the wires.

$$\boxed{\vec{\mathbf{B}}_1 = \frac{\mu_0 I_1}{2\pi d} \perp \text{ to the plane of the wires}}$$

(b) $\vec{\mathbf{F}} = I_2\vec{\mathbf{L}}\times\vec{\mathbf{B}}_1 = I_2 L\left(\dfrac{\mu_0 I_1}{2\pi d}\right)$ toward $I_1 = \boxed{\dfrac{\mu_0 I_1 I_2 L}{2\pi d} \text{ toward } I_1}$

(c) According to RHR 2, the direction of the field is perpendicular to the plane containing the wires but opposite to $\vec{\mathbf{B}}_1$, so $\boxed{\vec{\mathbf{B}}_2 = \dfrac{\mu_0 I_2}{2\pi d} \perp \text{ to the plane of the wires and opposite to } \vec{\mathbf{B}}_1}$.

(d) $\vec{\mathbf{F}} = I_1\vec{\mathbf{L}}\times\vec{\mathbf{B}}_2 = \boxed{\dfrac{\mu_0 I_1 I_2 L}{2\pi d} \text{ toward } I_2}$

(e) According to parts (b) and (d), parallel currents $\boxed{\text{attract}}$.

(f) Reversing one of the currents' directions causes the cross products to be oppositely directed (one $\vec{\mathbf{L}}$ is switched and one $\vec{\mathbf{B}}$), so the forces are away from the currents. Thus, antiparallel currents $\boxed{\text{repel}}$.

81. Strategy The currents have the same magnitude and are equidistant from the point at which the net magnetic field's direction is evaluated. So, the magnitudes of the four fields are the same.

Solution According to RHR 2 and symmetry, the field directions are the following:

	Current	Direction at the center of the square
1	top left	toward the top right wire
2	top right	toward the bottom right wire
3	bottom left	toward the bottom right wire
4	bottom right	toward the top right wire

Since the magnitudes of the fields are equal, the vertical component of the net field is zero and the horizontal component is to the right.

Each field is at a 45° angle to the horizontal (either above or below). The y-components cancel and the x-components add. The field for a long thin wire is $B = \mu_0 I/(2\pi r)$. Find the magnitude of the field.

$$B_x = \frac{\mu_0 I}{2\pi r}\cos 45° = \frac{\mu_0 I}{2\sqrt{2}\pi r}. \text{ Thus, } B_{\text{net},\,x} = \frac{4\mu_0 I}{2\sqrt{2}\pi r} = \frac{\sqrt{2}\mu_0 I}{\pi r} = B_{\text{net}}, \text{ where } r = \sqrt{\left(\frac{s}{2}\right)^2 + \left(\frac{s}{2}\right)^2} = \sqrt{\frac{s^2}{2}} = \frac{s}{\sqrt{2}}.$$

So, $B_{\text{net}} = \dfrac{2\mu_0 I}{\pi s} = \dfrac{2(4\pi\times10^{-7}\ \text{T}\cdot\text{m/A})(10.0\ \text{A})}{\pi(0.10\ \text{m})} = 80\ \mu\text{T}$, and $\vec{\mathbf{B}} = \boxed{80\ \mu\text{T to the right}}$.

83. Strategy Use RHR 2, symmetry, and the expression for a long thin wire.

Solution By symmetry and RHR 2, the fields due to the two currents on the left are directed to the right at point R. By symmetry and RHR 2, the vertical components of the fields due to the two currents on the right cancel at point R, and the horizontal components are equal in magnitude and are directed to the right.

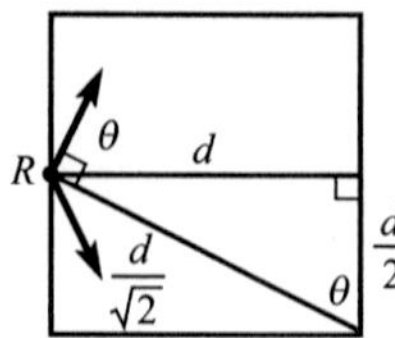

From the figure (which does not show the field vectors for the two currents on the left), we see that if $d = 0.10$ m, the distance from each of the two currents on the right to point P is $d/\sqrt{2}$. Find the angle θ.

$$\theta = \tan^{-1}\frac{d}{d/2} = \tan^{-1} 2 \approx 63.435°$$

The vertical components cancel; the horizontal components add. The field for a long thin wire is $B = \mu_0 I/(2\pi r)$. Find the magnitude of the field.

$$B_{\text{net},x} = 2B_{\text{left}} + 2B_{\text{right}} = 2\frac{\mu_0 I}{2\pi(d/2)} + 2\frac{\mu_0 I}{2\pi(d/\sqrt{2})}\cos 63.435° = \frac{2\mu_0 I}{\pi d} + \frac{\sqrt{2}\mu_0 I}{\pi d}\cos 63.435°$$

$$= \frac{\mu_0 I}{\pi d}(2 + \sqrt{2}\cos 63.435°) = \frac{(4\pi\times 10^{-7}\ \text{T}\cdot\text{m/A})(10.0\ \text{A})(2 + \sqrt{2}\cos 63.435°)}{\pi(0.10\ \text{m})} = 0.11\ \text{mT}$$

Thus, $\vec{\mathbf{B}} = \boxed{0.11\ \text{mT to the right}}$.

85. Strategy Use Newton's second law and Eqs. (19-12a) and (19-14). Let r be the distance between the wires.

Solution

(a) Let the tension in a string be T and θ be the angle the strings make with the vertical. F is the magnitude of the magnetic force. Use Newton's second law for one wire (and two strings).

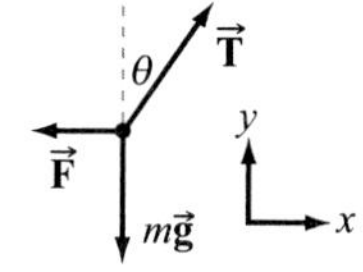

$$\Sigma F_x = 0 = 2T\sin\theta - F, \text{ so } T = \frac{F}{2\sin\theta}. \quad \Sigma F_y = 0 = 2T\cos\theta - mg, \text{ so } T = \frac{mg}{2\cos\theta}.$$

The currents must be in opposite directions since the force is repulsive. According to RHR 2, $\vec{\mathbf{B}}$ is perpendicular to $\vec{\mathbf{d}}$ for both wires, where $\vec{\mathbf{d}}$ is used for $\vec{\mathbf{L}}$ since L is used for the length of the strings.

The field due to a long straight wire is $B = \frac{\mu_0 I}{2\pi r}$ and the magnitude of the force is $F = IdB$, so $F = \frac{\mu_0 dI^2}{2\pi r}$.

From above, $T = \frac{F}{2\sin\theta} = \frac{mg}{2\cos\theta}$, so $F = mg\tan\theta$. Now, $\tan\theta \approx \sin\theta = \frac{\frac{1}{2}r}{L} = \frac{r}{2L}$ since θ is assumed small. Solve for r.

$$\frac{\mu_0 dI^2}{2\pi r} = mg\left(\frac{r}{2L}\right), \text{ so } r = \sqrt{\frac{\mu_0 LI^2}{\pi\left(\frac{m}{d}\right)g}} = \sqrt{\frac{(4\pi\times 10^{-7}\ \text{T}\cdot\text{m/A})(1.2\ \text{m})(50.0\ \text{A})^2}{\pi(0.050\ \text{kg/m})(9.80\ \text{m/s}^2)}} = \boxed{4.9\ \text{cm}}.$$

(b) According to the RHR and $\vec{\mathbf{F}} = I\vec{\mathbf{d}}\times\vec{\mathbf{B}}$ (see above for why $\vec{\mathbf{d}}$ is used instead of $\vec{\mathbf{L}}$), the currents must be in $\boxed{\text{opposite}}$ directions so that the force is repulsive. Currents in the same direction result in an attractive force.

89. Strategy Use Ampère's law, $\Sigma B_{\parallel} \Delta l = \mu_0 I$.

Solution Work in terms of N instead of n since N is constant and n depends upon r, since
$$n = \frac{N}{L} = \frac{N}{2\pi r}.$$
If a closed circular path of radius r is used, then $B_{\parallel} = B$. Then, $\Sigma B_{\parallel}\Delta l = B(2\pi r) = \mu_0 NI$ since N loops cut through

the area enclosed by the path. So, $\boxed{B = \dfrac{\mu_0 NI}{2\pi r}}$. $\boxed{\text{The field is not uniform since } B \propto \dfrac{1}{r}}$.

93. Strategy Use the definition of electric current.

Solution

(a) $\Delta q = e$ and $C =$ circumference $= v\Delta t$, so
$$I = \frac{\Delta q}{\Delta t} = \frac{e}{C/v} = \frac{ev}{2\pi r} = \frac{(1.602\times10^{-19}\ \text{C})(2.2\times10^6\ \text{m/s})}{2\pi(53\times10^{-12}\ \text{m})} = \boxed{1.1\ \text{mA}}.$$

(b) $IA = \dfrac{ev}{2\pi r}(\pi r^2) = \dfrac{1}{2}evr = \dfrac{1}{2}(1.602\times10^{-19}\ \text{C})(2.2\times10^6\ \text{m/s})(53\times10^{-12}\ \text{m}) = \boxed{9.3\times10^{-24}\ \text{A}\cdot\text{m}^2}$

(c) Comparing the orbital dipole moment found in part (b) and the intrinsic magnetic dipole moment, we see that $\boxed{\text{the orbital and intrinsic dipole moments are the same}}$.

97. Strategy The blood speed is equal to the Hall drift speed $v_D = E_H/B = V_H/(wB)$, where the width w is the diameter of the artery. The flow rate is equal to the drift speed times the cross-sectional area of the artery. The magnetic force on a moving charge is given by Eq. (19-5). The direction of the force is determined by the right-hand rule and the sign of the charge.

Solution

(a) Compute the drift speed.
$$v_D = \frac{V_H}{wB} = \frac{0.35\times10^{-3}\ \text{V}}{(0.0040\ \text{m})(0.25\ \text{T})} = \boxed{0.35\ \text{m/s}}$$

(b) Compute the flow rate.
$$\text{flow rate} = v_D A = (0.35\ \text{m/s})\pi(0.0020\ \text{m})^2 = \boxed{4.4\times10^{-6}\ \text{m}^3/\text{s}}$$

(c) The positive ions are deflected east (south$\times$down), so $\boxed{\text{the east lead}}$ is at the higher potential.

101. Strategy The maximum magnitude of the magnetic force on a charged particle is given by $F = |q|vB$. Thus, the maximum magnitude of the electric field is $E = vB$ (since $F = qE$). Treating the artery like a capacitor, the potential difference is $V = Ed$.

Solution Compute the greatest possible potential difference across the artery.
$$V = Ed = (vB)d = (4.25\ \text{m/s})(30\times10^{-6}\ \text{T})(0.010\ \text{m}) = \boxed{1.3\ \mu\text{V}}$$

105. Strategy Use Eqs. (19-12) and (19-14) and Newton's second law.

Solution

(a) According to Newton's second law, $\Sigma F_y = N - mg = N - \lambda Lg = 0$, so $N = \lambda Lg$ for each

wire, where λ is the mass per unit length. The force of static friction, $f_s = \mu_s N = \mu_s \lambda Lg$, opposes the magnetic force. The magnetic field due to each wire is perpendicular to the axis of the opposite wire, so the magnitude of the force on one wire due to the other is

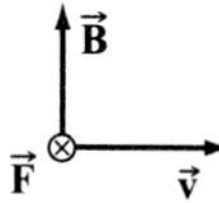

$$F = ILB = IL\frac{\mu_0 I}{2\pi r} = \frac{\mu_0 LI^2}{2\pi r}.$$

Set this result and the magnitude of the force due to static friction equal and solve for the minimum current necessary to make the wires start to move.

$$\frac{\mu_0 LI^2}{2\pi r} = \mu_s \lambda Lg, \text{ so } I = \sqrt{\frac{2\pi r \mu_s \lambda g}{\mu_0}} = \sqrt{\frac{2\pi(0.0025 \text{ m})(0.035)(0.0250 \text{ kg/m})(9.80 \text{ m/s}^2)}{4\pi \times 10^{-7} \text{ T}\cdot\text{m/A}}} = \boxed{10 \text{ A}}.$$

(b) The currents flow in opposite directions. According to $\vec{F} = I\vec{L} \times \vec{B}$ and the RHR, the force due to one wire on the other is repulsive, so the wires move $\boxed{\text{farther apart}}$.

109. Strategy v is that of the velocity selector, E_H / B, and $E_H = V_H/d$ for a uniform field.

Solution Compute the average speed.

$$v = \frac{V_H}{dB} = \frac{88.0 \times 10^{-6} \text{ V}}{(3.80 \times 10^{-3} \text{ m})(0.115 \text{ T})} = \boxed{20.1 \text{ cm/s}}$$

113. Strategy The velocity of the electrons is to the right. Use Eq. (19-5).

Solution The magnetic field points upward, toward the south pole of the magnet. According to $\vec{F} = -e\vec{v} \times \vec{B}$ and RHR 1, the electrons are deflected (and thus the beam moves) $\boxed{\text{into the page}}$.

117. Strategy Use Eqs. (19-12a) and (19-14).

Solution

(a) The field due to the long wire is given by $B = \mu_0 I_2/(2\pi r)$. The direction of the field is given by RHR 2; it is out of the page. The force on a side is $\vec{F} = I_1\vec{L} \times \vec{B}$. RHR 1 gives the following directions for $\vec{L} \times \vec{B}$:

Side	Current direction	Field direction	Force direction
top	right	out of the page	attracted to long wire
bottom	left	out of the page	repelled by long wire
left	up	out of the page	right
right	down	out of the page	left

(b) Due to symmetry, the magnitudes of the forces on the left and right sides are equal. The force directions are opposite, so they cancel. The force due to the bottom side is up, and since it is closer to the long wire, its magnitude is larger than that due to the top side; thus, the direction of the net force is up (away from the long wire). Calculate the magnitude.

$$F = I_1 L_{\text{bottom}} B_{\text{bottom}} - I_1 L_{\text{top}} B_{\text{top}} = I_1 L B_{\text{bottom}} - I_1 L B_{\text{top}} = I_1 L \frac{\mu_0 I_2}{2\pi}\left(\frac{1}{r_{\text{bottom}}} - \frac{1}{r_{\text{top}}}\right)$$

$$= \frac{(4\pi\times10^{-7}\ \text{T·m/A})(0.0020\ \text{A})(8.0\ \text{A})(0.090\ \text{m})}{2\pi}\left(\frac{1}{0.020\ \text{m}} - \frac{1}{0.070\ \text{m}}\right) = 1.0\times10^{-8}\ \text{N}$$

So, $\vec{\mathbf{F}} = \boxed{1.0\times10^{-8}\ \text{N away from the long wire}}$.

121. (a) Strategy From Example 19.5, $v = eBr/m$, and $f = v/C = v/(2\pi r)$.

Solution Find the frequency of oscillation.

$$f = \frac{eBr}{2\pi r m} = \frac{eB}{2\pi m} = \frac{(1.602\times10^{-19}\ \text{C})(1.3\ \text{T})}{2\pi(1.673\times10^{-27}\ \text{kg})} = \boxed{20\ \text{MHz}}$$

(b) Strategy Use Eq. (6-6) and $v = eBr/m$.

Solution Find the kinetic energy.

$$K = \frac{1}{2}mv^2 = \frac{1}{2}m\left(\frac{eBr}{m}\right)^2 = \frac{e^2 B^2 r^2}{2m} = \frac{(1.602\times10^{-19}\ \text{C})^2(1.3\ \text{T})^2(0.16\ \text{m})^2}{2(1.673\times10^{-27}\ \text{kg})} = \boxed{3.3\times10^{-13}\ \text{J}}$$

(c) Strategy $\Delta U = q\Delta V = e\Delta V$; equate this with the final kinetic energy and solve for ΔV.

Solution Find the equivalent voltage.

$$\Delta V = \frac{\Delta U}{e} = \frac{K}{e} = \frac{eB^2 r^2}{2m} = \frac{(1.602\times10^{-19}\ \text{C})(1.3\ \text{T})^2(0.16\ \text{m})^2}{2(1.673\times10^{-27}\ \text{kg})} = \boxed{2.1\ \text{MV}}$$

(d) Strategy and Solution The energy of a proton increases by $e\Delta V$ each time it crosses the gap, or $2e\Delta V$ for each revolution (two gap crossings), so its total energy is $2Ne\Delta V$, where N is the number of revolutions. Equate this with the final kinetic energy and solve for N.

$$2Ne\Delta V = K, \text{ so } N = \frac{K}{2e\Delta V} = \frac{3.32\times10^{-13}\ \text{J}}{2(1.602\times10^{-19}\ \text{C})(10.0\times10^3\ \text{V/rev})} = \boxed{100\ \text{rev}}.$$

Chapter 20

ELECTROMAGNETIC INDUCTION

Conceptual Questions

1. As the loop rotates, the magnitude of the flux increases as it approaches the maximum positive value with the loop perpendicular to the field and then decreases to zero as the loop becomes parallel to the field. As it turns through the maximum position and the flux changes from increasing to decreasing, the induced current reverses direction, because the sign of the change it's opposing has flipped. As the loop continues to rotate, the flux reaches its maximum negative value when the loop is once again perpendicular to the field but facing the opposite way. As it rotates through this position and the flux changes from decreasing to increasing, the induced current will once again reverse direction. Thus, the induced current reverses its direction twice per rotation.

5. (a) In position 1, there is an eddy current circulating clockwise to oppose the increase in magnetic flux as the plate enters the magnetic field.

   (b) In position 2, there is an eddy current circulating counterclockwise to oppose the decrease in magnetic flux as the plate leaves the magnetic field.

   (c) From the right hand rule, the induced magnetic field due to the induced current produces a magnetic force on the metal plate that opposes the motion of the plate. This force slows the pendulum each time it enters or leaves the magnetic field and rapidly brings the plate to rest.

9. No; knowing the flux through the surface and the area of the surface would be sufficient to calculate the average component of the magnetic field perpendicular to the surface only.

13. (a) A transformer only works for alternating currents because its operation is based upon the principle of magnetic induction. A current can only be induced by a changing magnetic flux which requires a changing primary current—impossible to achieve with a direct current.

   (b) The back emf in the primary coil limits the size of the current. For dc emf, no back emf exists, and thus, the current is much too large (limited only by the small resistance of the coils).

17. The back emf decreases as the mixer's motor slows and the current to the mixer therefore increases. As the current increases, resistive heating causes the motor to heat up. The motor will "burn out" if the temperature gets too high.

Problems

1. **Strategy** Use Eqs. (20-2a), (19-5), and (19-12a).

   Solution

   (a) The motional emf is $\mathscr{E} = vBL$, so $I = \dfrac{\mathscr{E}}{R} = \boxed{\dfrac{vBL}{R}}$.

   (b) By the RHR and $\vec{\mathbf{F}} = -e\vec{\mathbf{v}} \times \vec{\mathbf{B}}$, the direction of the force on the electrons in the rod is down. So, the direction of the current is $\boxed{\text{CCW}}$.

   (c) By the RHR and $\vec{\mathbf{F}} = I\vec{\mathbf{L}} \times \vec{\mathbf{B}}$, the direction of the magnetic force on the rod is $\boxed{\text{left}}$.

(d) $F = ILB = \dfrac{vBL}{R}LB = \boxed{\dfrac{vB^2L^2}{R}}$

3. **Strategy** Use the result of Problem 1d, $P = Fv$, and Eqs. (20-2a) and (18-21b).

Solution

(a) According to Problem 1d, the magnitude of the magnetic force on the rod is vB^2L^2/R. The net force must be

zero for constant velocity. So, $F_{\text{ext}} = F_{\text{B}} = \boxed{\dfrac{vB^2L^2}{R}}$.

(b) $\dfrac{\Delta W}{\Delta t} = P = Fv = \boxed{\dfrac{v^2B^2L^2}{R}}$

(c) $\mathscr{E} = vBL$, so $P = \dfrac{V^2}{R} = \boxed{\dfrac{v^2B^2L^2}{R}}$.

(d) $\boxed{\text{Energy is conserved since the rate at which the external force does work is equal to the power dissipated in the resistor.}}$

5. **(a) Strategy** Use Newton's second law, Eq. (19-12b), and Eq. (20-2a).

Solution $\Sigma F_y = F_{\text{B}} - mg = 0$, so $F_{\text{B}} = mg$ when the rod is falling with constant velocity. The magnitude of

the magnetic force is $F_{\text{B}} = ILB = \dfrac{\mathscr{E}}{R}LB = \dfrac{vBL}{R}LB = \dfrac{vL^2B^2}{R}$. Set this equal to mg and solve for the terminal

speed.

$\dfrac{vL^2B^2}{R} = mg$, so $v = \dfrac{mgR}{L^2B^2} = \dfrac{(0.0150\ \text{kg})(9.80\ \text{m/s}^2)(8.00\ \Omega)}{(1.30\ \text{m})^2(0.450\ \text{T})^2} = \boxed{3.44\ \text{m/s}}$.

(b) Strategy Use the potential energy in a uniform gravitational field and Eqs. (18-21b) and (20-2a).

Solution The change in gravitational energy per second is

$\dfrac{\Delta U}{\Delta t} = \dfrac{mg\Delta y}{\Delta t} = -mgv = -\dfrac{m^2g^2R}{L^2B^2} = -\dfrac{(0.0150\ \text{kg})^2(9.80\ \text{m/s}^2)^2(8.00\ \Omega)}{(1.30\ \text{m})^2(0.450\ \text{T})^2} = -0.505\ \text{W},$

so the magnitude of the change is 0.505 W. The power dissipated is

$P = \dfrac{\mathscr{E}^2}{R} = \dfrac{v^2B^2L^2}{R} = \left(\dfrac{mgR}{L^2B^2}\right)^2\dfrac{B^2L^2}{R} = \dfrac{m^2g^2R}{L^2B^2} = 0.505\ \text{W}.$

$\boxed{\text{The magnitude of the change in gravitational potential energy per second and the power dissipated in the resistor are the same, 0.505 W.}}$

9. **Strategy** Use Eqs. (19-5) and (20-2a).

   Solution

   (a) By $\vec{F} = -e\vec{v} \times \vec{B}$, the electrons are forced toward the center of the disk, so $\boxed{\text{positive}}$ charge accumulates on the edge of the disk.

   (b) The motional emf is $\mathscr{E} = vBL$ where the average speed is $v = \dfrac{1}{2}\omega R$ and L is the radius R.

   $$\Delta V = vBL = \left(\frac{\omega R}{2}\right)B(R) = \boxed{\frac{1}{2}\omega BR^2}$$

13. (a) **Strategy** The angle between the magnetic field and the normal of the loop is $\theta = 90.0° - 30.0° = 60.0°$. Use Eq. (20-5).

    Solution Compute the flux through the loop.
    $$\Phi_{\text{B}} = BA\cos\theta = (0.32\text{ T})(0.75\text{ m})^2\cos 60.0° = \boxed{0.090\text{ Wb}}$$

    (b) **Strategy** The new angle between the magnetic field and the normal of the loop is $\theta = 90.0° - 60.0° = 30.0°$. Use Eq. (20-5).

    Solution Compute the flux through the loop.
    $$\Phi_{\text{B}} = BA\cos\theta = (0.32\text{ T})(0.75\text{ m})^2\cos 30.0° = \boxed{0.16\text{ Wb}}$$

    (c) **Strategy** According to Lenz's law, the direction of an induced current in a loop always opposes the *change* in magnetic flux that induces the current.

    Solution As the angle is increasing, the flux is increasing, so the current will flow such that it decreases the net flux through the loop. The magnetic field must be in the negative x-direction, therefore, by the RHR, the current in the top side of the loop must flow in the $\boxed{-z\text{-direction}}$.

15. **Strategy** According to Lenz's law, the direction of an induced current in a loop always opposes the *change* in magnetic flux that induces the current. According to Faraday's law, the rate of change of the magnetic flux is equal to the induced emf.

    Solution

    (a) According to the RHR, $\vec{B}$ is directed into the page at the loop. Since the current is decreasing and $B \propto I$ for a long straight wire, $\vec{B}$ is decreasing, thus a $\boxed{\text{CW}}$ current will flow in the loop to generate a magnetic field directed into the page.

    (b) At the uppermost point of the loop, the current in it is parallel to the current in the wire. Parallel currents attract and opposite currents repel, but since the half of the loop with the parallel components of current is closest to the long straight wire, the force on the circle is toward the wire. Thus, the external force must be $\boxed{\text{away from the long straight wire}}$.

    (c) By Faraday's law, $\left|\dfrac{\Delta\Phi_{\text{B}}}{\Delta t}\right| = |\mathscr{E}| = iR = (84\times10^{-3}\text{ A})(24\text{ }\Omega) = \boxed{2.0\text{ Wb/s}}$.

17. **Strategy** According to Lenz's law, the direction of an induced current in a loop always opposes the *change* in magnetic flux that induces the current. According to Faraday's law, the rate of change of the magnetic flux is equal to the induced emf. Consider the magnetic force on a moving charge.

    Solution While I_1 is increasing, its field is increasing and, by the RHR, it is directed to the left. By Lenz's law, the changing magnetic flux is opposed by an induced current in loop 2 that flows counterclockwise as viewed from the right. The currents in the two loops flow in opposite directions, so they repel each other. The force on loop 2 is $\boxed{\text{to the right (away from loop 1)}}$. Alternatively, use Lenz's law. Loop 1's field is getting stronger, so the force on loop 2 will push it in the direction that tends to neutralize the change in flux—that is, away from loop 1 where its field is weaker.

21. **Strategy** Use the definition of resistance and Eqs. (20-5) and (20-6b).

    Solution

    (a) $\quad I = \dfrac{|\mathscr{E}|}{R} = \dfrac{1}{R}\left|-N\dfrac{\Delta\Phi_B}{\Delta t}\right| = \dfrac{1}{R}\left(\dfrac{NB_\perp A}{\Delta t}\right) = \dfrac{50(1.8\text{ T})\pi(0.050\text{ m})^2}{(2.8\text{ }\Omega)(3.6\text{ s})} = \boxed{0.070\text{ A}}$

    (b) The perpendicular component of $\vec{\mathbf{B}}$ is away from the viewer and increasing, so the induced field is toward the viewer. Thus, the current must flow $\boxed{\text{CCW}}$ according to the RHR.

25. (a) **Strategy** Use Faraday's law, Eq. (20-6a), and Eq. (20-5) for the flux.

    Solution In one rotation, the change in the flux is BA, and the frequency is $f = 1/\Delta t = \omega/(2\pi)$. Find the magnitude of the emf.

    $$|\mathscr{E}| = \left|\dfrac{\Delta\Phi_B}{\Delta t}\right| = \dfrac{BA}{2\pi/\omega} = \dfrac{B\omega(\pi R^2)}{2\pi} = \boxed{\dfrac{B\omega R^2}{2}}$$

    (b) **Strategy** The speed of the tip of the rod is $v = \omega R$.

    Solution Write the magnitude of the emf.

    $$|\mathscr{E}| = \dfrac{B\omega R^2}{2} = \dfrac{B(\omega R)R}{2} = \dfrac{BvR}{2}$$

    The motional emf of a rod of length R moving at constant velocity is $\mathscr{E} = vBR$, so our result

    > is half of the value of the motional emf of a rod moving at constant speed v; which is reasonable, since different points on the rotating rod have different speeds ranging from 0 to v.

29. (a) **Strategy** When the motor is running smoothly, the current and net emf are constant.

    Solution Compute the current when the motor is running smoothly.

    $$I = \dfrac{\mathscr{E}_{\text{ext}} - \mathscr{E}_{\text{back}}}{R} = \dfrac{24.00\text{ V} - 18.00\text{ V}}{8.00\text{ }\Omega} = \boxed{0.750\text{ A}}$$

    (b) **Strategy** Compute the current when the motor quits spinning and the back emf is zero. Consider how this may affect the motor.

    Solution Compute the current when there is no back emf.

    $$I = \dfrac{\mathscr{E}_{\text{ext}}}{R} = \dfrac{24.00\text{ V}}{8.00\text{ }\Omega} = \boxed{3.00\text{ A}}$$

    This is four times the normal current. This much current flowing though the motor may damage it.

    > Tim should shut the trimmer off because the wires in the motor were not meant to sustain this much current. The wires will burn up if this current flows through them for very long.

31. Strategy Use Eq. (20-9).

Solution Compute the secondary voltage amplitude.

$$\mathcal{E}_2 = \frac{N_2}{N_1}\mathcal{E}_1 = \frac{200}{4000}(2.2\times10^3 \text{ V}) = \boxed{110 \text{ V}}$$

33. Strategy Use Eq. (20-9).

Solution Compute the turns ratio and number of turns in the primary.

(a) $\dfrac{N_2}{N_1} = \dfrac{\mathcal{E}_2}{\mathcal{E}_1} = \dfrac{8.5 \text{ V}}{170 \text{ V}} = \boxed{\dfrac{1}{20}}$

(b) $N_1 = \dfrac{\mathcal{E}_1}{\mathcal{E}_2}N_2 = \dfrac{170 \text{ V}}{8.5 \text{ V}}(50) = \boxed{1000}$

35. Strategy Use Eq. (20-9).

Solution Compute the turns ratio.

$$\frac{N_2}{N_1} = \frac{\mathcal{E}_2}{\mathcal{E}_1} = \frac{10.0 \text{ V}}{5.00 \text{ V}} = \boxed{2.00}$$

37. Strategy Use Eq. (20-10).

Solution Compute the voltage amplitude in the secondary coil.

$$\frac{\mathcal{E}_1}{\mathcal{E}_2} = \frac{N_1}{N_2}, \text{ so } \mathcal{E}_2 = \frac{N_2}{N_1}\mathcal{E}_1 = \frac{300}{1800}(170 \text{ V}) = \boxed{28 \text{ V}}.$$

Compute the current amplitude in the primary coil.

$$\frac{I_2}{I_1} = \frac{N_1}{N_2}, \text{ so } I_1 = \frac{N_2}{N_1}I_2 = \frac{300}{1800}(3.2 \text{ A}) = \boxed{0.53 \text{ A}}.$$

39. Strategy The induced emf in a solid conductor subjected to a changing magnetic flux causes eddy currents to flow simultaneously along many different paths. These eddy currents dissipate energy. According to Lenz's law, the direction of an induced current in a loop always opposes the *change* in magnetic flux that induces the current.

Solution

(a) In the wall of the pipe, the magnetic field due to the falling magnet is directed primarily downward. Above the magnet, the field is decreasing as the magnet falls. According to Lenz's law, eddy currents are generated to oppose this decrease. By the right-hand rule, these eddy currents flow $\boxed{\text{CW}}$ when viewed from above.

(b) Below the magnet, the field is increasing, so to oppose this increase, the eddy currents flow $\boxed{\text{CCW}}$ as viewed from above.

(c) Since the eddy currents move in a resistive medium, they dissipate energy—the magnet's kinetic energy. Thus, the magnet's speed is reduced. The speed cannot be reduced to zero, though, since then there would be no eddy currents generated to slow the magnet (no changing flux). Therefore, the magnet must reach some terminal speed, like for a marble falling through honey. Below is a qualitative sketch of the speed of the magnet as a function of time.

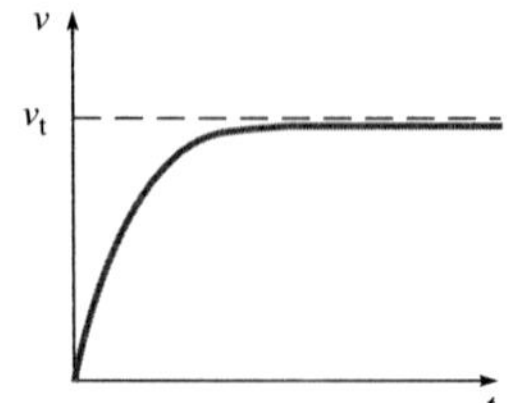

41. Strategy The magnitude of the magnetic field of a solenoid with N turns is given by $B = \mu_0 NI/L$. Use Eq. (20-7a) and Faraday's law.

Solution

(a) The fields due to each solenoid are

$$B_1 = \frac{\mu_0 N_1 I_1}{L} \text{ and } B_2 = \frac{\mu_0 N_2 I_2}{L}.$$

The flux linkage through N turns, each with cross-sectional area $A = \pi r^2$, is $N\Phi = NBA = NB\pi r^2$. Find the total flux through solenoid 2 due to the current in solenoid 1 as a function of time.

$$N_2 \Phi_{21} = N_2 B_1 \pi r^2 = N_2 \left(\frac{\mu_0 N_1 I_1}{L} \right) \pi r^2 = N_2 \frac{\mu_0 N_1 (I_{\mathrm{m}} \sin \omega t)}{L} \pi r^2 = \boxed{\frac{\mu_0 N_1 N_2 \pi r^2 I_{\mathrm{m}} \sin \omega t}{L}}$$

(b) If $\Phi(t) = \Phi_0 \sin \omega t$, then $\dfrac{\Delta \Phi}{\Delta t} = \omega \Phi_0 \cos \omega t$ (for small Δt).

So, by Faraday's law, $\mathscr{E}_{\max} = \left| -N_2 \dfrac{\Delta \Phi_{21}}{\Delta t} \right| = \boxed{\dfrac{\mu_0 N_1 N_2 \pi r^2 \omega I_{\mathrm{m}} \cos \omega t}{L}}$.

43. Strategy The total flux linkage is $N\Phi = LI$, so the flux through one winding is Φ. Use Eq. (20-15).

Solution Find the magnetic flux through one winding.

$$\Phi = \frac{LI}{N} = \frac{\mu_0 N^2 \pi \left(\frac{d}{2} \right)^2 I}{\ell N} = \frac{1}{4} \mu_0 \pi n d^2 I = \frac{1}{4} (4\pi \times 10^{-7} \text{ H/m}) \pi (160 \text{ cm}^{-1})(10^2 \text{ cm/m})(0.0075 \text{ m})^2 (0.20 \text{ A})$$

$$= \boxed{1.8 \times 10^{-7} \text{ Wb}}$$

45. Strategy The inductance of a solenoid is inversely proportional to its length.

Solution

$$L \propto \frac{1}{\ell}, \text{ so } \frac{L_2}{L_1} = \frac{\ell_1}{\ell_2} = \frac{\ell}{0.50\ell} = 2.0$$

The inductance is increased to 2.0 times its initial value.

49. Strategy $\mathscr{E}_{\mathrm{av}}$ is given by Faraday's law and $N\Phi = LI$.

Solution Find the average emf in the solenoid.

$$\mathscr{E}_{\mathrm{av}} = N \frac{\Delta \Phi}{\Delta t} = L \frac{\Delta I}{\Delta t} = (0.080 \text{ H}) \frac{(160.0 - 20.0) \times 10^{-3} \text{ A}}{7.0 \text{ s}} = \boxed{1.6 \text{ mV}}$$

51. Strategy Since the inductors are in parallel, the emfs induced in each must be equal. An equivalent inductor replacing L_1 and L_2 would have the same induced emf. The current through the equivalent emf must be the sum of the currents through the individual inductors. The relation between induced emf and a changing current is given by Faraday's law.

Solution Calculate the equivalent inductance of two ideal inductors in parallel.

$$\frac{\Delta I_{\mathrm{eq}}}{\Delta t} = \frac{\Delta I_1}{\Delta t} + \frac{\Delta I_2}{\Delta t}, \quad \mathscr{E}_{\mathrm{eq}} = \mathscr{E}_1 = \mathscr{E}_2, \text{ and } \mathscr{E} = -L \frac{\Delta I}{\Delta t}. \text{ Find } L_{\mathrm{eq}} \text{ in terms of } L_1 \text{ and } L_2.$$

$$L_{\text{eq}}\frac{\Delta I_{\text{eq}}}{\Delta t} = L_1\frac{\Delta I_1}{\Delta t} = L_2\frac{\Delta I_2}{\Delta t}, \text{ so } \frac{\Delta I_2}{\Delta t} = \frac{L_1}{L_2}\frac{\Delta I_1}{\Delta t}. \quad \frac{\Delta I_{\text{eq}}}{\Delta t} = \frac{\Delta I_1}{\Delta t} + \frac{\Delta I_2}{\Delta t} = \frac{\Delta I_1}{\Delta t}\left(1+\frac{L_1}{L_2}\right), \text{ so}$$

$$L_{\text{eq}}\frac{\Delta I_{\text{eq}}}{\Delta t} = L_{\text{eq}}\frac{\Delta I_1}{\Delta t}\left(1+\frac{L_1}{L_2}\right) = L_1\frac{\Delta I_1}{\Delta t}.$$

Solve for L_{eq}.

$$L_{\text{eq}}\left(1+\frac{L_1}{L_2}\right) = L_{\text{eq}}\left(\frac{L_1+L_2}{L_2}\right) = L_1, \text{ so } \boxed{L_{\text{eq}} = \frac{L_1 L_2}{L_1+L_2}}.$$

53. (a) Strategy Immediately after the switch is closed, the current through the inductor is zero, so the entire current flows through the 5.0-Ω and 10.0-Ω resistors as if they are in series. Use Kirchhoff's loop rule to find the current.

Solution Find the current.

$$6.0 \text{ V} - I(5.0\ \Omega) - I(10.0\ \Omega) = 0, \text{ so } I = \frac{6.0 \text{ V}}{15.0\ \Omega} = 0.40 \text{ A}.$$

Compute the voltages across the resistors.

$$V_{5.0} = (0.40 \text{ A})(5.0\ \Omega) = \boxed{2.0 \text{ V}} \text{ and } V_{10.0} = (0.40 \text{ A})(10.0\ \Omega) = \boxed{4.0 \text{ V}}.$$

(b) Strategy and Solution After the switch has been closed for a long time, the current in the inductor is no longer changing, so it acts like a short circuit. Thus, no current flows through the 10.0-Ω resistor, so the voltage across it is $\boxed{0}$. The entire voltage of the battery is dropped across the 5.0-Ω resistor, so the voltage across it is $\boxed{6.0 \text{ V}}$.

(c) Strategy Since the inductor acts like a short and the 10.0-Ω resistor is bypassed, the current through the inductor is the current that would flow through the circuit if it only contained the battery and the 5.0-Ω resistor.

Solution Compute the current.

$$I = \frac{6.0 \text{ V}}{5.0\ \Omega} = \boxed{1.2 \text{ A}}$$

57. (a) Strategy Use Eqs. (20-20) and (20-21).

Solution Find t when $I = 0.67 I_f$.

$$I = I_f(1-e^{-t/\tau})$$
$$0.67 I_f = I_f(1-e^{-tR/L})$$
$$\ln e^{-tR/L} = \ln 0.33$$
$$t = -\frac{L}{R}\ln 0.33 = -\frac{0.00067 \text{ H}}{130\ \Omega}\ln 0.33 = \boxed{5.7\times10^{-6} \text{ s}}$$

(b) Strategy The energy in the inductor is given by Eq. (20-16). This energy is at its maximum when the current flowing through the inductor is at its maximum; that is, when the current is equal to $\mathscr{E}/R$.

Solution Compute the maximum energy stored in the inductor.

$$U = \frac{1}{2}LI^2 = \frac{1}{2}L\frac{\mathscr{E}^2}{R^2} = \frac{1}{2}(0.00067 \text{ H})\left(\frac{24 \text{ V}}{130\ \Omega}\right)^2 = \boxed{1.1\times10^{-5} \text{ J}}$$

(c) Strategy Since $U = \frac{1}{2}LI^2$ and $I = I_f(1 - e^{-t/\tau})$, $U(t) = U_f(1 - e^{-t/\tau})^2$.

Solution Find t when $U = 0.67U_f$.

$$U = U_f(1 - e^{-t/\tau})^2$$
$$0.67U_f = U_f(1 - e^{-t/\tau})^2$$
$$\pm\sqrt{0.67} = 1 - e^{-t/\tau}$$
$$\ln e^{-tR/L} = \ln(1 \pm \sqrt{0.67})$$
$$t = -\frac{L}{R}\ln(1 \pm \sqrt{0.67}) = -\frac{0.00067 \text{ H}}{130 \ \Omega}\ln(1 \pm \sqrt{0.67}) = \boxed{8.8 \times 10^{-6} \text{ s}} \text{ or } -3.1 \times 10^{-6} \text{ s,}$$

which is extraneous since $t \geq 0$.

> This is more than in part (a) because the energy stored in the inductor is proportional to the current squared. It takes longer for the *square* of the current to be 67% of the maximum *square* of the current than for the current itself to be 67% of the maximum current.

61. (a) Strategy Use the definition of resistance and Eq. (20-16).

Solution Find the maximum current that flowed through the inductor.
$$I = \frac{\mathscr{E}}{R} = \frac{6.0 \text{ V}}{12 \ \Omega} = 0.50 \text{ A}$$
Calculate the stored energy.
$$U = \frac{1}{2}LI^2 = \frac{1}{2}(0.30 \text{ H})(0.50 \text{ A})^2 = \boxed{38 \text{ mJ}}$$

(b) Strategy $P = I\mathscr{E}$ and $I = 0.50$ A (at $t = 0$).

Solution Compute the instantaneous rate of change of the inductor's energy.
$\mathscr{E}$ = the voltage drop across the resistors = $(0.50 \text{ A})(30 \ \Omega) = 15$ V

Thus, $P = -(0.50 \text{ A})(15 \text{ V}) = \boxed{-7.5 \text{ W}}$, where $P < 0$ since the energy of the inductor is decreasing.

(c) Strategy Assume that $U_f \approx 0$. Use the definition of average power.

Solution Compute the average rate of change of the inductor's energy.
$$P_{av} = \frac{\Delta U}{\Delta t} = \frac{0 - 38 \times 10^{-3} \text{ J}}{1.0 \text{ s}} = \boxed{-38 \text{ mW}}$$

(d) Strategy Use Eq. (20-23).

Solution Solve for t when $I = 0.0010I_0$.
$$I = I_0 e^{-t/\tau}$$
$$\ln e^{t/\tau} = \ln \frac{I_0}{I}$$
$$t = \tau \ln \frac{I_0}{I} = \frac{L}{R_{eq}}\ln\frac{1}{0.0010} = -\frac{0.30 \text{ H}}{18 \ \Omega + 12 \ \Omega}\ln 0.0010 = \boxed{69 \text{ ms}}$$
Since 69 ms $\ll$ 1.0 s, the assumption in part (c) is valid.

63. (a) Strategy The current increases until it reaches a maximum. At this point, the emf in the inductor is zero.

Solution Compute the maximum current.
$$I_0 = \frac{\mathcal{E}}{R} = \frac{9.0 \text{ V}}{200.0 \ \Omega} = \boxed{45 \text{ mA}}$$

(b) Strategy The current increases from zero to I_0. Use Eq. (20-20).

Solution Solve for t when $I = \frac{1}{2}I_0$.
$$\frac{1}{2}I_0 = I_0(1 - e^{-t/\tau})$$
$$e^{-t/\tau} = 1 - \frac{1}{2}$$
$$\ln e^{-t/\tau} = -\frac{t}{\tau} = \ln \frac{1}{2}$$
$$t = \tau \ln 2 = \frac{L}{R} \ln 2 = \frac{0.30 \text{ H}}{200.0 \ \Omega} \ln 2 = \boxed{1.0 \text{ ms}}$$

(c) Strategy Use Eqs. (18-19), (18-21a), (20-16), and (20-22).

Solution Find the energy stored in the inductor when $I = \frac{1}{2}I_0$.
$$U = \frac{1}{2}LI^2 = \frac{1}{2}L\left(\frac{I_0}{2}\right)^2 = \frac{1}{8}L\left(\frac{\mathcal{E}_b}{R}\right)^2 = \frac{(0.30 \text{ H})(9.0 \text{ V})^2}{8(200.0 \ \Omega)^2} = \boxed{76 \ \mu\text{J}}$$

The rate at which energy is being stored in the inductor is equal to the current times the induced emf.
$$\frac{\Delta U_L}{\Delta t} = I\mathcal{E}_L = \frac{I_0}{2}(\mathcal{E}_b e^{-t/\tau}) = \frac{\mathcal{E}_b^2}{2R}e^{-(L\ln 2/R)/(L/R)} = \frac{\mathcal{E}_b^2}{2R}e^{-\ln 2} = \frac{(9.0 \text{ V})^2}{2(200.0 \ \Omega)}\left(\frac{1}{2}\right) = \boxed{0.10 \text{ W}}$$

Calculate the rate at which energy is dissipated in the resistor.
$$P = I^2R = \left(\frac{I_0}{2}\right)^2 R = \left(\frac{\mathcal{E}_b}{2R}\right)^2 R = \frac{\mathcal{E}_b^2}{4R} = \frac{(9.0 \text{ V})^2}{4(200.0 \ \Omega)} = \boxed{0.10 \text{ W}}$$

(d) Strategy Use the definition of resistance, Eq. (18-13), and the result of part (b).

Solution
$$I_0 = \frac{\mathcal{E}_b}{R_{\text{eq}}} = \frac{9.0 \text{ V}}{75 \ \Omega + 200.0 \ \Omega + 20.0 \ \Omega} = \boxed{31 \text{ mA}}; \ t = \frac{L}{R_{\text{eq}}}\ln 2 = \frac{0.30 \text{ H}}{75 \ \Omega + 200.0 \ \Omega + 20.0 \ \Omega}\ln 2 = \boxed{0.70 \text{ ms}}$$

65. Strategy According to Faraday's law, the magnitude of the induced emf around a loop is equal to the rate of change of the magnetic flux through the loop. According to Lenz's law, the direction of an induced current in a loop always opposes the *change* in magnetic flux that induces the current.

Solution

(a) The magnetic field is directed to the right inside the coil. As the magnet is pulled away to the left, the magnetic field decreases. A current in the coil is generated to oppose this decrease. The field due to the current is directed to the right, so according to the right-hand rule, the current flows CW as viewed from the left, or $\boxed{\text{to the left}}$ through the galvanometer.

(b) The magnetic field strength doubles, so the rate of change of the flux also doubles. This implies a doubling of the induced emf; thus, $\boxed{\text{the current doubles}}$.

(c) With the bar magnets held with the opposite poles together, the net magnetic flux through the coil is zero, so $\boxed{\text{the current is reduced to zero}}$.

69. (a) Strategy $\vec{F} = I\vec{L} \times \vec{B}$ and $\vec{L}$ and $\vec{B}$ are always perpendicular for both sides. In side 2, current flows into the page. In side 4, current flows out of the page.

Solution The magnitude of the force is

$$F = ILB = \frac{\omega BA}{R} LB \sin \omega t = \frac{\omega B^2 AL}{R} \sin \omega t$$

where, according to the RHR, $\vec{F}_2$ is down and $\vec{F}_4$ is up, or

$$\boxed{\vec{F}_2 = \frac{\omega B^2 AL}{R} \sin \omega t \text{ down and } \vec{F}_4 = \frac{\omega B^2 AL}{R} \sin \omega t \text{ up}}.$$

(b) Strategy and Solution $\boxed{\text{The magnetic forces on sides 1 and 3 are always parallel to the axis of rotation.}}$ Therefore, they do not cause a torque about the axis.

(c) Strategy Sum the torques using the forces found in part (a).

Solution $\Sigma \tau = (F_2 \sin \theta)r + (F_4 \sin \theta)r = 2Fr \sin \omega t$, since $F = \left| \vec{F}_2 \right| = \left| \vec{F}_4 \right|$ and $\theta = \omega t$.

Thus, $\tau(t) = 2 \left(\dfrac{\omega B^2 AL}{R} \sin \omega t \right) r \sin \omega t = \dfrac{2\omega r B^2 AL}{R} \sin^2 \omega t.$

Therefore, $\vec{\tau} = \boxed{\dfrac{2\omega r B^2 AL}{R} \sin^2 \omega t \text{ CCW}}.$

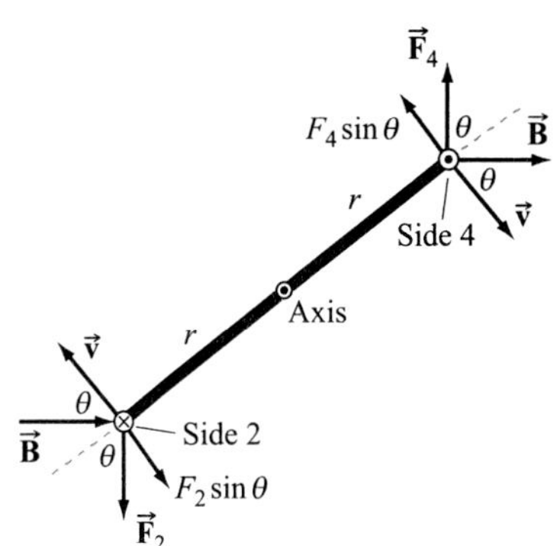

(d) Strategy and Solution

$\boxed{\text{The torque is counterclockwise, and the angular velocity is clockwise, so the magnetic torque would tend to decrease the angular velocity.}}$

73. Strategy Since $a \ll R$, the field inside the toroid can be considered uniform and perpendicular to the cross-sectional area a^2. So, we can consider the toroid to be a square solenoid with length $\ell = 2\pi R$. Use Eq. (20-15).

Solution Find the self inductance of the toroid.

$$L = \frac{\mu_0 N^2 A}{\ell} = \boxed{\frac{\mu_0 N^2 a^2}{2\pi R}}$$

77. Strategy The energy density in a magnetic field is $u_B = B^2 / (2\mu_0)$.

Solution Compute the energy.

$$U = u_B V = \frac{B^2 V}{2\mu_0} = \frac{(0.045 \times 10^{-3} \text{ T})^2 (1.0 \text{ m}^3)}{2(4\pi \times 10^{-7} \text{ T} \cdot \text{m/A})} = \boxed{0.81 \text{ mJ}}$$

81. Strategy and Solution

(a) According to Faraday's law, a changing magnetic flux through the coil induces an emf in the coil. The changing flux requires a changing magnetic field, and this in turn requires a changing current in the wire. Therefore, the induction ammeter will not work equally well for ac and dc currents; ac currents are required. The answer is no.

(b) Since equal magnitude currents flowing in opposite directions produce magnetic fields that cancel, there is no changing flux through the coil and, thus, no emf is measured by the induction ammeter. Therefore, the current drawn by the appliance cannot be measured. The answer is no.

85. (a) Strategy Since the inductor is ideal, $\mathscr{E}_L(t) = \mathscr{E}(t) = \mathscr{E}_m \sin \omega t$. Use Faraday's law and Eq. (20-7b).

Solution Write an expression for the current.

By Faraday's law, $\mathscr{E} = -N\dfrac{\Delta \Phi}{\Delta t} = -L\dfrac{\Delta I}{\Delta t}$. So, $\dfrac{\Delta I}{\Delta t} = -\dfrac{\mathscr{E}_m}{L}\sin \omega t$. Now, if $\Phi(t) = \Phi_0 \cos \omega t$, then

$\dfrac{\Delta \Phi}{\Delta t} = -\omega \Phi_0 \sin \omega t$. Comparing this to $\dfrac{\Delta I}{\Delta t}$, it is obvious that $\boxed{I(t) = \dfrac{\mathscr{E}_m}{\omega L}\cos \omega t}$.

(b) **Strategy** Divide the maximum emf by the maximum current.

Solution Compute the reactance.

$$\dfrac{\mathscr{E}_{max}}{I_{max}} = \dfrac{\mathscr{E}_m}{\dfrac{\mathscr{E}_m}{\omega L}} = \boxed{\omega L}$$

(c) **Strategy** $\cos \omega t$ and $\sin \omega t$ have a phase difference of a quarter period.

Solution Compute the phase difference.

$$T = \dfrac{1}{f} = \dfrac{2\pi}{\omega}, \text{ so } \dfrac{T}{4} = \boxed{\dfrac{\pi}{2\omega}}.$$

87. (a) Strategy We can model the airplane as a metal rod moving through a magnetic field. Then, the motional emf is $\mathscr{E} = vBL$, where B is the component of the field perpendicular to the motion of the plane and to the wing. Therefore, only the upwardly directed component of the magnetic field contributes to the motional emf.

Solution Compute the motional emf.

$\mathscr{E} = vBL = (180 \text{ m/s})(0.38 \times 10^{-3} \text{ T})(46 \text{ m}) = \boxed{3.1 \text{ V}}$.

(b) **Strategy** Consider the magnetic forces on electrons in the plane's wings.

Solution The force on an electron in the wing is $\vec{\mathbf{F}} = -e\vec{\mathbf{v}} \times \vec{\mathbf{B}}$. So, according to the RHR, the electrons are forced along the southernmost wing. Negative charge builds up on the southernmost wingtip and positive on the northernmost. Thus, the northernmost wingtip is positively charged.

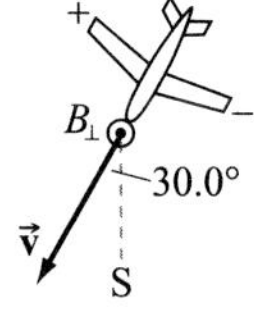

89. Strategy Use Eqs. (19-17), (20-5), and (20-6b). Φ_{12} is the flux through the inner solenoid (1) due to the outer solenoid (2).

Solution Find the magnitude of the induced emf in the inner solenoid.

$$\left|\mathscr{E}_{12}\right| = \left|-N_1 \dfrac{\Delta \Phi_{12}}{\Delta t}\right| = N_1 \dfrac{\Delta(B_2 A_1)}{\Delta t} = N_1 A_1 \dfrac{\Delta B_2}{\Delta t} = N_1 (\pi r_1^2) \dfrac{\Delta}{\Delta t}\left(\dfrac{\mu_0 N_2 I_2}{L_2}\right) = \boxed{\dfrac{\mu_0 N_1 N_2 \pi r_1^2}{L_2} \dfrac{\Delta I_2}{\Delta t}}$$

Chapter 21

ALTERNATING CURRENT

Conceptual Questions

1. When an ac generator is turned on within a circuit containing an uncharged capacitor, current begins to flow and charge begins to accumulate on the capacitor's plates. As charge accumulates, the potential across the capacitor increases, thereby inducing an emf directed opposite to the emf of the generator. The flow of current in the circuit therefore decreases as the capacitor's potential increases. When the direction of the emf from the generator changes, the potential across the capacitor decreases and the magnitude of the current in the circuit increases. Therefore, the current in an ac circuit and the potential difference across a capacitor in that circuit are always out of phase.

5. Appliances such as light bulbs and electric heaters operate by converting the energy stored in electric current into heat energy. The operation of such devices depends only upon the energy dissipating properties of their resistive elements and thus operate equally well with ac or dc. Appliances requiring a transformer of some type will only operate with ac as they work on the principle of induction and therefore require a changing current.

9. The 500 W reported on the appliance is the average power consumption, given by $P_{av} = I_{rms}V_{rms}\cos\phi$. The power factor $\cos\phi = R/Z$ depends on the capacitance, inductance, and resistance of the circuit. Only if the power factor is one, or $R = Z$, would the average power consumption be 600 W.

13. Inserting a coil of wire with a soft iron core into the circuit in series with a light bulb would add inductance and increase the overall impedance of the circuit. The amplitude of the current in the circuit would be reduced and the bulb would be dimmed. Moving the soft-iron core in and out of the coil would vary the amount of dimming by changing the inductance of the coil.

17. Since the power dissipated in a light bulb is given by $P_{av} = V_{rms}^2/R$, the bulb would consume considerably less power with an rms voltage of 120 V (in fact 1/4 as much) and would burn less brightly.

Problems

1. **Strategy and Solution** The current reverses direction twice per cycle and there are 60 cycles per second, so the current reverses direction $\boxed{120 \text{ times per second}}$.

3. **Strategy** 1500 W is the average power dissipated by the heater and $P_{av} = I_{rms}V_{rms}$.

   Solution Calculate I, the peak current.

   $$I = \sqrt{2}I_{rms} = \sqrt{2}\left(\frac{P_{av}}{V_{rms}}\right) = \sqrt{2}\left(\frac{1500 \text{ W}}{120 \text{ V}}\right) = \boxed{18 \text{ A}}$$

5. **Strategy** Find the resistance of the hair dryer. Then, compute the power dissipated in Europe.

   Solution

   $$P_{av} = \frac{V_{rms}^2}{R}, \text{ so } R = \frac{V_{US}^2}{P_{US}}. \text{ The power dissipated by the dryer in Europe is}$$

   $$P_R = \frac{V_E^2}{R} = \frac{V_E^2}{V_{US}^2}P_{US} = \left(\frac{240}{120}\right)^2 (1500 \text{ W}) = \boxed{6000 \text{ W}}.$$

   $\boxed{\text{The heating element of the hair dryer will burn out because it is not designed for this amount of power.}}$

7. **(a) Strategy** 4200 W is the average power drawn by the heater.

Solution Compute the rms current.

$$I_{rms} = \frac{P_{av}}{V_{rms}} = \frac{4200\text{ W}}{120\text{ V}} = \boxed{35\text{ A}}$$

(b) Strategy Since $P = V^2/R$ and $R_2 = R_1$, $P \propto V^2$. Form a proportion.

Solution Calculate P_2.

$$\frac{P_2}{P_1} = \left(\frac{V_2}{V_1}\right)^2, \text{ so } P_2 = \left(\frac{105}{120}\right)^2 (4200\text{ W}) = \boxed{3.2\text{ kW}}.$$

9. **Strategy** Use the definition of rms.

Solution Compute the amplitude.

$$\mathscr{E}_m = \sqrt{2}\mathscr{E}_{rms} = \sqrt{2}(4.0\text{ V}) = 5.7\text{ V}$$

The instantaneous sinusoidal emf oscillates between $\boxed{-5.7\text{ V and }5.7\text{ V}}$.

13. **Strategy** The reactance is $X_C = 1/(\omega C)$ where $\omega = 2\pi f$.

Solution Solve for f.

$$X_C = \frac{1}{2\pi f C}, \text{ so } f = \frac{1}{2\pi X_C C} = \frac{1}{2\pi(1.0\times10^3\text{ }\Omega)(6.0\times10^{-6}\text{ F})} = \boxed{27\text{ Hz}}.$$

15. **(a) Strategy** The reactance is $X_C = 1/(\omega C)$ where $\omega = 2\pi f$.

Solution Compute the reactance.

$$X_C = \frac{1}{\omega C} = \frac{1}{2\pi(50.0\text{ Hz})(0.250\times10^{-6}\text{ F})} = \boxed{12.7\text{ k}\Omega}$$

(b) Strategy The rms current is related to the reactance by $V_{rms} = I_{rms}X_C$.

Solution Solve for I_{rms}.

$$I_{rms} = \frac{V_{rms}}{X_C} = \omega C V_{rms} = 2\pi(50.0\text{ Hz})(0.250\times10^{-6}\text{ F})(220\text{ V}) = \boxed{17\text{ mA}}$$

17. **Strategy** Replace each quantity in $X_C = 1/(\omega C)$ with its SI units.

Solution Show that the units of capacitive reactance are ohms.

ω is in s^{-1} and C is in $\text{F} = \dfrac{\text{C}}{\text{V}}$, so X_C is in $\dfrac{1}{\text{s}^{-1}\cdot\frac{\text{C}}{\text{V}}} = \dfrac{\text{V}}{\frac{\text{C}}{\text{s}}} = \dfrac{\text{V}}{\text{A}} = \Omega.$

21. Strategy Since $R = X_C$, I is the amplitude of the current through R, also. The current through the capacitor leads the voltage by $\pi/2$ radians and that through the resistor is in phase with the voltage. Since the capacitor and resistor are in parallel, the voltage across each is the same. So, the current through the capacitor leads that through the resistor by $\pi/2$ radians, or $i_R(t) = I \sin(\omega t - \pi/2)$.

Solution Sketch the graphs of $i_C(t) = I \sin \omega t$ and $i_R(t) = I \sin(\omega t - \pi/2)$.

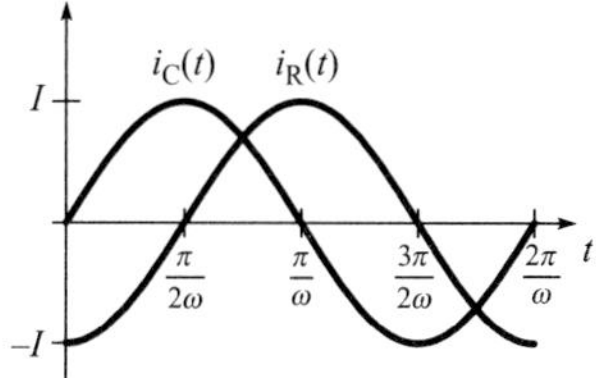

23. Strategy The inductive reactance is given by $X_L = \omega L$ where $\omega = 2\pi f$.

Solution Solve for f.
$$\omega L = 2\pi f L = X_L, \text{ so } f = \frac{X_L}{2\pi L} = \frac{18.8\ \Omega}{2\pi(20.0 \times 10^{-3}\ \text{H})} = \boxed{150\ \text{Hz}}.$$

25. (a) Strategy Use Eq. (21-9).

Solution Find the inductive reactance.
$$V_L = IX_L, \text{ so } X_L = \frac{V_L}{I} = \frac{15\ \text{V}}{3.5 \times 10^{-2}\ \text{A}} = \boxed{430\ \Omega}.$$

(b) Strategy Use Eqs. (21-10) and (20-14).

Solution Find the length of the solenoid.
$$X_L = \omega L = \omega \mu_0 n^2 \pi r^2 \ell = \frac{V_L}{I}, \text{ so } \ell = \frac{V_L}{\omega \mu_0 n^2 \pi r^2 I}$$
$$= \frac{15\ \text{V}}{2\pi(22 \times 10^3\ \text{Hz})(4\pi \times 10^{-7}\ \text{T} \cdot \text{m/A})(2.0 \times 10^4\ \text{m}^{-1})^2 \pi (8.0 \times 10^{-3}\ \text{m})^2 (3.5 \times 10^{-2}\ \text{A})} = \boxed{3.1\ \text{cm}}.$$

27. Strategy Use Eqs. (21-9) and (21-10).

Solution

(a) Form a proportion.
$$V = IX_L = I\omega L = 2\pi f I L, \text{ so } \frac{V_i}{V} = \frac{2\pi f I L_i}{2\pi f I L_{eq}} = \frac{L_i}{L_{eq}} = \frac{L_i}{L_1 + L_2}. \text{ Thus,}$$

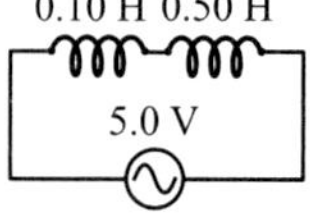

$$V_i = \frac{L_i}{L_1 + L_2} V = \frac{L_i}{0.10\ \text{H} + 0.50\ \text{H}}(5.0\ \text{V}) = \left(\frac{25}{3.0}\ \text{V/H}\right) L_i \text{ is the peak}$$
voltage across inductor i.
The peak voltages are given in the table below.

L (H)	V (V)
0.10	0.83
0.50	4.2

(b) Compute the peak current.

$$I = \frac{V}{X_{\mathrm{L}}} = \frac{V}{\omega L_{\mathrm{eq}}} = \frac{V}{2\pi f(L_1 + L_2)} = \frac{5.0\ \mathrm{V}}{2\pi(126\ \mathrm{Hz})(0.10\ \mathrm{H} + 0.50\ \mathrm{H})} = \boxed{11\ \mathrm{mA}}$$

29. Strategy Since the capacitor and inductor are in series, the same current flows through both.

Solution

(a) $v_{\mathrm{C}}(t)$ lags $i(t)$ by $90°$ and $v_{\mathrm{L}}(t)$ leads $i(t)$ by $90°$, so the phase difference is

$$\Delta\phi = \phi_{\mathrm{L}} - \phi_{\mathrm{C}} = 90° - (-90°) = \boxed{180°}.$$

(b) $v_{\mathrm{L}}(t)$ leads $v_{\mathrm{C}}(t)$ by $180°$, which means they are opposite in sign. The voltmeter reads the difference, $\boxed{4.0\ \mathrm{V}}$.

33. Strategy $Z = \sqrt{R^2 + X_{\mathrm{L}}^2} = 30.0\ \Omega$ where $X_{\mathrm{L}} = \omega L = 2\pi f L$.

Solution Find L.

$$R^2 + X_{\mathrm{L}}^2 = R^2 + \omega^2 L^2 = R^2 + (2\pi f)^2 L^2 = Z^2,\ \text{so } L = \frac{\sqrt{Z^2 - R^2}}{2\pi f} = \frac{\sqrt{(30.0\ \Omega)^2 - (20.0\ \Omega)^2}}{2\pi(50.0\ \mathrm{Hz})} = \boxed{71.2\ \mathrm{mH}}.$$

35. Strategy Use $\mathscr{E} = IZ$ where $Z = \sqrt{R^2 + X_{\mathrm{C}}^2}$.

Solution Find R.

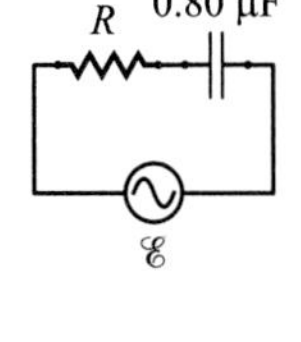

$$Z = \frac{\mathscr{E}}{I} = \sqrt{R^2 + X_{\mathrm{C}}^2}$$

$$\left(\frac{\mathscr{E}}{I}\right)^2 = R^2 + X_{\mathrm{C}}^2 = R^2 + \frac{1}{\omega^2 C^2}$$

$$R = \sqrt{\left(\frac{\mathscr{E}}{I}\right)^2 - \frac{1}{4\pi^2 f^2 C^2}} = \sqrt{\left(\frac{110\ \mathrm{V}}{0.0284\ \mathrm{A}}\right)^2 - \frac{1}{4\pi^2(60.0\ \mathrm{Hz})^2(0.80\times10^{-6}\ \mathrm{F})^2}} = \boxed{2\ \mathrm{k}\Omega}$$

37. Strategy Find the phase angle and the voltages across each device; then draw the phasor diagram.

Solution

(a) Find the reactances.

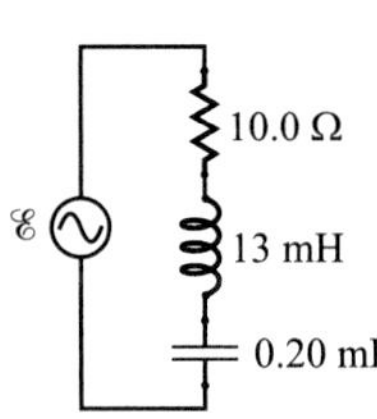

$$X_{\mathrm{L}} = \omega L = 2\pi(60\ \mathrm{Hz})(13\times10^{-3}\ \mathrm{H}) = 4.901\ \Omega\ \text{ and}$$

$$X_{\mathrm{C}} = \frac{1}{\omega C} = \frac{1}{2\pi(60\ \mathrm{Hz})(0.20\times10^{-3}\ \mathrm{F})} = 13.26\ \Omega.$$

Find the phase angle.

$$\phi = \tan^{-1}\frac{X_{\mathrm{L}} - X_{\mathrm{C}}}{R} = \tan^{-1}\frac{4.901\ \Omega - 13.26\ \Omega}{10.0\ \Omega} = \boxed{-40°}$$

Find the impedance.

$$Z = \frac{R}{\cos\phi} = \frac{10.0\ \Omega}{\cos(-39.9°)} = 13.035\ \Omega$$

Find the current in the circuit.

$$I = \frac{\mathscr{E}_{\mathrm{m}}}{Z} = \frac{9.0\ \mathrm{V}}{13.035\ \Omega} = 0.6904\ \mathrm{A}$$

Find the voltages across each device.

$$V_L = IX_L = (0.6904\ \text{A})(4.901\ \Omega) = \boxed{3.4\ \text{V}}, \quad V_C = IX_C = (0.6904\ \text{A})(13.26\ \Omega) = \boxed{9.2\ \text{V}}, \quad \text{and}$$

$$V_R = IR = (0.6904\ \text{A})(10.0\ \Omega) = \boxed{6.9\ \text{V}}.$$

(b) Draw a phasor diagram using the results of part (a).

$$V_L - V_C = 3.4\ \text{V} - 9.2\ \text{V} = -5.8\ \text{V}$$

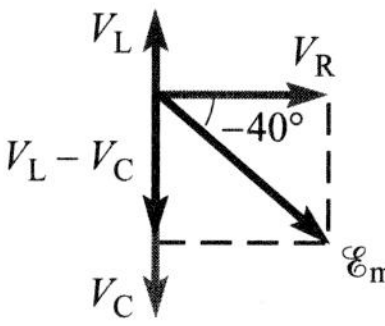

39. Strategy $P_{av} = I_{rms}\mathscr{E}_{rms}\cos\phi$ is the average power.

Solution Find the power factor and the phase difference.

(a) $\cos\phi = \dfrac{P_{av}}{I_{rms}\mathscr{E}_{rms}} = \dfrac{240\ \text{W}}{(2.80\ \text{A})(120\ \text{V})} = \boxed{0.71}$

(b) $\phi = \cos^{-1}\dfrac{P_{av}}{I_{rms}\mathscr{E}_{rms}} = \cos^{-1}\dfrac{240\ \text{W}}{(2.80\ \text{A})(120\ \text{V})} = \boxed{44°}$

41. (a) Strategy Use Eq. (21-17).

Solution Find the phase angle.

$$P_{av} = I_{rms}\mathscr{E}_{rms}\cos\phi, \text{ so } \phi = \cos^{-1}\dfrac{P_{av}}{I_{rms}\mathscr{E}_{rms}} = \cos^{-1}\dfrac{100\ \text{W}}{(2.0\ \text{A})(120\ \text{V})} = \boxed{65°}.$$

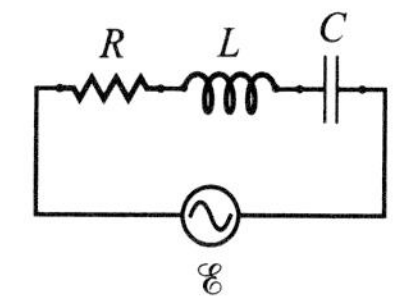

(b) Strategy Use Eqs. (21-4) and (21-15).

Solution Find the value of the resistor.

$$P_{av} = I_{rms}{}^2 R, \text{ so } R = \dfrac{P_{av}}{I_{rms}{}^2} = \dfrac{100\ \text{W}}{(2.0\ \text{A})^2} = \boxed{25\ \Omega}.$$

Find the value of the inductor.

$$\tan\phi = \dfrac{X_L - X_C}{R} = \dfrac{X_L - 0.50X_L}{R} = \dfrac{X_L}{2.0R} = \dfrac{\omega L}{2.0R}, \text{ so } L = \dfrac{2.0R}{\omega}\tan\phi = \dfrac{2.0(25\ \Omega)}{2\pi(60\ \text{Hz})}\tan 65.4° = \boxed{0.29\ \text{H}}.$$

Find the value of the capacitor.

$$X_C = \dfrac{1}{\omega C} = 0.50X_L = 0.50\omega L, \text{ so } C = \dfrac{2.0}{\omega^2 L} = \dfrac{2.0}{4\pi^2(60\ \text{Hz})^2(0.29\ \text{H})} = \boxed{4.9\times10^{-5}\ \text{F}}.$$

45. (a) Strategy Use Eqs. (21-14) with $X_C = 0$, Eqs. (21-9) and (21-10), and Ohm's law.

Solution Find I.

$$I = \dfrac{\mathscr{E}_m}{Z} = \dfrac{\mathscr{E}_m}{\sqrt{R^2 + X_L^2}} = \dfrac{\mathscr{E}_m}{\sqrt{R^2 + \omega^2 L^2}} = \dfrac{1.20\times10^3\ \text{V}}{\sqrt{(145.0\ \Omega)^2 + 4\pi^2(1250\ \text{Hz})^2(22.0\times10^{-3}\ \text{H})^2}}$$

$$= 5.32\ \text{A}$$

Calculate the voltage amplitudes.

$$V_L = IX_L = I\omega L = 2\pi(5.32\ \text{A})(1250\ \text{Hz})(22.0\times10^{-3}\ \text{H}) = \boxed{919\ \text{V}}$$

$$V_R = IR = (5.32\ \text{A})(145.0\ \Omega) = \boxed{771\ \text{V}}$$

(b) Strategy and Solution $\boxed{\text{No}}$, the voltage amplitudes do not add to give the amplitude of the source voltage; the voltages across the resistor and the inductor are 90° out of phase. Similar to the impedance, the source voltage is the square root of the sum of squares, $\boxed{\mathscr{E}_m = \sqrt{V_L^2 + V_R^2}}$, as can be seen in a phasor diagram.

(c) Strategy Use the results of parts (a) and (b).

Solution Sketch the phasor diagram.

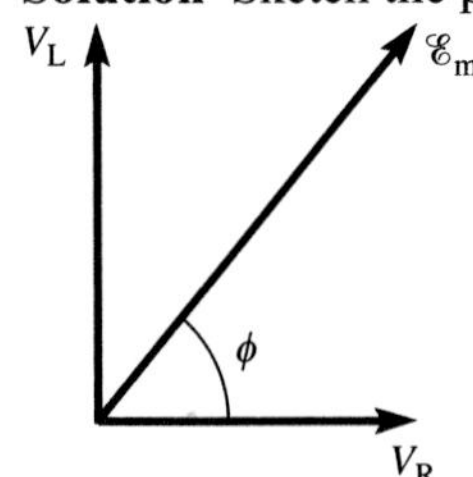

47. (a) Strategy Set $V_R = IR$ equal to $V_L = IX_L = I\omega L = 2\pi fIL$ and solve for the frequency.

Solution Solve for f.

$$V_R = IR = V_L = IX_L = I(2\pi fL),\ \text{so}\ f = \frac{R}{2\pi L} = \frac{150\ \Omega}{2\pi(0.75\ \text{H})} = \boxed{32\ \text{Hz}}.$$

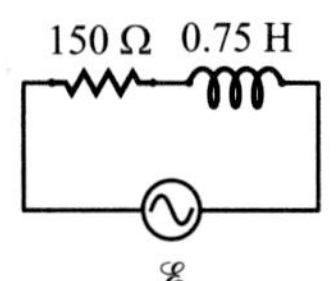

(b) Strategy In a phasor diagram, the source is the hypotenuse of a right triangle and V_R and V_L are the legs, so $\mathscr{E}_m = \sqrt{V_R^2 + V_L^2}$ according to the Pythagorean theorem.

Solution Find $V_L / \mathscr{E}_m = V_R / \mathscr{E}_m$.

$$\mathscr{E}_m^2 = V_R^2 + X_L^2 = V_R^2 + V_R^2 = 2V_R^2,\ \text{so}\ \frac{V_R^2}{\mathscr{E}_m^2} = \frac{1}{2}\ \text{and}\ \frac{V_R}{\mathscr{E}_m} = \frac{1}{\sqrt{2}} = \frac{V_L}{\mathscr{E}_m}.$$

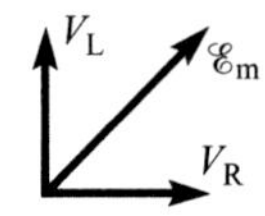

Therefore, the rms voltages across the components are not half of the rms voltage of the source: $\boxed{\dfrac{V_R}{\mathscr{E}_m} = \dfrac{V_L}{\mathscr{E}_m} = \dfrac{1}{\sqrt{2}}}.$

(c) Strategy The voltage across the inductor leads the current through it, so the source voltage leads the current. Use the power factor to find ϕ.

Solution Find the phase angle.

$$\cos\phi = \frac{R}{Z},\ \text{so}$$

$$\phi = \cos^{-1}\frac{R}{Z} = \cos^{-1}\frac{R}{\sqrt{R^2 + X_L^2}} = \cos^{-1}\frac{IR}{\sqrt{(IR)^2 + (IX_L)^2}} = \cos^{-1}\frac{V_R}{\sqrt{V_R^2 + V_L^2}} = \cos^{-1}\frac{V_R}{\mathscr{E}_m} = \cos^{-1}\frac{1}{\sqrt{2}} = \frac{\pi}{4}.$$

Therefore, $\boxed{\mathscr{E}\ \text{leads}\ I\ \text{by}\ \dfrac{\pi}{4}\ \text{rad} = 45°.}$

(d) Strategy Use the power factor to find Z.

Solution Find the impedance.
$$\cos\phi = \frac{R}{Z}, \text{ so } Z = \frac{R}{\cos\phi} = \frac{150\ \Omega}{\cos\frac{\pi}{4}} = \boxed{210\ \Omega}.$$

49. (a) Strategy Use Eq. (21-10).

Solution Compute the inductive reactance.
$$X_{\text{L}} = \omega L = 2\pi(250.0\text{ Hz})(10.0\times10^{-3}\text{ H}) = \boxed{15.7\ \Omega}$$

(b) Strategy Use Eq. (21-14b) with $X_{\text{C}} = 0$.

Solution Compute the impedance.
$$Z = \sqrt{R^2 + X_{\text{L}}^2} = \sqrt{(10.0\ \Omega)^2 + (15.71\ \Omega)^2} = \boxed{18.6\ \Omega}$$

(c) Strategy Use Eq. (21-14a).

Solution Compute the peak voltage.
$$I = \frac{\mathscr{E}_{\text{m}}}{Z} = \frac{1.00\text{ V}}{\sqrt{(10.0\ \Omega)^2 + (15.71\ \Omega)^2}} = \boxed{53.7\text{ mA}}$$

(d) Strategy Use Eq. (21-16).

Solution Compute the phase angle.
$$\cos\phi = \frac{R}{Z}, \text{ so } \phi = \cos^{-1}\frac{R}{Z} = \cos^{-1}\frac{10.0\ \Omega}{18.6\ \Omega} = \boxed{57.5^\circ}.$$

53. Strategy Use Eq. (21-18).

Solution Find the resonant frequency.
$$\omega_0 = \sqrt{\frac{1}{LC}} = \sqrt{\frac{1}{(40.0\times10^{-3}\text{ H})(0.0500\text{ F})}} = \boxed{22.4\text{ rad/s}}, \text{ or } f_0 = \frac{\omega_0}{2\pi} = \boxed{3.56\text{ Hz}}.$$

57. (a) Strategy and Solution At resonance, the phase difference between the voltages across the capacitor and the inductor is $\boxed{0^\circ}$, since the reactances are the same.

(b) Strategy At resonance, $Z = R$.

Solution Find the rms value of the applied emf.
$$\mathscr{E}_{\text{rms}} = I_{\text{rms}}Z = I_{\text{rms}}R = (0.12\text{ A})(20\ \Omega) = \boxed{2.4\text{ V}}$$

(c) Strategy and Solution If the frequency of the emf is changed without changing its rms value, the impedance increases and $\boxed{\text{the rms current decreases}}$.

59. (a) Strategy Use Eq. (21-18).

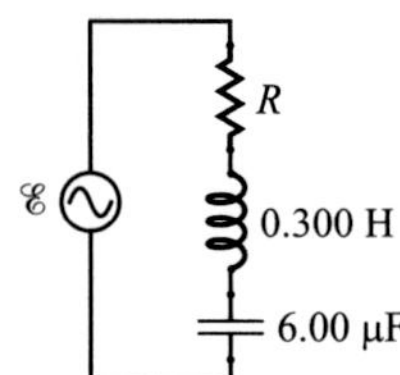

Solution Compute the resonant frequency.

$$\omega_0 = \sqrt{\frac{1}{LC}} = \sqrt{\frac{1}{(0.300 \text{ H})(6.00\times10^{-6} \text{ F})}} = \boxed{745 \text{ rad/s}}$$

(b) Strategy At resonance, $X_L = X_C$. Use Eq. (21-14a).

Solution Compute the resistance.

$$R = Z = \frac{\mathcal{E}_m}{I} = \frac{440 \text{ V}}{0.560 \text{ A}} = \boxed{790 \text{ }\Omega}$$

(c) Strategy At resonance, the voltages across the capacitor and inductor are 180° out of phase and equal in magnitude, so they cancel.

Solution The peak voltage across the resistor is $V_R = \mathcal{E}_m = \boxed{440 \text{ V}}$.

Across the inductor, the peak voltage is

$$V_L = IX_L = I\omega_0 L = \frac{IL}{\sqrt{LC}} = I\sqrt{\frac{L}{C}} = (0.560 \text{ A})\sqrt{\frac{0.300 \text{ H}}{6.00\times10^{-6} \text{ F}}} = \boxed{125 \text{ V}}.$$

Since $X_L = X_C$, $V_C = V_L = \boxed{125 \text{ V}}$.

61. Strategy Find the phase angle and the voltages across each device. Then, draw the phasor diagram and give the resonant frequency for the circuit.

Solution

(a) Find the reactances.

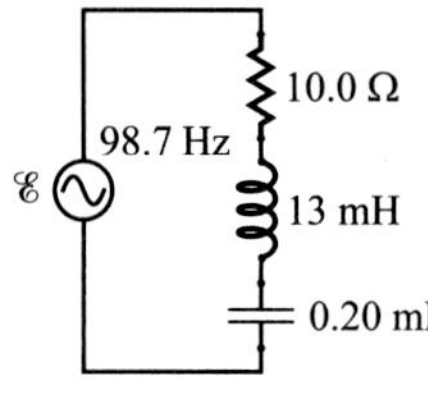

$$X_L = \omega L = 2\pi(98.7 \text{ Hz})(13\times10^{-3} \text{ H}) = 8.06 \text{ }\Omega \text{ and}$$

$$X_C = \frac{1}{\omega C} = \frac{1}{2\pi(98.7 \text{ Hz})(0.20\times10^{-3} \text{ F})} = 8.06 \text{ }\Omega.$$

Find the phase angle.

$$\phi = \tan^{-1}\frac{X_L - X_C}{R} = \tan^{-1}\frac{8.06 \text{ }\Omega - 8.06 \text{ }\Omega}{10.0 \text{ }\Omega} = \boxed{0°}$$

Since the phase angle is zero, the impedance is equal to the resistance. Find the current in the circuit.

$$I = \frac{\mathcal{E}_m}{R} = \frac{9.0 \text{ V}}{10.0 \text{ }\Omega} = 0.90 \text{ A}$$

Find the voltages across each device.

$$V_L = IX_L = (0.90 \text{ A})(8.06 \text{ }\Omega) = 7.3 \text{ V}, \ V_C = IX_C = (0.90 \text{ A})(8.06 \text{ }\Omega) = 7.3 \text{ V}, \text{ and}$$

$$V_R = IR = (0.90 \text{ A})(10.0 \text{ }\Omega) = 9.0 \text{ V} = \mathcal{E}_m.$$

(b) Draw a phasor diagram using the results of part (a). Note that $V_L - V_C = 0$ V.

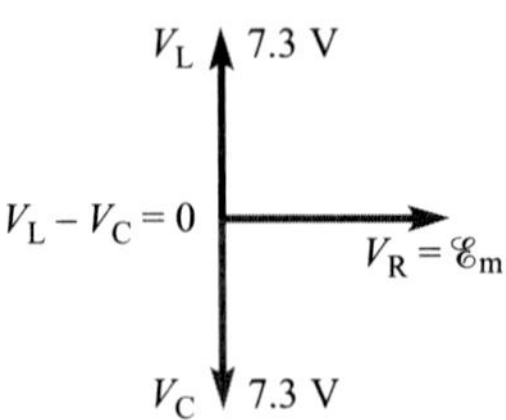

(c) Since the reactances are equal for a frequency of 98.7 Hz, $\boxed{98.7 \text{ Hz}}$ is the resonant frequency.

65. (a) Strategy Use Eq. (21-18).

Solution Compute the resonant angular frequency.

$$\omega_0 = \sqrt{\frac{1}{LC}} = \sqrt{\frac{1}{(0.800\text{ H})(2.22\times10^{-6}\text{ F})}} = \boxed{750\text{ rad/s}}$$

(b) Strategy At resonance, $X_L = X_C$, so $V_L = V_C$ and $Z = R$. Also,

$$V_R = IR = IZ = \mathcal{E}_m = \sqrt{2}\mathcal{E}_{rms} = \sqrt{2}(440\text{ V}) = 620\text{ V}.$$

Solution Find $V_L = V_C$.

$$V_L = V_C = IX_C = \left(\frac{\mathcal{E}_m}{R}\right)\left(\frac{1}{\omega_0 C}\right) = \frac{\mathcal{E}_m}{RC}\sqrt{LC} = \frac{\mathcal{E}_m}{R}\sqrt{\frac{L}{C}} = \frac{\sqrt{2}(440\text{ V})}{250\ \Omega}\sqrt{\frac{0.800\text{ H}}{2.22\times10^{-6}\text{ F}}} = 1.5\text{ kV}$$

Draw the phasor diagram.

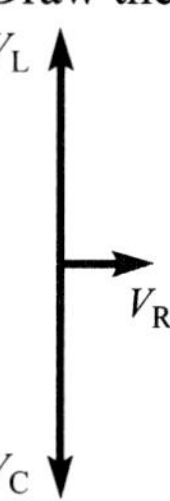

(c) Strategy Use Eq. (21-14a), Ohm's law, and Eq. (21-9) with rms values.

Solution Find the rms current.

$$I_{rms} = \frac{\mathcal{E}_{rms}}{Z} = \frac{\mathcal{E}_{rms}}{R}$$

Find the rms voltages.

$$V_{ab} = I_{rms}R = \mathcal{E}_{rms} = \boxed{440\text{ V}}$$

$$V_{bc} = I_{rms}X_L = \frac{\mathcal{E}_{rms}}{R}\omega_0 L = \frac{(440\text{ V})(750\text{ rad/s})(0.800\text{ H})}{250\ \Omega} = \boxed{1.1\text{ kV}}$$

Since $X_C = X_L$, $V_{cd} = V_{bc} = \boxed{1.1\text{ kV}}$.

$V_{bd} = \boxed{0}$, since the voltages across the inductor and capacitor oppose each other (180° out of phase).

V_{ad} must be equal to the source voltage, so $V_{ad} = \mathcal{E}_{rms} = \boxed{440\text{ V}}$.

(d) Strategy and Solution ω_0 is independent of R, so $\omega_0 = \boxed{750\text{ rad/s}}$.

(e) Strategy Use Ohm's law.

Solution Compute the rms current at resonance with the new resistor.

$$I_{rms} = \frac{\mathcal{E}_{rms}}{R} = \frac{440\text{ V}}{125\ \Omega} = \boxed{3.5\text{ A}}$$

69. Strategy Use Eqs. (18-8), (20-9), and (21-4).

Solution

(a) Find the total resistance in the copper wires.

$$R = \rho \frac{L}{A} = (1.67 \times 10^{-8} \ \Omega \cdot \text{m}) \frac{240 \times 10^3 \ \text{m}}{\pi (0.050 \ \text{m})^2} = \boxed{0.51 \ \Omega}$$

(b) The current in the wires is $I_{\text{rms}} = P_{\text{av, plant}} / V_{\text{rms}}$.

The average power dissipated by heating of the wires is

$$P_{\text{heat}} = I_{\text{rms}}^2 R = \left(\frac{P_{\text{av, plant}}}{V_{\text{rms}}} \right)^2 R = \left(\frac{2.5 \times 10^6 \ \text{W}}{1200 \ \text{V}} \right)^2 (0.51 \ \Omega) = \boxed{2.2 \ \text{MW}}.$$

(c) Find the number of turns in the secondary coil.

$$\frac{\mathscr{E}_2}{\mathscr{E}_1} = 150 = \frac{N_2}{N_1}, \text{ so } N_2 = 150 N_1 = 150(1000) = \boxed{150,000}.$$

(d) The new rms current in the wires is $I_{\text{rms}} = \dfrac{P_{\text{av, plant}}}{V_{\text{rms}}} = \dfrac{2.5 \times 10^6 \ \text{W}}{150(1200 \ \text{V})} = \boxed{14 \ \text{A}}.$

(e) The average power dissipated in the transmission lines using the transformer is

$$P_{\text{heat}} = I_{\text{rms}}^2 R = \left(\frac{P_{\text{av, plant}}}{V_{\text{rms}}} \right)^2 R = \left[\frac{2.5 \times 10^6 \ \text{W}}{150(1200 \ \text{V})} \right]^2 (0.51 \ \Omega) = \boxed{98 \ \text{W}}.$$

73. Strategy $P = [v(t)]^2 / R$ and $[v(t)]^2$ reaches its maximum value twice per cycle.

Solution The instantaneous power is at its maximum value two times per cycle, which is twice the frequency, so it is at maximum $2(60 \ \text{Hz}) = \boxed{120 \ \text{times per second}}$.

77. (a) Strategy Model the coil as an RL series circuit. Use Eq. (21-14b) with $X_C = 0$.

Solution Calculate the impedance.

$$Z = \sqrt{R^2 + X_L^2} = \sqrt{R^2 + \omega^2 L^2} = \sqrt{(120 \ \Omega)^2 + 4\pi^2 (60.0 \ \text{Hz})^2 (12.0 \ \text{H})^2} = \boxed{4.53 \ \text{k}\Omega}$$

(b) Strategy Use Eq. (21-14a).

Solution Calculate the current.

$$I_{\text{rms}} = \frac{\mathscr{E}_{\text{rms}}}{Z} = \frac{110 \ \text{V}}{4530 \ \Omega} = \boxed{24 \ \text{mA}}$$

81. Strategy Use Eqs. (21-7) and (21-10).

Solution

(a) A power factor of 1.00 implies that $X_L = X_C$. Find the inductive reactance.

$$X_L = X_C = \frac{1}{\omega C} = \frac{1}{2\pi (60.0 \ \text{Hz})(38.0 \times 10^{-6} \ \text{F})} = \boxed{69.8 \ \Omega}$$

(b) Find the inductance of the coil.

$$\omega L = X_L = X_C = \frac{1}{\omega C}, \text{ so } L = \frac{1}{\omega^2 C} = \frac{1}{4\pi^2 (60.0 \ \text{Hz})^2 (38.0 \times 10^{-6} \ \text{F})} = \boxed{185 \ \text{mH}}.$$

85. (a) Strategy Model the coil as an *RL* series circuit. Use Eq. (21-16).

Solution Find the power factor.

$$\cos\phi = \frac{R}{Z} = \frac{R}{\sqrt{R^2 + X_L^2}} = \frac{R}{\sqrt{R^2 + \omega^2 L^2}} = \frac{450\ \Omega}{\sqrt{(450\ \Omega)^2 + 4\pi^2(9.55\ \text{Hz})^2(2.47\ \text{H})^2}} = \boxed{0.95}$$

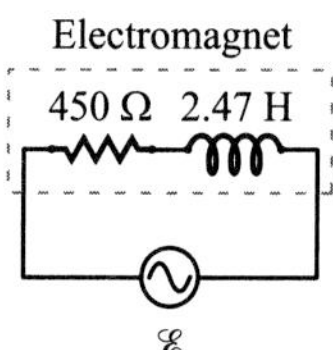

(b) Strategy Use Eq. (21-14b).

Solution Find the impedance.

$$Z = \sqrt{(450\ \Omega)^2 + 4\pi^2(9.55\ \text{Hz})^2(2.47\ \text{H})^2} = \boxed{470\ \Omega}$$

(c) Strategy Use Eq. (21-14a).

Solution Find the peak current.

$$I = \frac{\mathscr{E}_m}{Z} = \frac{2.0\times10^3\ \text{V}}{474\ \Omega} = \boxed{4.2\ \text{A}}$$

(d) Strategy Use Eq. (21-17).

Solution Find the average power.

$$P_{\text{av}} = I_{\text{rms}}V_{\text{rms}}\cos\phi = \frac{1}{2}IV\cos\phi = \frac{1}{2}(4.2\ \text{A})(2.0\times10^3\ \text{V})(0.95) = \boxed{4.0\ \text{kW}}$$

89. Strategy Use Eq. (21-4).

Solution

(a) The load for one house is $P_{\text{load, 1}} = I_{\text{rms}}V_{\text{rms}}$. So, for four houses,

$$P_{\text{load}} = 4I_{\text{rms}}V_{\text{rms}} = 4(60\ \text{A})(220\ \text{V}) = \boxed{53\ \text{kV}\cdot\text{A}}.$$

(b) $\boxed{\text{No}}$, the rating is inadequate, since $53\ \text{kV}\cdot\text{A} > 35\ \text{kV}\cdot\text{A}$.

(c) The load factor may be less than 1 since the transformer may have to supply a reactive load. Even though the load draws less power than a purely resistive load, the transformer must supply the same current and has the same heating in its windings and core.

93. (a) Strategy The capacitive reactance is $X_C = 1/(\omega C) = 1/(2\pi f C)$.

Solution Compute the reactances.

$$X_{C1} = \frac{1}{2\pi(12.0 \text{ Hz})(5.00\times10^{-6} \text{ F})} = \boxed{2.65 \text{ k}\Omega} \text{ and } X_{C2} = \frac{1}{2\pi(1.50\times10^3 \text{ Hz})(5.00\times10^{-6} \text{ F})} = \boxed{21.2 \text{ }\Omega}.$$

(b) Strategy The impedance is $Z = \sqrt{R^2 + X_C^2} = \sqrt{R^2 + 1/(2\pi f C)^2}$.

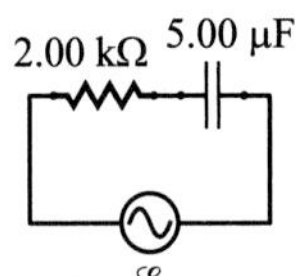

Solution Compute the impedances.

$$Z_1 = \sqrt{(2.00\times10^3 \text{ }\Omega)^2 + \frac{1}{4\pi^2(12.0 \text{ Hz})^2(5.00\times10^{-6} \text{ F})^2}} = \boxed{3.32 \text{ k}\Omega}$$

$$Z_2 = \sqrt{(2.00\times10^3 \text{ }\Omega)^2 + \frac{1}{4\pi^2(1.50\times10^3 \text{ Hz})^2(5.00\times10^{-6} \text{ F})^2}} = \boxed{2.00 \text{ k}\Omega}$$

(c) Strategy At high frequencies, $X_C \ll R$, so $Z \approx R$.

Solution Compute the maximum current.

$$I = \frac{\mathscr{E}_m}{Z} = \frac{\mathscr{E}_m}{R} = \frac{2.00 \text{ V}}{2.00\times10^3 \text{ }\Omega} = \boxed{1.00 \text{ mA}} \text{ is the maximum current } (Z \text{ is at its minimum value}).$$

(d) Strategy Use Eq. (21-16).

Solution Initially, when the ac source is first applied, there is no charge on the capacitor. Thus, $V_C = 0$ and the current is at its maximum value. Therefore, the current leads the voltage. Calculate the phase angle for each frequency using the power factor, $\cos\phi = R/Z$.

$f = 12.0$ Hz:

$$\phi = \cos^{-1}\frac{R}{Z} = \cos^{-1}\frac{R}{\sqrt{R^2 + X_C^2}} = \cos^{-1}\frac{1}{\sqrt{1 + \frac{X_C^2}{R^2}}} = \cos^{-1}\frac{1}{\sqrt{1 + \frac{1}{\omega^2 C^2 R^2}}}$$

$$= \cos^{-1}\frac{1}{\sqrt{1 + \frac{1}{4\pi^2(12.0 \text{ Hz})^2(5.00\times10^{-6} \text{ F})^2(2.00\times10^3 \text{ }\Omega)^2}}} = \boxed{53.0°}$$

$f = 1.50$ kHz:

$$\phi = \cos^{-1}\frac{1}{\sqrt{1 + \frac{1}{4\pi^2(1.50\times10^3 \text{ Hz})^2(5.00\times10^{-6} \text{ F})^2(2.00\times10^3 \text{ }\Omega)^2}}} = \boxed{0.608°}$$

Review Exercises

1. **Strategy** The maximum torque occurs when the plane of the loop is parallel to the magnetic field of the solenoid. Use Eqs. (19-13a) and (19-17).

   Solution Find the maximum possible magnetic torque on the loop.
   $$\tau = N_1 I_1 A_1 B_s = N_1 I_1 A_1 \mu_0 n_s I_s = (100)(2.20 \text{ A})\pi(0.0800 \text{ m})^2 (4\pi\times10^{-7} \text{ T}\cdot\text{m/A})(8500 \text{ m}^{-1})(25.0 \text{ A})$$
   $$= \boxed{1.2 \text{ N}\cdot\text{m}}$$
   Since the magnetic field inside the solenoid is along the axis of the solenoid, the magnetic torque on the loop is at its maximum value $\boxed{\text{when the plane of the loop is parallel to the axis of the solenoid}}$.

5. **Strategy** Use Eq. (19-14) and $\vec{F}_B = q\vec{v}\times\vec{B}$. Let the positive y-direction be up.

   Solution According to the RHR, the magnetic field generated by the power line at point P is in the positive y-direction. Looking at the side view, we see that the angle between the velocity of the muon and the magnetic field is $180° - 25° = 155°$. The charge of the muon is negative, so according to the RHR and $\vec{F}_B = q\vec{v}\times\vec{B}$, the direction of the force on the muon is out of the plane of the paper in the side view (or to the right in the end-on view). Compute the magnitude of the force.
   $$F_B = evB\sin\theta = ev\frac{\mu_0 I}{2\pi r}\sin\theta = \frac{(1.602\times10^{-19} \text{ C})(7.0\times10^7 \text{ m/s})(4\pi\times10^{-7} \text{ T}\cdot\text{m/A})(16.0 \text{ A})\sin155°}{2\pi(0.850 \text{ m})}$$
   $$= 1.8\times10^{-17} \text{ N}$$
   Thus, $\vec{F}_B = \boxed{1.8\times10^{-17} \text{ N out of the plane of the paper in the side view (or to the right in the end on view)}}$.

9. **(a) Strategy** Use $\vec{F} = I\vec{L}\times\vec{B}$.

   Solution The current flows into the page through the rod, so, by the RHR, the force on the rod is $\boxed{\text{to the right}}$.

   (b) Strategy Use $\vec{F} = I\vec{L}\times\vec{B}$, Newton's second law, and Eq. (4-5).

   Solution Find the normal force on the rod.
   $\Sigma F_y = N - mg = 0$, so $N = mg$.

   The force of kinetic friction is $f_k = \mu_k N = \mu_k mg$. Find the acceleration of the rod.
   $$\Sigma F_x = F_B - f_k = ILB - \mu_k mg = ma_x, \text{ so } a_x = \frac{ILB}{m} - \mu_k g.$$
   Find the speed of the rod.
   $$v_{fx}^2 - v_{ix}^2 = v^2 - 0 = 2a_x\Delta x = 2\left(\frac{ILB}{m} - \mu_k g\right)\Delta x, \text{ so}$$
   $$v = \sqrt{2\left(\frac{ILB}{m} - \mu_k g\right)\Delta x} = \sqrt{2\left[\frac{(2.00 \text{ A})(0.500 \text{ m})(0.750 \text{ T})}{0.0500 \text{ kg}} - 0.350(9.80 \text{ m/s}^2)\right](8.00 \text{ m})} = \boxed{13.6 \text{ m/s}}.$$

   (c) Strategy and Solution As the rod slides down the rails,
   $\boxed{\text{the applied emf has to increase because of the increased resistance in the longer rail lengths in the circuit and because of the increasingly large induced emf as the rod moves faster } (\mathscr{E} = vBL).}$

13. (a) Strategy Use Eq. (21-18).

Solution Compute the resonant frequency of the *RLC* series circuit.

$$\omega_0 = 2\pi f_0 = \frac{1}{\sqrt{LC}}, \text{ so } f_0 = \frac{1}{2\pi\sqrt{LC}} = \frac{1}{2\pi\sqrt{(0.146 \text{ H})(877\times10^{-9} \text{ F})}} = \boxed{445 \text{ Hz}}.$$

(b) Strategy and Solution At resonance the capacitive and inductive reactances are equal. If the frequency is less than the resonant frequency, the capacitive reactance will be greater than the inductive reactance because $X_C \propto f^{-1}$, while $X_L \propto f$. When $X_C > X_L$, the voltage lags the current. Therefore, the $\boxed{\text{current}}$ leads the voltage.

(c) Strategy Use Eq. (21-15) and the definitions of capacitive and inductive reactance.

Solution Find the phase angle of the circuit.

$$\tan\phi = \frac{X_L - X_C}{R} = \frac{\omega L - \frac{1}{\omega C}}{R} = \frac{0.50\omega_0 L - \frac{1}{0.50\omega_0 C}}{R} = \frac{\frac{0.50L}{\sqrt{LC}} - \frac{2.0}{\frac{1}{\sqrt{LC}}C}}{R} = \frac{0.50\sqrt{\frac{L}{C}} - 2.0\sqrt{\frac{L}{C}}}{R} = -\frac{1.5}{R}\sqrt{\frac{L}{C}}, \text{ so}$$

$$\phi = \tan^{-1}\left(-\frac{1.5}{R}\sqrt{\frac{L}{C}}\right) = \tan^{-1}\left[-\frac{1.5}{(255 \ \Omega)}\sqrt{\frac{0.146 \text{ H}}{877\times10^{-9} \text{ F}}}\right] = \boxed{-67°}.$$

(d) Strategy Use Eqs. (21-14a) and (21-16).

Solution Find the impedance.

$$\cos\phi = \frac{R}{Z}, \text{ so } Z = \frac{R}{\cos\phi}.$$

Find the rms current.

$$\mathscr{E}_m = IZ = \sqrt{2}I_{rms}Z = \sqrt{2}I_{rms}\frac{R}{\cos\phi}, \text{ so } I_{rms} = \frac{\mathscr{E}_m \cos\phi}{\sqrt{2}R} = \frac{(480 \text{ V})\cos(-67.4°)}{\sqrt{2}(255 \ \Omega)} = \boxed{0.51 \text{ A}}.$$

(e) Strategy Use Eq. (21-4).

Solution The average power dissipated in the circuit is

$$P_{av} = I_{rms}{}^2 R = (0.512 \text{ A})^2 (255 \ \Omega) = \boxed{67 \text{ W}}.$$

(f) Strategy Use Ohm's law and Eqs. (21-6) and (21-9).

Solution The maximum voltages are $V_R = IR = \sqrt{2}I_{rms}R = \sqrt{2}(0.51 \text{ A})(255 \ \Omega) = \boxed{180 \text{ V}}$,

$$V_L = IX_L = \sqrt{2}I_{rms}\omega L = \sqrt{2}(0.51 \text{ A})2\pi(0.50)(445 \text{ Hz})(0.146 \text{ H}) = \boxed{150 \text{ V}}, \text{ and}$$

$$V_C = IX_C = \sqrt{2}I_{rms}\frac{1}{\omega C} = \frac{\sqrt{2}(0.51 \text{ A})}{2\pi(0.50)(445 \text{ Hz})(877\times10^{-9} \text{ F})} = \boxed{590 \text{ V}}.$$

17. (a) Strategy Use Eq. (19-5) and $\vec{F}_E = q\vec{E}$.

Solution $\vec{F}_E$ must oppose $\vec{F}_B$ in the velocity selector. According to $\vec{F}_B = e\vec{v} \times \vec{B}$ and RHR 1, $\vec{F}_B$ is downward for a positive ion, so the electric force $\vec{F}_E$ opposing $\vec{F}_B$ is directed upward, or north in the figure. Since the ion is positive, the electric field must be upward. Set the magnitudes of the forces equal and solve for the magnitude of $\vec{E}$. $evB = eE$, so $E = vB$. Thus, $\vec{E} = \boxed{vB \text{ north}}$.

(b) Strategy Form a proportion to determine the direction of the deflection.

Solution

$$\frac{F_{B,\,\text{ion}}}{F_{E,\,\text{ion}}} = \frac{ev_{\text{ion}}B}{evB} = \frac{v_{\text{ion}}}{v} < 1, \text{ so } F_{B,\,\text{ion}} < F_{E,\,\text{ion}}, \text{ and the ions are deflected upward in the selector.}$$

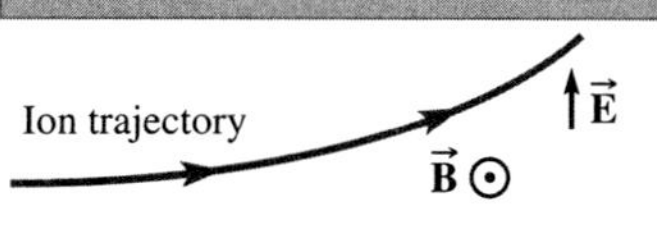

21. Strategy Use the relationships for power and fluid flow.

Solution If water were free to flow, we'd have

$$\rho g h = \frac{1}{2}\rho v^2$$

Let the volume flow rate be Q. The power drawn from the water is given by

$$P_w = Q\left(\rho g h - \frac{1}{2}\rho v^2\right)$$

The maximum power supplied is $P_{\max} = 0.80 P_w$. Find the peak current of a single generator.

$$\frac{1}{2}I_p V_p = 0.80 P_w, \text{ so}$$

$$
\begin{aligned}
I_{p,1} &= \frac{2(0.80)P_w}{V_p} = \frac{1.6}{V_p}Q\left(\rho g h - \frac{1}{2}\rho v^2\right) \\
&= \frac{1.6(100 \text{ m}^3/\text{s})(1000 \text{ kg}/\text{m}^3)}{(10{,}000 \text{ V})}\left[(9.8 \text{ m}/\text{s}^2)(120 \text{ m}) - \frac{(10 \text{ m}/\text{s})^2}{2}\right] = \boxed{18 \text{ kA}}
\end{aligned}
$$

MCAT Review

1. **Strategy and Solution** The current flows counterclockwise through the apparatus. By the RHR, the magnetic field generated by the current is directed upward. The correct answer is $\boxed{\text{A}}$.

2. **Strategy** Refer to the data in the table.

   Solution Two rows of data are given for a mass of 0.01 kg. The current is increased by 150% from 10.0 A to 15.0 A. Due to the increase in the current, the exit speed of the projectile increases by 150% from 2.0 km/s to 3.0 km/s. Thus, the exit speed is directly proportional to the current. So, if the current were decreased by a factor of two, the exit speed would be decreased by a factor of two, as well. The correct answer is $\boxed{\text{D}}$.

3. **Strategy and Solution** Since, for a given current, the force is constant along the entire length of the railgun, lengthening the rails would increase the exit speed because of the longer distance over which the force is present. The correct answer is $\boxed{\text{D}}$.

4. **Strategy and Solution** Since the resistance of the rails is directly proportional to their resistivity, lowering the resistivity of the rails would decrease the power required to maintain the current that flows through them. The correct answer is $\boxed{\text{B}}$.

5. **Strategy** The average power is equal to the change in kinetic energy divided by the time interval.

   Solution Find the average power supplied by the railgun.
   $$P_{\text{av}} = \frac{\Delta K}{\Delta t} = \frac{\frac{1}{2}mv^2 - 0}{\Delta t} = \frac{mv^2}{2\Delta t} = \frac{(0.10\ \text{kg})(10.0\ \text{m/s})^2}{2(2.0\ \text{s})} = 2.5\ \text{W}$$
   The correct answer is $\boxed{\text{B}}$.

6. **Strategy and Solution** $F = ma \propto I^2$, so $a \propto I^2/m$. The speed of the projectile is directly proportional to its acceleration, so $v \propto I^2/m$. Form a proportion using data from the table to find the approximate speed of the 0.08-kg projectile.
   $$\frac{v_2}{v_1} = \frac{I_2^2/m_2}{I_1^2/m_1} = \left(\frac{I_2}{I_1}\right)^2 \frac{m_1}{m_2}, \text{ so } v_2 = \left(\frac{I_2}{I_1}\right)^2 \frac{m_1}{m_2}v_1. \text{ For (1), we use the data given in the third row of the table.}$$
   $$v_2 = \left(\frac{I_2}{I_1}\right)^2 \frac{m_1}{m_2}v_1 = \left(\frac{20.0}{10.0}\right)^2 \frac{0.02}{0.08}(1.4\ \text{km/s}) = 1.4\ \text{km/s}$$
   The correct answer is $\boxed{\text{C}}$.

7. **Strategy and Solution** The power from the power plant is given by $P = IV$, so, for a given amount of power, increasing the voltage decreases the current required. Since the power lost as heat is given by $P = I^2R$, reducing the current reduces the power lost as heat. The correct answer is $\boxed{\text{A}}$.

8. **Strategy and Solution** The magnetic field lines due to the section of current-carrying wire are circles (which is the only possibility, given the symmetry of the situation). The direction of the field is determined by the RHR. Pointing the thumb of the right hand in the direction of the current in the wire, then curling the fingers inward toward the palm, the direction of the field is indicated by the direction of the curl of the fingers—in this case counterclockwise, as seen from above. The correct answer is $\boxed{\text{D}}$.

Chapter 22

ELECTROMAGNETIC WAVES

Conceptual Questions

1. Yes, the axis of the loop of a magnetic dipole antenna should be aligned with the magnetic field of the wave. The changing magnetic field of the wave induces an emf in the loop of the antenna. For best reception, the induced emf should be as large as possible, which occurs when the axis of the loop is aligned with the magnetic field.

5. The Ampère-Maxwell law tells us that a changing electric field induces a circulating magnetic field in the space around it. An electric wave with no magnetic component would be impossible because the changing electric field of the wave would itself induce a magnetic component.

9. At the point P shown in the figure, the magnetic field vector points in the $\pm z$-direction, parallel to the axis of the antenna, while the electric field vector points in the $\pm y$-direction, perpendicular to both the magnetic field and the direction of propagation of the wave.

13. Incandescent light bulbs, such as those used in greenhouses, emit a significant amount of infrared radiation. Additionally, plants and other warm bodies emit radiation in the infrared region of the spectrum. Infrared imaging is therefore an effective means of detecting biological activity such as the growing of plants.

Problems

1. **Strategy and Solution** At a point due south of the transmitter, the EM waves are traveling due south. The electric field at this point is oriented vertically, like the antenna. The magnetic field is perpendicular to both the electric field and the direction the EM waves travel, so the magnetic field must be oriented east-west.

3. **Strategy and Solution** At a point due north of the transmitter, the EM waves are traveling due north. The electric field at this point is oriented vertically, like the antenna. The magnetic field is perpendicular to both the electric field and the direction the EM waves travel, so the magnetic field is oriented east-west. The axis of a magnetic dipole antenna should be aligned with the orientation of the magnetic field; therefore, the antenna should be in the vertical plane defined by the vertical electric dipole antenna and the direction of wave propagation.

5. **Strategy and Solution** At a point due east of the transmitter, the EM waves are traveling due east. The electric field at this point is oriented horizontally north-south, like the antenna. The magnetic field is oriented perpendicular to both the direction of the wave and the electric field. Therefore, the axis of the magnetic dipole antenna should be oriented up-down.

7. **Strategy** Use Faraday's law, Eq. (20-5) for magnetic flux, and the expression for the amplitude of the emf found in Example 22.1.

   Solution

   (a) The axis of the coil should be aligned with the magnetic field, so that the flux through the coil will be greatest.

   (b) The magnitude of the maximum emf induced in the antenna due to the magnetic component of the wave is given by Faraday's law:

$$\mathscr{E}_{\mathrm{m}} = N \frac{\Delta \Phi_{\mathrm{B}}}{\Delta t}$$

The maximum magnetic flux is $\Phi_B = BA = B\pi r^2$. So, $\mathscr{E} = N\pi r^2 \dfrac{\Delta B}{\Delta t}$.

Assuming the field is sinusoidal, we let $B = B_m \sin(kz + \omega t)$. Then, $\dfrac{\Delta B}{\Delta t} = \omega B_m \cos(kz + \omega t)$.

The maximum value of $\dfrac{\Delta B}{\Delta t}$ is ωB_m. So, the amplitude of the induced emf is

$$\mathscr{E}_m = N\pi r^2 \omega B_m = 50\pi(0.050 \text{ m})^2 \, 2\pi(870{,}000 \text{ Hz})(1.7\times10^{-9} \text{ T}) = \boxed{3.6 \text{ mV}}.$$

(c) As in Example 22.1, the amplitude of the emf is $\mathscr{E}_m = E_m L = (0.50 \text{ V/m})(0.050 \text{ m}) = \boxed{25 \text{ mV}}$.

9. **Strategy** The frequency, wavelength, and speed of EM radiation are related by $\lambda f = c$.

Solution Compute the frequency of the microwaves.
$$f = \frac{c}{\lambda} = \frac{3.00\times10^8 \text{ m/s}}{0.12 \text{ m}} = 2.5\times10^9 \text{ Hz} = \boxed{2.5 \text{ GHz}}$$

11. **Strategy** The speed of light in matter is given by $v = c/n$, where n is the index of refraction.

Solution Compute the index of refraction of topaz.
$$n = \frac{c}{v} = \frac{3.00\times10^8 \text{ m/s}}{1.85\times10^8 \text{ m/s}} = \boxed{1.62}$$

13. **Strategy** The time of travel is equal to the distance traveled divided by the rate of travel.

Solution Compute the time it takes for light to travel 50.0 cm.
$$\Delta t = \frac{d}{c} = \frac{50.0\times10^{-2} \text{ m}}{3.00\times10^8 \text{ m/s}} = 1.67\times10^{-9} \text{ s} = \boxed{1.67 \text{ ns}}$$

15. (a) **Strategy** The wavelength is shorter in matter than it is in vacuum. Use Eq. (22-3).

Solution Compute the wavelength of light inside the glass.
$$\lambda_g = \frac{\lambda_v}{n} = \frac{692 \text{ nm}}{1.52} = \boxed{455 \text{ nm}}$$

(b) **Strategy** The frequency in glass is the same as the frequency in air. Use $c = f\lambda$.

Solution Compute the frequency of light inside the glass.
$$f_g = f_a = \frac{c}{\lambda_a} = \frac{3.00\times10^8 \text{ m/s}}{692\times10^{-9} \text{ m}} = \boxed{4.34\times10^{14} \text{ Hz}}$$

17. Strategy The frequency, wavelength, and speed of EM radiation are related by $\lambda f = c$. Use Figure 22.5 to determine the part of the EM spectrum to which the waves belong.

Solution

(a) Compute the wavelength.
$$\lambda = \frac{c}{f} = \frac{3.00\times10^8 \text{ m/s}}{60.0 \text{ Hz}} = \boxed{5.00\times10^6 \text{ m}}$$

(b) $\boxed{\text{The radius of the Earth is } 6.4\times10^6 \text{ m, which is close in value to the wavelength.}}$

(c) According to Figure 22.5, the waves are $\boxed{\text{radio waves}}$.

21. Strategy The time of transmission is equal to the distance divided by the speed of EM waves. For (c), use the definition of average velocity.

Solution

(a) A one-way transmission from the Rover *Spirit* on its landing day to the scientists on Earth took
$$\Delta t = \frac{\Delta x}{c} = \frac{170.2\times10^9 \text{ m}}{2.998\times10^8 \text{ m/s}} \times \frac{1 \text{ min}}{60 \text{ s}} = \boxed{9.462 \text{ min}}.$$

(b) A one-way transmission from the scientists on Earth to the Rover *Opportunity* on its landing day took
$$\Delta t = \frac{\Delta x}{c} = \frac{198.7\times10^9 \text{ m}}{2.998\times10^8 \text{ m/s}} \times \frac{1 \text{ min}}{60 \text{ s}} = \boxed{11.05 \text{ min}}.$$

25. Strategy The electric and magnetic fields oscillate with the same frequency, and their amplitudes are proportional to each other. Use Eq. (22-4).

Solution Compute the amplitude of the magnetic field.
$$B_{\text{m}} = \frac{E_{\text{m}}}{c} = \frac{0.60\times10^{-3} \text{ V/m}}{3.00\times10^8 \text{ m/s}} = \boxed{2.0\times10^{-12} \text{ T}}$$
The frequency of the magnetic field is $\boxed{30 \text{ GHz}}$, the same as that of the electric field.

27. Strategy The electric and magnetic fields oscillate with the same frequency, and their amplitudes are proportional to each other. Use Eqs. (22-4) and (22-5). The direction of propagation of EM waves is determined by $\vec{\mathbf{E}}\times\vec{\mathbf{B}}$.

Solution

(a) Compute the amplitude of the electric field.
$$E_{\text{m}} = cB_{\text{m}} = (3.00\times10^8 \text{ m/s})(2.5\times10^{-11} \text{ T}) = \boxed{7.5 \text{ mV/m}}$$
The frequency of the electric field is $\boxed{3.0 \text{ MHz}}$, the same as that of the magnetic field.

(b) Compute the magnitude of the electric field.
$$E = cB = (3.00\times10^8 \text{ m/s})(1.5\times10^{-11} \text{ T}) = \boxed{4.5 \text{ mV/m}}$$
Since the magnetic field is in the $+z$-direction and the wave is traveling in the $-y$-direction, by $\vec{\mathbf{E}}\times\vec{\mathbf{B}}$ and the RHR, the electric field at $y = 0$ and $t = 0$ must point $\boxed{\text{in the } +x\text{-direction}}$.

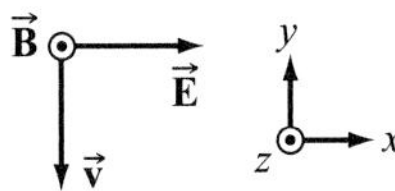

29. **(a) Strategy and Solution** Since the magnetic field depends on the value of z but not on the values of x or y, the wave moves parallel to the z-axis. The direction can be found by noting that as t increases in $kz + \omega t$, z must decrease to maintain the relative phase. So, the wave is moving in the $\boxed{-z\text{-direction.}}$

   (b) Strategy The amplitudes of the electric and magnetic fields are proportional to each other. Use Eq. (22-5). The direction of propagation of EM waves is determined by $\vec{E} \times \vec{B}$.

   Solution Since the magnetic field is in the $+y$-direction when $t = 0$ and $kz = \pi/2$ and the wave is traveling in the $-z$-direction, by $\vec{E} \times \vec{B}$ and the RHR, the electric field at $t = 0$ and $y = 0$ must point in the $-x$-direction. The components are $\boxed{E_x = -cB_\text{m}\sin(kz + \omega t),\ E_y = E_z = 0}$.

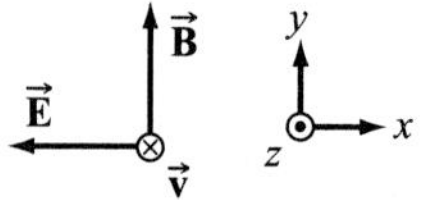

33. **Strategy** Use Eqs. (22-8), (22-10), and (22-11).

   Solution Find the intensity of the beam.
   $$I = \frac{\langle P \rangle}{A} = \frac{10.0\times10^{-3}\ \text{W}}{\frac{1}{4}\pi(0.85\times10^{-2}\ \text{m})^2} = 180\ \text{W/m}^2$$
   Find the rms value of the electric field.
   $$I = \langle u \rangle c = (\epsilon_0 E_\text{rms}{}^2)c,\ \text{so}\ E_\text{rms} = \sqrt{\frac{I}{\epsilon_0 c}} = \sqrt{\frac{180\ \text{W/m}^2}{[8.854\times10^{-12}\ \text{C}^2/(\text{N}\cdot\text{m}^2)](3.00\times10^8\ \text{m/s})}} = \boxed{260\ \text{V/m}}.$$

37. **Strategy** Intensity is related to the average power radiated by $I = \langle P \rangle / A$ where $A = 4\pi r^2$ and $r = 14\times10^6$ ly.

   Solution Find the rate at which the star radiates EM energy.
   $$\langle P \rangle = 4\pi r^2 I = 4\pi(14\times10^6\ \text{ly})^2(9.461\times10^{15}\ \text{m/ly})^2(4\times10^{-21}\ \text{W/m}^2) = \boxed{9\times10^{26}\ \text{W}}$$

41. **Strategy** Use Eqs. (22-5), (22-8), (22-11), and (22-12).

   Solution

   (a) Compute the average power incident on the telescope.
   $$\langle P \rangle = IA\cos\theta = (1.0\times10^{-26}\ \text{W/m}^2)\pi(305\ \text{m/2})^2\cos0° = \boxed{7.3\times10^{-22}\ \text{W}}$$

   (b) Compute the average power incident on the Earth's surface.
   $$\langle P \rangle = IA\cos\theta = (1.0\times10^{-26}\ \text{W/m}^2)\pi(6.371\times10^6\ \text{m})^2\cos0° = \boxed{1.3\times10^{-12}\ \text{W}}$$

   (c) Find the rms electric and magnetic fields.
   $$\langle u \rangle = \epsilon_0 E_\text{rms}^2 = \frac{I}{c},\ \text{so}\ E_\text{rms} = \sqrt{\frac{I}{\epsilon_0 c}} = \sqrt{\frac{1.0\times10^{-26}\ \text{W/m}^2}{[8.854\times10^{-12}\ \text{C}^2/(\text{N}\cdot\text{m}^2)](3.00\times10^8\ \text{m/s})}} = \boxed{1.9\times10^{-12}\ \text{V/m}}.$$
   Since $E_\text{rms} = cB_\text{rms}$, $B_\text{rms} = \dfrac{1.94\times10^{-12}\ \text{V/m}}{3.00\times10^8\ \text{m/s}} = \boxed{6.5\times10^{-21}\ \text{T}}.$

45. Strategy Use Eq. (22-14b).

Solution

(a) The intensity of the light passing through the first polarizing sheet is
$$I_1 = I_0 \cos^2 \theta_1 = I_0 \cos^2 \theta.$$
The intensity of the light passing through the second sheet is
$$I_2 = I_1 \cos^2 \theta_2 = I_1 \cos^2(90° - \theta) = I_0 \cos^2 \theta \cos^2(90° - \theta) = I_0 \cos^2 \theta \sin^2 \theta = \boxed{\frac{1}{4} I_0 \sin^2 2\theta},$$
since $\cos(90° - \theta) = \sin \theta$ and $2\cos\theta \sin\theta = \sin 2\theta$.

(b) The sine function reaches a maximum when its argument is $90°$.
$$2\theta = 90°, \text{ so } \theta = \boxed{45°}.$$

49. Strategy and Solution

(a) Microwaves that are transmitted are linearly polarized perpendicular to the strips. Since the microwaves are vertically polarized, a plate with horizontal strips will transmit the microwaves best. Plate $\boxed{\text{(a)}}$ will have the best transmission, since its strips are horizontal.

(b) The worst transmitter will be the best reflector. The strips in plate (c) are vertical, so plate $\boxed{\text{(c)}}$ will be the worst transmitter and the best reflector.

(c) The best transmitter is (a), so (a) transmits intensity I_1; the worst, is (c). Plate (b) is rotated $30.0°$ with respect to (a), so $I_{\text{second best}} = I_b = I_1 \cos^2 30.0° = \boxed{0.750 I_1}$.

53. Strategy and Solution The incident light is horizontal, as would occur shortly after sunrise. It is unpolarized, so the molecules in the atmosphere become oscillating dipoles, which oscillate in random directions perpendicular to the incident wave. As an oscillating dipole, the molecules radiate EM waves. Vertical oscillations of the dipoles do not radiate in the up-down direction. So, $\boxed{\text{yes}}$, the light is polarized in the $\boxed{\text{north-south}}$ direction (which is perpendicular to both the direction of the incident light and the direction of the scattered light).

55. Strategy First try Eq. (22-16). If the resulting velocity is small compared to c, we are done. If the velocity is not small compared to c, we will have to use Eq. (22-15).

Solution Find the speed of the star with respect to Earth.
$$f_o \approx f_s\left(1 + \frac{v_{rel}}{c}\right)$$
$$1 + \frac{v_{rel}}{c} \approx \frac{f_o}{f_s} = \frac{\lambda_s}{\lambda_o}$$
$$v_{rel} \approx c\left(\frac{\lambda_s}{\lambda_o} - 1\right) = (3.00 \times 10^8 \text{ m/s})\left(\frac{659.6 \text{ nm}}{661.1 \text{ nm}} - 1\right) = -680 \text{ km/s}$$

This velocity is small compared to c, so the use of Eq. (22-16) was justified. $v_{rel} < 0$, so $\boxed{\text{the star is moving at 680 km/s away from the Earth.}}$

57. Strategy Use the Doppler shift formula, Eq. (22-15). The cars are approaching, so the relative speed is positive.

Solution

(a) Since the cars are approaching each other, the frequency of the microwaves observed by the speeder is greater than that of the emitted microwaves. Therefore, $\boxed{f_2}$ is larger.

(b) Find the frequency difference.

$$f_2 - f_1 = f_1\sqrt{\frac{1+v_{\text{rel}}/c}{1-v_{\text{rel}}/c}} - f_1 = f_1\left(\sqrt{\frac{1+v_{\text{rel}}/c}{1-v_{\text{rel}}/c}}-1\right) = (7.50\times10^9 \text{ Hz})\left(\sqrt{\frac{1+48.0/2.998\times10^8}{1-48.0/2.998\times10^8}}-1\right) = \boxed{1.2 \text{ kHz}}$$

59. Strategy A Doppler shift of this magnitude almost certainly requires a relativistic relative velocity. Use Eq. (22-15).

Solution $v_{\text{rel}} > 0$ since the source of the light is stationary (observer approaching the source). Let $v_{\text{rel}} = v$ for simplicity. Find v.

$$f_o = f_s\sqrt{\frac{1+\frac{v}{c}}{1-\frac{v}{c}}}$$

$$\left(\frac{f_o}{f_s}\right)^2 = \frac{1+\frac{v}{c}}{1-\frac{v}{c}}$$

$$\left(\frac{f_o}{f_s}\right)^2 - \left(\frac{f_o}{f_s}\right)^2\frac{v}{c} = 1+\frac{v}{c}$$

$$\left(\frac{f_o}{f_s}\right)^2 - 1 = \frac{v}{c}\left[1+\left(\frac{f_o}{f_s}\right)^2\right]$$

$$v = c\frac{(f_o/f_s)^2-1}{(f_o/f_s)^2+1} = c\frac{(\lambda_s/\lambda_o)^2-1}{(\lambda_s/\lambda_o)^2+1} = (3.00\times10^8 \text{ m/s})\frac{(630/530)^2-1}{(630/530)^2+1} = \boxed{5\times10^7 \text{ m/s}}$$

61. Strategy Use the Doppler shift formula, Eq. (22-15), and the binomial approximations, since the relative speed is small compared to c. The cars are receding, so the relative speed is negative.

Solution

(a) Since the speeder is moving away from the police car, the frequency of the microwaves observed by the police is less than that of the emitted microwaves. The microwaves must travel farther for a fixed speed. Therefore, $\boxed{f_1}$ is larger.

(b) Find the frequency of the pulse as received by the speeder.

$$f_{\text{at speeder}} = f_{as} = f_1\sqrt{\frac{1+v_{\text{rel}}/c}{1-v_{\text{rel}}/c}}$$

Find the frequency of the pulse as received by the police.

$$f_2 = f_{as}\sqrt{\frac{1+v_{\text{rel}}/c}{1-v_{\text{rel}}/c}} = f_1\left(\frac{1+v_{\text{rel}}/c}{1-v_{\text{rel}}/c}\right) \approx f_1\left(1+\frac{v_{\text{rel}}}{c}\right)\left(1+\frac{v_{\text{rel}}}{c}\right) \approx f_1\left[1+\left(\frac{v_{\text{rel}}}{c}\right)^2\right] \approx f_1\left(1+\frac{2v_{\text{rel}}}{c}\right), \text{ so}$$

$$f_2 - f_1 \approx f_1\left(1+\frac{2v_{\text{rel}}}{c}\right) - f_1 = f_1\left(1+\frac{2v_{\text{rel}}}{c}-1\right) = \frac{2v_{\text{rel}}f_1}{c} = \frac{2(-43.0 \text{ m/s})(36.0\times10^9 \text{ Hz})}{2.998\times10^8 \text{ m/s}} = \boxed{-10.3 \text{ kHz}}.$$

65. Strategy Since the normal to the surface of the detector makes an angle with the rays of the Sun, the effective area of the detector is reduced by a factor of $\cos 30.0°$. Use the definition of power and the relationship between intensity and power.

Solution Find how long it takes for the detector to measure 420 kJ of energy.

$$\langle P \rangle = IA = \frac{\Delta U}{\Delta t}, \text{ so } \Delta t = \frac{\Delta U}{IA} = \frac{420,000 \text{ J}}{(1.00\times10^3 \text{ W/m}^2)(5.00 \text{ m})^2 \cos 30.0°} = \boxed{19 \text{ s}}.$$

69. Strategy (a) Refer to Figure 22.5. (b) The length of the pulse is equal to the distance EM radiation travels in 10.0 ps. (c) Divide the length of a pulse by the wavelength to find the number that fit in one pulse.

Solution

(a) According to Figure 22.5, the pulse is in the $\boxed{\text{UV}}$ part of the EM spectrum.

(b) $\text{length} = c\Delta t = (3.00\times10^8 \text{ m/s})(10.0\times10^{-12} \text{ s}) = \boxed{0.300 \text{ cm}}$

(c) $\dfrac{\text{length of pulse}}{\text{wavelength}} = \dfrac{0.300\times10^{-2} \text{ m}}{193\times10^{-9} \text{ m}} = \boxed{15,500}$

73. Strategy To find the intensity of the transmitted light, use Eq. (22-14b) to find the transmitted intensity of each sheet and combine the results.

Solution The angle of polarization of the first sheet is θ_1. The transmitted intensity is $I_1 = I_0 \cos^2 \theta_1$. Light exits the first polarizer at an angle θ_1 relative to the original polarization direction. The polarization angle of the second sheet is θ_2 relative to the original polarization direction, so the polarization angle of the second sheet relative to the first polarizer is $\theta_2 - \theta_1$. The transmitted intensity for the second sheet is

$I_2 = I_1 \cos^2(\theta_2 - \theta_1) = I_1 \cos^2(\theta_1 - \theta_2)$. Combining the equations gives $I_2 = I_0 \cos^2 \theta_1 \cos^2(\theta_1 - \theta_2) = I$, the final transmitted intensity.

77. Strategy The rms value for the electric field is $E_{\text{rms}} = E_{\text{m}}/\sqrt{2}$. The average energy density is related to the electric field by $\langle u \rangle = \epsilon_0 E_{\text{rms}}^2 = \epsilon_0 E_{\text{m}}^2/2$. The intensity is $I = \langle u \rangle c$.

Solution Find the average energy density.

$$\langle u \rangle = \frac{1}{2}[8.854\times10^{-12} \text{ C}^2/(\text{N}\cdot\text{m}^2)](32.0\times10^{-3} \text{ V/m})^2 = \boxed{4.53\times10^{-15} \text{ J/m}^3}$$

Compute the intensity.

$$I = \langle u \rangle c = (4.53\times10^{-15} \text{ J/m}^3)(3.00\times10^8 \text{ m/s}) = \boxed{1.36\times10^{-6} \text{ W/m}^2}$$

Chapter 23

REFLECTION AND REFRACTION OF LIGHT

Conceptual Questions

1. In specular reflection, radiation incident on a surface at a given angle always reflects at the same angle. Specular reflection occurs for materials with surface features that are small compared to the wavelength of the radiation. For visible light, such materials include polished mirrors, lenses, and metals. In diffuse reflection, radiation incident on a surface at a given angle is not necessarily reflected at the same angle. Diffuse reflection occurs for materials with surface features that are large compared to the wavelength of the reflected radiation. Examples of such materials for visible light include rock, wood, and most other items encountered in the natural world.

5. The rough surface of "no glare glass" results in diffuse reflection instead of specular reflection that is indicative of smooth glass. As a result, light rays from a bright source are reflected in various directions so that no focused image of the source is formed.

9. **(a)** The image is upright.

(b) The image is inverted.

(c) The real image is formed at a distance greater than $2f$, and therefore, cannot be seen.

13. A virtual image would be formed on the left side of the lens, while a real image would be formed on the right side.

17. An image formed by a projector needs to be located on the viewing screen, where the light is diffusely reflected allowing the image to be seen from any angle. Since the light from the projector actually comes from the location of the image, it is a real image. For a camera, the image must be focused on the film so that the exposed film forms a copy of the actual image. Thus, the camera must form a real image as well. The lens in the eye works like a camera, forming a real image on the retina.

21. Because the index of refraction of air is less than the index of refraction of the slide, refraction occurs. Thus, less light can be collected than if the gap is filled with oil with the same index of refraction as the slide.

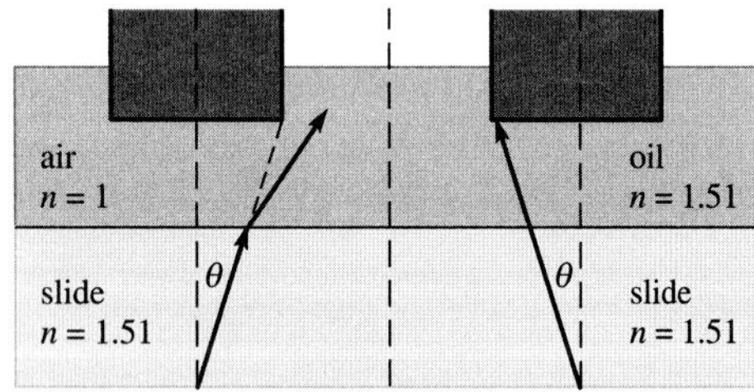

Problems

1. **Strategy** Every point on a wavefront is considered a source of spherical wavelets. A surface tangent to the wavelets at a later time is the wavefront at that time.

   Solution The planar wavefront incident on the reflecting wall at normal incidence is transmitted through the opening and reflected at the edges of the opening.

   > On the transmitted side is one 4-cm-long planar wave.
   > On the incident side are two hemispherical waves.

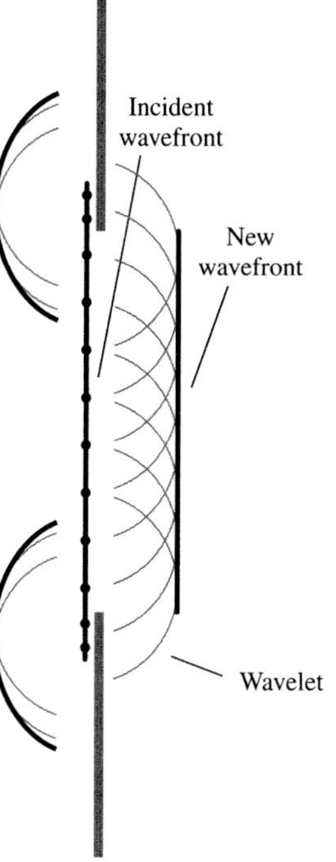

3. **Strategy** With respect to the normal at the point of incidence, the angle of incidence equals the angle of reflection and the reflected ray lies in the same plane, the plane of incidence, as the incident ray and the normal. Every point on a wavefront is considered a source of spherical wavelets. A surface tangent to the wavelets at a later time is the wavefront at that time.

   Solution The ray diagram and a wavefront for a plane wave reflected from the surface of a sphere.

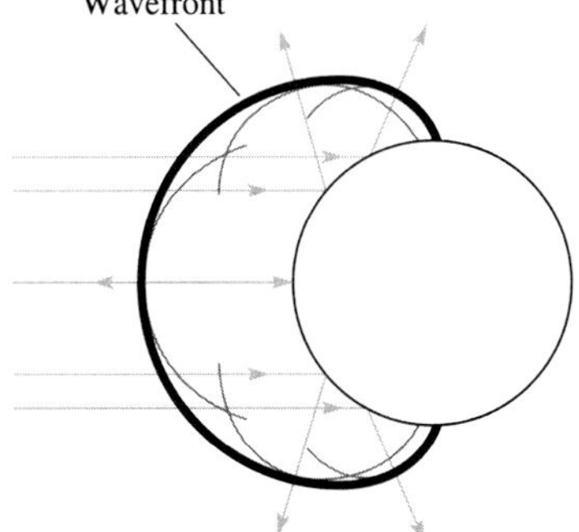

5. **Strategy** The normal is perpendicular to the surface of the pond. Use the laws of reflection.

   Solution

   (a) The angle of incidence is $\theta_i = 90° - 35° = \boxed{55°}$.

   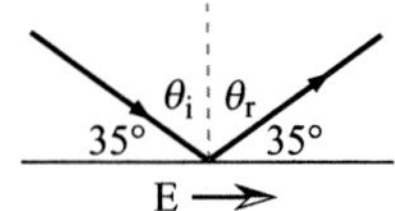

   (b) The angle of reflection is $\theta_r = \theta_i = \boxed{55°}$.

   (c) $55°$ from the normal is equivalent to $35°$ above the horizontal. The sun is above the western horizon. So, the reflected rays are traveling at an angle $\boxed{35° \text{ above the surface of the pond to the east}}$.

9. **Strategy** Use Snell's law, Eq. (23-4).

   Solution Find the angle with respect to the normal at which the fish sees the Sun.
   $$n_i \sin \theta_i = n_t \sin \theta_t, \text{ so } \theta_t = \sin^{-1}\left(\frac{n_i}{n_t}\sin \theta_i\right) = \sin^{-1}\left(\frac{1.000}{1.333}\sin 30.0°\right) = \boxed{22.0°}.$$

11. **Strategy** Use Snell's law, Eq. (23-4).

    Solution Find the angle with respect to the normal at which light is traveling in the aqueous fluid.
    $n_{air} \sin \theta_{air} = n_{cornea} \sin \theta_{cornea}$ at the air-cornea interface.
    $n_{cornea} \sin \theta_{cornea} = n_{aq} \sin \theta_{aq}$ at the cornea-aqueous fluid interface.
    So, $n_{aq} \sin \theta_{aq} = n_{air} \sin \theta_{air}$. Solve for θ_{aq}.
    $$\theta_{aq} = \sin^{-1}\left(\frac{n_{air}}{n_{aq}}\sin \theta_{air}\right) = \sin^{-1}\left(\frac{1.000}{1.336}\sin 17.5°\right) = \boxed{13.0°}$$

13. **Strategy** Use Snell's law, Eq. (23-4). Let the index of refraction for air be n, and let n_1, n_2, n_3, and n_4 be 1.20, 1.40, 1.32, and 1.28, respectively. Let θ be the angle of emergence.

    Solution Find the angle the beam makes with the normal when it emerges into the air after passing through the entire stack of four flat transparent materials.
    $$n \sin 60.0° = n_1 \sin \theta_1 = n_2 \sin \theta_2 = n_3 \sin \theta_3 = n_4 \sin \theta_4 = n \sin \theta, \text{ so } \theta = \boxed{60.0°}.$$

15. **Strategy** Draw a diagram. Use Snell's law, Eq. (23-4).

    Solution Find θ_2.
    Relate θ_1 and θ_3.
    $n_1 \sin(90° - \theta_1) = n_3 \sin \theta_3$
    Relate θ_2 and θ_3.
    $n_2 \sin(90° - \theta_2) = n_3 \sin \theta_3$
    Eliminate $n_3 \sin \theta_3$ and solve for θ_2.
    $n_2 \sin(90° - \theta_2) = n_1 \sin(90° - \theta_1)$

    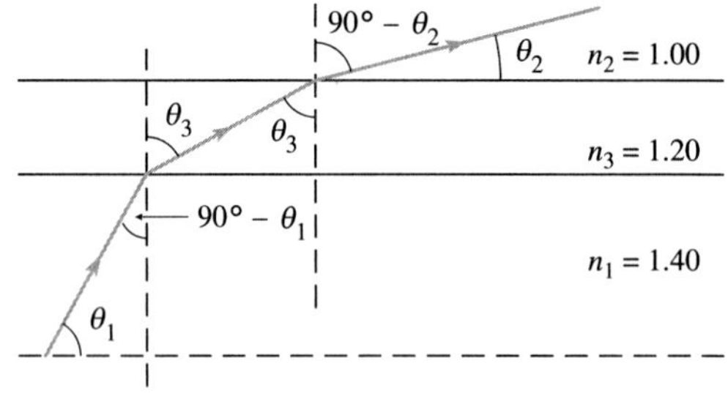

    $$90° - \theta_2 = \sin^{-1}\left[\frac{n_1}{n_2}\sin(90° - \theta_1)\right]$$

    $$\theta_2 = 90° - \sin^{-1}\left[\frac{1.40}{1.00}\sin(90° - 49.0°)\right] = \boxed{23.3°}$$

17. Strategy Use Snell's law, Eq. (23-4), and Eq. (22-2). Let $\theta_2 = \theta_t$.

Solution Find the angle of the transmitted ray.

$$n_1 \sin\theta_1 = n_2 \sin\theta_2, \text{ so } \sin\theta_2 = \frac{n_1}{n_2}\sin\theta_1 = \frac{c/v_1}{c/v_2}\sin\theta_1 = \frac{v_2}{v_1}\sin\theta_1 \text{ and } \theta_2 = \sin^{-1}\left[\frac{0.67c}{0.80c}\sin 12.0^\circ\right] = \boxed{10^\circ}.$$

19. Strategy Draw a diagram. Use geometry and Snell's law, Eq. (23-4).

Solution Find α and θ_1.

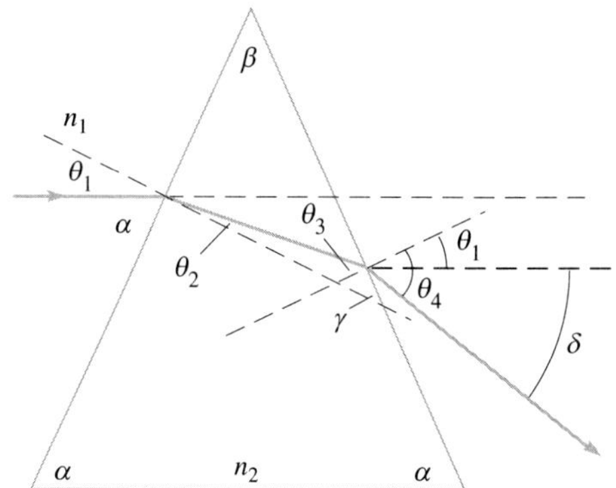

$$\beta + 2\alpha = 180^\circ \text{ implies that } \alpha = \frac{180^\circ - \beta}{2} = \frac{180^\circ - 30.0^\circ}{2} = 75.0^\circ.$$

$\theta_1 = 90^\circ - \alpha = 90^\circ - 75.0^\circ = 15.0^\circ$ is the angle of incidence.

Find θ_2. Here, $n_1 = 1.000$ and $n_2 = 1.517$.

$n_1 \sin\theta_1 = n_2 \sin\theta_2$, so

$$\theta_2 = \sin^{-1}\left(\frac{n_1}{n_2}\sin\theta_1\right) = \sin^{-1}\left(\frac{1.000}{1.517}\sin 15.0^\circ\right).$$

Find θ_3.

From the figure, $90^\circ + \beta + \gamma = 180^\circ$, which implies $\gamma = 90^\circ - \beta$.

$\gamma + (90^\circ + \theta_3) + \theta_2 = 90^\circ - \beta + 90^\circ + \theta_3 + \theta_2 = 180^\circ$, so $\theta_3 = \beta - \theta_2$.

Find θ_4.

$$n_2 \sin\theta_3 = n_1 \sin\theta_4, \text{ so } \theta_4 = \sin^{-1}\left[\frac{n_2}{n_1}\sin\theta_3\right] = \sin^{-1}\left[\frac{n_2}{n_1}\sin(\beta - \theta_2)\right].$$

Next, find δ.

$\delta + \theta_1 = \theta_4$, so

$$\delta = \theta_4 - \theta_1 = \sin^{-1}\left\{\frac{n_2}{n_1}\sin\left[\beta - \sin^{-1}\left(\frac{n_1}{n_2}\sin\theta_1\right)\right]\right\} - \theta_1$$

$$= \sin^{-1}\left\{\frac{1.517}{1.000}\sin\left[30.0^\circ - \sin^{-1}\left(\frac{1.000}{1.517}\sin 15.0^\circ\right)\right]\right\} - 15.0^\circ = \boxed{16.5^\circ}.$$

21. Strategy The speed of light in matter is given by $v = c/n$. Use Snell's law, Eq. (23-4).

Solution

(a) Find the indices of refraction.

$$n_1 \sin\theta_1 = n_2 \sin\theta_2, \text{ so } n_2 = n_1 \frac{\sin\theta_1}{\sin\theta_2}.$$

For red light: $n_{\text{red}} = 1.000\dfrac{\sin 26.00^\circ}{\sin 10.48^\circ} = \boxed{2.410}$

For blue light: $n_{\text{blue}} = 1.000\dfrac{\sin 26.00^\circ}{\sin 10.33^\circ} = \boxed{2.445}$

(b) To get the ratio of the speeds, express the speeds using the indices of refraction.

$$\frac{v_{\text{red}}}{v_{\text{blue}}} = \frac{c/n_{\text{red}}}{c/n_{\text{blue}}} = \frac{n_{\text{blue}}}{n_{\text{red}}} = \frac{\sin 10.48^\circ}{\sin 10.33^\circ} = \boxed{1.014}$$

(c) With no dispersion, all colors would undergo the same refraction. $\boxed{\text{The diamond would be clear, with no color.}}$

23. **Strategy** From Table 23.1, the index of refraction for air is 1.00 and for sapphire it is 1.77. Use Eq. (23-5a).

Solution Calculate the critical angle for sapphire surrounded by air.

$$\theta_c = \sin^{-1}\frac{n_t}{n_i} = \sin^{-1}\frac{1.00}{1.77} = \boxed{34.4^\circ}$$

25. **Strategy** Total internal reflection occurs when the angle of incidence is greater than or equal to the critical angle. Use Eqs. (23-5).

Solution The critical angle is given by $\sin\theta_c = n_t/n_i$. Since the sine of any angle is less than or equal to 1, the index of refraction of the media in which the transmitted rays travel must be less than or equal to that in which the incident rays travel. Since $n_t = 1.4 > 1.2 = n_i$, the answer is $\boxed{\text{no}}$, there is no critical angle.

27. **Strategy** Total internal reflection occurs when the angle of incidence is greater than or equal to the critical angle. The angle of incidence on the back of the prism is 45°. Use Eqs. (23-5).

Solution Compute the critical angle.

$$\theta_c = \sin^{-1}\frac{n_t}{n_i} = \sin^{-1}\frac{1.0}{1.6} = 39^\circ$$

Since the angle of incidence (45°) is greater than the critical angle (39°), no light exits the back of the prism. The light is totally reflected downward, and then passes through the bottom surface ($\theta_i < \theta_c$), with a small amount reflected back into the prism.

29. **Strategy** Total internal reflection occurs when the angle of incidence is greater than or equal to the critical angle. Use the identities (1) $\cos^2\theta + \sin^2\theta = 1$ and (2) $\sin(90^\circ - \theta) = \cos\theta$.

Solution From the figure and identity 2, we see that
$$n\sin\theta_{\max} = n_1\sin(90^\circ - \theta_c) = n_1\cos\theta_c.$$
From the figure, we see that the critical angle is given by $\sin\theta_c = n_2/n_1$.
Using this and identity 1, we have
$$\cos\theta_c = \sqrt{1 - \sin^2\theta_c} = \sqrt{1 - \frac{n_2^2}{n_1^2}}.$$

Substituting for $\cos\theta_c$ in the first equation gives
$$n\sin\theta_{\max} = n_1\cos\theta_c = n_1\sqrt{1 - \frac{n_2^2}{n_1^2}} = \sqrt{n_1^2 - n_2^2},\ \text{which is the numerical aperture.}$$

31. **Strategy** Total internal reflection occurs when the angle of incidence is greater than or equal to the critical angle. Use Eqs. (23-5) and Snell's law, Eq. (23-4).

Solution When the light is incident on the Plexiglas tank, some is transmitted at angle θ_1, so $n\sin\theta_i = n_1\sin\theta_1$ where $n = 1.00$ for air and $n_1 = 1.51$ for Plexiglas. At the Plexiglas carbon disulfide interface, θ_1 is the incident angle and θ_2 is the transmitted angle, so $n_2\sin\theta_2 = n_1\sin\theta_1$ where $n_2 = 1.628$ for carbon disulfide. The ray passes through the carbon disulfide and is incident on the bottom tank-liquid interface at angle θ_2. Here the light must experience total internal reflection, so $\theta_2 = \theta_c = \sin^{-1}\frac{n_1}{n_2}$, or $\frac{n_1}{n_2} = \sin\theta_2$. Find θ_i.

$$n\sin\theta_i = n_1\sin\theta_1 = n_2\sin\theta_2 = n_2\frac{n_1}{n_2} = n_1,\ \text{so}\ \theta_i = \sin^{-1}\frac{n_1}{n} = \sin^{-1}\frac{1.51}{1.00} = \sin^{-1}1.51,\ \text{or}\ \sin\theta_i = 1.51,$$

which is impossible since $\sin\theta \le 1$ for all θ. Thus, there is $\boxed{\text{no}}$ angle θ_i for which light is transmitted into the carbon disulfide but not into the Plexiglas at the bottom of the tank.

33. Strategy Use Eq. (23-5a).

Solution Find the index of refraction of the core of the optical fiber.

$$\theta_c = \sin^{-1}\frac{n_t}{n_i}, \text{ so } \sin\theta_c = \frac{n_t}{n_i} \text{ and } n_i = \frac{n_t}{\sin\theta_c} = \frac{1.20}{\sin 45.0°} = \boxed{1.70}.$$

37. (a) Strategy The reflected light is totally polarized when the angle of incidence equals Brewster's angle. Use Eq. (23-6).

Solution Compute Brewster's angle.

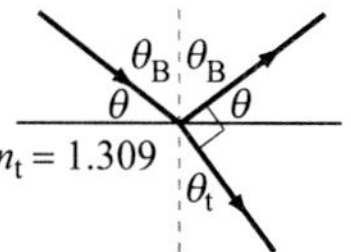

$$\theta_B = \tan^{-1}\frac{n_t}{n_i} = \tan^{-1}\frac{1.309}{1.000} = 52.62°$$

The angle with respect to the horizontal is the complement of this angle.

$$\theta = 90° - 52.62° = \boxed{37.38°}$$

(b) Strategy and Solution For Brewster's angle, the reflected light is polarized $\boxed{\text{perpendicular to the plane of incidence.}}$

(c) Strategy When the angle of incidence is Brewster's angle, the incident and transmitted rays are complementary.

Solution Find the angle of transmission.

$$\theta_t = 90° - \theta_i = 90° - 52.62° = 37.38°$$

The angle below the horizontal is the complement of this angle.

$$90° - 37.38° = \boxed{52.62°}$$

39. Strategy The equation derived in Example 23.4 can be used for this problem (with $n_w = n_d$) since $n_d > n_a$.

Solution Find the depth of the defect.

$$\frac{\text{apparent depth}}{\text{actual depth}} = \frac{n_a}{n_d} = \frac{1.000}{2.419}, \text{ so actual depth = apparent depth} \cdot 2.419 = 2.0 \text{ mm} \cdot 2.419 = \boxed{4.8 \text{ mm}}.$$

41. Strategy Use Snell's law with $n_a = 1.000$, $n_w = 1.333$, and $\theta_a = 90°$.

Solution Find θ_w.

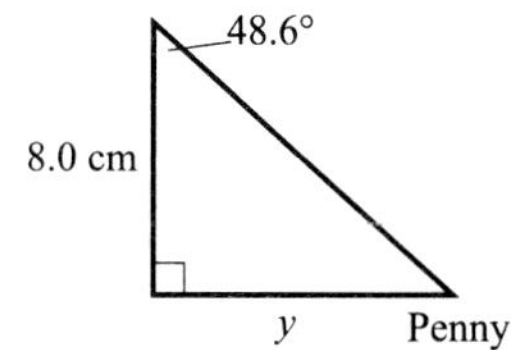

$$n_w \sin\theta_w = n_a \sin\theta_a, \text{ so}$$

$$\theta_w = \sin^{-1}\left(\frac{n_a}{n_w}\sin 90°\right) = \sin^{-1}\left[\frac{n_a}{n_w}(1)\right] = \sin^{-1}\left(\frac{1.000}{1.333}\right) = 48.6°.$$

A right triangle is formed by the actual location of the penny, the apparent location of the penny, and the bottom of the bowl directly below the apparent location. Let y be the distance from the latter location to the actual location of the penny, then $y = (8.0 \text{ cm}) \tan 48.6° = 9.1 \text{ cm}$. Since the penny appears to be 3.0 cm from the edge of the bowl, the horizontal distance between the penny and the edge of the bowl is 9.1 cm + 3.0 cm = $\boxed{12.1 \text{ cm}}$.

43. Strategy For a plane mirror, a point source and its image are at the same distance from the mirror (on opposite sides) and both lie on the same normal line. Draw a ray diagram. Use geometry and the laws of reflection.

Solution Suppose the mirror is hung at the proper height (see the figure). The top of Daniel's head is at point A, his eyes are at point B, and his shoes are at point D. Lines AD and CE are perpendicular, and $\theta_i = \theta_r$,

so triangles BCE and DCE are congruent and $BC = CD = \dfrac{1}{2}BD$.

Similarly, triangles CDE and FED are congruent, so

$$EF = CD = \frac{1}{2}BD = \frac{1}{2} \cdot 1.82 \text{ m} = \boxed{0.91 \text{ m}}.$$

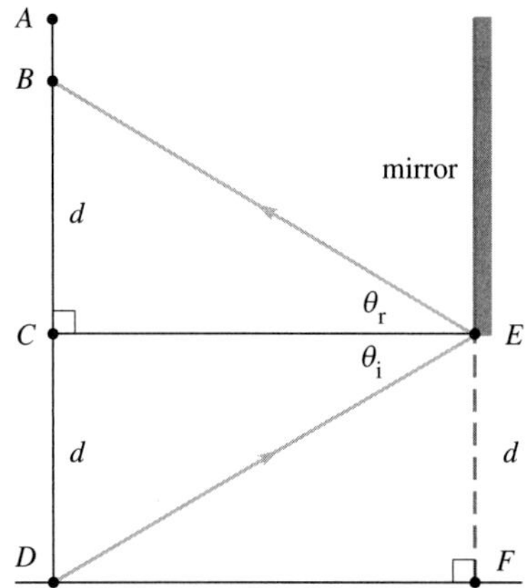

45. Strategy For a plane mirror, a point source and its image are at the same distance from the mirror (on opposite sides) and both lie on the same normal line.

Solution Since Gustav is holding the match, the distance to the image is twice the distance from Gustav to the mirror, so the distance from Gustav to the mirror is half this distance.

$$\frac{1}{2} \cdot 4 \text{ m} = \boxed{2 \text{ m}}$$

47. Strategy For a plane mirror, a point source and its image are at the same distance from the mirror (on opposite sides) and both lie on the same normal line.

Solution Maurizio sees three images by looking straight into each mirror. He sees three other images by looking where each pair of mirrors meet (left wall and right wall, left wall and ceiling, right wall and ceiling). He sees one more image by looking at the corner where all three mirrors meet. $\boxed{\text{He sees 7 images total.}}$

49. Strategy (a) Use Figure 23.27a and similar triangles. (b) Use any pair of rays in Figure 23.27a. Draw a diagram.

Solution

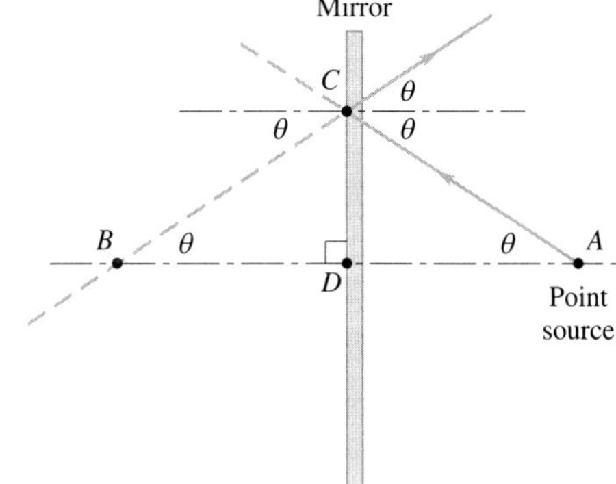

(a) In the figure, a ray from the point source strikes the mirror at point C and reflects. The line $\overrightarrow{AD}$ is normal to the surface of the mirror. When the reflected ray is traced back behind the mirror, it crosses $\overrightarrow{AD}$ at point B. By geometry and the laws of reflection, $\angle CAB$ and $\angle ABC$ have the same measure. Thus, the two right triangles CAD and CBD are similar and $AD = BD$. Since AD is the same for all rays, we conclude that all rays, when extended behind the mirror, will meet at point B.

(b) Point B is the image point, since all rays appear to come from it. Thus, the image point lies on a line through the object and perpendicular to the mirror, $\overrightarrow{AD}$. Since $BD = AD$ as shown in part (a), the object and image distances are equal.

51. Strategy The mirror is concave, so $p = 20.0$ cm and $f = 5.00$ cm. Find the image distance q using the mirror equation.

Solution Find the position of the image.

$$\frac{1}{p} + \frac{1}{q} = \frac{1}{f}, \text{ so } q = \frac{pf}{p-f} = \frac{(20.0 \text{ cm})(5.00 \text{ cm})}{20.0 \text{ cm} - 5.00 \text{ cm}} = 6.67 \text{ cm}.$$

The image is formed $\boxed{6.67 \text{ cm in front of the mirror}}$.

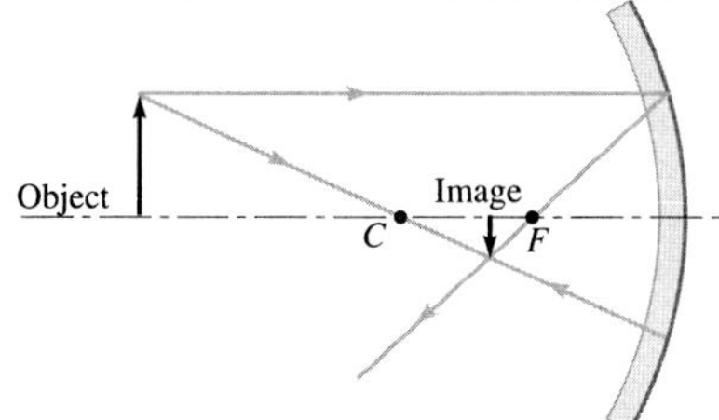

53. Strategy Since the image is real, the image distance is positive and the image is inverted (negative). The image is twice the size of the object, so $h' = -2h$. Use the magnification and mirror equations.

Solution Relate the object and image locations.

$$\frac{h'}{h} = \frac{-2h}{h} = -\frac{q}{p}, \text{ so } q = 2p.$$

Find the distance of the object from the mirror.

$$\frac{1}{p} + \frac{1}{q} = \frac{1}{p} + \frac{1}{2p} = \frac{3}{2p} = \frac{1}{f}, \text{ so } p = \frac{3}{2}f = \frac{3}{2}\left(\frac{R}{2}\right) = \frac{3}{4}(25.0 \text{ cm}) = 18.8 \text{ cm}.$$

The object is $\boxed{18.8 \text{ cm in front of the mirror}}$.

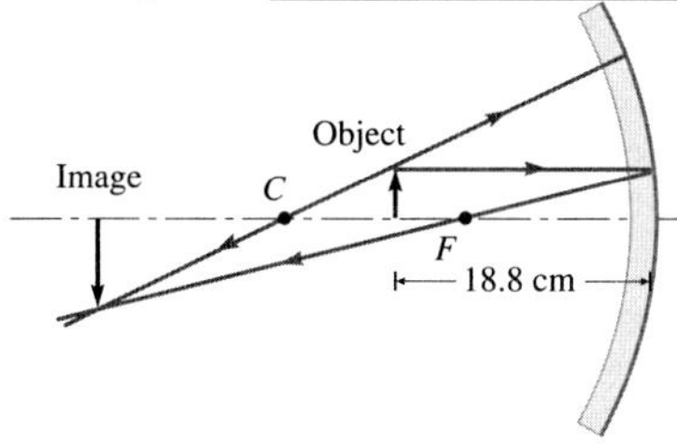

57. Strategy Refer to the derivation of the magnification for a concave mirror. Draw a ray diagram using a ray that is not one of the three principal rays.

Solution The ray diagram.

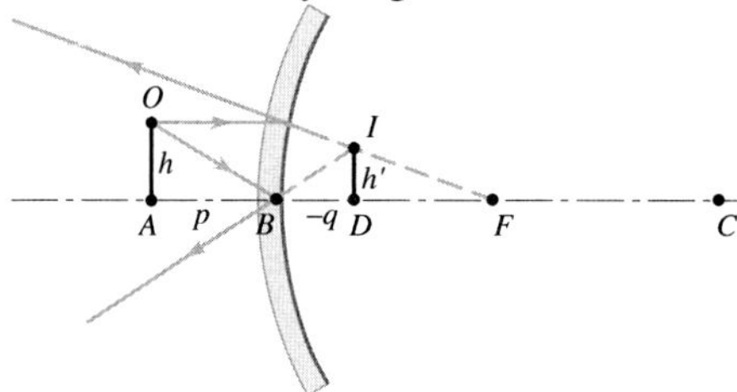

In the diagram, the two right triangles OAB and IDB are similar, so $\dfrac{h}{p} = \dfrac{h'}{-q}$ or $\dfrac{h'}{h} = \dfrac{-q}{p}$. The negative sign is included since $h/p > 0$; $q < 0$ since the image is behind the mirror, so $h'/q < 0$, and $-h'/q > 0$. Combining this with the magnification definition $h' = mh$ gives $m = \dfrac{h'}{h} = -\dfrac{q}{p}$.

61. (a) Strategy Use the thin lens equation to find the object distance. Then, draw the ray diagram.

Solution Find the object distance.

$$\frac{1}{p}+\frac{1}{q}=\frac{1}{f}, \text{ so } p=\frac{qf}{q-f}=\frac{(5.00 \text{ cm})(3.50 \text{ cm})}{5.00 \text{ cm}-3.50 \text{ cm}}=\boxed{11.7 \text{ cm}}.$$

The diagram is shown.

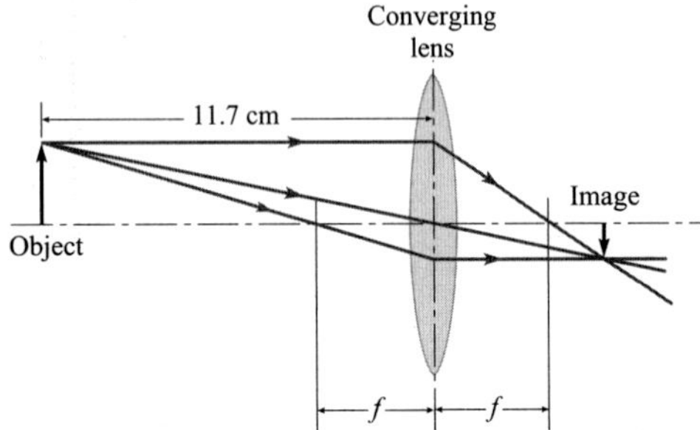

(b) Strategy and Solution Light rays actually pass through the image location, so the image is $\boxed{\text{real}}$.

(c) Strategy Use the magnification equation.

Solution Compute the magnification of the image.

$$m=-\frac{q}{p}=-\frac{5.00 \text{ cm}}{11.667 \text{ cm}}=\boxed{-0.429}$$

63. Strategy Sketch a ray diagram for a converging lens using the three principal rays. Place the object twice the focal length from the lens.

Solution As can be seen in the figure, when an object is placed twice the focal length away from a converging lens, an inverted, real, and same-sized image is formed.

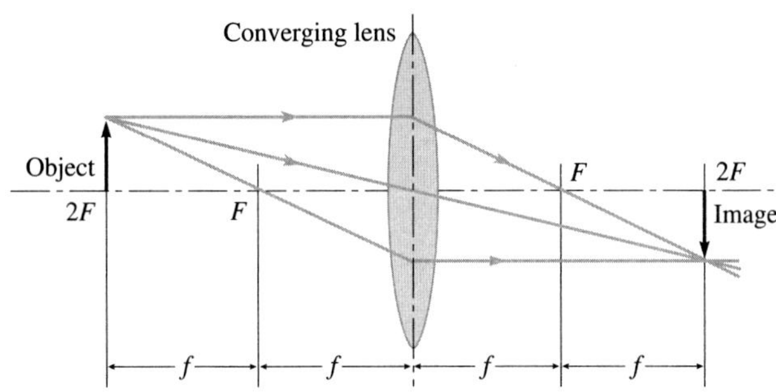

65. Strategy Sketch a ray diagram for a converging lens using the three principal rays. Place the object a distance equal to the focal length from the lens.

Solution As can be seen in the figure, when an object is placed at the focal point of a converging lens, the rays emerging from the lens are parallel to each other, so the image is at infinity.

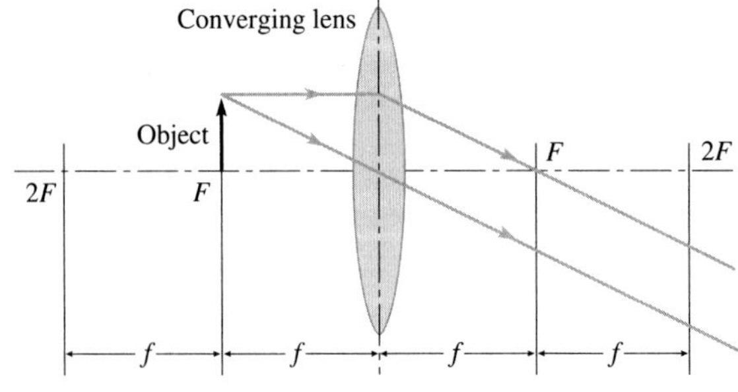

67. Strategy Draw a ray diagram using the three principal rays.

Solution Find the height and position of the image.

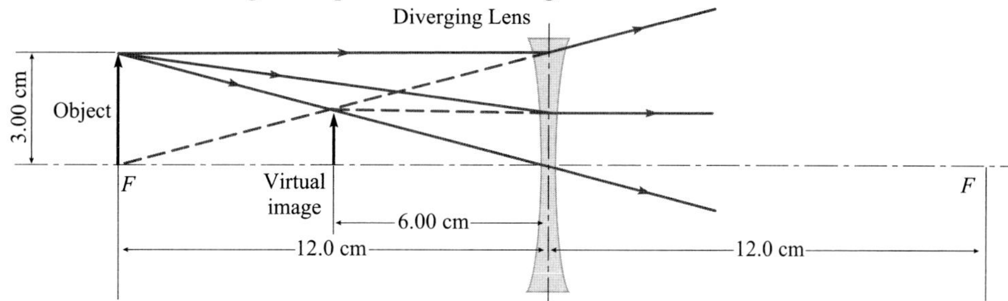

The image is located 6.00 cm from the lens on the same side as the object and has a height of 1.50 cm.

69. Strategy Use the thin lens and magnification equations.

Solution

(a) Solve the thin lens equation for q.

$$\frac{1}{p}+\frac{1}{q}=\frac{1}{f}, \text{ so } q=\frac{pf}{p-f}.$$

For $f = 8.00$ cm and $p = 5.00$ cm , we have

$$q=\frac{(5.00 \text{ cm})(8.00 \text{ cm})}{5.00 \text{ cm}-8.00 \text{ cm}}=-13.3 \text{ cm}.$$

Since q is negative, the image is virtual and on the same side of the lens as the object. The magnification is

$$m=-\frac{q}{p}=-\frac{-13.33 \text{ cm}}{5.00 \text{ cm}}=2.67.$$

Since m is positive and greater than 1, the image is upright and enlarged. The same process is used for the other object distances, and the results are summarized in the table on the following page.

p (cm)	q (cm)	m	Real or virtual	Orientation	Relative size
5.00	−13.3	2.67	virtual	upright	enlarged
14.0	18.7	−1.33	real	inverted	enlarged
16.0	16.0	−1.00	real	inverted	same
20.0	13.3	−0.667	real	inverted	diminished

(b) The image height can be found from the magnification equation.
$h' = mh$

For $p = 5.00$ cm , $h = 4.00$ cm , and $m = 2.67$ we have $h' = 2.67 \cdot 4.00$ cm $= \boxed{10.7 \text{ cm}}$.

For $p = 20.0$ cm , $h = 4.00$ cm , and $m = -0.667$ we have $h' = -0.667 \cdot 4.00$ cm $= \boxed{-2.67 \text{ cm}}$.

73. Strategy Use the thin lens and magnification equations. First find p and q. In this arrangement, p and q are positive so $m = -q/p$ is negative and the image is inverted.

Solution

(a) We calculate m using the given heights.

$$m = \frac{h'}{h} = \frac{-60.0 \text{ cm}}{2.40 \text{ cm}} = -25.0 = \frac{-q}{p}, \text{ so } q = 25.0p.$$

Substitute this expression for q and $f = 12.0$ cm into the thin lens equation.

$$\frac{1}{p} + \frac{1}{q} = \frac{1}{p} + \frac{1}{25.0p} = \frac{1.04}{p} = \frac{1}{f} = \frac{1}{12.0 \text{ cm}}, \text{ so } p = 1.04(12.0 \text{ cm}) = 12.48 \text{ cm and } q = 25.0p = 312 \text{ cm}.$$

The distance from the slide to the screen is $p + q = 12.48 \text{ cm} + 312 \text{ cm} = \boxed{3.24 \text{ m}}$

(b) Since $\dfrac{1}{f} = \dfrac{1}{p} + \dfrac{1}{q}$, moving the screen away makes q larger, so p must be smaller. To maintain focus, the lens is moved $\boxed{\text{closer}}$ to the slide.

77. Strategy Redraw the diagram, labeling the vertices of similar triangles.

Solution In the figure, triangle ABF and triangle ACG are similar, so

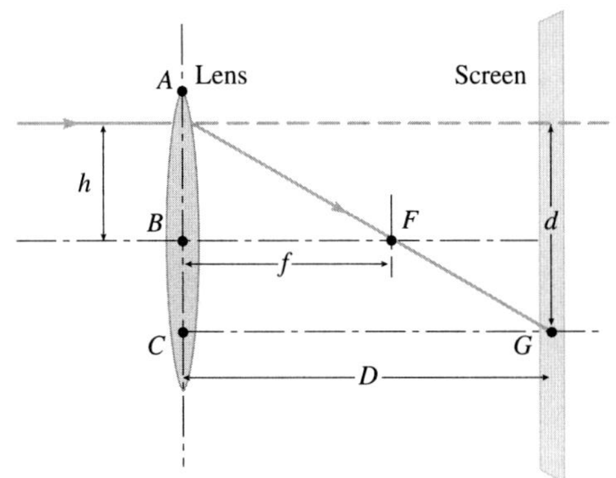

$$\frac{h}{f} = \frac{-d}{D}, \text{ so } f = -\frac{h}{d}D.$$

For paraxial rays, the slope of the d vs. h graph is constant. The middle three data points reflect this case. Find the slope.

$$m = \frac{\Delta d}{\Delta h} = \frac{-1.0 - 1.0}{0.5 - (-0.5)} = \frac{-2.0}{1.0} = -2.0$$

For a constant slope, $\dfrac{\Delta d}{\Delta h} = \dfrac{d}{h}$, so

$$f = -\frac{1}{-2.0}D = 0.50(1.0 \text{ m}) = \boxed{50 \text{ cm}}.$$

81. (a) Strategy The image is virtual, so the image distance is negative. Use the magnification equation.

Solution Find the object distance.

$$m = \frac{h'}{h} = -\frac{q}{p}, \text{ so } p = -\frac{h}{h'}q = -\frac{8.0 \text{ cm}}{3.5 \text{ cm}}(-4.0 \text{ cm}) = \boxed{9.1 \text{ cm}}.$$

(b) Strategy and Solution The image is upright, virtual, smaller than the object, and closer to the mirror than the object. The mirror is $\boxed{\text{convex}}$.

(c) Strategy The radius of curvature is twice the absolute value of the focal length. Use the mirror equation.

Solution Find the focal length of the mirror.

$$\frac{1}{p} + \frac{1}{q} = \frac{1}{f}, \text{ so } f = \left(\frac{1}{p} + \frac{1}{q}\right)^{-1} = \left(-\frac{h'}{hq} + \frac{1}{q}\right)^{-1} = q\left(1 - \frac{h'}{h}\right)^{-1} = (-4.0 \text{ cm})\left(1 - \frac{3.5}{8.0}\right)^{-1} = \boxed{-7.1 \text{ cm}}.$$

Compute the radius of curvature.

$$R = 2|f| = 2(7.1 \text{ cm}) = \boxed{14 \text{ cm}}$$

85. Strategy Use the mirror and magnification equations.

Solution

(a) The image appears to be behind the mirror, where no light passes through it, so it is a $\boxed{\text{virtual}}$ image.

(b) The image is virtual, so $q = -|q| < 0$. The image is upright and diminished, so $0 < m < 1$. This implies $|q| < p$ since $m = -\dfrac{q}{p} = \dfrac{|q|}{p} < 1$. Using the mirror equation, we have

$$\frac{1}{q} + \frac{1}{p} = -\frac{1}{|q|} + \frac{1}{p} = \frac{1}{f}, \text{ so } \frac{p}{|q|} - 1 = -\frac{p}{f}.$$

Since $\dfrac{|q|}{p} < 1$, $\dfrac{p}{|q|} > 1$. So, $\dfrac{p}{|q|} - 1 > 0$, and $-\dfrac{p}{f} > 0$, or $f < 0$ since $p > 0$. Since the focal length is negative, the mirror is $\boxed{\text{convex}}$.

(c) In a convex mirror, the image is upright and smaller than the object, so $m < 1$ and $pm < p$. But $pm = -q$, so $-q < p$. Since q is negative, $|q| < p$ and the image is closer to the mirror than the object. $\boxed{\text{The image seems to be farther away than the object because its angular size is smaller than that of the object.}}$

87. Strategy and Solution For a plane mirror, the image distance equals the object distance. Since speed is distance divided by time, the image speed equals the object speed: $\boxed{0.8 \text{ m/s}}$.

89. Strategy Use the laws of reflection.

Solution In (1), a ray strikes a plane mirror. The angles of incidence and reflection are both θ. In (2), the mirror has been rotated relative to the normal by an angle α. The angle of incidence must equal the angle of reflection, so $\theta_2 = \theta + \alpha$. Then $\theta_3 = \alpha + \theta_2 = \alpha + \theta + \alpha = \theta + 2\alpha$.

Since the angle of the original reflected ray was θ, the reflected ray has rotated through an angle of 2α.

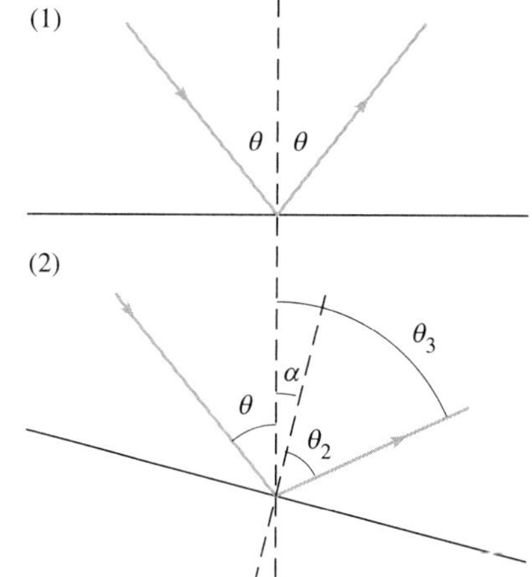

93. Strategy Since the image is virtual, q must be negative. Use the thin lens equation.

Solution Find the focal length.
$$\frac{1}{p} + \frac{1}{q} = \frac{1}{f}, \text{ so } f = \frac{pq}{p+q} = \frac{(10.0 \text{ cm})(-30.0 \text{ cm})}{10.0 \text{ cm} - 30.0 \text{ cm}} = \boxed{15.0 \text{ cm}}.$$
The focal length is positive, so the lens is $\boxed{\text{converging}}$.

97. Strategy Use Eq. (23-5a) to determine the critical angles for each color of light.

Solution Compute the critical angles.

$$\theta_{c\,(red)} = \sin^{-1}\frac{n_t}{n_i} = \sin^{-1}\frac{1.0003}{1.6182} = 38.182°, \quad \theta_{c\,(yellow)} = \sin^{-1}\frac{1.0003}{1.6276} = 37.922°, \text{ and}$$

$$\theta_{c\,(blue)} = \sin^{-1}\frac{1.0003}{1.6523} = 37.258°.$$

$\theta_{c\,(red)}$ and $\theta_{c\,(yellow)}$ are greater than $\theta_i = 37.5°$, so $\boxed{\text{red and yellow}}$ reach the detector.

101. Strategy Use the thin lens and magnification equations.

Solution

(a) Find the image location.

$$\frac{1}{p}+\frac{1}{q}=\frac{1}{f}, \text{ so } q = \frac{1}{\frac{1}{f}-\frac{1}{p}} = \frac{1}{\frac{1}{-20.0 \text{ cm}} - \frac{1}{50.0 \text{ cm}}} = -14.3 \text{ cm}.$$

Find the image height.

$$\frac{h'}{h}=-\frac{q}{p}, \text{ so } h' = -\frac{qh}{p} = -\frac{-14.3 \text{ cm}\cdot 5.0 \text{ cm}}{50.0 \text{ cm}} = \boxed{1.4 \text{ cm}}.$$

(b) Since $h' > 0$, the image is $\boxed{\text{upright}}$.

105. Strategy Use geometry and Snell's law, Eq. (23-4).

Solution From Snell's law,

$n_{air}\sin i = n\sin r$ and $n\sin i' = n_{air}\sin r'$.

Assuming $i = r'$, then $n_{air}\sin i = n_{air}\sin r'$, and we have

$n\sin r = n\sin i'$, which implies $r = i'$.

In the small isosceles triangle in the figure with angles r and i', there is a third unlabeled angle; call it θ. We know that $\theta + r + i' = 180°$. Also, $A + 180° + \theta = 360°$, or $\theta + A = 180°$.

So, $A = r + i'$. In addition, $r = i'$, thus $A = 2r$, or $r = \frac{1}{2}A$. The deviation angle δ is given by $\delta = \delta_1 + \delta_2$, where $\delta_1 = i - r$ and $\delta_2 = r' - i'$. So, $\delta = \delta_1 + \delta_2 = i - r + r' - i'$.

For the minimum deviation, substitute $\delta_{min} = D$, $i = r'$, $r = i'$, and $r = \frac{1}{2}A$, and solve for i.

$$D = i - \frac{1}{2}A + i - \frac{1}{2}A = 2i - A, \text{ so } i = \frac{1}{2}(A + D).$$

Substitute for i and $r = \frac{1}{2}A$ into Snell's law and solve for n.

$$n_{air}\sin i = 1.00\sin\frac{1}{2}(A+D) = n\sin r = n\sin\frac{1}{2}A, \text{ so } n = \frac{\sin\frac{1}{2}(A+D)}{\sin\frac{1}{2}A}.$$

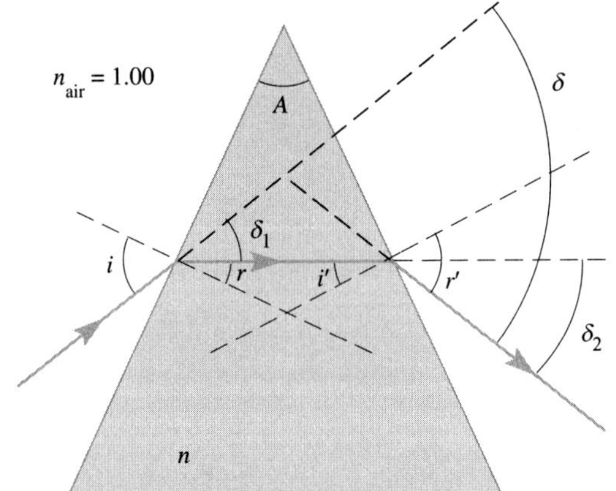

Chapter 24

OPTICAL INSTRUMENTS

Conceptual Questions

1. A camera or slide projector needs to form a real image of a real object. This can only be done with a converging lens. Similarly, the objective lens in a microscope or telescope must be converging so that it can form a real image for viewing with the eyepiece. The eyepiece used to view the image forms a virtual image at a large distance (usually infinity) for ease of viewing. Either a converging or diverging lens can be used as the eyepiece, since both are capable of forming a virtual image.

5. One of the greatest factors limiting the resolution of Earth-based telescopes is the variation in air temperature and density in the atmosphere, which limits the amount of detail that can be seen in astronomical objects. Putting telescopes on mountaintops helps reduce this effect by decreasing the amount of atmosphere that the light rays must pass through.

9. Chromatic aberration is caused by dispersion—the varying value of the index of refraction as a function of frequency. Chromatic aberration in lenses may be reduced by the use of a compound lens made of several lenses with different dispersion relations so that the aberrations caused by one lens offset those caused by another.

13.

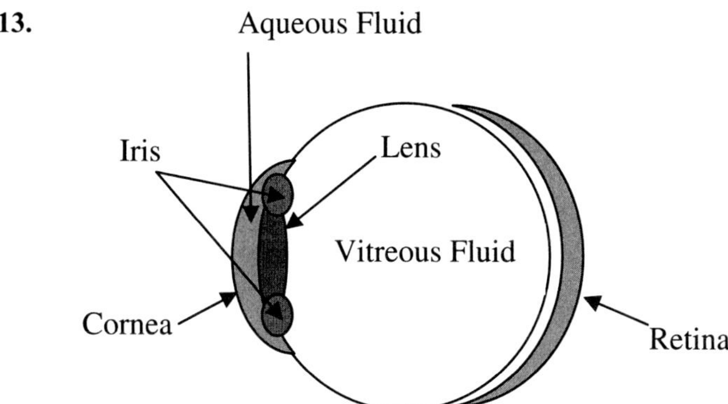

17. The primary advantage of building large astronomical telescopes with mirrors instead of lenses is that it is much easier to construct a precisely built large mirror than a precisely built large lens. Additionally, for an equally sized lens and mirror, the telescope built with a mirror is much smaller because reflections can be used to "fold up" the light path.

Problems

1. **Strategy** Use the lens and total transverse magnification equations.

   Solution

   (a) Find the image due to the first lens.
   $$\frac{1}{p_1}+\frac{1}{q_1}=\frac{1}{f_1}, \text{ so } q_1 = \frac{1}{\frac{1}{f_1}-\frac{1}{p_1}} = \frac{1}{\frac{1}{5.0\text{ cm}}-\frac{1}{12.0\text{ cm}}} = 8.6\text{ cm.}$$
   Find the object distance for the second lens.
   $$p_2 = s - q_1 = 2.0\text{ cm} - 8.6\text{ cm} = -6.6\text{ cm}$$
   Find the location of the final image.
   $$\frac{1}{p_2}+\frac{1}{q_2}=\frac{1}{f_2}, \text{ so } q_2 = \frac{1}{\frac{1}{f_2}-\frac{1}{p_2}} = \frac{1}{\frac{1}{4.0\text{ cm}}-\frac{1}{-6.6\text{ cm}}} = 2.5\text{ cm.}$$
   The final image is $\boxed{2.5\text{ cm past the 4.0-cm lens}}$. The image is $\boxed{\text{real}}$ since q_2 is positive.

   (b) Compute the overall magnification.
   $$m = m_1 m_2 = -\frac{q_1}{p_1}\left(-\frac{q_2}{p_2}\right) = -\frac{8.6\text{ cm}}{12.0\text{ cm}}\times -\frac{2.5\text{ cm}}{-6.6\text{ cm}} = \boxed{-0.27}$$

3. (a) **Strategy** Use the lens equations.

   Solution Find the image location of the first lens.
   $$\frac{1}{p_1}+\frac{1}{q_1}=\frac{1}{f_1}, \text{ so } q_1 = \left(\frac{1}{f_1}-\frac{1}{p_1}\right)^{-1}. \text{ So, the object distance for the second lens is}$$
   $$p_2 = s - q_1 = s - \left(\frac{1}{f_1}-\frac{1}{p_1}\right)^{-1} = 0.880\text{ m} - \left(\frac{1}{0.250\text{ m}}-\frac{1}{1.100\text{ m}}\right)^{-1} = 0.556\text{ m.}$$
   Find the focal length of the second lens.
   $$\frac{1}{p_2}+\frac{1}{q_2}=\frac{1}{f_2}, \text{ so } f_2 = \left(\frac{1}{p_2}+\frac{1}{q_2}\right)^{-1} = \left(\frac{1}{0.556\text{ m}}+\frac{1}{0.150\text{ m}}\right)^{-1} = 0.118\text{ m} = \boxed{11.8\text{ cm}}.$$

   (b) **Strategy** Use the magnification and total transverse magnification equations.

   Solution Find the total magnification of this lens combination.
   $$m = m_1 \times m_2 = -\frac{q_1}{p_1}\times -\frac{q_2}{p_2} = \frac{(0.3235)(0.150)}{(1.100)(0.5565)} = \boxed{0.0793}$$

5. **Strategy** Draw a ray diagram for the system of lenses. Use the lens equations.

Solution The ray diagram:

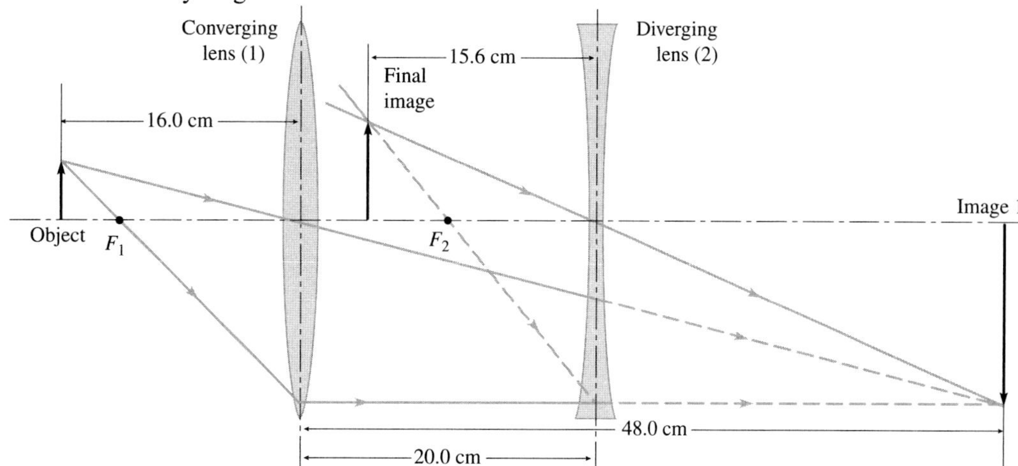

From the figure, the final image is about 15.6 cm left of lens 2. Verify using the lens equations. Find the image due to the first lens (1).

$$\frac{1}{p_1} + \frac{1}{q_1} = \frac{1}{f_1}, \text{ so } q_1 = \frac{1}{\frac{1}{f_1} - \frac{1}{p_1}} = \frac{1}{\frac{1}{12.0 \text{ cm}} - \frac{1}{16.0 \text{ cm}}} = 48.0 \text{ cm}.$$

For the diverging lens (2), the object distance is $p_2 = s - q_1 = 20.0 \text{ cm} - 48.0 \text{ cm} = -28.0 \text{ cm}$. Find q_2.

$$q_2 = \frac{1}{\frac{1}{f_2} - \frac{1}{p_2}} = \frac{1}{\frac{1}{-10.0 \text{ cm}} - \frac{1}{-28.0 \text{ cm}}} = -15.6 \text{ cm}$$

The final image is located $\boxed{15.6 \text{ cm to the left of the diverging lens}}$.

9. **Strategy** Use the lens equations and transverse and total transverse magnification equations.

Solution

(a) Find the image location of the first lens.

$$\frac{1}{p_1} + \frac{1}{q_1} = \frac{1}{f_1}, \text{ so } q_1 = \left(\frac{1}{f_1} - \frac{1}{p_1}\right)^{-1} = \left(\frac{1}{3.70 \text{ cm}} - \frac{1}{6.00 \text{ cm}}\right)^{-1} = 9.65 \text{ cm}.$$

So, the object distance for the second lens is $p_2 = s - q_1 = (24.65 \text{ cm} - 6.00 \text{ cm}) - 9.65 \text{ cm} = 9.00 \text{ cm}$. Find the focal length of the second lens.

$$\frac{1}{p_2} + \frac{1}{q_2} = \frac{1}{f_2}, \text{ so } f_2 = \left(\frac{1}{p_2} + \frac{1}{q_2}\right)^{-1} = \left(\frac{1}{9.00 \text{ cm}} + \frac{1}{32.0 \text{ cm} - 24.65 \text{ cm}}\right)^{-1} = \boxed{4.05 \text{ cm}}.$$

(b) Since its focal length is positive, the lens is $\boxed{\text{converging}}$.

(c) Find the total magnification of this system.

$$m = m_1 \times m_2 = -\frac{q_1}{p_1} \times -\frac{q_2}{p_2} = \frac{(9.65)(32.0 - 24.65)}{(6.00)(9.00)} = \boxed{1.31}$$

(d) Find the image height.

$$m = \frac{h'}{h}, \text{ so } h' = mh = 1.314(12.0 \text{ cm}) = \boxed{15.8 \text{ cm}}.$$

13. Strategy Use the thin lens and transverse magnification equations.

Solution

(a) Find the image distance.
$$\frac{1}{p}+\frac{1}{q}=\frac{1}{f},\ \text{so}\ q=\frac{1}{\frac{1}{f}-\frac{1}{p}}=\frac{1}{\frac{1}{50.0\ \text{mm}}-\frac{1}{3.0\times10^{3}\ \text{mm}}}=\boxed{50.8\ \text{mm}}.$$

(b) Compute the magnification.
$$m=-\frac{q}{p}=-\frac{50.8\ \text{mm}}{3.0\times10^{3}\ \text{mm}}=\boxed{-0.0169}$$

(c) Compute the image height.
$$h'=mh=-0.0169(1.2\ \text{m})=-0.0203\ \text{m}=-20.3\ \text{mm}$$
The image height is $\boxed{20.3\ \text{mm}}$.

15. Strategy The largest possible image size is 7.2 mm. To find p, solve the transverse magnification equation for q and substitute this into the thin lens equation.

Solution Find the image location in terms of the object location.
$$-\frac{q}{p}=\frac{h'}{h},\ \text{so}\ q=-\frac{h'p}{h}.$$
Substitute for q.
$$\frac{1}{p}+\frac{1}{q}=\frac{1}{p}-\frac{h}{h'p}=\frac{1}{p}\left(1-\frac{h}{h'}\right)=\frac{1}{f},\ \text{so}\ p=f\left(1-\frac{h}{h'}\right).$$
Using $f=50.0$ mm, $h=52$ m, and $h'=-7.2$ mm (inverted image) yields
$$p=(50.0\ \text{mm})\left(1-\frac{52\ \text{m}}{-7.2\ \text{mm}}\right)=(0.0500\ \text{m})\left(1+\frac{52\ \text{m}}{0.0072\ \text{m}}\right)=\boxed{360\ \text{m}}.$$

17. Strategy The slide is inverted with respect to the image, so h is negative. Use the thin lens and transverse magnification equations.

Solution Find object location.
$$-\frac{q}{p}=\frac{h'}{h},\ \text{so}\ p=-\frac{qh}{h'}=-\frac{(12.0\ \text{m})(-36\ \text{mm})}{1.50\ \text{m}}=290\ \text{mm}.$$
Find the focal length.
$$\frac{1}{f}=\frac{1}{p}+\frac{1}{q},\ \text{so}\ f=\frac{1}{\frac{1}{p}+\frac{1}{q}}=\frac{1}{\frac{1}{290\ \text{mm}}+\frac{1}{12{,}000\ \text{mm}}}=\boxed{280\ \text{mm}}.$$

19. Strategy Use the lens equations.

Solution

(a) Find the image location of the first lens.

$$\frac{1}{p_1}+\frac{1}{q_1}=\frac{1}{f_1}, \text{ so } q_1=\left(\frac{1}{f_1}-\frac{1}{p_1}\right)^{-1}=\frac{f_1 p_1}{p_1-f_1}=\frac{(3.00 \text{ cm})(4.00 \text{ cm})}{4.00 \text{ cm}-3.00 \text{ cm}}=12.0 \text{ cm}.$$

This is a real image located between the two lenses; so, to display an image, a screen can be placed $\boxed{12.0 \text{ cm to the right of the converging lens}}$.

(b) The object distance for the second lens is $p_2=s-q_1=10.0 \text{ cm}-12.0 \text{ cm}=-2.0 \text{ cm}.$

Find the final image location.

$$\frac{1}{p_2}+\frac{1}{q_2}=\frac{1}{f_2}, \text{ so } q_2=\left(\frac{1}{f_2}-\frac{1}{p_2}\right)^{-1}=\frac{f_2 p_2}{p_2-f_2}=\frac{(-5.00 \text{ cm})(-2.0 \text{ cm})}{-2.0 \text{ cm}-(-5.00 \text{ cm})}=3.3 \text{ cm}.$$

This is a real image located 3.3 cm to the right of the diverging lens; so, to display an image, a screen can be placed $\boxed{3.3 \text{ cm to the right of the diverging lens}}$.

21. Strategy Use the thin lens equation.

Solution Solve for the focal length.

$$\frac{1}{f}=\frac{1}{p}+\frac{1}{q}=\frac{p+q}{pq}, \text{ so } f=\frac{pq}{p+q}.$$

For $p=25.0 \text{ cm}$ and $q=2.00 \text{ cm}$, the focal length is $f=\dfrac{(25.0 \text{ cm})(2.00 \text{ cm})}{25.0 \text{ cm}+2.00 \text{ cm}}=1.85 \text{ cm}.$

For $p=\infty$, $\dfrac{1}{p}=0$, so $\dfrac{1}{f}=\dfrac{1}{q}$. With $q=2.00 \text{ cm}$, $f=2.00 \text{ cm}.$

Thus, the focal length of the lens system must vary between 1.85 cm and 2.00 cm to see objects from 25.0 cm to infinity.

25. (a) Strategy and Solution A focal length range of 1.85 cm to 2.00 cm corresponds to a 2.00 cm lens-retina distance in a normal eye. For a distant object, the focus of the eye can be adjusted to 1.90 cm so that the object is seen clearly, but for close objects, there is not much room for adjustment by the eye muscle to accommodate. Thus, this eye is $\boxed{\text{farsighted}}$.

(b) Strategy Use the thin lens equation.

Solution Solve for the object location.

$$\frac{1}{p}+\frac{1}{q}=\frac{1}{f}, \text{ so } p=\frac{1}{\frac{1}{f}-\frac{1}{q}}.$$

Find p for $q=1.90 \text{ cm}$ and f between 1.85 cm and 1.90 cm.

$$p=\frac{1}{\frac{1}{1.85 \text{ cm}}-\frac{1}{1.90 \text{ cm}}}=70 \text{ cm and } p=\frac{1}{\frac{1}{1.90 \text{ cm}}-\frac{1}{1.90 \text{ cm}}}=\infty.$$

The eye can focus from $\boxed{70 \text{ cm to infinity}}$.

27. Strategy The refractive power of a lens is the reciprocal of the focal length. Use the thin lens equation. Let $p = \infty$ for distant objects and $q = -2.0$ m for a virtual image at Colin's far point.

Solution Find the required refractive power.

$$P = \frac{1}{f} = \frac{1}{q} + \frac{1}{p} = \frac{1}{-2.0 \text{ m}} + \frac{1}{\infty} = \boxed{-0.50 \text{ D}}$$

29. (a) Strategy A hyperopic (farsighted) eye has too large a near point; the refractive power is too small.

Solution Sketch a qualitative ray diagram for Anne's eye.

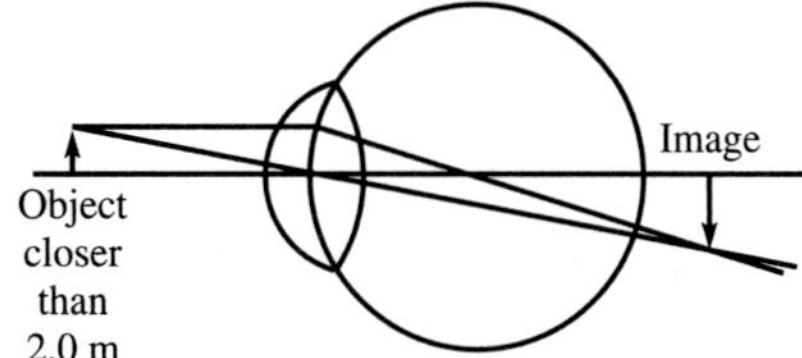

(b) Strategy An object at 20.0 cm should have a virtual image at her near point, 2.0 m. Use $p = 20.0$ cm and $q = -2.0$ m in the thin lens equation.

Solution Find the focal length of the contact lenses.

$$\frac{1}{f} = \frac{1}{p} + \frac{1}{q}, \text{ so } f = \left(\frac{1}{p} + \frac{1}{q} \right)^{-1} = \left(\frac{1}{20.0 \text{ cm}} + \frac{1}{-200 \text{ cm}} \right)^{-1} = \boxed{22.2 \text{ cm}}.$$

33. Strategy When the object is placed at the focal point of the magnifier, the angular magnification is given by N/f. When the object is placed closer to the magnifier ($p < f$), the angular magnification is given by N/p.

Solution Compute the angular magnifications.

(a) $\dfrac{N}{f} = \dfrac{25 \text{ cm}}{15 \text{ cm}} = 1.67$; (b) $\dfrac{N}{p} = \dfrac{25 \text{ cm}}{10 \text{ cm}} = 2.5$; (c) $\dfrac{25}{10} = 2.5$; (d) $\dfrac{25}{20} = 1.25$; (e) $\dfrac{25}{15} = 1.67$

Ranking the angular magnifications, greatest to least, we have $\boxed{\text{(b) = (c), (a) = (e), (d)}}$.

37. Strategy A refractive power of $+40.0$ D means $P = 1/f = 40.0 \text{ m}^{-1}$. Use the thin lens equation and the equation for the angular magnification found in Example 24.6.

Solution

(a) The image location is $q = -25.0$ cm $= -0.250$ m. Find the object distance; that is the distance between the stamp and the magnifier.

$$\frac{1}{f} = \frac{1}{p} + \frac{1}{q}, \text{ so } p = \frac{1}{\frac{1}{f} - \frac{1}{q}} = \frac{1}{40.0 \text{ m}^{-1} - \frac{1}{-0.250 \text{ m}}} = \boxed{2.27 \text{ cm}}.$$

(b) Assuming the nearpoint $N = 25$ cm, the angular magnification is $M = \dfrac{N}{p} = \dfrac{25 \text{ cm}}{2.27 \text{ cm}} = \boxed{11}$.

(c) The size of the image is given by M times the object height, so $11(3.00 \text{ cm}) = \boxed{33 \text{ cm}}$.

41. (a) Strategy The image distance is $q = -N$. Use the thin lens equation.

Solution Find the object distance.

$$\frac{1}{p} + \frac{1}{q} = \frac{1}{f}, \text{ so } p = \frac{fq}{q-f} = \frac{-Nf}{-N-f} = \boxed{\frac{Nf}{N+f}}.$$

(b) Strategy The angular size of the image is equal to the ratio of the object height h to the object distance p, assuming $p \gg h$.

Solution

$$\theta = \frac{h}{p} = \frac{h}{\dfrac{Nf}{N+f}} = \frac{h(N+f)}{Nf}$$

(c) Strategy Use Equation (24-6) and the equation for the angular magnification found in Example 24.6.

Solution The angular magnification when the image is at the near point is

$$M = \frac{N}{p} = \frac{N}{\dfrac{Nf}{N+f}} = \frac{N(N+f)}{Nf} = \frac{N+f}{f} = \boxed{\frac{N}{f}+1}.$$

For an image at infinity, the angular magnification is $M_\infty = \dfrac{N}{f}$, so $\boxed{M = M_\infty + 1}$.

45. Strategy and Solution At the near point, the wing subtends an angle of $\theta_1 \approx \dfrac{h}{N} = \dfrac{0.10 \text{ cm}}{25 \text{ cm}} = 0.0040$. The image subtends an angle of $\theta_2 \approx \dfrac{h'}{q} = \dfrac{1.0 \text{ m}}{5.0 \text{ m}} = 0.20$. The angular magnification is $M = \dfrac{\theta_2}{\theta_1} = \dfrac{0.20}{0.0040} = \boxed{50}$.

47. Strategy Since the final image is not at infinity, the image from the objective lens is not at the focal point of the eyepiece. The angular magnification due to the eyepiece is equal to the near point divided by the object distance for the eyepiece. Use the thin lens equation and Figure 24.16.

Solution Find the object distance for the eyepiece.

$$M_e = \frac{N}{p_e}, \text{ so } p_e = \frac{N}{M_e} = \frac{25.0 \text{ cm}}{5.00} = 5.00 \text{ cm}.$$

Find the focal length of the eyepiece.

$$\frac{1}{p_e} + \frac{1}{q_e} = \frac{1}{f_e}, \text{ so } f_e = \frac{1}{\dfrac{1}{p_e} + \dfrac{1}{q_e}} = \frac{1}{\dfrac{1}{5.00 \text{ cm}} + \dfrac{1}{-25.0 \text{ cm}}} = 6.25 \text{ cm}.$$

Find the image distance for the objective lens.

$$q_o + p_e = f_o + L + f_e, \text{ so } q_o = f_o + L + f_e - p_e = 1.50 \text{ cm} + 16.0 \text{ cm} + 6.25 \text{ cm} - 5.00 \text{ cm} = 18.8 \text{ cm}.$$

Find the object distance for the objective lens.

$$\frac{1}{p_o} + \frac{1}{q_o} = \frac{1}{f_o}, \text{ so } p_o = \frac{1}{\dfrac{1}{f_o} - \dfrac{1}{q_o}} = \frac{1}{\dfrac{1}{1.50 \text{ cm}} - \dfrac{1}{18.8 \text{ cm}}} = \boxed{1.63 \text{ cm}}.$$

49. (a) Strategy Assuming the final image is at infinity, the image from the objective will be at the focal point of the eyepiece. The distance between the lenses is equal to the sum of the focal length of the eyepiece and the image distance for the objective.

Solution Compute the distance between the lenses.

$$q_o + f_e = 16.5 \text{ cm} + 2.80 \text{ cm} = \boxed{19.3 \text{ cm}}$$

(b) Strategy Use Eqs. (24-7) and (24-8).

Solution Find the angular magnification.

$$M_{\text{total}} = -\frac{L}{f_o} \times \frac{N}{f_e} = -\frac{q_o - f_o}{f_o} \times \frac{N}{f_e} = -\frac{16.5 \text{ cm} - 0.500 \text{ cm}}{0.500 \text{ cm}} \times \frac{25.0 \text{ cm}}{2.80 \text{ cm}} = \boxed{-286}$$

(c) Strategy Use the thin lens equation.

Solution Find the object distance for the objective.

$$\frac{1}{p_o} + \frac{1}{q_o} = \frac{1}{f_o}, \text{ so } p_o = \frac{1}{\frac{1}{f_o} - \frac{1}{q_o}} = \frac{1}{\frac{1}{5.00 \text{ mm}} - \frac{1}{165 \text{ mm}}} = \boxed{5.16 \text{ mm}}.$$

53. Strategy Use the thin lens, transverse magnification, and tube length equations.

Solution The transverse magnification due to the objective lens is

$$m_o = -\frac{q_o}{p_o} = -q_o\left(\frac{1}{p_o}\right).$$

From the thin lens equation, we have $\frac{1}{p_o} = \frac{1}{f_o} - \frac{1}{q_o}$. Substituting this for $\frac{1}{p_o}$ in the previous equation yields

$$m_o = -q_o\left(\frac{1}{f_o} - \frac{1}{q_o}\right) = -q_o\left(\frac{q_o - f_o}{f_o q_o}\right) = -\frac{q_o - f_o}{f_o}.$$

If the image of the objective is at the focal point of the eyepiece, as in Fig. 24.16, the tube length is $L = q_o - f_o$. Combining this with the previous result yields

$$m_o = -\frac{L}{f_o}.$$

57. Strategy Use Eq. (24-9) for the barrel length of a telescope.

Solution Find the distance between the objective and the eyepiece—the barrel length.

$$\text{barrel length} = f_o + f_e = 19.8 \text{ m} + 0.0390 \text{ m} = \boxed{19.8 \text{ m}}$$

61. (a) Strategy The moon is far enough away that we can approximate its distance as infinite. Then we can use the equation for the barrel length of a telescope, Eq. (24-9), for the distance between the lenses.

Solution Compute the barrel length of the telescope.

$$\text{barrel length} = f_o + f_e = 2.40 \text{ m} + 0.160 \text{ m} = \boxed{2.56 \text{ m}}$$

(b) Strategy Use the transverse magnification equation. $h = 3474$ km, $p = 384,500$ km, and $q \approx f_o = 2.40$ m.

Solution Find the diameter of the image produced.

$$\frac{h'}{h} = -\frac{q}{p}, \text{ so } h' = -\frac{qh}{p} = -\frac{(2.40 \text{ m})(3474 \text{ km})}{384,500 \text{ km}} = -0.0217 \text{ m} = -2.17 \text{ cm}.$$

The diameter of the image is $\boxed{2.17 \text{ cm}}$.

(c) Strategy Use Eq. (24-10).

Solution Compute the angular magnification.

$$M = -\frac{f_o}{f_e} = -\frac{2.40 \text{ m}}{0.160 \text{ m}} = \boxed{-15}$$

65. Strategy The refractive power of a lens is the reciprocal of the focal length. Use the thin lens equation.

Solution

(a) Since the eye is relaxed, the image is formed at the man's far point. $p = 40.0 \text{ cm} - 2.0 \text{ cm} = 38.0 \text{ cm}$ and $P = 1/f = 2.0 \text{ m}^{-1}$. Find the uncorrected far point.

$$q = \left(\frac{1}{f} - \frac{1}{p}\right)^{-1} = \left(2.0 \text{ m}^{-1} - \frac{1}{0.380 \text{ m}}\right)^{-1} = -1.6 \text{ m}$$

The man's uncorrected far point is $1.6 \text{ m} + 0.020 \text{ m} = \boxed{1.6 \text{ m}}$.

(b) For distant objects, his lenses should form an image at his far point.
$p = \infty$ and $q = -1.603 \text{ m} + 0.020 \text{ m} = -1.583 \text{ m}$.
Compute the refractive power required for distance vision.

$$P = \frac{1}{f} = \frac{1}{p} + \frac{1}{q} = 0 + \frac{1}{-1.583 \text{ m}} = \boxed{-0.63 \text{ D}}$$

(c) Now, $p = 0.25 \text{ m} - 0.020 \text{ m} = 0.23 \text{ m}$ and $q = -1.0 \text{ m} + 0.020 \text{ m} = -0.98 \text{ m}$.
Compute the refractive power for close-up vision.

$$P = \frac{1}{f} = \frac{1}{p} + \frac{1}{q} = \frac{1}{0.23 \text{ m}} + \frac{1}{-0.98 \text{ m}} = 3.3 \text{ D}$$

The refractive power of the two lenses in his bifocals should be $\boxed{-0.63 \text{ D and } 3.3 \text{ D}}$.

69. Strategy Use the angular magnification for an astronomical telescope and the fact that the distance separating the lenses is the sum of the focal lengths.

Solution

(a) Find the focal length of the objective lens in terms of that of the eyepiece.

$$M = -\frac{f_o}{f_e}, \text{ so } f_o = -M f_e = 5.0 f_e.$$

The focal length of the objective must be 5 times the focal length of the eyepiece. The focal length of lens 1 is 5 times the focal length of lens 2, so $\boxed{\text{lens 1 is the objective and lens 2 is the eyepiece}}$.

(b) Find the distance between the lenses.

$$25.0 \text{ cm} + 5.0 \text{ cm} = \boxed{30.0 \text{ cm}}$$

73. (a) Strategy and Solution A large magnitude magnification is desired. Since $M = -f_o/f_e$, the objective lens should have the longer focal length. So, $\boxed{\text{the lens with the 30.0-cm focal length}}$ should be the objective.

(b) **Strategy** Use Eq. (24-10).

Solution Compute the angular magnification.

$$M = -\frac{f_o}{f_e} = -\frac{30.0 \text{ cm}}{3.0 \text{ cm}} = \boxed{-10}$$

(c) **Strategy** The distance between the lenses is the sum of the focal lengths.

Solution Compute the distance between the lenses.

$$30.0 \text{ cm} + 3.0 \text{ cm} = \boxed{33.0 \text{ cm}}$$

77. Strategy For a distant mountain range, the object distance can be approximated as infinite, so the lens-sensor distance must equal the focal length of the lens, 50.0 mm. For the flower bed at 1.5 m, the lens-sensor distance is found by solving the thin lens equation for q, using $p = 1.5 \text{ m} = 1500 \text{ mm}$ and $f = 50.0 \text{ mm}$.

Solution Find the distance the lens moves with respect to the sensor.

$$\frac{1}{p} + \frac{1}{q} = \frac{1}{f}, \text{ so } q = \left(\frac{1}{f} - \frac{1}{p}\right)^{-1} = \left(\frac{1}{50.0 \text{ mm}} - \frac{1}{1500 \text{ mm}}\right)^{-1} = 51.7 \text{ mm}.$$

The lens must move a distance of $51.7 \text{ mm} - 50.0 \text{ mm} = \boxed{1.7 \text{ mm}}$.

81. (a) Strategy The refractive power of a lens is the reciprocal of the focal length. The distance between the lenses equals the sum of the focal lengths of the lenses and the tube length L.

Solution Compute the focal lengths of the lenses.
$$f_o = f_e = \frac{1}{18\ \text{D}} = 0.056\ \text{m} = 5.6\ \text{cm}$$
Find the tube length.
$$\text{lens distance} = f_o + f_e + L$$
$$28\ \text{cm} = 5.6\ \text{cm} + 5.6\ \text{cm} + L$$
$$L = \boxed{17\ \text{cm}}$$

(b) Strategy Use Eq. (24-8).

Solution Calculate the angular magnification for the microscope.
$$M_{\text{total}} = -\frac{L}{f_o} \times \frac{N}{f_e} = -L \times P_o \times N \times P_e = -0.17\ \text{m} \times 18\ \text{D} \times 0.25\ \text{m} \times 18\ \text{D} = \boxed{-14}$$

(c) Strategy Use the thin lens equation with $q_o = L + f_o$ and $1/f_o = P_o = 18\ \text{D}$.

Solution Find the object distance for the objective.
$$\frac{1}{p_o} + \frac{1}{q_o} = \frac{1}{f_o},\ \text{so}\ p_o = \left(\frac{1}{f_o} - \frac{1}{q_o}\right)^{-1} = \left(P_o + \frac{1}{L + f_o}\right)^{-1} = \left(18\ \text{m}^{-1} - \frac{1}{0.17\ \text{m} + 0.056\ \text{m}}\right)^{-1} = \boxed{7.4\ \text{cm}}.$$

85. Strategy Refer to the solution for Problem 84. Draw ray diagrams for the two images. Start with a fairly large object and draw three rays to determine each image.

Solution

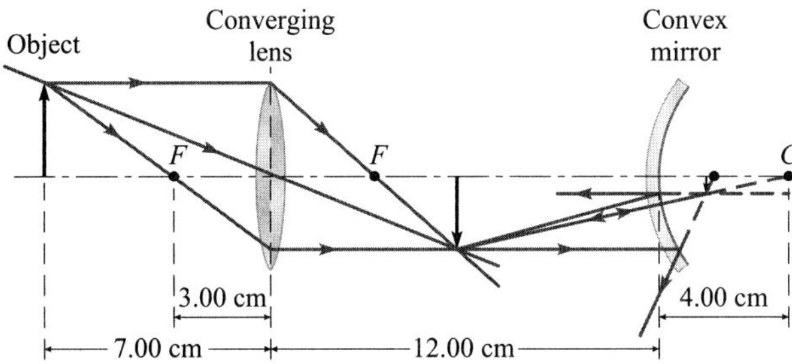

Chapter 25

INTERFERENCE AND DIFFRACTION

Conceptual Questions

1. Coherent waves have a definite fixed phase relationship. Two waves of different frequencies do not have a fixed phase relationship, and are therefore incoherent. Nevertheless, if the two frequencies are nearly identical, the waves can be nearly coherent, and maintain an approximately fixed phase relationship over a large distance. If the two frequencies are significantly different, the phase relationship is not even approximately constant over any appreciable distance.

5. The magnitude of diffractive effects is proportional to the wavelength of the diffracted wave. The wavelengths of sound waves are several orders of magnitude larger than those of visible light, so diffractive effects are much more pronounced for sound.

9. The smallest distance that a microscope can resolve is about the same size as the wavelength of light used to illuminate the object. For visible light, the wavelength is about 400–700 nm, which is far too large to be able to resolve an object that is 0.1 nm in width.

13. As the width of the slit decreases, the diffraction pattern spreads out and becomes dimmer.

17. The first ray incident from the air ($n = 1.00$) reflects from the layer of MgF_2 ($n = 1.38$), while the second ray travels into the MgF_2 and is reflected from the lens ($n = 1.51$). Both of these rays are incident in a material with lower n and reflect from a material with higher n, so both reflected rays undergo phase shifts of $180°$. If the antireflective coating had $n = 1.62$ instead, then the first ray would be phase shifted while the second would not.

Problems

1. **(a)** **Strategy** The wavelength and frequency are related by $\lambda f = c$.

   Solution Compute the wavelength of a 60-kHz EM wave.
   $$\lambda = \frac{c}{f} = \frac{3.00 \times 10^8 \text{ m/s}}{6.0 \times 10^4 \text{ Hz}} = \boxed{5.0 \text{ km}}$$

   (b) **Strategy** Compute the path difference to determine the phase difference.

   Solution The path difference is $\Delta l = (19 \text{ km} + 12 \text{ km}) - 21 \text{ km} = 10 \text{ km}$, which is equal to two wavelengths. So, the path difference results in constructive interference, but the reflection of the signal from the helicopter results in a half-wavelength phase difference, which cause destructive interference. To summarize: $\boxed{\text{Destructive interference occurs, since the path difference is 10 km and there is a } \lambda/2 \text{ phase shift}}$.

3. **Strategy** Since the radio waves interfere destructively and the reflections at the planes introduce $180°$ phase shifts, the path differences must be equal to integral numbers of wavelengths.

Solution The path differences are $d + h - 102$ km, where $d = \sqrt{(102 \text{ km})^2 + h^2}$. The wavelength is $\lambda = c/f$. Find the wavelength in terms of the path differences.

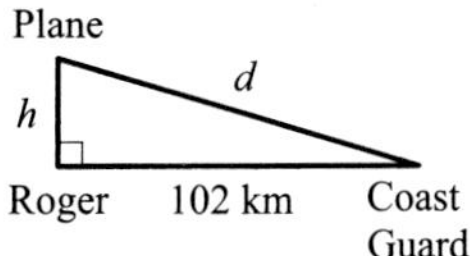

$$\Delta l = \sqrt{(102 \text{ km})^2 + h^2} + h - 102 \text{ km} = m\lambda, \text{ so}$$

$$\lambda = \frac{\sqrt{(102 \text{ km})^2 + h^2} + h - 102 \text{ km}}{m}.$$ Now, the heights of the planes are $h = 780$ m, 975 m, and 1170 m, so we

have $\lambda = \dfrac{783 \text{ m}}{m_1} = \dfrac{980 \text{ m}}{m_2} = \dfrac{1177}{m_3}$ for the respective heights. Form ratios of the integers m.

$\dfrac{m_2}{m_1} = \dfrac{980}{783} = 1.25 = \dfrac{5}{4}$ and $\dfrac{m_3}{m_1} = \dfrac{1177}{783} = 1.50 = \dfrac{6}{4}$. So, $m_1 = 4$ and $\lambda = \dfrac{783 \text{ m}}{4} = 196$ m. Therefore, the

frequency of the signal is $f = \dfrac{c}{\lambda} = \dfrac{3.00 \times 10^8 \text{ m/s}}{196 \text{ m}} = \boxed{1530 \text{ kHz}}$.

5. **(a) Strategy** Sketch four sinusoidal waves, each with wavelength 6 cm. From top to bottom, the amplitudes are 2 cm, 2 cm, 3 cm, and 1 cm. The third wave is $180°$ out of phase with the other three.

Solution The waves are shown.

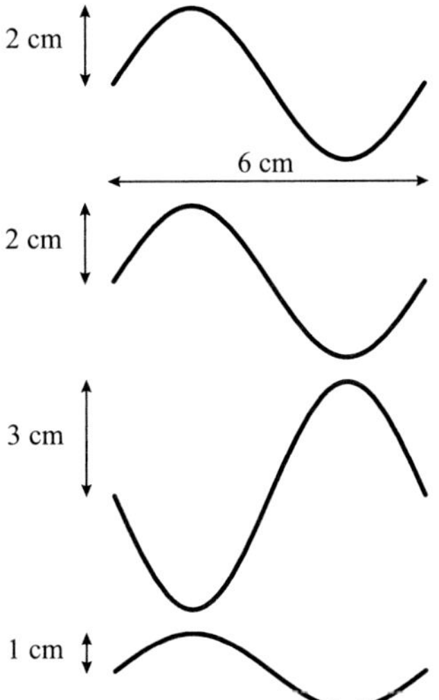

Three of the waves are in phase, so their amplitudes add. The amplitude of the third wave is $180°$ out of phase, so its amplitude is subtracted. The amplitude of the sum is $2 \text{ cm} + 2 \text{ cm} - 3 \text{ cm} + 1 \text{ cm} = \boxed{2 \text{ cm}}$.

(b) Strategy The first two waves interfere constructively, so use Eq. (25-3) to find the intensity of the new wave. The second two waves interfere destructively, so use Eq. (25-6) to find the intensity of the new wave; this wave will also be $180°$ out of phase with the sum of the first two, since the third wave's amplitude is greater than the fourth's, so use Eq. (25-6) again to find the final intensity. The first two waves each have intensity I_0. Use the fact that intensity is directly proportional to amplitude to find the intensity of the third and fourth waves in terms of I_0.

Solution Find the intensity of the combined first two waves.

$$I = I_1 + I_2 + 2\sqrt{I_1 I_2} = I_0 + I_0 + 2\sqrt{I_0 I_0} = 4I_0$$

Find the intensities of the third and fourth waves in terms of I_0. Form proportions.

$$\frac{I_3}{I_0} = \frac{3^2}{2^2} = \frac{9}{4}, \text{ so } I_3 = \frac{9}{4}I_0. \quad \frac{I_4}{I_0} = \frac{1^2}{2^2} = \frac{1}{4}, \text{ so } I_4 = \frac{1}{4}I_0.$$

Find the intensity of the combined third and fourth waves.

$$I = I_3 + I_4 - 2\sqrt{I_3 I_4} = \frac{9}{4}I_0 + \frac{1}{4}I_0 - 2\sqrt{\frac{9}{4}I_0 \frac{1}{4}I_0} = \frac{5}{2}I_0 - 2\sqrt{\frac{9}{16}I_0^2} = I_0$$

Find the intensity of the combination of all four waves.

$$I = 4I_0 + I_0 - 2\sqrt{4I_0 I_0} = 5I_0 - 4I_0 = \boxed{I_0}$$

(c) **Strategy** Three of the waves are in phase, so their amplitudes add. The amplitude of the fourth wave is $180°$ out of phase, so its amplitude is subtracted.

Solution The amplitude of the sum is $2 \text{ cm} + 2 \text{ cm} + 3 \text{ cm} - 1 \text{ cm} = \boxed{6 \text{ cm}}$.

(d) **Strategy** Repeat the process of part (b), but this time, the combination of the second two waves will be in phase with the combination of the first two.

Solution Find the intensity of the combined first two waves.

$$I = I_1 + I_2 + 2\sqrt{I_1 I_2} = I_0 + I_0 + 2\sqrt{I_0 I_0} = 4I_0$$

Find the intensities of the third and fourth waves in terms of I_0. Form proportions.

$$\frac{I_3}{I_0} = \frac{3^2}{2^2} = \frac{9}{4}, \text{ so } I_3 = \frac{9}{4}I_0. \quad \frac{I_4}{I_0} = \frac{1^2}{2^2} = \frac{1}{4}, \text{ so } I_4 = \frac{1}{4}I_0.$$

Find the intensity of the combined third and fourth waves.

$$I = I_3 + I_4 - 2\sqrt{I_3 I_4} = \frac{9}{4}I_0 + \frac{1}{4}I_0 - 2\sqrt{\frac{9}{4}I_0 \frac{1}{4}I_0} = \frac{5}{2}I_0 - 2\sqrt{\frac{9}{16}I_0^2} = I_0$$

Find the intensity of the combination of all four waves.

$$I = 4I_0 + I_0 + 2\sqrt{4I_0 I_0} = 5I_0 + 4I_0 = \boxed{9I_0}$$

7. **Strategy** Since the light from each lamp is incoherent, the intensity of the lamps together is just the sum of the intensities.

Solution Find the combined intensity of the lamps.

$$I = I_0 + 4I_0 = \boxed{5I_0}$$

9. (a) **Strategy** As in Example 25.1, the formula relating the path length difference to the wavelength is $2\Delta x = \lambda$.

Solution The maxima in the figure are at $x = 0.7$ cm and $x = 2.3$ cm, so $\Delta x = 1.6$ cm and the wavelength is $\lambda = 2 \cdot 1.6 \text{ cm} = \boxed{3.2 \text{ cm}}$.

(b) **Strategy** Amplitude is proportional to the square root of intensity and intensity is proportional to power, so amplitude is proportional to the square root of power.

Solution The maxima have power readings of 3.4 units and 2.6 units. The ratio of the amplitudes is $\sqrt{3.4/2.6} = \boxed{1.1}$.

13. **(a) Strategy and Solution** Since the wavelength of light is shorter in glass than in vacuum, it will take a larger number of wavelengths to pass through the glass. To increase the number of wavelengths in the other arm, the mirror must be moved out.

(b) Strategy Let the original length of each arm be D. Let T be the thickness of the glass. Then, the total distance traveled by the light in the glass is $2T$, and the distance traveled in the same arm but outside of the glass is $2(D-T)$. If the vacuum wavelength of one of the components of the white light is λ_0, then the wavelength in the glass is $\lambda = \lambda_0/n$.

Solution Find the thickness of the slab of glass.

The number of wavelengths traveled in the glass is $\dfrac{2T}{\lambda} = \dfrac{2T}{\lambda_0/n} = \dfrac{2Tn}{\lambda_0}$.

The number of wavelengths traveled in the same arm but outside of the glass is $\dfrac{2(D-T)}{\lambda_0}$.

The total number of wavelengths traveled in this arm is $\dfrac{2D-2T}{\lambda_0} + \dfrac{2Tn}{\lambda_0} = \dfrac{2D+2T(n-1)}{\lambda_0} = \dfrac{2D+2T(0.46)}{\lambda_0}$.

Let d be the distance the mirror moved. Then, the number of wavelengths traveled in this arm is $\dfrac{2D+2d}{\lambda_0}$.

The number of wavelengths in both cases must be equal.

$\dfrac{2D+2T(0.46)}{\lambda_0} = \dfrac{2D+2d}{\lambda_0}$, so $T = \dfrac{d}{0.46} = \dfrac{6.73 \text{ cm}}{0.46} = \boxed{15 \text{ cm}}$.

15. **Strategy** At the air-oil boundary, reflected light is $180°$ out of phase, since $n_{\text{oil}} > n_{\text{air}}$. For transmitted light reflected at the oil-water boundary, the reflective ray is not phase shifted since $n_{\text{water}} < n_{\text{oil}}$. The two reflected rays are $180°$ out of phase, so constructive interference occurs when the relative path length difference, $2t$, is an odd multiple of one half the wavelength in the oil, $\lambda = \lambda_0 / n_{\text{oil}}$.

Solution Find the wavelength.

$2t = \left(m + \dfrac{1}{2}\right)\lambda = \left(m + \dfrac{1}{2}\right)\dfrac{\lambda_0}{n_{\text{oil}}}$. Solving for λ_0 yields $\lambda_0 = \dfrac{2tn_{\text{oil}}}{m + \frac{1}{2}}$.

$n_{\text{air}} = 1.00$
$n_{\text{oil}} = 1.50 \quad t = 0.40 \ \mu\text{m}$
$n_{\text{water}} = 1.33$
(Rays are normally incident)

Substituting $t = 0.40 \ \mu\text{m}$ and $n_{\text{oil}} = 1.50$ gives $\lambda_0 = \dfrac{1200 \text{ nm}}{m + \frac{1}{2}}$.

Evaluating this for $m = 0, 1, 2,$ and 3 gives $\lambda_0 = 2400$ nm, 800 nm, 480 nm, and 343 nm, respectively. Of these values, only $\boxed{480 \text{ nm}}$ is in the visible spectrum.

17. **Strategy** The wavelength most strongly transmitted will be the wavelength whose reflected rays combine destructively (minimum reflection). At both boundaries, the reflected rays are inverted, since $n_{\text{air}} < n_{\text{film}}$ and $n_{\text{film}} < n_{\text{glass}}$. So, the phase difference between the reflected rays is due only to the thickness of the film. Destructive interference occurs when the path length difference, $2t$, is equal to an odd multiple of half the wavelength in the film, $\lambda = \lambda_0 / n_{\text{film}}$.

Solution Find the wavelength.

$2t = \left(m + \dfrac{1}{2}\right)\lambda = \left(m + \dfrac{1}{2}\right)\dfrac{\lambda_0}{n_{\text{film}}}$. Solving for λ_0, and substituting $n_{\text{film}} = 1.38$, t

$n_{\text{air}} = 1.00$
$n_{\text{film}} = 1.38 \quad t = 90.0 \text{ nm}$
$n_{\text{glass}} = 1.50$
(Rays are normally incident)

$= 90.0$ nm, and $m = 0$ yields $\lambda_0 = \dfrac{2tn_{\text{film}}}{m + \frac{1}{2}} = \dfrac{2 \cdot 90.0 \text{ nm} \cdot 1.38}{0 + \frac{1}{2}} = \boxed{497 \text{ nm}}$.

All other values of m give wavelengths not in the visible spectrum.

19. **(a) Strategy** The weakest wavelengths in transmitted light correspond to the strongest wavelengths in reflected light. Rays reflected off the front of the film will be inverted since $n_{air} < n_{film}$, but rays reflected off the back of the film are not inverted, so the reflected rays will be 180° out of phase plus the phase shift caused by the path length difference. The condition for constructive interference is used to find the strongest wavelengths in reflected light. This occurs when the path length difference, $2t$, is an odd multiple of half the wavelength in the film.

Solution Find the weakest wavelengths.

$$2t = \left(m + \frac{1}{2}\right)\lambda = \left(m + \frac{1}{2}\right)\frac{\lambda_0}{n_{film}}$$

Solving for λ_0 with $t = 910.0$ nm and $n_{film} = 1.50$ gives

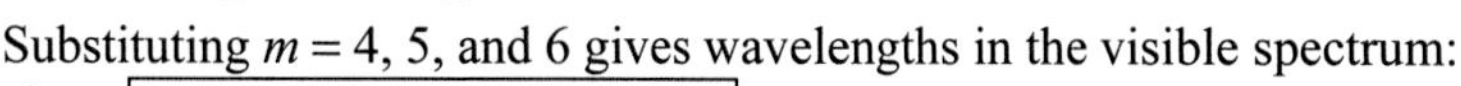

$$\lambda_0 = \frac{2tn_{film}}{m + \frac{1}{2}} = \frac{2730 \text{ nm}}{m + \frac{1}{2}}.$$

Substituting $m = 4, 5,$ and 6 gives wavelengths in the visible spectrum:
$\lambda_0 = \boxed{607 \text{ nm}, 496 \text{ nm, and } 420 \text{ nm}}$.

(b) Strategy The strongest wavelengths is transmitted light correspond to the weakest wavelengths in reflected light. Destructive interference in the reflected light occurs when the path length difference, $2t$, is equal to an integral number of wavelengths in the film, $\lambda = \lambda_0 / n_{film}$.

Solution Find the strongest wavelengths.

$$2t = m\lambda = m\frac{\lambda_0}{n_{film}}$$

Solving for λ_0 with $t = 910.0$ nm and $n_{film} = 1.50$ gives

$$\lambda_0 = \frac{2tn_{film}}{m} = \frac{2730 \text{ nm}}{m}.$$

Substituting $m = 4, 5,$ and 6 gives wavelengths in the visible spectrum: $\lambda_0 = \boxed{683 \text{ nm}, 546 \text{ nm, and } 455 \text{ nm}}$.

21. **(a) Strategy** When light reflects from a boundary with a medium with a higher index of refraction, it is inverted (180° phase change). Inversion alone will result in destructive interference.

Solution If there is no gap between the plates of glass—they are $\boxed{\text{touching}}$—the light is inverted at the boundary and destructive interference occurs. Therefore, the minimum distance between the two glass plates for one of the dark regions is $\boxed{\text{zero}}$.

(b) Strategy There must be some nonzero thickness of air between the plates for constructive interference to occur. Let this distance between the plates be t.

Solution The light is inverted after is reflects of the bottom layer of glass after it has passed through the thin film of air between the plates, so the path difference due to the light traveling through the thin film of air must be equal to one half wavelength to compensate. The light passes through the thickness of the air twice, before and after reflection. Find t.

$$2t = \frac{\lambda}{2}, \text{ so } t = \frac{\lambda}{4} = \frac{550 \text{ nm}}{4} = \boxed{140 \text{ nm}}.$$

(c) Strategy Refer to parts (a) and (b). For destructive interference, the path difference due to the light traveling through the thin film of air must be equal to one full wavelength.

Solution Find the next largest distance between the plates resulting in a dark region.

$$2t = \lambda, \text{ so } t = \frac{\lambda}{2} = \frac{550 \text{ nm}}{2} = \boxed{280 \text{ nm}}.$$

25. Strategy Constructive interference occurs when the rays are in phase. Destructive interference occurs when they are $180°$ out of phase. Consider phase shifts due to reflections and path-length differences.

Solution Ray 1 is inverted upon reflection and ray 2 is not. Constructive interference occurs for rays 1 and 2 when the path length difference, $2t$, is equal to an odd multiple of the wavelength in the film, $\lambda = \lambda_0 / n$.

$$2t = \left(m + \frac{1}{2} \right)\lambda = \left(m + \frac{1}{2} \right)\frac{\lambda_0}{n_2} \qquad (1)$$

The path length difference between rays 3 and 4 is $2t$, and ray 4 is not inverted by either of its two reflections within the film. So, destructive interference occurs if $2t$ is equal to an odd multiple of the wavelength in the film.

Thus, $2t = \left(m + \frac{1}{2} \right)\dfrac{\lambda_0}{n_2}$, which is the same as the condition for constructive interference for rays 1 and 2.

For rays 1 and 2 to interfere destructively, $2t$ must equal an integral number of wavelengths in the film, so

$$2t = \frac{m\lambda_0}{n_2} \qquad (2).$$

For rays 3 and 4 to interfere constructively, $2t$ must equal an integral number of wavelengths in the film, so

$2t = \dfrac{m\lambda_0}{n_2}$, which is the same as the condition for destructive interference for rays 1 and 2.

29. Strategy Bright fringes occur at angles θ according to Eq. (25-10). Assume θ is small.

Solution In the figure, a bright fringe of order m appears on the screen at a distance x from the central bright fringe. D, x, and the distance to the mth fringe from the slits form a right triangle. So, $x = D\tan\theta$.

When θ is small ($x \ll D$), the approximation $\sin\theta \approx \tan\theta$ may be used, yielding $x = D\sin\theta$. Solving the double-slit maxima equation for $\sin\theta$ gives $\sin\theta = \dfrac{m\lambda}{d}$. Substituting this into the previous equation gives $x = m\dfrac{D\lambda}{d}$. D, λ, and d are constant. m is an integer representing the fringe order. Thus, for $x \ll D$, the fringes are equally spaced.

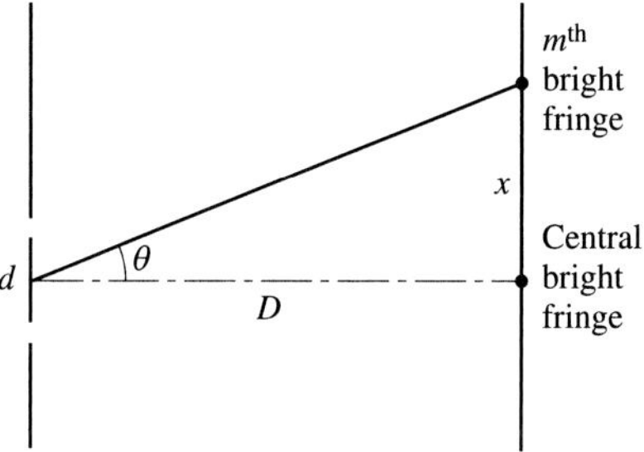

31. Strategy Draw a diagram of the situation and use the double-slit maxima equation. Assume θ is small.

Solution In the figure, a bright fringe of order m appears on the screen at a distance x from the central bright fringe. D, x, and the distance to the mth fringe from the slits form a right triangle. So, $x = D\tan\theta$. When θ is small ($x \ll D$), the approximation $\sin\theta \approx \tan\theta$ may be used, yielding $x = D\sin\theta$. Solving the double-slit maxima equation for $\sin\theta$ gives $\sin\theta = \dfrac{m\lambda}{d}$. Substituting this into the previous equation gives $x = m\dfrac{D\lambda}{d}$. So, the linear distance

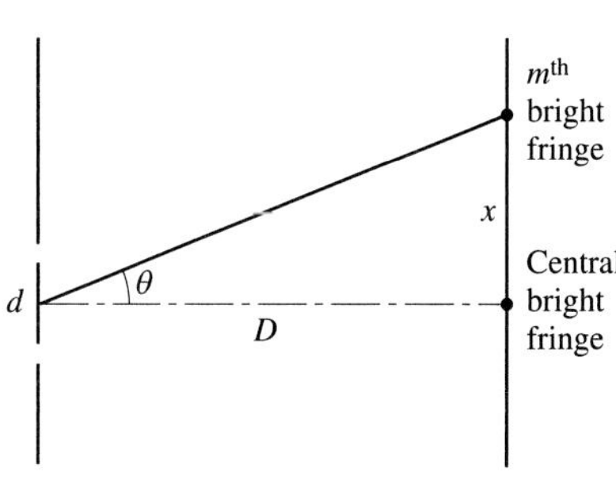

between adjacent maxima on the screen is $\dfrac{D\lambda}{d} = \dfrac{(0.368 \text{ m})(475 \times 10^{-9} \text{ m})}{0.120 \times 10^{-3} \text{ m}} = \boxed{1.46 \text{ mm}}$.

Since $D = 0.368 \text{ m} \gg x = 0.00146 \text{ m}$, the small angle approximation is justified.

33. Strategy To display five interference maxima, the screen must be twice as wide as the distance from the central bright fringe (zeroth-order maximum) to the second-order interference minimum (which defines the edge of the second-order maximum). Let x be the distance from the central bright fringe to the second-order interference minimum. Then, the minimum width of the screen is $2x$. Use the double-slit minima equation.

Solution The double-slit minima equation gives

$$d \sin \theta = (m + 1/2)\lambda = (2 + 1/2)\lambda = 5\lambda/2$$

for the second-order interference minimum.

The hypotenuse of the right triangle formed by x and D is $\sqrt{x^2 + D^2}$.

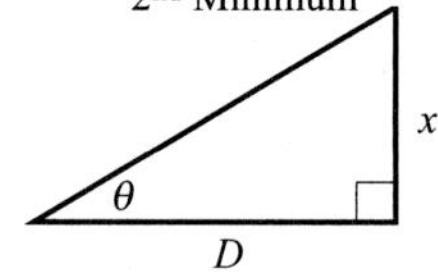

Therefore, $\sin \theta = \dfrac{x}{\sqrt{x^2 + D^2}} = \dfrac{5\lambda}{2d} = \dfrac{5(547 \times 10^{-9}\text{ m})}{2(0.00150\text{ m})} = 9.117 \times 10^{-4}.$

Solve for $2x$, the minimum width of the screen.

$$\frac{x^2}{x^2 + D^2} = (9.117 \times 10^{-4})^2$$

$$1 + \frac{D^2}{x^2} = (9.117 \times 10^{-4})^{-2}$$

$$2x = 2D[(9.117 \times 10^{-4})^{-2} - 1]^{-1/2} = 2(0.900\text{ m})[(9.117 \times 10^{-4})^{-2} - 1]^{-1/2} = \boxed{1.64\text{ mm}}$$

35. Strategy $x \ll D$ (distant screen), so using the small angle approximation is justified. Then, from Example 25.4, $d = \lambda D / x$. So, for fixed D and d, $\lambda \propto x$.

Solution Form a proportion to find the wavelength of the second source.

$$\frac{\lambda_2}{\lambda_1} = \frac{x_2}{x_1}, \text{ so } \lambda_2 = \frac{0.640\text{ cm}}{0.530\text{ cm}}(589\text{ nm}) = \boxed{711\text{ nm}}.$$

37. Strategy Solve Eq. (25-10) for θ and use the fact that $d = 1/N$.

Solution Find the expected angle.

$$d \sin \theta = m\lambda, \text{ so } \theta = \sin^{-1}\frac{m\lambda}{d} = \sin^{-1}(m\lambda N) = \sin^{-1}\left[3(546 \times 10^{-7}\text{ cm})\frac{8000\text{ slits}}{2.54\text{ cm}}\right] = \boxed{31.1^\circ}.$$

39. Strategy Since the third-order line is the last to appear, the angle for this line must be approaching 90°. Use Eq. (25-10) and $d = 1/N$.

Solution Compute the slit separation.

$$d \sin \theta = m\lambda, \text{ so } d = \frac{m\lambda}{\sin \theta} \approx \frac{3(650 \times 10^{-9}\text{ m})}{\sin 90^\circ} = 2 \times 10^{-6}\text{ m}.$$

Compute the number of slits per cm.

$$N = \frac{1}{d} = \frac{1}{2 \times 10^{-6}\text{ m}} = \frac{1}{2 \times 10^{-4}\text{ cm}} = \boxed{5000\text{ slits/cm}}$$

41. (a) Strategy Solve Eq. (25-10) for m. $d = 1.50\ \mu\text{m}$, $\lambda = 0.600\ \mu\text{m}$, and $\theta = 90°$.

Solution Find m.

$$d\sin\theta = m\lambda, \text{ so } m = \frac{d\sin\theta}{\lambda} = \frac{(1.50\ \mu\text{m})\sin 90°}{0.600\ \mu\text{m}} = 2.50.$$

The maxima visible on the screen will be associated with $m = 0$, ± 1, and ± 2, so $\boxed{\text{five}}$ maxima are seen.

(b) Strategy Draw a diagram of the situation; use it and Eq. (25-10) to find the locations of the maxima.

Solution Find θ_1.

$$d\sin\theta_1 = \lambda \text{ for } m = 1, \text{ so } \theta_1 = \sin^{-1}\frac{\lambda}{d}.$$

Substitute for θ_1 and solve for x_1.

$$x_1 = D\tan\left(\sin^{-1}\frac{\lambda}{d}\right) = (3.0\ \text{m})\tan\left(\sin^{-1}\frac{0.600\ \mu\text{m}}{1.50\ \mu\text{m}}\right) = 1.3\ \text{m}$$

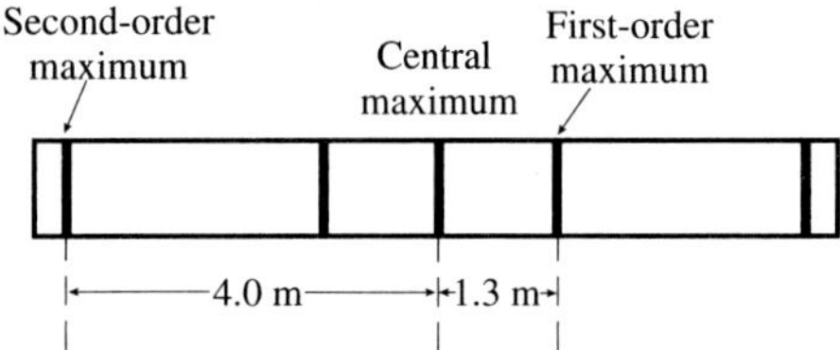

Solve for x_2.

$$\theta_2 = \sin^{-1}\frac{2\lambda}{d} \text{ since } m = 2.$$

$$x_2 = (3.0\ \text{m})\tan\left[\sin^{-1}\frac{2(0.600\ \mu\text{m})}{1.50\ \mu\text{m}}\right] = 4.0\ \text{m}$$

A sketch of the pattern is shown.

43. (a) Strategy Use Eq. (25-10) to find each angle. The slit separation is given by $d = 1/N$.

Solution Find the slit separation.

$$d = \frac{1}{5000.0\ \text{slits/cm}} = 2.0000\times10^{-6}\ \text{m}$$

The line at $12.98°$ must be a first-order line, since it is the smallest angle. Find the wavelength.

$$d\sin\theta = m\lambda, \text{ so } \lambda = \frac{d\sin\theta}{m} = \frac{(2.0000\times10^{-6}\ \text{m})\sin 12.98°}{1} = 449.2\ \text{nm}.$$

Now, the angles for the second-, third-, and fourth-order lines for this wavelength can be found.

$$d\sin\theta = m\lambda, \text{ so } \theta = \sin^{-1}\frac{m\lambda}{d}.$$

$$\theta_2 = \sin^{-1}\left(\frac{2\cdot449.2\times10^{-9}\ \text{m}}{2.0000\times10^{-6}\ \text{m}}\right) = 26.69°, \ \theta_3 = \sin^{-1}\left(\frac{3\cdot449.2\times10^{-9}\ \text{m}}{2.0000\times10^{-6}\ \text{m}}\right) = 42.36°, \text{ and}$$

$$\theta_4 = \sin^{-1}\left(\frac{4\cdot449.2\times10^{-9}\ \text{m}}{2.0000\times10^{-6}\ \text{m}}\right) = 63.95°.$$

Next, the line at $19.0°$ must be a first-order line for a different wavelength. Find the wavelength.

$$\lambda = \frac{d\sin\theta}{m} = \frac{(2.0000\times10^{-6}\ \text{m})\sin 19.0°}{1} = 651\ \text{nm}$$

Now the angles for the second- and third-order lines for this wavelength can be found.

$$\theta_2 = \sin^{-1}\left(\frac{2\cdot 651\times 10^{-9}\ \text{m}}{2.0000\times 10^{-6}\ \text{m}}\right) = 40.6°\ \text{and}\ \theta_3 = \sin^{-1}\left(\frac{3\cdot 651\times 10^{-9}\ \text{m}}{2.0000\times 10^{-6}\ \text{m}}\right) = 77.6°.$$

All the spectral lines are accounted for by $\boxed{\text{two}}$ wavelengths, $\boxed{449.2\ \text{nm and}\ 651\ \text{nm}}$.

(b) Strategy To find the total number of lines seen on the screen to one side of the central maximum, solve Eq. (25-10) for m (for both wavelengths) using $\theta = 90°$ and $d = 1/(2000.0\ \text{slits/cm}) = 5.0000\times 10^{-6}\ \text{m}$, making sure to round m down to the nearest whole number.

Solution Find the number of spectral lines for each wavelength.

$$d\sin\theta = m\lambda,\ \text{so}\ m = \frac{d\sin\theta}{\lambda} = \frac{(5.0000\times 10^{-6}\ \text{m})\sin 90°}{449.2\times 10^{-9}\ \text{m}} = 11\ \text{and}\ m = \frac{(5.0000\times 10^{-6}\ \text{m})\sin 90°}{651\times 10^{-9}\ \text{m}} = 7.$$

The total is $11+7 = \boxed{18\ \text{lines}}$.

45. (a) Strategy Solve Eq. (25-10) for m using $\lambda = 440.936$ nm, $\theta = 90°$, and $d = 1.600\ \text{cm}/(12,000\ \text{slits})$.

Solution Find the number of orders that can be seen.

$$d\sin\theta = m\lambda,\ \text{so}\ m = \frac{d\sin\theta}{\lambda} = \frac{1.600\times 10^{-2}\ \text{m}}{12,000}\cdot\frac{\sin 90°}{440.936\times 10^{-9}\ \text{m}} = 3.024.$$

Since m must be an integer, we have $m = 3$, and $\boxed{\text{three}}$ orders of lines can be seen.

(b) Strategy For order 0, both angles are at $0°$, so the separation is $0 - 0 = 0$.
For orders 1, 2, and 3, solve Eq. (25-10) for θ for each pair of lines, and subtract to find the difference.

Solution Find the angular separation.

$$d\sin\theta = m\lambda,\ \text{so}\ \theta = \sin^{-1}\left(m\lambda\frac{1}{d}\right).$$

Here, $\dfrac{1}{d} = \dfrac{12,000\ \text{slits}}{1.600\times 10^{-2}\ \text{m}} = 7.500\times 10^{5}\ \text{slits/m}$.

Compute the angles.

$$\theta_{a1} = \sin^{-1}(1\cdot 440.000\times 10^{-9}\ \text{m}\cdot 7.500\times 10^{5}\ \text{m}^{-1}) = 19.27°$$
$$\theta_{b1} = \sin^{-1}(1\cdot 440.936\times 10^{-9}\ \text{m}\cdot 7.500\times 10^{5}\ \text{m}^{-1}) = 19.31°$$
$$\theta_{a2} = \sin^{-1}(2\cdot 440.000\times 10^{-9}\ \text{m}\cdot 7.500\times 10^{5}\ \text{m}^{-1}) = 41.30°$$
$$\theta_{b2} = \sin^{-1}(2\cdot 440.936\times 10^{-9}\ \text{m}\cdot 7.500\times 10^{5}\ \text{m}^{-1}) = 41.41°$$
$$\theta_{a3} = \sin^{-1}(3\cdot 440.000\times 10^{-9}\ \text{m}\cdot 7.500\times 10^{5}\ \text{m}^{-1}) = 81.89°$$
$$\theta_{b3} = \sin^{-1}(3\cdot 440.936\times 10^{-9}\ \text{m}\cdot 7.500\times 10^{5}\ \text{m}^{-1}) = 82.80°$$

The angular separations are: $\boxed{\theta_{b1} - \theta_{a1} = 0.04°,\ \theta_{b2} - \theta_{a2} = 0.11°,\ \theta_{b3} - \theta_{a3} = 0.91°}$.

(c) $\boxed{\text{The two lines are best resolved by the third-order spectra, since the angular separation is the largest.}}$

49. Strategy The distance s is given by the position of the second minima. To find this position, combine Eq. (25-12) with the equation relating s, θ, and D (the screen-slit distance), $\tan\theta = s/D$. Assume θ is a small angle.

Solution First solve Eq. (25-12) for $\sin\theta$.

$$a\sin\theta = m\lambda, \text{ so } \sin\theta = \frac{m\lambda}{a}.$$

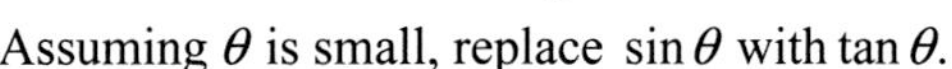

Assuming θ is small, replace $\sin\theta$ with $\tan\theta$.

$$\tan\theta = \frac{m\lambda}{a}$$

Substitute this expression into the equation relating s, θ, and D.

$$\frac{m\lambda}{a} = \frac{s}{D}$$

Now solve for s using $m = 2$, $\lambda = 630$ nm, $a = 0.40$ mm, and $D = 2.0$ m.

$$s = \frac{m\lambda D}{a} = \frac{2\cdot 630\times10^{-9} \text{ m}\cdot 2.0 \text{ m}}{0.40\times10^{-3} \text{ m}} = \boxed{6.3 \text{ mm}}$$

$s \ll D$, so using the small angle approximation was justified.

51. (a) Strategy Eq. (25-12) is $d\sin\theta = m\lambda$. Since we are concerned about the width of the central maximum, $m = 1$. For a fixed d, $\sin\theta \propto \lambda$, so larger wavelengths give larger angles for the first minima.

Solution Since the width of the central bright fringe is twice the distance from the center of the screen to the first minimum, the fringe is wider for greater wavelengths. Since $\lambda_{\text{red}} > \lambda_{\text{blue}}$, the central maximum is $\boxed{\text{wider}}$ for red light than it is for blue.

(b) Strategy Call the width of the central maximum W. Referring to Figure 25.33, we have $W = 2x$, and by trigonometry, $x = D\tan\theta$, so $W = 2x = 2D\tan\theta$. Assume θ is a small angle.

Solution Since θ is assumed to be small, we have $W = 2D\tan\theta \approx 2D\sin\theta$.

From Eq. (25-12), we have $\sin\theta = \dfrac{\lambda}{a}$ since $m = 1$, so W becomes $W = \dfrac{2L\lambda}{a}$.

Applying this equation to the current problem gives $W_{\text{R}} = \dfrac{2L\lambda_{\text{R}}}{a}$ and $W_{\text{B}} = \dfrac{2L\lambda_{\text{B}}}{a}$.

Dividing these equations gives $\dfrac{W_{\text{R}}}{W_{\text{B}}} = \dfrac{\frac{2L\lambda_{\text{R}}}{a}}{\frac{2L\lambda_{\text{B}}}{a}} = \dfrac{\lambda_{\text{R}}}{\lambda_{\text{B}}}$.

Now solve for W_{R} when $W_{\text{B}} = 2.0$ cm, $\lambda_{\text{R}} = 0.70$ μm, and $\lambda_{\text{B}} = 0.43$ μm.

$$W_{\text{R}} = W_{\text{B}}\frac{\lambda_{\text{R}}}{\lambda_{\text{B}}} = (2.0 \text{ cm})\frac{0.70 \text{ μm}}{0.43 \text{ μm}} = \boxed{3.3 \text{ cm}}$$

53. Strategy Use Eq. (25-12) and the small angle approximation $\sin\theta \approx \tan\theta = x/L$, where x is the half-width of the maximum and L is the distance to the screen. The first minimum occurs at $m = 1$.

Solution Find the thickness of the hair.

$$m\lambda = (1)\lambda = a\sin\theta \approx ax/L, \text{ so } a = hair\,thickness = \frac{\lambda L}{x} = \frac{(632.8\times10^{-9}\text{ m})(2.0\text{ m})}{0.015/2\text{ m}} = \boxed{170\ \mu\text{m}}.$$

57. Strategy Let a be the diameter of the hole in the pinhole camera, and let x be the diameter of the central maximum produced on the film by light entering through the hole. The value x is related to the length of the camera L and the angular size of the first minimum θ by the formula $\tan\theta = (x/2)/L$. Assume θ is a small angle. Use Eq. (25-13).

Solution Assuming θ is small, we have $\tan\theta = x/(2L) \approx \sin\theta$. Combining this with Eq. (25-13)

$(a\sin\theta = 1.22\lambda)$ yields $a\dfrac{x}{2L} = 1.22\lambda$ or $ax = 2.44\lambda L$. The hole will be the optimum size when the diameter of the

hole equals the diameter of the central maximum, or $x = a$. Substitute this into the previous result and solve for a.
$a^2 = 2.44\lambda L$, so $a = \sqrt{2.44\lambda L}$.
Evaluating this using $\lambda = 560$ nm and $L = 16.0$ cm gives the optimum value for the diameter a.

$$a = \sqrt{2.44\cdot 560\times10^{-9}\text{ m}\cdot 0.160\text{ m}} = \boxed{0.47\text{ mm}}$$

61. Strategy Assume that the two light sources are just far enough apart to be resolvable according to Rayleigh's criterion. For the smallest angular separation, use the smallest visible wavelength, which is approximately 400 nm.

Solution Compute the angular separation.

$$a\sin\Delta\theta = 1.22\lambda_0, \text{ so } \Delta\theta = \sin^{-1}\frac{1.22\lambda_0}{a} = \sin^{-1}\frac{1.22(400\times10^{-9}\text{ m})}{5\times10^{-3}\text{ m}} = 0.0056°.$$

This angular separation corresponds to a distance x on the fovea given by $x = L\tan\Delta\theta$.
Using $L = 25$ mm (the diameter of the eye) and $\Delta\theta = 0.0056°$ gives a separation of

$x = (25\times10^{-3}\text{ m})\tan 0.0056° = 2\times10^{-6}\text{ m} = 2\ \mu\text{m}.$

If this is the distance between two non-adjacent cones with a single intervening cone, then the distance between

adjacent cones will be $2\ \mu\text{m}/2 = \boxed{1\ \mu\text{m}}$.

> Our vision would not be better if the cones were closer together, since diffraction would blur images enough that the extra cones wouldn't do any good.

63. (a) Strategy and Solution All points on the line $\theta = 0$ are equidistant from the towers. Since the radio waves started out in phase, they will arrive in phase and combine constructively at all points on $\theta = 0$. The power will be a $\boxed{\text{maximum}}$.

(b) Strategy This can be treated as a double-slit problem with the antennas playing the role of the slits. Solving Eq. (25-10) for θ, and using $d = \lambda$ and $m = 0$ and ± 1 yields the maxima.

Solution Solve for the angle.

$d \sin \theta = \lambda \sin \theta = m\lambda$, so $\theta = \sin^{-1} m$.

Substitute the values of m.

$\theta_0 = \sin^{-1}(0) = 0$ and $\theta_{\pm 1} = \sin^{-1}(\pm 1) = \pm 90°$.

The situation for $90° < \theta < 180°$ and $-180° < \theta < -90°$ is the mirror image of the maxima found above. All together, the maxima occur at $\theta = -180°,\ -90°,\ 0°,\ 90°,$ and $180°,$ where $\pm 180°$ is the same location. The minima are midway between the maxima: $\pm 45°$ and $\pm 135°$. A qualitative graph is shown.

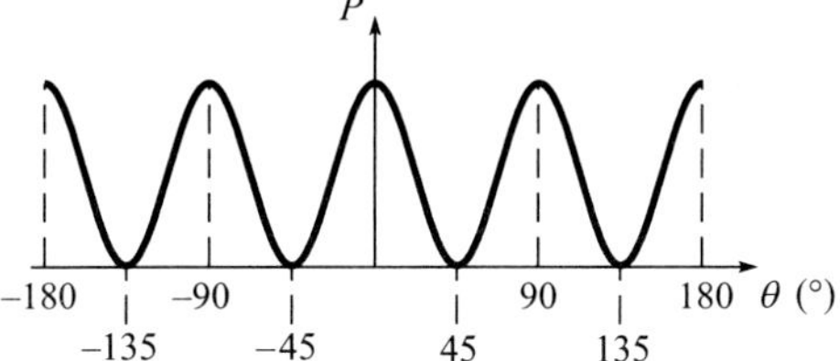

(c) Strategy Solve Eq. (25-10) for θ using $d = \lambda/2$ and $m = 0$.

Solution Solve for the angle.

$d \sin \theta = \dfrac{\lambda}{2} \sin \theta = m\lambda$, so $\theta = \sin^{-1}(2m)$.

For $m = 0$, we have $\theta_0 = \sin^{-1}(0) = 0$.

By symmetry, $\theta = \pm 180°$ will also be a maximum. Since $d = \lambda/2$, destructive interference will occur at $\theta = \pm 90°$.

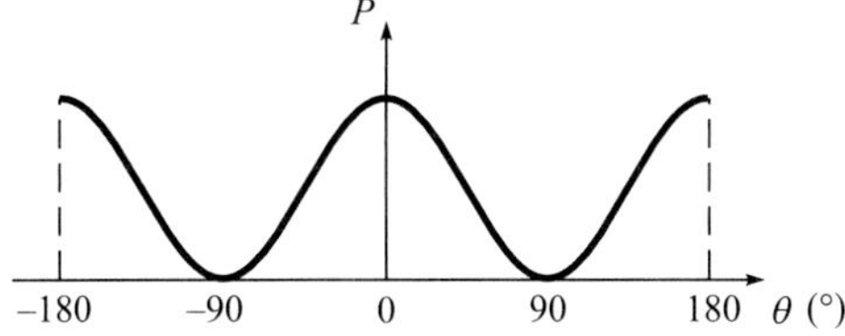

65. Strategy Let r be the radius of the circular aperture, and let x be the radius of the central maximum produced on the moon by light passing through the circular aperture. The value x is related to the Earth-Moon distance L and the angular size θ of the first minimum by the formula $\tan \theta = x/L$. Assume θ is a small angle. Use Eq. (25-13).

Solution Since the angle is small,

$\sin \theta \approx \tan \theta = \dfrac{x}{L}$.

Combine this result with Eq. (25-13).

$1.22\lambda = a \sin \theta = 2r \sin \theta \approx 2r\dfrac{x}{L}$ or $rx = 0.610\lambda L$, where r is the radius of the spot.

The spot on the moon will be smallest when the radius of the aperture equals the radius of the central maximum, i.e., $r = x$. Solve for x.

$x^2 = 0.610\lambda L$, so $x = \sqrt{0.610\lambda L} = \sqrt{0.610(0.60\times 10^{-6}\ \text{m})(3.845\times 10^{8}\ \text{m})} = \boxed{12\ \text{m}}$.

69. Strategy Points A, C, and E are on maxima and B and D are on minima. Use Eqs. (25-9) through (25-11).

Solution Point A is at the center of the central bright fringe (zeroth-order maximum), so the path length difference is $\Delta l = d \sin\theta = m\lambda = (0)\lambda = \boxed{0}$. Point B is at the zeroth-order minimum, so the path length difference is $\Delta l = d \sin\theta = (m+1/2)\lambda = (0+1/2)\lambda = 660\text{ nm}/2 = \boxed{330\text{ nm}}$. Point C is at the first-order maximum, so the path length difference is $\Delta l = m\lambda = (1)(660\text{ nm}) = \boxed{660\text{ nm}}$. Point D is at the first-order minimum, so the path length difference is $\Delta l = (m+1/2)\lambda = (1+1/2)\lambda = 3(660\text{ nm})/2 = \boxed{990\text{ nm}}$. Point E is at the second-order maximum, so the path length difference is $\Delta l = m\lambda = (2)(660\text{ nm}) = \boxed{1.3\ \mu\text{m}}$.

73. Strategy The first single-slit diffraction minimum is coincident with the fifth-order double-slit interference maximum. Use the equations for the double-slit interference maxima and single-slit diffraction minima. The distance to the first diffraction minimum is half the width of the central maximum.

Solution

(a) Find the width of the slits.

$$a \sin\theta = m\lambda = (1)\lambda = \lambda,\text{ so } a = \frac{\lambda}{\sin\theta}.$$

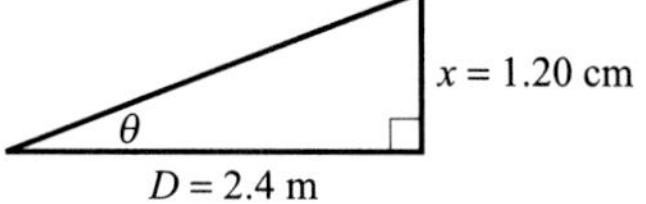

Since $x = (2.40\text{ cm})/2 = 0.0120\text{ m} \ll D = 2.4\text{ m}$, $\sin\theta \approx \tan\theta = \dfrac{x}{D}$ and

$$a = \frac{\lambda D}{x} = \frac{(510 \times 10^{-9}\text{ m})(2.4\text{ m})}{0.0120\text{ m}} = \boxed{0.10\text{ mm}}.$$

(b) Find the relationship between the slit width a and the distance between the slits d using the fifth-order double-slit interference maximum and the first single-slit diffraction minimum.

$$d \sin\theta = 5\lambda \text{ and } a \sin\theta = \lambda,\text{ so } \frac{d\sin\theta}{a\sin\theta} = \frac{d}{a} = \frac{5\lambda}{\lambda} = 5 \text{ and } d = 5a = 5(0.102\text{ mm}) = \boxed{0.51\text{ mm}}.$$

77. Strategy Rays reflected at both boundaries will be inverted, since $n_{\text{air}} < n_{\text{coating}} < n_{\text{glass}}$, so the phase difference in the two reflected rays depends only on the thickness of the coating. For enhanced reflection, constructive interference is needed. This occurs when the path length difference in the coating, $2t$, is equal to the wavelength in the coating, $\lambda_{\text{coating}} = \lambda / n_{\text{coating}}$ (for minimum thickness).

Solution Find the minimum thickness of the coating.

$$2t = \lambda_{\text{coating}} = \frac{\lambda}{n_{\text{coating}}},\text{ so } \boxed{t = \frac{\lambda}{2n_{\text{coating}}}}.$$

81. Strategy Use Eq. (25-10). The relationship between the diffraction angle θ, the grating-screen distance D, and the position x on the screen is $\tan\theta = x/D$. The slit separation is given by $d = 1/N$.

Solution Compute the slit separation.

$$d = \frac{1}{10{,}000.0 \ \text{slits/cm}} = 1.00000\times10^{-6} \ \text{m}$$

Solve for θ.

$$d\sin\theta = m\lambda, \ \text{so} \ \theta = \sin^{-1}\frac{m\lambda}{d}.$$

Solve for x.

$$\tan\theta = \frac{x}{D}, \ \text{so} \ x = D\tan\theta = D\tan\left(\sin^{-1}\frac{m\lambda}{d}\right) = (2.0 \ \text{m})\tan\left(\sin^{-1}\frac{m\lambda}{1.00000\times10^{-6} \ \text{m}}\right).$$

Using $\lambda = 690$ nm and $m = 0, \pm1$ gives the locations of the red lines.

$$x_{R0} = (2.0 \ \text{m})\tan\left(\sin^{-1}\frac{0\cdot\lambda}{1.00000\times10^{-6} \ \text{m}}\right) = 0 \ \text{and} \ x_{R\pm1} = (2.0 \ \text{m})\tan\left(\sin^{-1}\frac{\pm1\cdot690\times10^{-9} \ \text{m}}{1.00000\times10^{-6} \ \text{m}}\right) = \pm1.9 \ \text{m}.$$

Using $\lambda = 460$ nm and $m = 0, \pm1$, and ±2 gives the locations of the blue lines.

$$x_{B0} = 0, \ x_{B\pm1} = (2.0 \ \text{m})\tan\left(\sin^{-1}\frac{\pm1\cdot460\times10^{-9} \ \text{m}}{1.00000\times10^{-6} \ \text{m}}\right) = \pm1.0 \ \text{m, and}$$

$$x_{B\pm2} = (2.0 \ \text{m})\tan\left(\sin^{-1}\frac{\pm2\cdot460\times10^{-9} \ \text{m}}{1.00000\times10^{-6} \ \text{m}}\right) = \pm4.7 \ \text{m}.$$

The only overlap is at $x = 0$, where the lines combine to make purple. The pattern is shown below.

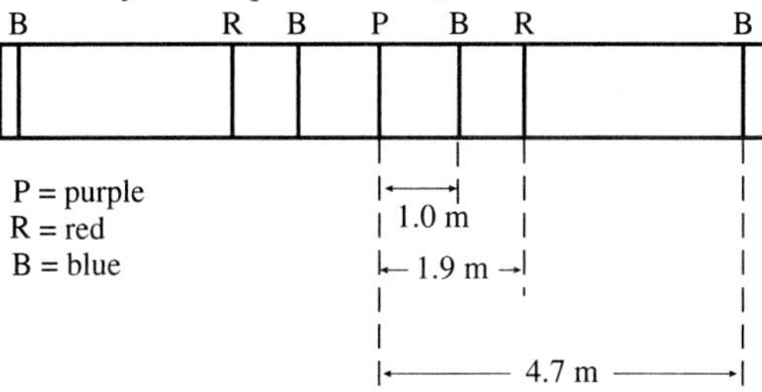

85. Strategy Use Eq. (25-15).

Solution

(a) Since the "reflected" beam makes an angle of 2θ with the incident beam, $2\theta = 1.3$ rad and $\theta = 0.65$ rad. Find the crystal plane separation.

$$2d\sin\theta = m\lambda, \ \text{so} \ d = \frac{m\lambda}{2\sin\theta} = \frac{(1)(0.18\times10^{-9} \ \text{m})}{\sin0.65} = \boxed{0.15 \ \text{nm}}.$$

(b) Check to see if there is a second maximum.

$$\sin\theta = \frac{m\lambda}{2d} = \frac{(2)(0.18\times10^{-9} \ \text{m})}{2(0.15\times10^{-9} \ \text{m})} = 1.2 > 1$$

Since $\sin\theta > 1$, there are no other maxima produced by this plane separation. The answer is $\boxed{\text{no}}$.

Review Exercises

1. **Strategy** The sound of the bat hitting the ball travels 22 m from the bat to the microphone in a time Δt_1. Then, the EM wave travels 4500 km from the stadium to the television set in a time Δt_2. Finally, the sound travels 2.0 m from the television set to your ears in a time Δt_3. Compute the time it takes for each wave, sound and EM, to travel the stated distances using $\Delta x = v\Delta t$.

   Solution Compute the minimum time it takes for you to hear the crack of the bat after the batter hit the ball.

   $$\Delta t_1 + \Delta t_2 + \Delta t_3 = \frac{22\ \text{m}}{343\ \text{m/s}} + \frac{4.50\times10^6\ \text{m}}{3.00\times10^8\ \text{m/s}} + \frac{2.0\ \text{m}}{343\ \text{m/s}} = \boxed{85\ \text{ms}}.$$

5. **Strategy** Since the radio waves interfere destructively and the reflection at the plane introduces a 180° phase shift, the minimum path difference must be equal to one wavelength.

   Solution

   (a) Find the wavelength of the radio waves.

   $$\lambda = \frac{c}{f} = \frac{2.998\times10^8\ \text{m/s}}{1408\times10^3\ \text{Hz}} = 212.93\ \text{m}$$

   $2\lambda = 425.9$ m and $3\lambda = 638.8$ m. If Juanita is correct that the plane is at least 500 m over her head, $\boxed{\text{the plane must be at least } 3\lambda = 638.8\ \text{m over Juanita's head}}$.

   (b) The lower estimate is $2\lambda = \boxed{425.9\ \text{m}}$ and the higher estimate is $4\lambda = \boxed{851.7\ \text{m}}$.

9. **Strategy and Solution** You don't see an interference pattern on your desk when light from two different lamps are illuminating the surface because $\boxed{\text{the lamps are not emitting coherent light}}$.

13. **Strategy** The distance between adjacent slits is the reciprocal of the number of slits per unit length. Use Eq. (25-10).

   Solution

   (a) Compute the distance between adjacent slits.

   $$d = \frac{1}{5550\ \text{slits/cm}} = \frac{1\ \text{cm}}{5550\ \text{slits}} = \frac{10^{-2}\ \text{m}}{5550\ \text{slits}} = \boxed{1.80\times10^{-6}\ \text{m}}$$

   (b) Find the distance between the central bright spot and the first-order maximum. Find θ.

   $d\sin\theta = m\lambda = (1)\lambda = \lambda$, so $\theta = \sin^{-1}\dfrac{\lambda}{d}$. Find x.

   $\tan\theta = \dfrac{x}{D}$, so

   $$x = D\tan\theta = D\tan\left(\sin^{-1}\frac{\lambda}{d}\right) = (5.50\ \text{m})\tan\left[\sin^{-1}\frac{0.680\times10^{-6}\ \text{m}}{1.8018\times10^{-6}\ \text{m}}\right] = \boxed{2.24\ \text{m}}.$$

(c) Find the distance between the central bright spot and the second-order maximum. Find θ.

$$d \sin \theta = m\lambda = 2\lambda, \text{ so } \theta = \sin^{-1} \frac{2\lambda}{d}. \text{ Find } x.$$

$$\tan \theta = \frac{x}{D}, \text{ so } x = D \tan \theta = D \tan\left(\sin^{-1}\frac{2\lambda}{d} \right) = (5.50 \text{ m}) \tan\left[\sin^{-1} \frac{2(0.680 \times 10^{-6} \text{ m})}{1.8018 \times 10^{-6} \text{ m}} \right] = \boxed{6.33 \text{ m}}.$$

(d) $\boxed{\text{The assumption that } \sin \theta \approx \tan \theta \text{ is not valid because the angles are not small.}}$

17. Strategy Geraldine want to display the zeroth- and first-order interference maxima across the full width of a 20.0-cm screen. To do so, the first-order minima must be at the edge of the screen. Use Eq. (25-11).

Solution Find D, the distance between the double slit and the screen.

$$d \sin \theta = d \frac{x}{\sqrt{x^2 + D^2}} = \left(m + \frac{1}{2} \right)\lambda = \left(1 + \frac{1}{2} \right)\lambda = \frac{3}{2}\lambda$$

Solve for D.

$$\frac{x}{\sqrt{x^2 + D^2}} = \frac{3\lambda}{2d}$$

$$\frac{x^2}{x^2 + D^2} = \left(\frac{3\lambda}{2d} \right)^2$$

$$x^2 + D^2 = \left(\frac{2d}{3\lambda} \right)^2 x^2$$

$$D = x\sqrt{\left(\frac{2d}{3\lambda} \right)^2 - 1} = \frac{0.200 \text{ m}}{2} \sqrt{\left[\frac{2(20.0 \times 10^{-6} \text{ m})}{3(423 \times 10^{-9} \text{ m})} \right]^2 - 1} = \boxed{3.15 \text{ m}}$$

21. Strategy Use Eq. (25-10). Let the subscript g stand for the green light and the subscript v stand for the violet light. Assume that the angles are small. Then $\sin \theta \approx \tan \theta = x/D$.

Solution

(a) Since $\sin \theta \propto \lambda$, the green maximum will be farther than the violet maximum from the central maximum. Find the separation $x_g - x_v$.

$$d \sin \theta_g = d\frac{x_g}{D} = \lambda_g \text{ and } d \sin \theta_v = d\frac{x_v}{D} = \lambda_v, \text{ so}$$

$$x_g - x_v = \frac{D}{d}\lambda_g - \frac{D}{d}\lambda_v = \frac{D}{d}(\lambda_g - \lambda_v) = \frac{0.720 \text{ m}}{0.020 \times 10^{-3} \text{ m}}(520 \times 10^{-9} \text{ m} - 412 \times 10^{-9} \text{ m}) = \boxed{3.9 \text{ mm}}.$$

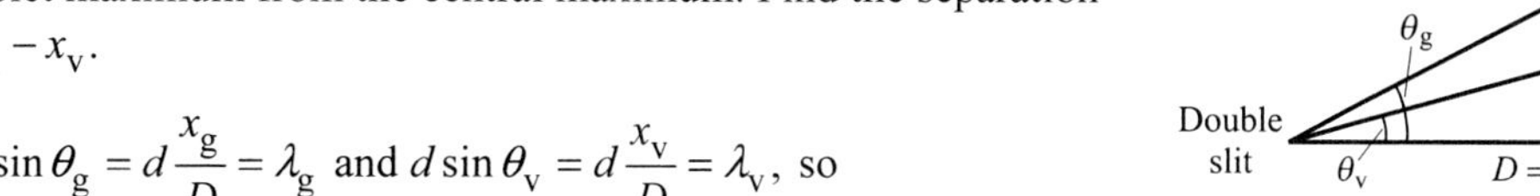

(b) Find the separation.

$$d \sin \theta_g = d\frac{x_g}{D} = 2\lambda_g \text{ and } d \sin \theta_v = d\frac{x_v}{D} = 2\lambda_v, \text{ so}$$

$$x_g - x_v = \frac{D}{d}(2\lambda_g) - \frac{D}{d}(2\lambda_v) = \frac{2D}{d}(\lambda_g - \lambda_v) = \frac{2(0.720 \text{ m})}{0.020 \times 10^{-3} \text{ m}}(520 \times 10^{-9} \text{ m} - 412 \times 10^{-9} \text{ m}) = \boxed{7.8 \text{ mm}}.$$

MCAT Review

1. **Strategy** Use the transverse magnification equation.

   Solution Find the ratio of the height of the image to the height of the object.
   $$m = \frac{h'}{h} = -\frac{q}{p} = -\frac{\frac{3}{2}f}{3f} = -\frac{1}{2}$$
   The correct answer is $\boxed{A}$.

2. **Strategy** The object and image distances are the same. Use the mirror equation and the fact that $f = R/2$.

   Solution Find p, the object location.
   $$\frac{1}{p} + \frac{1}{q} = \frac{1}{p} + \frac{1}{p} = \frac{2}{p} = \frac{1}{f} = \frac{2}{R}, \text{ so } p = R = 50 \text{ cm.}$$
   The correct answer is $\boxed{C}$.

3. **Strategy and Solution** When the telescope is focused on a very distant object, the image is located at the focal point of the mirror. The image is formed in front of the mirror, so it is real. The image and object distances are positive, so according to the magnification equation, $m = -q/p < 0$, therefore, the image is inverted.

   The correct answer is $\boxed{B}$.

4. **Strategy** The magnification is the equal to the ratio of the focal length of the mirror to that of the eyepiece.

   Solution The magnification of the Hubble is approximately
   $$\frac{f_{\text{mirror}}}{f_{\text{eyepiece}}} = \frac{13 \text{ m}}{2.5 \times 10^{-2} \text{ m}} = 520. \text{ The correct answer is } \boxed{C}.$$

5. **Strategy** For a very distant object, p is very large, so $1/p$ is approximately zero. Use the mirror equation.

   Solution Find the image location.
   $$\frac{1}{f} = \frac{1}{q} + \frac{1}{p} \approx \frac{1}{q} + 0 = \frac{1}{q}, \text{ so } q \approx f.$$
   The image location is very close to the focal point of the mirror. The correct answer is $\boxed{C}$.

6. **Strategy and Solution** Visible light has wavelengths greater than ultraviolet light. The wavelengths of visible light are large compared to atoms and molecules, so visible light is not that easily absorbed by the atmosphere. The wavelengths of ultraviolet light are closer to the size of atoms and molecules in the atmosphere, so ultraviolet waves are more readily absorbed by the atmosphere. The correct answer is $\boxed{A}$.

Chapter 26

RELATIVITY

Conceptual Questions

1. Whether something is obvious depends on our common sense about it. In everyday experience we only observe objects in motion with speeds much smaller than the speed of light. We have no experience with, and hence no common sense about, objects moving close to the speed of light. Therefore, it is not obvious that moving clocks don't run slow or that lengths don't shorten when the speeds involved approach the speed of light.

5. The astronaut would measure about 52 beats per minute. He is in his own rest frame, so he measures the proper time between heartbeats—the same as if he were taking his pulse back on Earth.

9. Celia will find that the flash from the left will reach her ship before the flash from the right. According to Abe, this occurs because her ship is moving toward the left flash. According to Celia, however, the two pulses travel equal distances to reach her, so the flash from the left must have been emitted earlier than the flash from the right.

Problems

1. **Strategy** Compute the times required for the optical signal and the train to reach the station and find the difference.

   Solution The time required for the optical signal to reach the station, as measured by an observer at rest relative to the station (the stationmaster) is

   $$\frac{d}{c} = \frac{1.0 \text{ km}}{3.00 \times 10^5 \text{ km/s}} = 3.3 \ \mu s.$$

   Measured by the same observer, the time needed for the train to arrive is

   $$\frac{d}{v} = \frac{d}{0.60c} = \frac{1.0 \text{ km}}{0.60 \cdot 3.00 \times 10^5 \text{ km/s}} = 5.6 \ \mu s.$$

   The difference in the arrival times is $5.56 \ \mu s - 3.33 \ \mu s = \boxed{2.2 \ \mu s}$.

5. **Strategy** The proper time interval Δt_0 is the time interval measured by the Rolex. The time interval measured at mission control is $\Delta t = 12.0$ h. The time-dilation equation can be used to find Δt_0.

   Solution

   $$\Delta t_0 = \frac{\Delta t}{\gamma} = \Delta t \sqrt{1 - \frac{v^2}{c^2}} = (12.0 \text{ h}) \sqrt{1 - \frac{(2.0 \times 10^8 \text{ m/s})^2}{(3.00 \times 10^8 \text{ m/s})^2}} = \boxed{8.9 \text{ h}}$$

9. **(a) Strategy** The age of the passenger at the end of the trip is found by adding the elapsed time Δt_0, as measured by the ship's clock, to the passenger's age at the start of the trip. To find the elapsed time Δt_0, use the time dilation equation with $\Delta t = \dfrac{\text{distance relative to Earth}}{\text{speed relative to Earth}} = \dfrac{710 \text{ ly}}{0.9999c} = 710 \text{ yr.}$

Solution Find Δt.

$$\Delta t_0 = \frac{\Delta t}{\gamma} = \Delta t\sqrt{1 - v^2/c^2} = (710 \text{ yr})\sqrt{1 - 0.9999^2} = 10 \text{ yr}$$

The passenger's age at the end of the trip is 20 years old + 10 years = $\boxed{30 \text{ years old}}$.

(b) Strategy and Solution The spaceship takes slightly more than 710 years to reach its destination, as measured by an observer on Earth. The radio signal takes 710 years to reach Earth. The total elapsed time between the departure of the ship and the arrival of the signal back on Earth is 710 years + 710 years = 1420 years. The year will be 2000 + 1420 = $\boxed{3420}$.

11. **Strategy** The proper time interval is 8.0 h. The time interval Δt measured by the clock on the ground is given by the time dilation equation $\Delta t = \gamma \Delta t_0$. Use the binomial approximation.

Solution The time difference is

$$\Delta t - \Delta t_0 = \gamma \Delta t_0 - \Delta t_0 = (\gamma - 1)\Delta t_0 = \left[(1 - v^2/c^2)^{-1/2} - 1\right]\Delta t_0.$$

Since $v \ll c$, by the binomial approximation, $\gamma = (1 - v^2/c^2)^{-1/2} \approx 1 + \dfrac{v^2}{2c^2}.$

The time difference is $\Delta t - \Delta t_0 \approx \left(1 + \dfrac{v^2}{2c^2} - 1\right)\Delta t_0 = \dfrac{v^2}{2c^2}\Delta t_0 = \dfrac{(220 \text{ m/s})^2}{2(3.00 \times 10^8 \text{ m/s})^2}(8.0 \text{ h}) = \boxed{7.7 \text{ ns}}.$

13. **(a) Strategy** The observers on Earth see the contracted heights of the occupants of the spaceship, since the heights are along the direction of motion. Use Eq. (26-4).

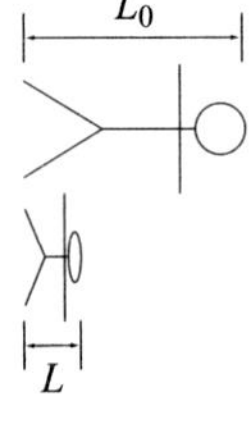

Solution Compute the approximate heights of the occupants as viewed by others on the ship.

$$L = \frac{L_0}{\gamma}, \text{ so } L_0 = \gamma L = \frac{L}{\sqrt{1 - v^2/c^2}} = \frac{0.50 \text{ m}}{\sqrt{1 - 0.97^2}} = \boxed{2 \text{ m}}.$$

(b) Strategy The widths of the occupants are perpendicular to their direction of motion, so the widths are not contracted.

Solution The others on the ship see the same widths as the observers on Earth, $\boxed{0.50 \text{ m}}$.

15. **(a) Strategy** The proper length is $L_0 = 91.5$ m. The length measured in the rest frame of the particle is given by Eq. (26-4).

Solution

$$L = \frac{L_0}{\gamma} = L_0\sqrt{1 - v^2/c^2} = (91.5 \text{ m})\sqrt{1 - 0.50^2} = \boxed{79 \text{ m}}$$

(b) Strategy Since the speed of the particle is constant relative to Earth, the time is found using the equation $\Delta x = v\Delta t$.

Solution

$$\Delta t = \frac{\Delta x}{v} = \frac{91.5 \text{ m}}{0.50(3.00\times10^8 \text{ m/s})} = 6.1\times10^{-7} \text{ s} = \boxed{610 \text{ ns}}$$

(c) Strategy In the rest frame of the particle, the distance traveled is 79 m and the speed is $0.50c$.

Solution The time required is

$$\Delta t = \frac{\Delta x}{v} = \frac{79 \text{ m}}{0.50(3.00\times10^8 \text{ m/s})} = 5.3\times10^{-7} \text{ s} = \boxed{530 \text{ ns}}.$$

17. **Strategy** Use Eqs. (26-2) and (26-4).

Solution The length measured from the other ship is contracted by a factor

$$\gamma = (1 - v^2/c^2)^{-1/2} = (1 - 0.90^2)^{-1/2} = 2.3. \text{ The contracted length is } L = \frac{L_0}{\gamma} = \frac{30.0 \text{ m}}{2.3} = \boxed{13 \text{ m}}.$$

19. **(a) Strategy and Solution** Length contraction occurs only along the direction of motion. Since the rod is held perpendicular to the direction of motion, the Earth observer will measure the same length for the rod as the pilot of the ship did, $\boxed{1.0 \text{ m}}$.

(b) Strategy The pilot still measures the proper length of the rod as $L_0 = 1.0$ m. To the Earth observer, the rod is now contracted. Use Eq. (26-4).

Solution Find the length of the rod according to the Earth observer.

$$L = \frac{L_0}{\gamma} = L_0\sqrt{1 - v^2/c^2} = (1.0 \text{ m})\sqrt{1 - 0.40^2} = \boxed{0.92 \text{ m}}$$

21. **Strategy** The observer on the ground measures the proper length between the towers as $L_0 = 3.0$ km. The distance is contracted for a passenger on the train. Use Eq. (26-4).

Solution

(a) Relative to the passenger on the train, the distance between the towers is

$$L = \frac{L_0}{\gamma} = L_0\sqrt{1 - v^2/c^2} = (3.0 \text{ km})\sqrt{1 - 0.80^2} = 1.8 \text{ km}.$$

The time interval measured by the passenger is $\Delta t = \dfrac{L}{v} = \dfrac{1.8 \text{ km}}{0.80(3.00\times10^5 \text{ km/s})} = 7.5\times10^{-6} \text{ s} = \boxed{7.5 \text{ μs}}.$

(b) The observer on the ground measures the time interval

$$\Delta t = \frac{L_0}{v} = \frac{3.0 \text{ km}}{0.80(3.00\times10^5 \text{ km/s})} = 1.3\times10^{-5} \text{ s} = \boxed{13 \text{ μs}}.$$

25. Strategy According to the postulates of special relativity, the speed of light must be the same in all inertial reference frames.

Solution The Earth and the spaceship are both inertial reference frames (no acceleration), so Siu-Ling will measure a speed of light of $\boxed{3.00\times10^8 \text{ m/s}}$.

27. Strategy Since the ships are approaching the Moon from opposite directions, we use a coordinate system relative to the Moon such that one ship (ship A) approaches from the left at a speed $0.60c$, and the other ship (ship B) approaches from the right at speed $0.80c$. The velocity of ship A relative to ship B is written v_{AB}. If the velocity of ship A relative to the Moon is $v_{AM} = 0.60c$, then the velocity of ship B relative to the Moon is $v_{BM} = -0.80c$; so $v_{MB} = -v_{BM} = 0.80c$. Use the relativistic velocity transformation equation.

Solution Find the velocity of ship A relative to ship B.

$$v_{AB} = \frac{v_{AM} + v_{MB}}{1 + v_{AM}v_{MB}/c^2} = \frac{0.60c + 0.80c}{1 + (0.60)(0.80)} = 0.946c$$

$v_{BA} = -v_{AB} = -0.946c$, so the relative speed measured by a passenger on either ship is $\boxed{0.946c}$.

29. Strategy The velocity of the electron relative to the lab is v_{EL}. $v_{PL} = \frac{4}{5}c$ is the velocity of the proton relative to the lab. $v_{EP} = -\frac{5}{7}c$ is the velocity of the electron relative to the proton. The velocity of the electron relative to the proton is negative since the electron moves to the left and the positive direction is chosen to be to the right. Use Eq. (26-5).

Solution

$$v_{EL} = \frac{v_{EP} + v_{PL}}{1 + v_{EP}v_{PL}/c^2} = \frac{-\frac{5}{7}c + \frac{4}{5}c}{1 + \left(-\frac{5}{7}\right)\left(\frac{4}{5}\right)} = \frac{1}{5}c$$

The speed of the electron relative to the lab is $\boxed{\frac{1}{5}c}$.

31. Strategy The speed of electron B in any frame of reference in which electron A is at rest is the same as the speed of electron B relative to electron A. The coordinate system relative to the lab is chosen so that objects traveling west have a positive velocity. Then the velocity of electron B relative to the lab is $v_{BL} = \frac{4}{5}c$ and the velocity of electron A relative to the lab is $v_{AL} = \frac{3}{5}c$, so $v_{LA} = -v_{AL} = -\frac{3}{5}c$. Use Eq. (26-5).

Solution

$$v_{BA} = \frac{v_{BL} + v_{LA}}{1 + v_{BL}v_{LA}/c^2} = \frac{\frac{4}{5}c + \left(-\frac{3}{5}c\right)}{1 + \left(\frac{4}{5}\right)\left(-\frac{3}{5}\right)} = \frac{5}{13}c$$

The speed of electron B in a frame of reference in which electron A is at rest is $\boxed{\frac{5}{13}c}$.

33. Strategy The magnitude of the momentum is given by $p = \gamma mv$.

Solution Find the speed of the electron.

$$p = \gamma mv$$

$$p\sqrt{1 - v^2/c^2} = mv$$

$$p^2(1 - v^2/c^2) = m^2 v^2$$

$$p^2 = m^2 v^2 + \frac{p^2 v^2}{c^2} = v^2\left(m^2 + \frac{p^2}{c^2}\right)$$

$$v = \frac{p}{\sqrt{m^2 + \frac{p^2}{c^2}}} = \frac{2.4\times10^{-22}\ \text{kg}\cdot\text{m/s}}{\sqrt{(9.109\times10^{-31}\ \text{kg})^2 + \frac{(2.4\times10^{-22}\ \text{kg}\cdot\text{m/s})^2}{(3.00\times10^8\ \text{m/s})^2}}} \times \frac{c}{3.00\times10^8\ \text{m/s}} = \boxed{0.66c}$$

35. (a) Strategy The magnitude of the momentum is given by $p = \gamma mv$.

Solution Compute the magnitude of the body's momentum.

$$p = \gamma mv = \frac{mv}{\sqrt{1 - v^2/c^2}} = \frac{(12.6\ \text{kg})0.87(3.00\times10^8\ \text{m/s})}{\sqrt{1 - 0.87^2}} = \boxed{6.7\times10^9\ \text{kg}\cdot\text{m/s}}$$

(b) Strategy Use the impulse-momentum theorem.

Solution Find the time required to bring the body to rest.

$$\Delta p = F\Delta t,\ \text{so}\ \Delta t = \frac{\Delta p}{F} = \frac{0 - 6.7\times10^9\ \text{kg}\cdot\text{m/s}}{-424.6\ \text{N}} \times \frac{1\ \text{yr}}{3.156\times10^7\ \text{s}} = \boxed{0.50\ \text{yr}}.$$

37. (a) Strategy Use Eq. (26-8) for the kinetic energy.

Solution Find the speed of the electrons.

$$K = (\gamma - 1)mc^2 \Rightarrow 1 + \frac{K}{mc^2} = \frac{1}{\sqrt{1 - \frac{v^2}{c^2}}} \Rightarrow 1 - \frac{v^2}{c^2} = \left(1 + \frac{K}{mc^2}\right)^{-2} \Rightarrow v = c\sqrt{1 - \left(1 + \frac{K}{mc^2}\right)^{-2}}$$

The speed is $v = c\sqrt{1 - \left(1 + \dfrac{25\ \text{MeV}}{0.5110\ \text{MeV}}\right)^{-2}} = \boxed{0.99980c}$.

(b) Strategy The time it takes the electrons to exit the accelerator and reach the patient in the reference frame of the hospital is $\Delta t = \Delta x/v$, where Δx is 15 cm and the speed v is that found in part (a). In the reference frame of the electrons, the time interval is shorter by a factor of $1/\gamma$.

Solution Compute the time interval in the reference frame of the electrons.

$$\Delta t_0 = \frac{\Delta t}{\gamma} = \frac{\Delta x/v}{\gamma} = \frac{\Delta x}{\gamma v} = \frac{\Delta x\sqrt{1 - v^2/c^2}}{v} = \frac{(0.15\ \text{m})\sqrt{1 - 0.99980^2}}{0.99980(2.998\times10^8\ \text{m/s})} = \boxed{0.010\ \text{ns}}$$

41. (a) Strategy and Solution The kinetic energy is $K = U = qV = +6e \times 125 \text{ MV} = \boxed{750 \text{ MeV}}$.

(b) Strategy Use Eq. (26-8) for the kinetic energy.

Solution Find the speed of the ions.

$$K = (\gamma - 1)mc^2 \Rightarrow 1 + \frac{K}{mc^2} = \gamma = \frac{1}{\sqrt{1 - \frac{v^2}{c^2}}} \Rightarrow 1 - \frac{v^2}{c^2} = \left(1 + \frac{K}{mc^2}\right)^{-2} \Rightarrow v = c\sqrt{1 - \left(1 + \frac{K}{mc^2}\right)^{-2}}$$

The speed is $v = c\sqrt{1 - \left(1 + \dfrac{750 \text{ MeV}}{11{,}172 \text{ MeV}}\right)^{-2}} = \boxed{0.34908c}$.

(c) Strategy In the reference frame of the laboratory, the length of the accelerator is 7.50 m. In the reference frame of the ions, the length is contracted by a factor of $1/\gamma$.

Solution Compute the length of the accelerator in the reference frame of the ions.

$$L = \frac{L_0}{\gamma} = \frac{L_0}{1 + K/(mc^2)} = \frac{7.50 \text{ m}}{1 + (750 \text{ MeV})/(11{,}172 \text{ MeV})} = \boxed{7.03 \text{ m}}$$

43. Strategy Since no energy is lost to the environment, the change in the kinetic energy of the two lumps can be equated to a change in the rest energy of the system.

Solution $\Delta E_0 = (\Delta m)c^2$, so

$$\Delta m = \frac{\Delta E_0}{c^2} = \frac{\Delta K_1 + \Delta K_2}{c^2} = \frac{\frac{1}{2}m_1 v_1^2 + \frac{1}{2}m_2 v_2^2}{c^2}.$$

The nonrelativistic form of the kinetic energy is used since the speeds are very small compared to c. Using $m_1 = m_2 = 1.00 \text{ kg}$ and $v_1 = v_2 = 30.0 \text{ m/s}$ gives

$$\Delta m = \frac{\frac{1}{2}(1.00 \text{ kg})(30.0 \text{ m/s})^2 + \frac{1}{2}(1.00 \text{ kg})(30.0 \text{ m/s})^2}{(3.00 \times 10^8 \text{ m/s})^2} = 1.00 \times 10^{-14} \text{ kg}.$$

The mass of the system $\boxed{\text{increased by } 1.00 \times 10^{-14} \text{ kg}}$.

45. Strategy The mass of the Sun is 1.987×10^{30} kg. 80.0% of the limiting mass of the white dwarf is converted to energy. Use Eq. (26-7).

Solution Compute the energy released by the explosion of the white dwarf.

$$E_0 = mc^2 = 0.800(1.4)(1.987 \times 10^{30} \text{ kg})(3.00 \times 10^8 \text{ m/s})^2 = \boxed{2.0 \times 10^{47} \text{ J}}$$

47. Strategy Total energy must be conserved.

Solution Find the energy released in the decay.

rest energy before decay $= m_{\text{Rn}}c^2 =$ rest energy after decay $+$ energy released $= m_{\text{Po}}c^2 + m_\alpha c^2 + E$, so

$$E = (m_{\text{Rn}} - m_{\text{Po}} - m_\alpha)c^2 = (221.970\,39\,\text{u} - 217.962\,89\,\text{u} - 4.001\,51\,\text{u})(931.494 \text{ MeV/u}) = \boxed{5.58 \text{ MeV}}.$$

49. Strategy Use Eq. (26-9).

Solution Find the electron's speed.

$$E = \gamma mc^2 = (1 - v^2/c^2)^{-1/2} mc^2$$

$$\sqrt{1 - \frac{v^2}{c^2}} = \frac{mc^2}{E}$$

$$\frac{v^2}{c^2} = 1 - \frac{m^2 c^4}{E^2}$$

$$v = c\sqrt{1 - \frac{m^2 c^4}{E^2}} = c\sqrt{1 - \frac{(9.109 \times 10^{-31}\ \text{kg})^2 (3.00 \times 10^8\ \text{m/s})^4}{(1.02 \times 10^{-13}\ \text{J})^2}} = \boxed{0.595c}$$

51. Strategy The object is moving at $1.80/3.00 = 0.600$ times c. Use Eq. (26-8).

Solution Find the kinetic energy of the object.

$$K = (\gamma - 1)mc^2 = [(1 - v^2/c^2)^{-1/2} - 1]mc^2 = [(1 - 0.600^2)^{-1/2} - 1](0.12\ \text{kg})(3.00 \times 10^8\ \text{m/s})^2 = \boxed{2.7 \times 10^{15}\ \text{J}}$$

53. Strategy Solve Eq. (26-10) for p using $E = 5.0mc^2$ and $E_0 = mc^2$.

Solution Find the magnitude of the electron's momentum as observed in the laboratory.

$$E^2 = (5.0mc^2)^2 = 25m^2 c^4 = E_0^2 + (pc)^2 = m^2 c^4 + p^2 c^2,\ \text{so}\ p = \sqrt{24}mc = \boxed{4.9mc}.$$

57. Strategy Convert 1 MeV to joules. Then divide the result by c.

Solution

$$1\ \text{MeV} = (1 \times 10^6\ \text{eV})\frac{1.602 \times 10^{-19}\ \text{J}}{1\ \text{eV}} = 1.602 \times 10^{-13}\ \text{J, so}\ \frac{1\ \text{MeV}}{c} = \frac{1.602 \times 10^{-13}\ \text{J}}{2.998 \times 10^8\ \text{m/s}} = 5.344 \times 10^{-22}\ \text{kg} \cdot \text{m/s}.$$

The conversion is $\boxed{1\ \text{MeV}/c = 5.344 \times 10^{-22}\ \text{kg} \cdot \text{m/s}}$.

61. Strategy Use Eq. (26-10) and the definition of total energy.

Solution Show that $(pc)^2 = K^2 + 2KE_0$.

The total energy is $E = K + E_0$.

Squaring both sides gives $E^2 = (K + E_0)^2 = K^2 + 2KE_0 + E_0^2$.

Combining this with the energy-momentum relation $E^2 = E_0^2 + (pc)^2$ yields

$E_0^2 + (pc)^2 = K^2 + 2KE_0 + E_0^2$, or $(pc)^2 = K^2 + 2KE_0$, which is Eq. (26-11).

65. Strategy Refer to Example 26.2. Use the frame of reference of an observer on the ground.

Solution

(a) For an observer on the ground, the distance traveled by the muons is just the difference between the initial and final altitudes.

$$4500 \text{ m} - 0 \text{ m} = \boxed{4500 \text{ m}}$$

(b) For an observer on the ground, the time of flight is

$$\Delta t = \frac{\Delta x}{v} = \frac{4500 \text{ m}}{0.9950(3.00 \times 10^8 \text{ m/s})} = 1.5 \times 10^{-5} \text{ s} = \boxed{15 \text{ μs}}$$

(c) The proper time interval for the half-life of a muon is measured in the reference frame of the muon itself, so $\Delta t_0 = 1.5$ μs. An observer on the ground sees the half-life as

$$\Delta t = \gamma \Delta t_0 = (1 - v^2/c^2)^{-1/2} \Delta t_0 = (1 - 0.9950^2)^{-1/2}(1.5 \text{ μs}) = \boxed{15 \text{ μs}}.$$

(d) The time of flight is equal to the half-life, so during the flight half of the muons decay. So, $\boxed{500{,}000}$ muons reach sea level.

69. Strategy Use the length contraction equation to find γ. Then, use the time dilation equation to find how long the astronaut would say that she exercised.

Solution Find γ.

$$L = \frac{L_0}{\gamma}, \text{ so } \gamma = \frac{L_0}{L}.$$

Find Δt_0.

$$\Delta t = \gamma \Delta t_0 = \frac{L_0}{L} \Delta t_0, \text{ so } \Delta t_0 = \frac{L}{L_0} \Delta t = \frac{30.5 \text{ m}}{35.2 \text{ m}}(22.2 \text{ min}) = \boxed{19.2 \text{ min}}.$$

73. Strategy Solve the time dilation equation for v. $\Delta t_0 = 3.0$ days and $\Delta t = 4.0$ days.

Solution Find the speed of the starship relative to the space station.

$$\Delta t = \gamma \Delta t_0$$
$$4.0 \text{ d} = (1 - v^2/c^2)^{-1/2}(3.0 \text{ d})$$
$$\sqrt{1 - v^2/c^2} = 0.75$$
$$\frac{v^2}{c^2} = 1 - 0.75^2$$
$$v = c\sqrt{1 - 0.75^2} = \boxed{0.66c}$$

77. (a) Strategy Let L_{sy} and L_{sE} be the lengths of the ship in your and the Earth's reference frames, respectively. Find the proper length of the ship. Then, find its length as viewed by observers on Earth. From Problem 76, the speed of the ship relative to the Earth is $0.966c$. Use Eq. (26-4).

Solution Find the proper length L_0.

$$L_{sy} = L_0\sqrt{1 - v_{sy}^2/c^2}, \text{ so } L_0 = \frac{L_{sy}}{\sqrt{1 - v_{sy}^2/c^2}} = \frac{24 \text{ m}}{\sqrt{1 - 0.50^2}} = 28 \text{ m}.$$

Find the length of the ship as measured by observers on Earth.

$$L_{sE} = L_0\sqrt{1 - v_{sE}^2/c^2} = (28 \text{ m})\sqrt{1 - 0.966^2} = \boxed{7.2 \text{ m}}$$

(b) Strategy Your speed relative to the Earth is $0.90c$. Use Eq. (26-4).

Solution Find the length of the rod as measured by observers on Earth.

$$L = L_0\sqrt{1 - v^2/c^2} = (24 \text{ m})\sqrt{1 - 0.90^2} = \boxed{10 \text{ m}}$$

(c) Strategy Your speed relative to the other ship is $0.50c$. Use Eq. (26-4).

Solution Find the length of the rod as measured by observers on the other ship.

$$L = L_0\sqrt{1 - v^2/c^2} = (24 \text{ m})\sqrt{1 - 0.50^2} = \boxed{21 \text{ m}}$$

81. Strategy Use the definition of total energy and Eq. (26-10).

Solution If K is much larger than E_0, then E is also much larger than E_0, since $E = K + E_0 > K \gg E_0$. Now, $E^2 = E_0^2 + (pc)^2$, so $E^2 - E_0^2 = (pc)^2$. Since $E \gg E_0$, $E^2 \gg E_0^2$, so $E^2 - E_0^2 \approx E^2$. Therefore, $E^2 \approx (pc)^2$, or $E \approx pc$.

85. (a) Strategy Use the conversion between eV and J.

Solution

$$K = (2.0\times10^{20} \text{ eV})\frac{1.602\times10^{-19} \text{ J}}{\text{eV}} = \boxed{32 \text{ J}}$$

(b) Strategy The amount of energy needed to lift an object is the same as the work done on the object.

Solution Find how far the object could be lifted.

$$W = Fd = mgd, \text{ so } d = \frac{W}{mg} = \frac{32 \text{ J}}{(1.0 \text{ kg})(9.80 \text{ m/s}^2)} = \boxed{3.3 \text{ m}}.$$

(c) Strategy Use Eqs. (26-2) and (26-8) and the binomial approximation.

Solution Find γ.

$$K = (\gamma - 1)mc^2, \text{ so } \gamma = \frac{K}{mc^2} + 1 = \frac{32 \text{ J}}{(1.673\times10^{-27} \text{ kg})(3.00\times10^8 \text{ m/s})^2} + 1 = 2.1\times10^{11}.$$

Find the speed of the proton.

$$\gamma = (1 - v^2/c^2)^{-1/2}$$

$$1 - \frac{v^2}{c^2} = \frac{1}{\gamma^2}$$

$$\frac{v}{c} = \left(1 - \frac{1}{\gamma^2}\right)^{1/2}$$

Since $\gamma \gg 1$, we know $1/\gamma^2 \ll 1$, so the binomial expansion can be used.

$$\frac{v}{c} \approx 1 - \frac{1}{2\gamma^2}, \text{ so}$$

$$v \approx \left(1 - \frac{1}{2\gamma^2}\right)c = \left[1 - \frac{1}{2(2.1 \times 10^{11})^2}\right]c \approx (1 - 1.1 \times 10^{-23})c = \boxed{0.999\,999\,999\,999\,999\,999\,999\,989\,c}.$$

89. Strategy Use the time dilation equation and the binomial approximation.

Solution Find an approximate form of the time dilation equation. Then, find the time the astronaut spent on the shuttle Δt_0.

$$\Delta t = \gamma \Delta t_0 = \frac{\Delta t_0}{\sqrt{1 - v^2/c^2}} \approx \Delta t_0 \left[1 - \left(-\frac{1}{2}\right)\frac{v^2}{c^2}\right] = \Delta t_0 + \Delta t_0 \frac{v^2}{2c^2}, \text{ so}$$

$$\Delta t_0 \approx \frac{2c^2}{v^2}(\Delta t - \Delta t_0) = \frac{2(3.00 \times 10^8 \text{ m/s})^2}{(7.860 \times 10^3 \text{ m/s})^2}(1.0 \text{ s}) \times \frac{1 \text{ yr}}{3.156 \times 10^7 \text{ s}} = \boxed{92 \text{ yr}}.$$

91. Strategy Use conservation of momentum and total energy with the *nonrelativistic* expressions $p = mv$ and $K = mv^2/2$.

Solution By conservation of momentum, the α particle and Po nucleus have equal momentum magnitudes: $p_\alpha = p_{\text{Po}}$. The kinetic energies are related to the momenta by $K = p^2/(2m)$. The total energy of the system is conserved, so

$$m_{\text{Ra}}c^2 = m_\alpha c^2 + m_{\text{Po}}c^2 + K_\alpha + K_{\text{Po}}$$

$$(m_{\text{Ra}} - m_\alpha - m_{\text{Po}})c^2 = K_\alpha + K_{\text{Po}} = \frac{p^2}{2m_\alpha} + \frac{p^2}{2m_{\text{Po}}}$$

where $p \equiv p_\alpha = p_{\text{Po}}$ and the mass defect is
$\Delta m = m_{\text{Ra}} - m_\alpha - m_{\text{Po}} = 221.970\,39 \text{ u} - 4.001\,51 \text{ u} - 217.962\,89 \text{ u} = 0.00599 \text{ u}.$
Solve for p.

$$p = \sqrt{\frac{2(\Delta m)c^2}{\frac{1}{m_\alpha} + \frac{1}{m_{\text{Po}}}}} = c\sqrt{\frac{2(0.00599 \text{ u})}{\frac{1}{4.001\,51 \text{ u}} + \frac{1}{217.962\,89 \text{ u}}}} = (0.217 \text{ u})c$$

Now we can find the speeds using $p = mv$.

(a) $v_{\text{Po}} = \dfrac{p}{m_{\text{Po}}} = \dfrac{(0.217 \text{ u})(2.998 \times 10^8 \text{ m/s})}{217.962\,89 \text{ u}} = \boxed{2.98 \times 10^5 \text{ m/s}}$

(b) $v_\alpha = \dfrac{p}{m_\alpha} = \dfrac{(0.217 \text{ u})(2.998 \times 10^8 \text{ m/s})}{4.001\,51 \text{ u}} = \boxed{1.63 \times 10^7 \text{ m/s}}$

Both speeds are much less than c, so the nonrelativistic approximation was valid.

Chapter 27

EARLY QUANTUM PHYSICS AND THE PHOTON

Conceptual Questions

1. The photoelectric effect describes the mechanism by which electrons are ejected from materials when certain frequencies of electromagnetic radiation are incident upon them. The first puzzling feature of this effect was that brighter light increases the current of electrons but does not affect their kinetic energy. This is understood with the photon model of light because increasing the intensity of the radiation simply increases the number of photons colliding with the material but does not increase their energy. The second puzzling feature was that the maximum kinetic energy of the electrons depends upon the wavelength of the photons. This too is explained by the photon theory because a shorter wavelength photon has a greater energy, and therefore, will provide a larger "kick" to the photoelectron. The third puzzling feature was that for a given metal there is a threshold frequency below which no electrons will be emitted. This is an understandable phenomenon because some minimum amount of energy is required to set the electrons free from the material to which they are bound, and if the photon energy is too low, an electron will not be ejected after absorption of a photon. The final puzzling feature of the photoelectric effect was that electrons are emitted virtually instantaneously after turning on a photon source. This too makes sense as a result of the photon theory because each individual photon carries enough energy to cause a photoelectron to be emitted—only a single photon is required to start a photocurrent.

5. According to the Bohr model, the difference in energy between two levels of the hydrogen atom is proportional to the value of the electron charge. If the electron charge varied from particle to particle, the photons emitted from a hydrogen atom would not all have the same energy. Furthermore, when the emission spectrum of hydrogen was observed, photons from the same transition would have slightly different wavelengths and would therefore not form sharp lines.

9. The Bohr model contains four assumptions: 1) The electron can exist without radiating only in certain circular orbits; 2) The laws of Newtonian mechanics apply to the motion of the electron in any of these stationary states; 3) The electron can make a transition between stationary states through the emission or absorption of a single photon; 4) The stationary states are those circular orbits in which the electron's angular momentum is quantized in integral multiples of $h/(2\pi)$.

13. The process of pair production becomes especially important for photons with energies in excess of 1.02 MeV.

17. In a hydrogen gas discharge tube, the hydrogen atoms are excited to higher energy states in a random process of collisions with energetic electrons. Once in an excited state, the atoms decay back to the ground state, possibly passing through several intermediate states, and emitting one or more photons. Because of the random nature of the excitations in this process, the atoms in the tube are excited, on average, to every possible state, and pass through every possible intermediate state during the decays. The emission spectrum of hydrogen therefore contains a line for every possible atomic transition. In an absorption spectrum, the atoms absorb incident photons in a similarly random process that puts them on average in every excited state. Once excited, however, the atom quickly decays back to the ground state by emitting one or more photons. The excited states do not live long enough to absorb a significant amount of the incident radiation and undergo transitions from intermediate states to higher energy states. For this reason, an absorption spectrum shows lines corresponding to transitions from the ground state only, while an emission spectrum contains lines from transitions between any two states.

21. Since UV light has a higher frequency than blue, and both metals produce photoelectrons for blue light, both metals will also produce photoelectrons for UV light. Metal 1 produces photoelectrons for red light, so it *might* produce them for infrared as well. Metal 2 will not produce them for infrared light, since it doesn't produce them for red light. Photoelectrons are produced whenever the photon energy is sufficient to overcome the work function of the metal. Therefore, the metal with the higher threshold frequency has the larger work function, which is metal 2.

Problems

1. **(a) Strategy and Solution** The energy of a photon is inversely proportional to its wavelength, so the photons with the shorter wavelength have the greater energy. Therefore, a single $\boxed{\text{ultraviolet}}$ photon has the greater energy.

(b) Strategy The energy of a photon of EM radiation with frequency f is $E = hf$. The frequency and wavelength are related by $\lambda f = c$.

Solution Compute the energy of a single infrared photon.
$$E_{\text{infrared}} = hf = \frac{hc}{\lambda} = \frac{(6.626\times10^{-34}\ \text{J}\cdot\text{s})(3.00\times10^{8}\ \text{m/s})}{2.0\times10^{-6}\ \text{m}} = \boxed{9.9\times10^{-20}\ \text{J}}$$
Compute the energy of a single ultraviolet photon.
$$E_{\text{ultraviolet}} = hf = \frac{hc}{\lambda} = \frac{(6.626\times10^{-34}\ \text{J}\cdot\text{s})(3.00\times10^{8}\ \text{m/s})}{7.0\times10^{-8}\ \text{m}} = \boxed{2.8\times10^{-18}\ \text{J}}$$

(c) Strategy Use the definition of power. Let N be the number of photons. The total energy is NE, where E is the energy of a single photon.

Solution Find the number of infrared photons emitted per second.
$$P = \frac{NE}{\Delta t},\ \text{so}\ \frac{N_{\text{infrared}}}{\Delta t} = \frac{P}{E} = \frac{200\ \text{W}}{9.9\times10^{-20}\ \text{J/photon}} = \boxed{2.0\times10^{21}\ \text{photons/s}}.$$
Find the number of ultraviolet photons emitted per second.
$$\frac{N_{\text{ultraviolet}}}{\Delta t} = \frac{P}{E} = \frac{200\ \text{W}}{2.84\times10^{-18}\ \text{J/photon}} = \boxed{7.0\times10^{19}\ \text{photons/s}}$$

3. **Strategy** The energy of a photon of EM radiation with frequency f is $E = hf$. The frequency and wavelength are related by $\lambda f = c$.

Solution

(a) Calculate the wavelength of a photon with energy 3.1 eV.
$$E = hf = \frac{hc}{\lambda},\ \text{so}\ \lambda = \frac{hc}{E} = \frac{1240\ \text{eV}\cdot\text{nm}}{3.1\ \text{eV}} = \boxed{400\ \text{nm}}.$$

(b) Calculate the frequency of a photon with energy 3.1 eV.
$$f = \frac{c}{\lambda} = \frac{3.00\times10^{8}\ \text{m/s}}{400\times10^{-9}\ \text{m}} = \boxed{7.5\times10^{14}\ \text{Hz}}$$

5. **(a) Strategy** Use Einstein's photoelectric equation.

Solution Calculate the maximum kinetic energy.
$$K_{\text{max}} = hf - \phi = \frac{hc}{\lambda} - \phi = \frac{1240\ \text{eV}\cdot\text{nm}}{413\ \text{nm}} - 2.16\ \text{eV} = \boxed{0.84\ \text{eV}}$$

(b) Strategy The threshold wavelength is related to the threshold frequency by $\lambda_0 f_0 = c$. Use Eq. (27-8).

Solution Find the threshold wavelength.
$$hf_0 = \frac{hc}{\lambda_0} = \phi,\ \text{so}\ \lambda_0 = \frac{hc}{\phi} = \frac{1240\ \text{eV}\cdot\text{nm}}{2.16\ \text{eV}} = \boxed{574\ \text{nm}}.$$

7. Strategy The work function for the metal is $\phi = 2.60$ eV. Use Eq. (27-8) to find the maximum wavelength of the photons that will eject electrons from the metal. The frequency and wavelength are related by $\lambda f = c$.

Solution Find the longest wavelength.
$$f_0 = \frac{\phi}{h} = \frac{c}{\lambda}, \text{ so } \lambda = \frac{hc}{\phi} = \frac{1240 \text{ eV} \cdot \text{nm}}{2.60 \text{ eV}} = \boxed{477 \text{ nm}}.$$

9. Strategy The maximum kinetic energy of an emitted electron is the difference between the photon energy and the work function. The maximum kinetic energy of an emitted electron is also equal to the increase in potential energy for an electron moving through a potential difference $-V_s$, or eV_s. The maximum kinetic energy of the emitted electrons is independent of the intensity of the light, so the values of the intensity are irrelevant.

Solution Find the stopping potential V_s.
$$K_{max} = \frac{hc}{\lambda} - \phi = eV_s, \text{ so } V_s = \frac{hc/e}{\lambda} - \phi/e.$$
The stopping potential is inversely proportional to the wavelength of the incident photons and does not vary with intensity; thus, ranking the wavelengths from smallest to largest is equivalent to ranking the stopping potentials from largest to smallest. Ranking the stopping potentials, largest to smallest, we have $\boxed{\text{(e), (a) = (f), (b) = (c), (d)}}$.

13. Strategy The frequency and wavelength are related by $\lambda f = c$. Use Eq. (27-8).

Solution

(a) The three relevant frequencies are
$$f_{yellow} = \frac{c}{\lambda} = \frac{3.00 \times 10^8 \text{ m/s}}{580 \times 10^{-9} \text{ m}} = 5.2 \times 10^{14} \text{ Hz}, \quad f_{violet} = \frac{c}{\lambda} = \frac{3.00 \times 10^8 \text{ m/s}}{425 \times 10^{-9} \text{ m}} = 7.06 \times 10^{14} \text{ Hz, and}$$
$$f_0 = 6.20 \times 10^{14} \text{ Hz}.$$
$\boxed{\text{No}}$; only the source of $\boxed{\text{violet light}}$ can eject electrons from the metal surface. Since $f_{yellow} < f_0$, the yellow light source cannot eject electrons from the metal. Electrons are ejected from the metal by the violet light, since $f_{violet} > f_0$.

(b) The amount of energy required to eject an electron from the metal is the work function for the metal.
$$\phi = hf_0 = (4.136 \times 10^{-15} \text{ eV} \cdot \text{s})(6.20 \times 10^{14} \text{ Hz}) = \boxed{2.56 \text{ eV}}$$

17. Strategy The minimum potential difference is found by equating the energy of an x-ray photon and the electron's kinetic energy, which is equal to the electric potential energy of the electron, eV_{min}. The energy of a photon of EM radiation with frequency f is $E = hf$. The frequency and wavelength are related by $\lambda f = c$.

Solution Find the minimum potential difference.
$$hf = \frac{hc}{\lambda} = K = eV_{min}, \text{ so } V_{min} = \frac{hc}{\lambda e} = \frac{1240 \text{ eV} \cdot \text{nm}}{0.250 \text{ nm} \cdot e} = \boxed{4.96 \text{ kV}}.$$

19. Strategy The minimum wavelength can be found by equating the photon's energy and the electron's kinetic energy. The kinetic energy of the electron is equal to the electric potential energy of the electron, eV. Use Eq. (27-9).

Solution Find the minimum wavelength.
$$hf_{max} = \frac{hc}{\lambda_{min}} = K = eV, \text{ so } \lambda_{min} = \frac{hc}{eV} = \frac{1240 \text{ eV} \cdot \text{nm}}{40.0 \times 10^3 \text{ eV}} = \boxed{31.0 \text{ pm}}.$$

21. **Strategy** Let V be the potential difference in the x-ray tube. The maximum frequency (or cutoff frequency) occurs when the photon's energy is equal to the electron's kinetic energy. The kinetic energy of the electron is equal to the electric potential energy of the electron, eV. Use Eq. (27-9).

Solution

$$hf_{max} = K = eV, \text{ so } f_{max} = \frac{e}{h}V.$$

The cutoff frequency is proportional to the potential difference V, since e and h are constant.

23. **Strategy** Use the Compton shift, Eq. (27-14).

Solution

(a) Find the Compton shift in wavelength.

$$\lambda' - \lambda = \frac{h}{m_e c}(1 - \cos\theta) = (2.426 \text{ pm})(1 - \cos 80.0°) = \boxed{2.00 \text{ pm}}$$

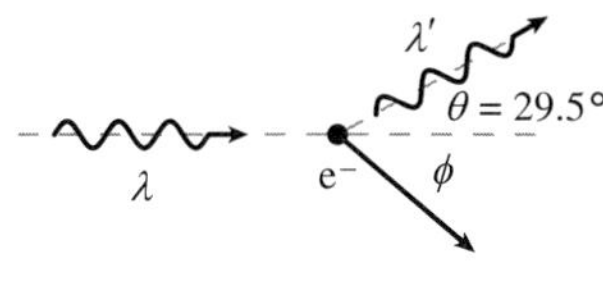

(b) Find the wavelength of the scattered photon.

$$\lambda' - \lambda = \Delta\lambda, \text{ so } \lambda' = \lambda + 2.00 \text{ pm} = 1.50\times10^2 \text{ pm} + 2.00 \text{ pm} = \boxed{152 \text{ pm}}.$$

25. **Strategy** Use the Compton shift, Eq. (27-14).

Solution The wavelength of the scattered x-rays is given by
$$\lambda' - \lambda = (2.426 \text{ pm})(1 - \cos\theta), \text{ so } \lambda' = \lambda + 2.426 \text{ pm} - (2.426 \text{ pm})\cos\theta = 2.426 \text{ pm} + \lambda - (2.426 \text{ pm})\cos\theta.$$
Compute $\lambda - (2.426 \text{ pm})\cos\theta$ for each case to find the change in wavelength of the scattered rays from 2.426 pm.
(a) $1.0 \text{ pm} - (2.426 \text{ pm})\cos 90° = 1.0 \text{ pm}$; (b) $1.0 \text{ pm} - (2.426 \text{ pm})\cos 60° = -0.213 \text{ pm}$;
(c) $4.0 \text{ pm} - (2.426 \text{ pm})\cos 120° = 5.213 \text{ pm}$; (d) $1.6 \text{ pm} - (2.426 \text{ pm})\cos 60° = 0.387 \text{ pm}$;
(e) $1.6 \text{ pm} - (2.426 \text{ pm})\cos 120° = 2.813 \text{ pm}$; (f) $4.0 \text{ pm} - (2.426 \text{ pm})\cos 2.0° = 1.6 \text{ pm}$
Ranking the scattered x-rays from largest to smallest wavelength, we have $\boxed{\text{(c), (e), (f), (a), (d), (b)}}$.

29. (a) **Strategy** Use the Compton shift, Eq. (27-14).

Solution Find the wavelength of the incident photon.

$$\lambda' - \lambda = \frac{h}{m_e c}(1 - \cos\theta), \text{ so } \lambda = \lambda' - \frac{h}{m_e c}(1 - \cos\theta)$$

$$= 2.81\times10^{-12} \text{ m} - \frac{6.626\times10^{-34} \text{ J}\cdot\text{s}}{(9.109\times10^{-31} \text{ kg})(2.998\times10^8 \text{ m/s})}(1 - \cos 29.5°)$$

$$= \boxed{2.50\times10^{-12} \text{ m}}.$$

(b) **Strategy** The kinetic energy of the electron plus the energy of the scattered photon is equal to the energy of the incident photon. Use Eq. (27-10).

Solution Find the kinetic energy of the electron.
$$K_e + E' = E, \text{ so}$$
$$K_e = E - E' = \frac{hc}{\lambda} - \frac{hc}{\lambda'} = (1240\times10^{-9} \text{ eV}\cdot\text{m})\left(\frac{1}{2.4954\times10^{-12} \text{ m}} - \frac{1}{2.81\times10^{-12} \text{ m}}\right) = \boxed{55.6 \text{ keV}}.$$

31. Strategy The change in kinetic energy of the electron is the difference in the energies of the incident photon and the scattered photon.

Solution Find the change in kinetic energy.

$$\Delta K = E - E' = \frac{hc}{\lambda} - \frac{hc}{\lambda'} = \frac{1240 \text{ eV} \cdot \text{nm}}{0.0100 \text{ nm}} - \frac{1240 \text{ eV} \cdot \text{nm}}{0.0124 \text{ nm}} = \boxed{2.4 \times 10^4 \text{ eV}}$$

33. Strategy The energy for a hydrogen atom in the nth stationary state is given by $E_n = (-13.6 \text{ eV})/n^2$.

Solution Find the energy.

$$E_4 = \frac{-13.6 \text{ eV}}{4^2} = \boxed{-0.850 \text{ eV}}$$

35. Strategy The energy supplied by the photon raises the atom from the ground state to a higher energy level. The energy for a hydrogen atom in the nth stationary state is given by $E_n = (-13.6 \text{ eV})/n^2$.

Solution Find the energy level to which the atom is excited.

$$E_n - E_1 = \frac{E_1}{n^2} - E_1 = 12.1 \text{ eV, so } n = \sqrt{\frac{E_1}{12.1 \text{ eV} + E_1}} = \sqrt{\frac{-13.6 \text{ eV}}{12.1 \text{ eV} - 13.6 \text{ eV}}} = 3.$$

The atom is excited to the $\boxed{n = 3}$ energy level.

37. Strategy The minimum energy for an ionized atom is $E_{\text{ionized}} = 0$. Use Eq. (27-24).

Solution The energy needed to ionize a hydrogen atom initially in the $n = 2$ state is

$$E = E_{\text{ionized}} - E_2 = E_{\text{ionized}} - \frac{E_1}{2^2} = 0 - \left(\frac{-13.6 \text{ eV}}{4} \right) = \boxed{3.40 \text{ eV}}.$$

39. Strategy The energy of a photon is related to its wavelength by $E = hc/\lambda$. Use Eq. (27-24) to find the energy of the transition.

Solution The amount of energy released when a hydrogen atom makes a transition from the $n = 6$ to the $n = 3$

state is $E = E_6 - E_3 = \dfrac{-13.6 \text{ eV}}{6^2} - \dfrac{-13.6 \text{ eV}}{3^2} = 1.13 \text{ eV}.$

The wavelength of the radiation emitted is $\lambda = \dfrac{hc}{E} = \dfrac{1240 \text{ eV} \cdot \text{nm}}{1.133 \text{ eV}} = \boxed{1.09 \text{ μm}}.$

41. Strategy and Solution

(a) The electron can return to the ground state directly, emitting one photon; thus, the minimum number of photons is $\boxed{\text{one}}$.

(b) The electron can return to the ground state by stopping at each intermediate state: $n = 5$ to $n = 4$, 4 to 3, 3 to 2, and then finally, from $n = 2$ to the ground state, $n = 1$. One photon is emitted for each transition, for a total of $\boxed{\text{four}}$ photons.

45. Strategy Use Eq. (27-20) and the values of the fundamental constants.

Solution Compute the numerical value of the Bohr radius.

$$a_0 = \frac{\hbar^2}{m_e k e^2} = \frac{h^2}{4\pi^2 m_e k e^2} = \frac{(6.626 \times 10^{-34} \text{ J} \cdot \text{s})^2}{4\pi^2 (9.109 \times 10^{-31} \text{ kg})(8.988 \times 10^9 \text{ N} \cdot \text{m}^2/\text{C}^2)(1.602 \times 10^{-19} \text{ C})^2} = 5.29 \times 10^{-11} \text{ m}$$

49. Strategy The orbital radius in the nth state is given by $r_n = n^2 a_0$.

Solution Compute the orbital radius.
$$r_3 = 3^2 \cdot 52.9 \text{ pm} = 476 \text{ pm} = \boxed{0.476 \text{ nm}}$$

53. Strategy The energy of a photon with the given wavelength is equal to the energy difference between the two helium levels. The energy of a photon of EM radiation with wavelength λ is $E = hc/\lambda$.

Solution Find the energy difference.
$$E = \frac{hc}{\lambda} = \frac{1240 \text{ eV} \cdot \text{nm}}{587.6 \text{ nm}} = \boxed{2.11 \text{ eV}}$$

57. Strategy For a photon, the maximum wavelength corresponds to the minimum energy. The minimum energy needed is the rest energy of an electron-positron pair. Use $E_0 = mc^2$ and $E = hc/\lambda$.

Solution Compute the required energy.
$$E = 2m_e c^2 = 2(9.109 \times 10^{-31} \text{ kg})(3.00 \times 10^8 \text{ m/s})^2 = 1.64 \times 10^{-13} \text{ J}$$
Compute the wavelength.
$$\lambda = \frac{hc}{E} = \frac{(6.626 \times 10^{-34} \text{ J} \cdot \text{s})(3.00 \times 10^8 \text{ m/s})}{1.64 \times 10^{-13} \text{ J}} = \boxed{1.21 \text{ pm}}$$

61. Strategy The rest energy of the muon and antimuon are converted into the energy of the two photons of equal energy. Since the photons have equal energy, their wavelengths are the same. The energy of a photon of EM radiation with wavelength λ is $E = hc/\lambda$. Use $E_0 = mc^2$.

Solution Find the wavelength.
$$2E_{\text{photon}} = E_{\text{muon}} + E_{\text{antimuon}} = 207 m_e c^2 + 207 m_e c^2 = 414 m_e c^2 = 2\frac{hc}{\lambda}, \text{ so}$$
$$\lambda = \frac{hc}{207 m_e c^2} = \frac{1240 \times 10^{-15} \text{ MeV} \cdot \text{m}}{207(0.511 \text{ MeV})} = \boxed{1.17 \times 10^{-14} \text{ m}}.$$

65. (a) Strategy The energy per second incident is the power incident, which is equal to the intensity times the area.

Solution Compute the energy per second falling on the atom.
$$P = IA = (0.01 \text{ W/m}^2)(0.1 \times 10^{-9} \text{ m})^2 = 1 \times 10^{-22} \text{ W} = \boxed{1 \times 10^{-22} \text{ J/s}}$$

(b) Strategy Since $P = \Delta E / \Delta t$, and $\Delta E = \phi$ in this case, $\Delta t = \phi / P$.

Solution Compute the time lag.
$$\Delta t = \frac{\phi}{P} = \frac{(2.0 \text{ eV})(1.6 \times 10^{-19} \text{ J/eV})}{1 \times 10^{-22} \text{ J/s}} = \boxed{3.2 \times 10^3 \text{ s}}$$

(c) Strategy and Solution Whether or not an electron is ejected has nothing to do with the intensity of the light; ejection of the electron depends upon the energy of individual photons. $\boxed{\text{An electron is ejected immediately when a single photon of sufficient energy strikes it.}}$

69. Strategy Power is equal to intensity times area. The energy of a photon of EM radiation with wavelength λ is $E = hc/\lambda$. The minimum detected number of photons per second is given by the ratio of the power to the energy per photon.

Solution The minimum amount of power that an owl's eye can detect is
$$P = IA = I\pi r^2 = (5.0\times10^{-13}\ \text{W/m}^2)\pi(4.25\times10^{-3}\ \text{m})^2 = 2.84\times10^{-17}\ \text{W}.$$
The energy of a photon of wavelength 510 nm is
$$E = \frac{hc}{\lambda} = \frac{(6.626\times10^{-34}\ \text{J}\cdot\text{s})(3.00\times10^{8}\ \text{m/s})}{510\times10^{-9}\ \text{m}} = 3.9\times10^{-19}\ \text{J}.$$
Compute the minimum number of photons per second that an owl eye can detect.
$$\frac{2.84\times10^{-17}\ \text{W}}{3.9\times10^{-19}\ \text{J}} = \boxed{73\ \text{photons/s}}$$

73. Strategy The shortest wavelength x-ray corresponds to the highest frequency, which occurs when the energy of the photon is equal to the electron's kinetic energy.

Solution Find the wavelength.
$$E_{\text{photon}} = \frac{hc}{\lambda} = K_{\text{electron}} = eV,\ \text{so}\ \lambda = \frac{hc}{eV} = \frac{1240\ \text{eV}\cdot\text{nm}}{e\times0.20\times10^{6}\ \text{V}} = 6.2\times10^{-3}\ \text{nm} = \boxed{6.2\ \text{pm}}.$$

77. (a) Strategy Plot the data and draw a best-fit line.

Solution

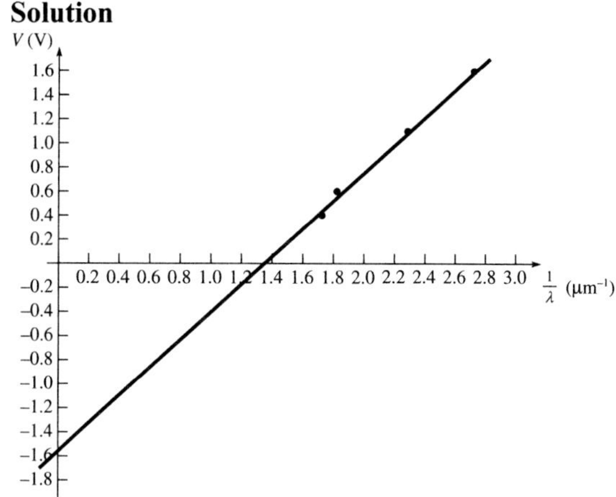

(b) Strategy The intercept of the stopping potential times e gives the work function. The $1/\lambda$-intercept gives the threshold wavelength.

Solution The V-intercept is about -1.57 V. This equates to a work function of $\boxed{1.57\ \text{eV}}$.

The $1/\lambda$-intercept is about $1.35\ \mu\text{m}^{-1}$. The corresponding value for the threshold wavelength is
$$\lambda = \frac{1}{1.35\ \mu\text{m}^{-1}} = \boxed{741\ \text{nm}}.$$

(c) **Strategy** Use Einstein's photoelectric equation.

Solution The slope is approximately
$$\frac{\Delta V}{\Delta\left(\frac{1}{\lambda}\right)} = \frac{1.57\ \text{V}}{1.35\ \mu\text{m}^{-1}} = 1.16\ \text{V}\cdot\mu\text{m} = \boxed{1160\ \text{V}\cdot\text{nm}}.$$
To calculate the theoretical value of the slope, start with Einstein's photoelectric equation.
$$K_{\max} = hf - \phi$$
Substitute $K_{\max} = eV$ and $f = c/\lambda$.
$$eV = \frac{hc}{\lambda} - \phi,\ \text{so}\ V = \frac{hc}{e}\cdot\frac{1}{\lambda} - \frac{\phi}{e}.$$
This equation relates V to $1/\lambda$, as does the graph. The theoretical slope is
$$\frac{hc}{e} = \frac{1240\ \text{eV}\cdot\text{nm}}{e} = 1240\ \text{V}\cdot\text{nm}.$$
This value is reasonably close to the measured value.

81. **Strategy** The energy of a photon of EM radiation with wavelength λ is $E = hc/\lambda$. Use the Compton shift, Eq. (27-14).

Solution The wavelength of an incident photon is
$$\lambda = \frac{hc}{E} = \frac{1240\ \text{eV}\cdot\text{nm}}{186\times10^3\ \text{eV}} = 6.667\ \text{pm}.$$
Calculate the Compton shifts for $\theta = 90.0°$ and $180.0°$.
$$\Delta\lambda_{90} = \frac{h}{m_e c}(1 - \cos 90.0°) = 2.426\ \text{pm and } \Delta\lambda_{180} = (2.426\ \text{pm})(1 - \cos 180.0°) = 4.852\ \text{pm}.$$
The wavelengths of the scattered x-rays are
$$\lambda'_{90} = \lambda + \Delta\lambda = 6.667\ \text{pm} + 2.426\ \text{pm} = 9.09\ \text{pm and } \lambda'_{180} = 6.667\ \text{pm} + 4.852\ \text{pm} = 11.52\ \text{pm}.$$
The energies of the scattered x-rays are
$$E_{90} = \frac{hc}{\lambda} = \frac{1240\ \text{eV}\cdot\text{nm}}{0.00909\ \text{nm}} = \boxed{136\ \text{keV}}\ \text{and } E_{180} = \frac{1240\ \text{eV}\cdot\text{nm}}{0.01152\ \text{nm}} = \boxed{108\ \text{keV}}.$$

85. **(a)** **Strategy** Substitute $K_{\max} = eV_s$ into Einstein's photoelectric equation and solve for V_s.

Solution Calculate the stopping potential.
$$K_{\max} = eV_s = hf - \phi = \frac{hc}{\lambda} - \phi,\ \text{so}\ V_s = \frac{hc}{e\lambda} - \frac{\phi}{e} = \frac{1240\ \text{eV}\cdot\text{nm}}{e\times0.20\times10^3\ \text{nm}} - \frac{4.5\ \text{eV}}{1\ e} = \boxed{1.7\ \text{V}}.$$

(b) **Strategy** Compute the energy of a photon and compare it to the work function.

Solution The energy of a photon with $\lambda = 400$ nm is
$$E = hf = \frac{hc}{\lambda} = \frac{1240\ \text{eV}\cdot\text{nm}}{400\ \text{nm}} = 3.1\ \text{eV}.$$
This energy is less than the work function for tungsten, $\phi = 4.5$ eV, so the current is $\boxed{\text{zero}}$.

89. Strategy After the collision the electron recoils in the direction of motion of the incident photon. Its motion has no component perpendicular to the motion of the incident photon. Conservation of momentum requires the scattering angle must be 180°. Use Eq. (27-10) and the Compton shift, Eq. (27-14).

Solution Solve the Compton shift formula for λ.

Before:

$$\lambda' - \lambda = \frac{h}{m_e c}(1 - \cos\theta), \text{ so}$$

After:

$$\lambda = \lambda' - \frac{h}{m_e c}(1 - \cos 180°) = \lambda' - 2(2.426 \text{ pm}) = \lambda' - 4.852 \text{ pm}.$$

The kinetic energy of the electron is $K_e = E - E' = \dfrac{hc}{\lambda} - \dfrac{hc}{\lambda'}$.

Substituting $\lambda = \lambda' - 4.852$ pm and $K_e = 0.20$ keV, and solving for λ' gives the wavelength of the scattered x-ray.

$$\frac{K_e}{hc} = \frac{1}{\lambda' - 4.852 \text{ pm}} - \frac{1}{\lambda'}$$

$$\frac{0.20 \times 10^3 \text{ eV}}{(1240 \text{ eV} \cdot \text{nm})(1000 \text{ pm/nm})} = \frac{1}{\lambda' - 4.852 \text{ pm}} - \frac{1}{\lambda'}$$

$$(1.613 \times 10^{-4} \text{ pm}^{-1})(\lambda' - 4.852 \text{ pm})\lambda' = \lambda' - (\lambda' - 4.852 \text{ pm})$$

$$\lambda'^2 - (4.852 \text{ pm})\lambda' = 3.008 \times 10^4 \text{ pm}^2$$

$$\lambda'^2 - (4.852 \text{ pm})\lambda' - 3.008 \times 10^4 \text{ pm}^2 = 0$$

Solve for λ' using the quadratic formula.

$$\lambda' = \frac{4.852 \text{ pm} \pm \sqrt{(-4.852 \text{ pm})^2 - 4 \cdot 1 \cdot (-3.008 \times 10^4 \text{ pm}^2)}}{2 \cdot 1} = \boxed{176 \text{ pm}} \text{ or } -171 \text{ pm, which is extraneous.}$$

The wavelength of the incident x-ray can now be calculated.

$$\lambda = \lambda' - 4.852 \text{ pm} = 176 \text{ pm} - 4.852 \text{ pm} = \boxed{171 \text{ pm}}$$

Chapter 28

QUANTUM PHYSICS

Conceptual Questions

1. We know the wavelengths of the electrons and x-rays are the same because a diffraction pattern depends only on the wavelength of the incident particles or radiation and the interatomic spacing of the sample. They would not give the same pattern if their energies were the same because then they would have different wavelengths. For a photon and an electron with the same energy E, $\lambda_{\text{photon}} = hc/E$, while $\lambda_{\text{electron}} = hc/\sqrt{E^2 - (mc^2)^2}$.

5. If we are only concerned with the energy of the state, then we need only specify the principal quantum number, n. This would be the case, for example, if we were talking about the energies of photons emitted during atomic transitions. If we are interested in describing the state more completely, then we need to include more quantum numbers such as the spin magnetic number, m_s.

9. The light that causes optical pumping must excite the atoms in the laser to a higher energy state than the one from which they decay when they emit the laser light. Therefore, the energy of the photons of the optical pumping light must be greater than the energy of the photons of the emitted laser light. This means the wavelength of the laser beam must be greater than the wavelength of the optical pumping light.

13. Photons incident on the photoconducting material transfer their kinetic energy to atoms in the semiconductor. Electrons in these atoms jump into the conduction band, and therefore, increase the conductivity of the material. The band gap for a good photoconductor should be approximately the same size as the energy of the incident photons. If the drum gets hot, the contrast between light and dark areas of the image is degraded because some of the thermal energy can cause electrons to jump into the conduction band even if no photons were incident upon them.

17. Increasing the temperature of a semiconductor increases the number of electrons excited into the conduction band. The increased number of conduction electrons increases the semiconductor's ability to conduct electricity, and therefore, decreases its resistivity.

Problems

1. **Strategy** Find the de Broglie wavelength using Eq. (28-1). Compare the result to the diameter of the hoop.

   Solution
   $$\lambda = \frac{h}{p} = \frac{h}{mv} = \frac{6.626\times10^{-34}\ \text{J}\cdot\text{s}}{(0.50\ \text{kg})(10\ \text{m/s})} = \boxed{1.3\times10^{-34}\ \text{m}}$$
   The wavelength is much too small compared to the diameter of the hoop for any appreciable diffraction to occur—for a diameter of ~ 1 m, it's a factor of 10^{-34} smaller!

5. **Strategy** The electron is relativistic, so use $p = \gamma mv$ to find the magnitude of its momentum. Find the de Broglie wavelength using Eq. (28-1).

   Solution
   $$\lambda = \frac{h}{p} = \frac{h}{\gamma mv} = \frac{h\sqrt{1-v^2/c^2}}{mv} = \frac{(6.626\times10^{-34}\ \text{J}\cdot\text{s})\sqrt{1-(3/5)^2}}{(9.109\times10^{-31}\ \text{kg})(\frac{3}{5}\times3.00\times10^8\ \text{m/s})} = \boxed{3.23\ \text{pm}}$$

9. **Strategy** The electron's kinetic energy is small compared to its rest energy, so the electron is nonrelativistic and we can use $p = mv$ and $K = \frac{1}{2}mv^2$ to find the momentum. Find the de Broglie wavelength using Eq. (28-1) and the wavelength of the photon using $E = hc/\lambda$.

Solution First solve for p in terms of K.

$$K = \frac{1}{2}mv^2, \text{ so } v = \sqrt{\frac{2K}{m}}.$$

The momentum is then $p = mv = m\sqrt{\frac{2K}{m}} = \sqrt{2Km}$.

Find the de Broglie wavelength of the electron.

$$\lambda_e = \frac{h}{p} = \frac{h}{\sqrt{2Km_e}}$$

The wavelength of a photon is $\lambda_p = \dfrac{hc}{E_p}$. Compute the ratio of the wavelengths.

$$\frac{\lambda_p}{\lambda_e} = \frac{hc/E_p}{h/\sqrt{2Km}} = \frac{c\sqrt{2Km}}{E_p} = \frac{(3.00\times10^8 \text{ m/s})\sqrt{2(0.100\times10^3 \text{ eV})(1.602\times10^{-19} \text{ J/eV})(9.109\times10^{-31} \text{ kg})}}{(0.100\times10^3 \text{ eV})(1.602\times10^{-19} \text{ J/eV})} = \boxed{101}$$

13. **(a) Strategy** The energy of a photon is given by $E = hc/\lambda$. The minimum energy corresponds to the maximum wavelength of a photon that is capable of resolving the details of the molecule.

     Solution Find the minimum photon energy.

     $$E = \frac{hc}{\lambda} = \frac{1240 \text{ eV} \cdot \text{nm}}{20 \text{ nm}} = \boxed{62 \text{ eV}}$$

   (b) Strategy The minimum kinetic energy corresponds to the maximum wavelength of an electron that is capable of resolving the details of the molecule. Assume that the electrons are nonrelativistic. Use the de Broglie wavelength, $K = \frac{1}{2}mv^2$, and $p = mv$.

     Solution Find the kinetic energy of the electrons.

     $$K = \frac{1}{2}mv^2 = \frac{(mv)^2}{2m} = \frac{p^2}{2m} = \frac{(h/\lambda)^2}{2m} = \frac{h^2}{2m\lambda^2} = \frac{(6.626\times10^{-34} \text{ J}\cdot\text{s})^2}{2(9.109\times10^{-31} \text{ kg})(20\times10^{-9} \text{ m})^2} \times \frac{1}{1.602\times10^{-19} \text{ J/eV}}$$

     $$= \boxed{0.0038 \text{ eV}}$$

     The kinetic energy is small compared to the rest energy of an electron, so the use of the nonrelativistic equations for momentum and kinetic energy was valid.

   (c) Strategy Use conservation of energy.

     Solution Find the required potential difference.

     $$\Delta U = -e\Delta V = -\Delta K = -K, \text{ so } \Delta V = \frac{K}{e} = \frac{0.0038 \text{ eV}}{e} = \boxed{0.0038 \text{ V}}.$$

17. **Strategy** Use the position-momentum uncertainty principle, Eq. (28-2).

   Solution Find the uncertainty in the position of the basketball.

   $$\Delta x \Delta p \geq \frac{1}{2}\hbar, \text{ so } \Delta x \geq \frac{\hbar}{2\Delta p} = \frac{\hbar}{2p \cdot 10^{-6}} = \frac{10^6 h}{4\pi mv} = \frac{10^6 (6.626\times10^{-34} \text{ J}\cdot\text{s})}{4\pi(0.50 \text{ kg})(10 \text{ m/s})} = \boxed{1\times10^{-29} \text{ m}}.$$

21. Strategy Use the position-momentum uncertainty principle, Eq. (28-2), and $p = mv$.

Solution

(a) Estimate the minimum uncertainty in the position of the bullet.

$$\Delta x \Delta p_x \geq \frac{1}{2}\hbar, \text{ so } \Delta x \geq \frac{\hbar}{2\Delta p_x} = \frac{\hbar}{2m\Delta v_x}. \text{ Thus,}$$

$$(\Delta x)_{\min} = \frac{\hbar}{2m\Delta v_x} = \frac{\hbar}{2m\left(\frac{\Delta v_x}{v_x}\right)v_x} = \frac{6.626\times10^{-34} \text{ J}\cdot\text{s}}{4\pi(10.000\times10^{-3} \text{ kg})(0.0004)(300.00 \text{ m/s})} = \boxed{4\times10^{-32} \text{ m}}.$$

(b) $(\Delta x)_{\min} = \dfrac{6.626\times10^{-34} \text{ J}\cdot\text{s}}{4\pi(9.109\times10^{-31} \text{ kg})(0.0004)(300.00 \text{ m/s})} = \boxed{0.5 \text{ mm}}$

(c) The uncertainty in the bullet's position is far too small to be observable. The uncertainty in the electron's position is huge on the atomic scale. Thus, the uncertainty principle can be neglected in the macroscopic world, but not on the atomic scale.

23. Strategy The single-slit diffraction minima are given by $a\sin\theta = m\lambda$. The edge of the central fringe corresponds to $m = 1$. Since the width of the central fringe is small compared to the slit-screen distance, use small angle approximations. Use the de Broglie wavelength with $p = \sqrt{2Km_e}$ for the electrons.

Solution Since $\theta \approx \tan\theta = x/D$ and $\theta \approx \sin\theta = \lambda/a$, we have $\lambda = ax/D$, where x is half the width of the central fringe and D is the distance from the slit to the screen. Solve for p.

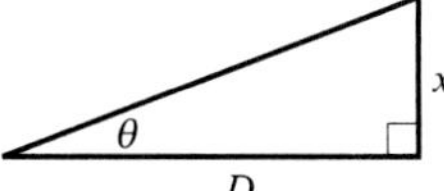

$$\lambda = \frac{h}{p} = \frac{ax}{D}, \text{ so } p = \frac{Dh}{ax}.$$

Find the kinetic energy.

$$p = \frac{Dh}{ax} = \sqrt{2Km_e}, \text{ so}$$

$$K = \frac{D^2h^2}{2a^2x^2m_e} = \frac{(1.0 \text{ m})^2(6.626\times10^{-34} \text{ J}\cdot\text{s})^2}{2(40.0\times10^{-9} \text{ m})^2(0.031 \text{ m})^2(9.109\times10^{-31} \text{ kg})}\times\frac{1 \text{ eV}}{1.602\times10^{-19} \text{ J}} = \boxed{0.98 \text{ eV}}.$$

25. Strategy Use the energy-time uncertainty principle, Eq. (28-3).

Solution Find the uncertainty in the energy.

$$\Delta E \Delta t \geq \frac{1}{2}\hbar, \text{ so } \Delta E \geq \frac{\hbar}{2\Delta t} \approx \frac{1\times10^{-34} \text{ J}\cdot\text{s}}{2(0.1\times10^{-9} \text{ s})} = 5\times10^{-25} \text{ J}.$$

Calculate the fractional uncertainty.

$$\frac{\Delta E}{E} \approx \frac{5\times10^{-25} \text{ J}}{(1672\times10^{6} \text{ eV})(1.602\times10^{-19} \text{ J/eV})} = \boxed{2\times10^{-15}}$$

27. Strategy The kinetic energy of the electron is given by $K = p_n^2/(2m)$, where $p_n = nh/(2L)$. The minimum kinetic energy is the $n = 1$ state.

Solution Find the minimum kinetic energy.

$$K = \frac{p_n^2}{2m} = \frac{n^2 h^2}{8mL^2} = \frac{1^2(6.626\times10^{-34}\text{ J·s})^2}{8(9.109\times10^{-31}\text{ kg})(1.0\times10^{-15}\text{ m})^2}\times\frac{1\text{ eV}}{1.602\times10^{-19}\text{ J}} = \boxed{380\text{ GeV}}$$

29. Strategy Use Eq. (28-8) and the ground-state energy of the hydrogen atom.

Solution Compute the ground-state energy of the one-dimensional-box "atom".

$$E_1 = \frac{h^2}{8mL^2} = \frac{h^2}{8m_e(2a_0)^2} = \frac{(6.626\times10^{-34}\text{ J·s})^2}{32(9.109\times10^{-31}\text{ kg})(5.292\times10^{-11}\text{ m})^2(1.602\times10^{-19}\text{ J/eV})} = \boxed{33.57\text{ eV}}$$

Compare this to the actual ground-state energy.

$$\frac{33.57\text{ eV}}{13.6\text{ eV}} = 2.47$$

This is about 2.47 times the actual ground-state energy of 13.6 eV. This is within an order of magnitude of the actual number, which is fairly close considering that the atom is three dimensional, not one dimensional.

33. Strategy A particle in a box can make a transition from an excited state n to a lower state m by radiating a photon with energy equal to the energy difference between the states. Use Eq. (28-8) to find the length of the box.

Solution

(a) Find the ground-state energy.

$$E = E_2 - E_1 = 2^2 E_1 - E_1 = 3E_1, \text{ so } E_1 = \frac{E}{3} = \frac{1.2\text{ eV}}{3} = \boxed{0.40\text{ eV}}.$$

(b) There are 2 ways the electron can go from the $n = 3$ state to the $n = 1$ state.

$3 \to 1$: 1 photon;

$$E_{31} = E_3 - E_1 = 9E_1 - E_1 = 8E_1 = 8\times0.40\text{ eV} = \boxed{3.2\text{ eV}}$$

$3 \to 2 \to 1$: 2 photons;

$$E_{32} = E_3 - E_2 = 9E_1 - 4E_1 = 5E_1 = 5\times0.40\text{ eV} = \boxed{2.0\text{ eV}} \text{ and}$$

$$E_{21} = E_2 - E_1 = \boxed{1.2\text{ eV}}.$$

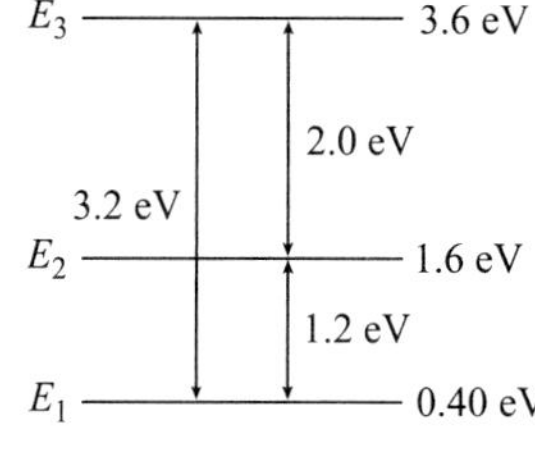

(c) Find the length of the box.

$$E_1 = \frac{h^2}{8mL^2}, \text{ so } L = \sqrt{\frac{h^2}{8mE_1}} = \frac{h}{2\sqrt{2mE_1}} = \frac{6.626\times10^{-34}\text{ J·s}}{2\sqrt{2(9.109\times10^{-31}\text{ kg})(0.40\text{ eV})(1.602\times10^{-19}\text{ J/eV})}} = \boxed{0.97\text{ nm}}.$$

35. Strategy There are $4\ell + 2$ electron states in a subshell. Use Table 28.1.

Solution Since $\ell = 1$, there are $4(1) + 2 = \boxed{6}$ electron states. For $\ell = 1$, $m_\ell = -1$, 0, and 1; for each m_ℓ, $m_s = \pm 1/2$. The states are given in the table.

n	3	3	3	3	3	3
ℓ	1	1	1	1	1	1
m_ℓ	-1	-1	0	0	1	1
m_s	$-\frac{1}{2}$	$+\frac{1}{2}$	$-\frac{1}{2}$	$+\frac{1}{2}$	$-\frac{1}{2}$	$+\frac{1}{2}$

37. Strategy and Solution For each pair of values of n and ℓ, the orbital magnetic quantum number can take $2\ell + 1$ values, and the spin magnetic quantum number can take 2 values. The number of possible states is the product, $\boxed{2(2\ell + 1)}$.

39. Strategy Nickel (Ni) has atomic number 28, so it has 28 electrons in a neutral atom. It is not an exception to the subshell order, so subshells are filled in the order listed in Eq. (28-16).

Solution The ground-state electron configuration of nickel is

$$\boxed{1s^2 2s^2 2p^6 3s^2 3p^6 4s^2 3d^8}.$$

41. Strategy Bromine has atomic number 35, so there are 35 electrons in the neutral atom. It is not an exception to the subshell order, so subshells are filled in the order listed in Eq. (28-16). Find the ground-state electron configuration for bromine, and then use it to find the principal and orbital angular momentum quantum numbers for an outer electron in the ground state of a bromine atom. Use Eq. (28-12).

Solution The electron configuration of bromine is $1s^2 2s^2 2p^6 3s^2 3p^6 4s^2 3d^{10} 4p^5$, so the principal quantum number of an outer electron is $n = 4$, and the largest value of the orbital angular momentum quantum number is $n - 1 = 4 - 1 = 3$. The maximum possible value of angular momentum for the electron is

$$L = \sqrt{\ell(\ell+1)}\hbar = \sqrt{3(3+1)}\hbar = \boxed{2\sqrt{3}\hbar = 3.655\times10^{-34}\ \text{kg}\cdot\text{m}^2/\text{s}}, \text{ where } \hbar = 1.055\times10^{-34}\ \text{kg}\cdot\text{m}^2/\text{s}.$$

43. Strategy Fluorine and chlorine have 9 and 17 electrons in a neutral atom, respectively. Neither of these appear in the list of exceptions to the subshell order, so subshells are filled in the order listed in Eq. (28-16). The Periodic Table of the elements is arranged in columns by electron configuration.

Solution

(a) The ground-state electron configurations are the following:

$$\boxed{\text{F: } 1s^2 2s^2 2p^5; \text{ Cl: } 1s^2 2s^2 2p^6 3s^2 3p^5}.$$

(b) Both neutral atoms have valence -1. Their outermost electrons are in the $\boxed{p^5 \text{ subshell}}$. This is why they are placed in the same column.

45. (a) Strategy Use Eq. (28-12).

Solution Find the angular momentum of the electron.

$$L = \sqrt{\ell(\ell+1)}\hbar = \sqrt{1(1+1)}\hbar = \boxed{\sqrt{2}\hbar}$$

(b) Strategy Use Table 28.1 and Eq. (28-13).

Solution Since $n = 2$, ℓ can be 0 or 1, and m_ℓ can have values $-1, 0,$ and 1 (for $\ell = 1$). $L_z = m_\ell \hbar$, so the allowed values of L_z are $\boxed{-\hbar,\, 0,\, \text{and } \hbar}$.

(c) Strategy The magnitude of the total angular momentum and the z-component are related by $L_z = L\cos\theta$, where θ is the angle between the positive z-axis and $\vec{L}$.

Solution Compute the angles between the positive z-axis and $\vec{L}$.

$$\theta = \cos^{-1}\frac{L_z}{L}, \text{ so the angles are } \cos^{-1}\frac{\hbar}{\sqrt{2}\hbar} = \boxed{45°},\ \cos^{-1}\frac{0}{\sqrt{2}\hbar} = \boxed{90°},\ \text{and } \cos^{-1}\frac{-\hbar}{\sqrt{2}\hbar} = \boxed{135°}.$$

49. Strategy The wavelength of a photon is related to its energy by $E = hc/\lambda$. Stimulated emission can occur between the energy levels 20.66 eV and 18.70 eV.

Solution The associated wavelength is

$$\lambda = \frac{hc}{\Delta E} = \frac{1240 \text{ eV}\cdot\text{nm}}{20.66 \text{ eV} - 18.70 \text{ eV}} = \boxed{633 \text{ nm}}.$$

53. Strategy The tunneling probability is proportional to $e^{-2\kappa a}$, where κ is given by Eq. (28-18).

Solution

(a) $\kappa \propto \sqrt{m}$ and $e^{-2\kappa a}$ is greater for smaller κ, so the particle with the smaller mass has the higher tunneling probability. That particle is the $\boxed{\text{proton}}$.

(b) Find the ratio of the tunneling probabilities.

$$\frac{P_\text{p}}{P_\text{d}} = \frac{e^{-2\kappa_\text{p}a}}{e^{-2\kappa_\text{d}a}} = e^{2a(\kappa_\text{d}-\kappa_\text{p})}, \text{ where } a = 10.0 \text{ fm and } \kappa_\text{d} - \kappa_\text{p} \text{ is}$$

$$\sqrt{\frac{2m_\text{d}}{\hbar^2}(U_0 - E)} - \sqrt{\frac{2m_\text{p}}{\hbar^2}(U_0 - E)} = \sqrt{\frac{2}{\hbar^2}(U_0 - E)}\left(\sqrt{2.0m_\text{p}} - \sqrt{m_\text{p}}\right) = (\sqrt{2.0}-1)\sqrt{\frac{2m_\text{p}}{\hbar^2}(U_0 - E)}, \text{ so}$$

$$\frac{P_\text{p}}{P_\text{d}} = e^{2(10.0\times10^{-15} \text{ m})(\sqrt{2.0}-1)\sqrt{\frac{8\pi^2(1.673\times10^{-27}\text{ kg})}{(6.626\times10^{-34}\text{ J}\cdot\text{s})^2}(10.0\text{ MeV}-3.0\text{ MeV})\left(\frac{1.602\times10^{-13}\text{ J}}{1\text{ MeV}}\right)}} = \boxed{120}.$$

57. Strategy Use the energy-time uncertainty principle, Eq. (28-3). The rest energy is given by $E_0 = mc^2$.

Solution Calculate the minimum uncertainty in the energy.

$$\Delta E \Delta t \geq \frac{1}{2}\hbar, \text{ so } (\Delta E)_{\min} = \frac{\hbar}{2\Delta t} = \frac{6.626 \times 10^{-34} \text{ J} \cdot \text{s}}{4\pi(15 \times 60 \text{ s})} = 5.86 \times 10^{-38} \text{ J.}$$

Find the inherent uncertainty in the mass of a free neutron.

$$\Delta m = \frac{\Delta E_{\min}}{c^2} = \frac{5.86 \times 10^{-38} \text{ J}}{(3.00 \times 10^8 \text{ m/s})^2} = \boxed{6.5 \times 10^{-55} \text{ kg}}$$

Compare to the average neutron mass.

$$\frac{6.5 \times 10^{-55} \text{ kg}}{1.67 \times 10^{-27} \text{ kg}} = 3.9 \times 10^{-28}, \text{ so } \boxed{\frac{\Delta m}{m} = 3.9 \times 10^{-28}}.$$

61. Strategy Use the energy-time uncertainty principle. The rest energy of a particle is mc^2.

Solution

(a) Compute the lifetime of an electron created from the vacuum.

$$\Delta E \Delta t = \frac{1}{2}\hbar, \text{ so } \Delta t = \frac{\hbar}{2\Delta E} = \frac{\hbar}{2m_e c^2} = \frac{6.582 \times 10^{-16} \text{ eV} \cdot \text{s}}{2(0.5110 \times 10^6 \text{ eV})} = \boxed{6.440 \times 10^{-22} \text{ s}}.$$

(b) Compute the lifetime of a shot put created from the vacuum.

$$\Delta t = \frac{\hbar}{2\Delta E} = \frac{\hbar}{2mc^2} = \frac{1.055 \times 10^{-34} \text{ J} \cdot \text{s}}{2(7 \text{ kg})(3.00 \times 10^8 \text{ m/s})^2} = \boxed{8 \times 10^{-53} \text{ s}}$$

65. Strategy The wavelength of the electrons must equal the wavelength of the photons to give the same diffraction pattern. The wavelength of a photon is related to its energy by $E = hc/\lambda$. The wavelength of the electrons is given by the de Broglie wavelength with $p = \sqrt{2Km}$, since the electrons are nonrelativistic $[K = p^2/(2m)]$.

Solution Find the kinetic energy of the electrons.

$$\lambda_{\text{ph}} = \frac{hc}{E} = \lambda_e = \frac{h}{p} = \frac{h}{\sqrt{2Km}}, \text{ so } 2Km = \frac{E^2}{c^2} \text{ or } K = \frac{E^2}{2mc^2} = \frac{(2.0 \text{ eV})^2}{2(511 \times 10^3 \text{ eV})} = \boxed{3.9 \times 10^{-6} \text{ eV}}.$$

69. Strategy Find the de Broglie wavelength using Eq. (28-1) and $p = mv$. Divide the wavelength by the diameter of a proton to compare.

Solution

(a) Calculate the de Broglie wavelength of the bullet.

$$\lambda = \frac{h}{p} = \frac{h}{mv} = \frac{6.626 \times 10^{-34} \text{ J} \cdot \text{s}}{(10.0 \times 10^{-3} \text{ kg})(300.0 \text{ m/s})} = \boxed{2.21 \times 10^{-34} \text{ m}}$$

(b) Compare the wavelength to the diameter of a proton.

$$\frac{\lambda}{d_{\text{proton}}} \approx \frac{\lambda}{1 \text{ fm}} = \frac{2.21 \times 10^{-34} \text{ m}}{1 \times 10^{-15} \text{ m}} \sim 10^{-19}$$

The wavelength of the bullet is $\boxed{\text{about } 10^{-19} \text{ smaller}}$ than the diameter of a proton.

(c) For diffraction to occur, the wavelength must be roughly the same size as the width of the aperture through which the wave passes. An aperture small enough to cause diffraction would be too small for the bullet to pass through. In addition, it would be exceedingly difficult, if not impossible, to find or create such a small aperture. So, no; the wavelength is so much smaller than any aperture that diffraction is negligible.

73. (a) **Strategy** The electrons have kinetic energy of 54.0 eV due to the accelerating potential. Since the kinetic energy is so much smaller than the rest energy of the electron, use the nonrelativistic forms of the momentum and kinetic energy. Find the de Broglie wavelength using Eq. (28-1).

Solution Find the momentum.

$$K = \frac{p^2}{2m}, \text{ so } p = \sqrt{2mK}.$$

Calculate the de Broglie wavelength.

$$\lambda = \frac{h}{p} = \frac{h}{\sqrt{2mK}} = \frac{6.626\times10^{-34} \text{ J}\cdot\text{s}}{\sqrt{2(9.109\times10^{-31} \text{ kg})(54.0 \text{ eV})(1.602\times10^{-19} \text{ J/eV})}} = \boxed{167 \text{ pm}}$$

(b) **Strategy** Use Bragg's law.

Solution Find the Bragg angle for $m = 1$.

$$2d\sin\theta = m\lambda = (1)\lambda, \text{ so } \theta = \sin^{-1}\frac{\lambda}{2d} = \sin^{-1}\frac{1.669\times10^{-10} \text{ m}}{2(0.091\times10^{-9} \text{ m})} = \boxed{66.5°}.$$

(c) **Strategy** Make a sketch to show the relationship between the Bragg angle and the scattering angle.

Solution In the figure to the right, $\theta_i = \theta_r$ and $\theta_{\text{Bragg}} = \theta_r + \alpha$; but

$\theta_i = \alpha$, since they are opposite angles, so $\theta_{\text{Bragg}} = \theta_r + \theta_i = 2\theta_i$,

or $\theta_{\text{Bragg}} = 2(66.5°) = 133°$, which equals $130°$ to two significant figures.

Yes, the results agree.

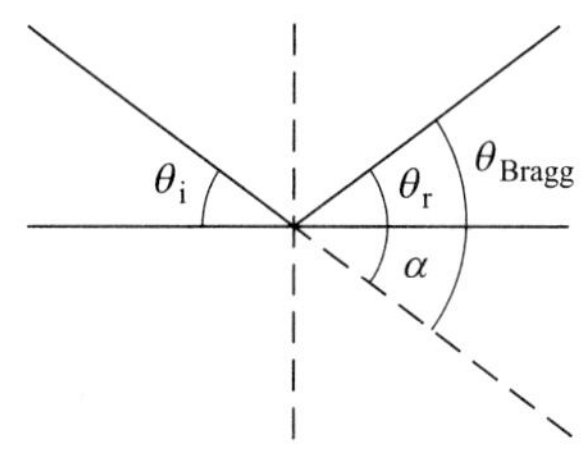

75. **Strategy** Use Figure 28.13 as a guide to make a qualitative sketch of the wave function for the $n = 5$ state. Estimate the number of bound states for the *finite* box by using the energy states for an *infinite* box.

Solution

(a) The figure shows the wave function for the $n = 5$ state of an electron in a finite box of length L.

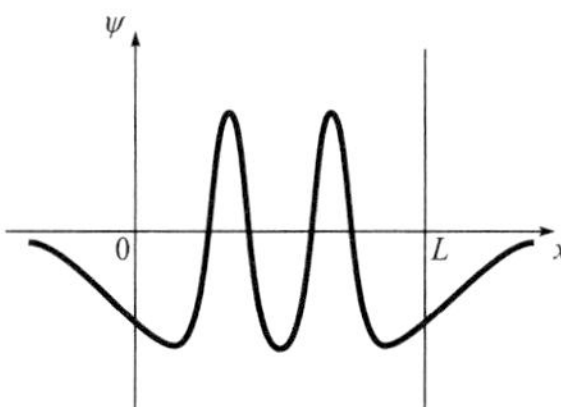

(b) The energy states for an infinite box are given by $E_n = n^2h^2/(8mL^2)$. Set $E_n = U_0$ and solve for n.

$$n = \sqrt{\frac{8mL^2U_0}{h^2}} = \sqrt{\frac{8(9.109\times10^{-31} \text{ kg})(1.0\times10^{-9} \text{ m})^2(1.0\times10^3 \text{ eV})(1.602\times10^{-19} \text{ J/eV})}{(6.626\times10^{-34} \text{ J}\cdot\text{s})^2}} = 51.6$$

$n = 52$ would give an energy greater than U_0, so there are approximately $\boxed{51}$ bound states.

77. (a) Strategy The length of the box can be found by using the equation for the ground-state energy, Eq. (28-8).

Solution Find the length.

$$E_1 = \frac{h^2}{8mL^2}, \text{ so } L = \frac{h}{2\sqrt{2mE_1}} = \frac{6.626\times10^{-34}\ \text{J}\cdot\text{s}}{2\sqrt{2(9.109\times10^{-31}\ \text{kg})(0.010\ \text{eV})(1.602\times10^{-19}\ \text{J/eV})}} = \boxed{6.1\ \text{nm}}.$$

(b) Strategy Use Figure 28.10 to sketch the wave functions for the three lowest energy states of the electron confined to a one-dimensional box of length $L = 6.1$ nm.

Solution The wave functions:

Lowest energy state:

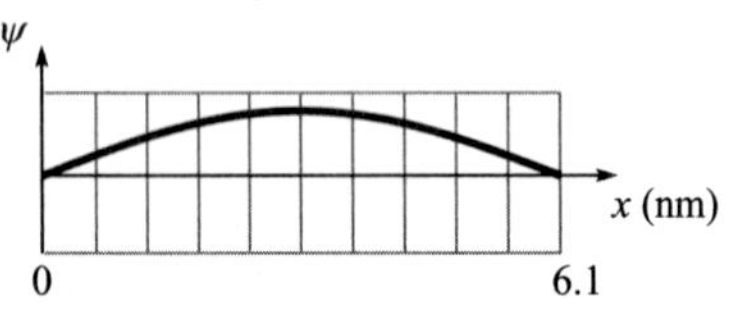

Second lowest energy state:

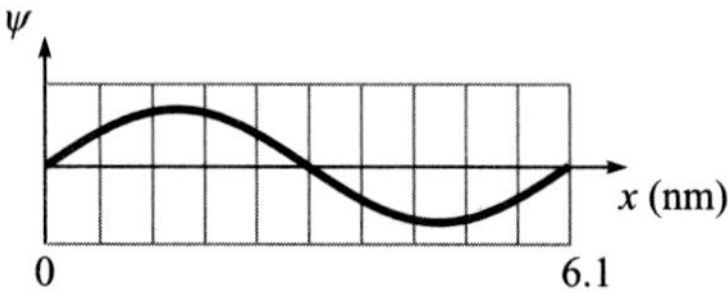

Third lowest energy state:

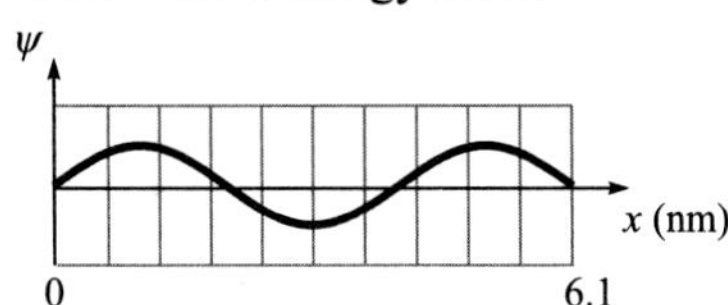

(c) Strategy The kinetic energy of the electron in the second excited state $(n = 3)$ is given by $E_3 = 3^2 h^2/(8mL^2)$. The momentum is related to the kinetic energy by $p = \sqrt{2mK} = \sqrt{2mE_3}$, so the wavelength is given by $\lambda = h/p = h/\sqrt{2mE_3}$.

Solution Find the wavelength.

$$\lambda = \frac{h}{\sqrt{2mE_3}} = \frac{h}{\sqrt{2m\frac{3^2 h^2}{8mL^2}}} = \frac{2L}{3} = \frac{2}{3}(6.1\ \text{nm}) = \boxed{4.1\ \text{nm}}$$

(d) Strategy The energy of a photon is related to its wavelength by $E = hc/\lambda$. Use Eqs. (28-9) and (28-10) to find the energies of the transitions.

Solution Find the energy of the photon.

$$E = \frac{hc}{\lambda} = \frac{1240\ \text{eV}\cdot\text{nm}}{15,500\ \text{nm}} = 0.0800\ \text{eV}$$

Find the energy state of the electron after absorbing the photon.

$$E = E_n - E_1 = n^2 E_1 - E_1, \text{ so } n = \sqrt{\frac{E + E_1}{E_1}} = \sqrt{\frac{0.0800\ \text{eV} + 0.010\ \text{eV}}{0.010\ \text{eV}}} = 3.$$

The possible transitions (and wavelengths) are

$3 \rightarrow 1$: one photon; $\lambda_{31} = \boxed{15.5\ \mu\text{m}}$

$3 \rightarrow 2 \rightarrow 1$: two photons;

$$\lambda_{32} = \frac{hc}{E_3 - E_2} = \frac{1240\ \text{eV}\cdot\text{nm}}{9(0.010\ \text{eV}) - 4(0.010\ \text{eV})} = \boxed{25\ \mu\text{m}}$$

$$\lambda_{21} = \frac{hc}{E_2 - E_1} = \frac{1240\ \text{eV}\cdot\text{nm}}{4(0.010\ \text{eV}) - 0.010\ \text{eV}} = \boxed{41\ \mu\text{m}}$$

Chapter 29

NUCLEAR PHYSICS

Conceptual Questions

1. Alpha, beta, and gamma rays were distinguished by their varying abilities to penetrate matter before their masses and electric charges were determined.

5. There is a mass defect due to the binding energy of atomic electrons but it is several orders of magnitude smaller than the binding energy of the nucleus and may thus be omitted without significant loss of accuracy.

9. If none of the iodine were eliminated from the body by biological processes, it would take 8 days for only half as much ^{131}I to be left in the body. However, at least some of the iodine is eliminated from the body by biological means before it has decayed, so it should take less than 8 days for there to be only half as much ^{131}I left in the body.

13. The products of a typical fusion reaction are light, non-radioactive elements. During a fusion reaction, electrons or positrons and photons are typically emitted. These by-products have little tendency to induce radioactivity in materials they come into contact with. A fission reaction, on the other hand, typically has radioactive daughter nuclei as the end products, constituting radioactive waste. Furthermore, fission reactions usually give off neutrons, which tend to induce radioactivity (neutron activation) in materials they come into contact with. The material of the fission reactor itself, the moderator substance, and the control rods all become radioactive as a result of neutron bombardment. As such, they must eventually be discarded as nuclear waste.

Problems

1. **Strategy** Nucleons have masses of approximately 1 u, so dividing the mass of the person by 1 u will give an estimate of the total number of nucleons in the person's body.

   Solution Estimate the number of nucleons.
   $$\frac{75 \text{ kg}}{1 \text{ u}} = \frac{75 \text{ kg}}{1.660\,539 \times 10^{-27} \text{ kg}} = \boxed{4.5 \times 10^{28}}$$

5. **Strategy** The nucleon number A is the sum of the total number of protons Z and neutrons N. Use the Periodic Table of the elements to find the number of protons.

   Solution Find the number of protons.
   Potassium has atomic number 19, so $Z = 19$.
   $N = \#$ of neutrons $= 21$
   Find the nucleon number.
   $A = Z + N = 19 + 21 = 40$
   So, the symbol is $\boxed{^{40}_{19}\text{K}}$.

7. **Strategy** Use the Periodic Table of the elements to find the number of protons.

   Solution Xe has atomic number 54, so the number of protons is $\boxed{54}$.

9. Strategy $A = 107$ for technetium-107. Find the radius of the nucleus using Eq. (29-4). Use the volume of a sphere to find the volume of the nucleus.

Solution Find the radius.
$$r = r_0 A^{1/3} = (1.2 \text{ fm})(107)^{1/3} = \boxed{5.7 \text{ fm}}$$
Find the volume.
$$V = \frac{4}{3}\pi r^3 = \frac{4}{3}\pi r_0^3 A = \frac{4}{3}\pi(1.2 \times 10^{-15} \text{ m})^3(107) = \boxed{7.7 \times 10^{-43} \text{ m}^3}$$

11. Strategy A deuteron has 1 proton and 1 neutron. Use Eqs. (29-7) and (29-8) to find the mass defect and binding energy.

Solution Find the mass defect.
$$\Delta m = 1 \times 1.007\ 276\ 5 \text{ u} + 1 \times 1.008\ 664\ 9 \text{ u} - 2.013\ 553 \text{ u} = 0.002\ 388 \text{ u}$$

The binding energy is $E_B = (\Delta m)c^2 = 0.002\ 388 4 \text{ u} \times 931.494 \text{ MeV/u} = \boxed{2.225 \text{ MeV}}$.

13. Strategy The nucleon number A is the sum of the total number of protons Z and neutrons N. Use Eqs. (29-7) and (29-8) to find the mass defect and binding energy. The binding energy per nucleon is the binding energy divided by the total number of nucleons in the nucleus.

Solution

(a) The ^{16}O atom has 8 protons, 8 neutrons, and 8 electrons. Its mass is 15.994 914 6 u. Find the mass defect.
$$\Delta m = (\text{mass of 8 } ^1\text{H atoms and 8 neutrons}) - (\text{mass of } ^{16}\text{O atom})$$
$$= 8 \times 1.007\ 825\ 0 \text{ u} + 8 \times 1.008\ 664\ 9 \text{ u} - 15.994\ 914\ 6 \text{ u} = 0.137\ 004\ 6 \text{ u}$$

The binding energy is $E_B = (\Delta m)c^2 = 0.137\ 004\ 6 \text{ u} \times 931.494 \text{ MeV/u} = \boxed{127.619 \text{ MeV}}$.

(b) Calculate the average binding energy per nucleon.
$$\frac{E_B}{A} = \frac{127.619 \text{ MeV}}{16 \text{ nucleons}} = \boxed{7.976\ 19 \text{ MeV/nucleon}}$$
This result matches the value given in Figure 29.2.

17. Strategy Use Eq. (29-8) to find the mass defect. Compare the mass defect to the mass of an electron.

Solution

(a) Calculate the mass defect.
$$\Delta m = \frac{E_B}{c^2} = \frac{13.6 \text{ eV}}{931.494 \times 10^6 \text{ eV/u}} = \boxed{1.46 \times 10^{-8} \text{ u}}$$

(b) The ratio of this mass defect to the mass of the electron is
$$\frac{1.46 \times 10^{-8} \text{ u}}{5.486 \times 10^{-4} \text{ u}} = 2.66 \times 10^{-5}.$$

This is small enough to be ignored. There is $\boxed{\text{no}}$ reason to worry when calculating the mass of the ^{1}H nucleus, especially since its mass is 1836 times that of the electron.

21. Strategy In beta-minus decay, the atomic number Z increases by 1 while the mass number A remains constant. Use Eq. (29-11).

Solution

For the parent $\left(^{40}_{19}\text{K}\right)$ $Z = 19$, so the daughter nuclide will have $Z = 19 + 1 = 20$, which is the element Ca. The symbol for the daughter is $\boxed{^{40}_{20}\text{Ca}}$.

23. Strategy In electron-capture decay, the atomic number Z is decreased by 1 while the mass number A stays the same.

Solution In this case, the parent nuclide has $Z = 11$ and $A = 22$, so the daughter nuclide will have $Z = 10$ and $A = 22$, which is the element neon. Write out the reaction.

$^{A}_{Z}P + ^{\ 0}_{-1}\text{e} \rightarrow ^{\ \ A}_{Z-1}D + ^{0}_{0}\nu$, so $\boxed{^{22}_{11}\text{Na} + ^{\ 0}_{-1}\text{e} \rightarrow ^{22}_{10}\text{Ne} + ^{0}_{0}\nu}$. The daughter nuclide is $\boxed{^{22}_{10}\text{Ne}}$.

25. Strategy The kinetic energy of the decay products equals the energy associated with the change in mass.

Solution Find the change in mass during the decay using atomic masses.
$$\Delta m = (\text{mass of } ^{222}_{86}\text{Rn} + \text{mass of } ^{4}_{2}\text{He}) - (\text{mass of } ^{226}_{88}\text{Ra}) = (222.017\ 577\ 7\ \text{u} + 4.002\ 603\ 2\ \text{u}) - 226.025\ 409\ 8\ \text{u}$$
$$= -0.005\ 228\ 9\ \text{u}$$
Find the kinetic energy.
$$K = E = |\Delta m|c^2 = 0.005\ 228\ 9\ \text{u} \times 931.494\ \text{MeV/u} = 4.8707\ \text{MeV}$$

Assuming the $^{222}_{86}\text{Rn}$ nucleus takes away an insignificant fraction of the kinetic energy, the alpha particle's kinetic energy will be $\boxed{4.8707\ \text{MeV}}$.

27. Strategy In Problem 21, the daughter nuclide in this decay was found to be $^{40}_{20}\text{Ca}$. The maximum kinetic energy of the beta particle is equal to the disintegration energy.

Solution The reaction is
$$^{40}_{19}\text{K} \rightarrow ^{40}_{20}\text{Ca} + ^{\ 0}_{-1}\text{e} + ^{0}_{0}\bar{\nu}$$

The atomic masses of $^{40}_{19}\text{K}$ and $^{40}_{20}\text{Ca}$ are 39.963 998 5 u and 39.962 591 0 u, respectively. To get the masses of the nuclei, we subtract Zm_e from each. The mass of the electron is 0.000 548 6 u, and the neutrino's mass is negligible. The mass difference is
$$\Delta m = [(M_\text{Ca} - 20m_\text{e}) + m_\text{e}] - (M_\text{K} - 19m_\text{e}) = M_\text{Ca} - M_\text{K} = 39.962\ 591\ 0\ \text{u} - 39.963\ 998\ 5\ \text{u} = -0.001\ 407\ 5\ \text{u}.$$

The disintegration energy is $E = |\Delta m|c^2 = 0.001\ 407\ 5\ \text{u} \times 931.494\ \text{MeV/u} = 1.3111\ \text{MeV}$.

The maximum kinetic energy of the β^- particle is $\boxed{1.3111\ \text{MeV}}$.

29. Strategy In a spontaneous decay, the mass of the decay products must be less than the mass of the parent.

Solution Alpha decay of $^{19}_{8}\text{O}$ would have a daughter nuclide of $^{15}_{6}\text{C}$.

$$^{19}_{8}\text{O} \rightarrow ^{15}_{6}\text{C} + ^{4}_{2}\text{He}$$

The mass of atomic $^{19}_{8}\text{O}$ is 19.003 578 7 u. The combined mass of $^{15}_{6}\text{C}$ and $^{4}_{2}\text{He}$ is

15.010 599 3 u + 4.002 603 2 u = 19.013 202 5 u.

Since 19.013 202 5 u > 19.003 578 7 u, the spontaneous alpha decay of $^{19}_{8}\text{O}$ is not possible.

33. Strategy The activity is reduced by a factor of two for each half-life. Use Eqs. (29-18), (29-20), and (29-22) to find the initial number of nuclei and the probability of decay per second.

Solution

(a) Find the number of half-lives.
$$600.0 \text{ s} = \frac{600.0 \text{ s}}{200.0 \text{ s/half-life}} = 3.000 \text{ half-lives}$$

The activity after 3.000 half-lives will be $R = \left(\frac{1}{2}\right)^{3.000} \times R_0 = \frac{1}{8.000} \times 80,000.0 \text{ s}^{-1} = \boxed{10,000 \text{ s}^{-1}}$.

(b) Find the initial number of nuclei.
$$N_0 = \frac{1}{\lambda} R_0 = \tau R_0 = \frac{R_0 T_{1/2}}{\ln 2} = 80,000.0 \text{ s}^{-1} \times \frac{200.0 \text{ s}}{\ln 2} = \boxed{2.308 \times 10^7}$$

(c) The probability per second is $\lambda = \frac{1}{\tau} = \frac{\ln 2}{T_{1/2}} = \frac{\ln 2}{200.0 \text{ s}} = \boxed{3.466 \times 10^{-3} \text{ s}^{-1}}$.

35. Strategy The activity as a function of time is given by $R = R_0 e^{-t/\tau}$. Use Eq. (29-22) to find the time constant. Assume that the original activity was 0.25 Bq per g of carbon.

Solution Find the age of the bones.
$$e^{-t/\tau} = \frac{R}{R_0}, \text{ so } -\frac{t}{\tau} = \ln\frac{R}{R_0} \text{ or } t = -\tau \ln\frac{R}{R_0} = -\frac{5730 \text{ yr}}{\ln 2} \times \ln\frac{0.242}{0.25} = \boxed{270 \text{ yr}}.$$

37. Strategy The half-life of $^{214}_{83}\text{Bi}$ is 19.9 min. The activity is given by $R = R_0(2^{-t/T_{1/2}})$.

Solution Find the activity.
$$R = R_0(2^{-t/T_{1/2}}) = 0.058 \text{ Ci} \times \left(\frac{1}{2}\right)^{60/19.9} = \boxed{0.0072 \text{ Ci}}$$

39. Strategy Use Eq. (29-22) to find the time constant. The number of nuclei is related to the mass by $N = mN_A/M$, where N_A is Avogadro's number and M is the molar mass. The activity is equal to the number of nuclei divided by the time constant.

Solution Convert the half-life of radium-226 to seconds.
$$1600 \text{ yr} \cdot \frac{3.156 \times 10^7 \text{ s}}{\text{yr}} = 5.0496 \times 10^{10} \text{ s}$$

The time constant τ is $\tau = \frac{T_{1/2}}{\ln 2} = \frac{5.0496 \times 10^{10} \text{ s}}{\ln 2} = 7.285 \times 10^{10} \text{ s}.$
Find the number of nuclei in 1.0 g of radium.
$$N = \frac{mN_A}{M} = \frac{1.0 \text{ g}}{226.0254 \text{ g/mol}} \times 6.022 \times 10^{23} \text{ nuclei/mol} = 2.6643 \times 10^{21} \text{ nuclei}$$

The activity is $R = \frac{N}{\tau} = \frac{2.6643 \times 10^{21} \text{ nuclei}}{7.285 \times 10^{10} \text{ s}} \times \frac{1 \text{ Ci}}{3.7 \times 10^{10} \text{ Bq}} = \boxed{0.99 \text{ Ci}}.$

41. Strategy Use Eqs. (29-20) and (29-22) to find the decay constant. The total number of nuclei in 1.00 g of carbon times the relative abundance gives the number of carbon-14 nuclei. Use the results of parts (a) and (b) and Eq. (29-18) to find the activity per gram in a living sample.

Solution

(a) Calculate the decay constant.
$$\lambda = \frac{1}{\tau} = \frac{\ln 2}{T_{1/2}} = \frac{\ln 2}{5730 \text{ yr} \times 3.156 \times 10^7 \text{ s/yr}} = \boxed{3.83 \times 10^{-12} \text{ s}^{-1}}$$

(b) Calculate the number of ^{14}C atoms.
$$N = \frac{\text{mass}}{\text{mass per mol}} \times N_A \times (\text{relative abundance}) = \frac{1.00 \text{ g}}{12.011 \text{ g/mol}} \times 6.022 \times 10^{23} \text{ atoms/mol} \times 1.3 \times 10^{-12}$$
$$= \boxed{6.5 \times 10^{10} \text{ atoms}}$$

(c) Calculate the activity per gram.
$$\frac{R}{1.00 \text{ g}} = \lambda \frac{N}{1.00 \text{ g}} = 3.83 \times 10^{-12} \text{ s}^{-1} \times \frac{6.5 \times 10^{10} \text{ atoms}}{1.00 \text{ g}} = \boxed{0.25 \text{ Bq/g}}$$

45. Strategy Look up the half-lives in Appendix B. Use Eq. (29-23).

Solution The numbers of oxygen-15 nuclei and oxygen-19 nuclei at time t are given by $N_{15} = N_0 \times 2^{-t/T_{15}}$ and $N_{19} = N_0 \times 2^{-t/T_{19}}$, respectively. There are twice as many oxygen-15 nuclei as oxygen-19 nuclei at time t. Find t.
$$N_0 \times 2^{-t/T_{15}} = 2N_0 \times 2^{-t/T_{19}}$$
$$2^{-t/T_{15}} = 2^{1-t/T_{19}}$$
$$-\frac{t}{T_{15}} = 1 - \frac{t}{T_{19}}$$
$$t\left(\frac{1}{T_{19}} - \frac{1}{T_{15}}\right) = 1$$
$$t = \frac{T_{15}T_{19}}{T_{15} - T_{19}} = \frac{(122.24 \text{ s})(26.88 \text{ s})}{122.24 \text{ s} - 26.88 \text{ s}} = \boxed{34.46 \text{ s}}$$
Compute the percent of oxygen-19 nuclei that have decayed.
$$\frac{N_0 - N_{19}}{N_0} \times 100\% = (1 - 2^{-t/T_{19}}) \times 100\% = (1 - 2^{-34.457/26.88}) \times 100\% = \boxed{58.87\%}$$

49. Strategy The absorbed dose of ionizing radiation is the amount of radiation energy absorbed per unit mass of tissue. Use Eq. (29-28a) to find the biologically equivalent dose in sieverts.

Solution Find the total energy absorbed by the tumor.
$$E = (1.16 \times 10^{17} \text{ protons})(950 \times 10^3 \text{ eV/proton})(1.602 \times 10^{-19} \text{ J/eV})$$
Find the biologically equivalent dose.
$$\text{absorbed dose} \times \text{QF} = \frac{(1.16 \times 10^{17} \text{ protons})(950 \times 10^3 \text{ eV/proton})(1.602 \times 10^{-19} \text{ J/eV})}{3.82 \times 10^{-6} \text{ kg}} \times 3.0 = \boxed{1.4 \times 10^{10} \text{ Sv}}$$

53. Strategy The unknown nuclide can be found by balancing the reaction. The total charge and total number of nucleons must remain the same.

Solution The reaction is $_0^1 n + _b^a(?) \to _{80}^{198}\text{Hg}^* + _{-1}^0 e + \bar{\nu}$.

Find a and b.

$1 + a = 198 \Rightarrow a = 197$ and $0 + b = 80 - 1 \Rightarrow b = 79$.

Therefore (?) is $\boxed{_{79}^{197}\text{Au}}$.

57. Strategy From Figure 29.2, the binding energies per nucleon for $_{92}^{235}\text{U}$, $_{54}^{139}\text{Xe}$, and $_{38}^{95}\text{Sr}$ are approximately 7.6 MeV, 8.3 MeV, and 8.7 MeV, respectively. The energy released is equal to the increase in the binding energy.

Solution The binding energies are as follows:

$_{92}^{235}\text{U}$: $235 \times 7.6 \text{ MeV} = 1786 \text{ MeV}$, $_{54}^{139}\text{Xe}$: $139 \times 8.3 \text{ MeV} = 1154 \text{ MeV}$, $_{38}^{95}\text{Sr}$: $95 \times 8.7 \text{ MeV} = 827 \text{ MeV}$.

Find the energy released.

$1154 \text{ MeV} + 827 \text{ MeV} - 1786 \text{ MeV} = 195 \text{ MeV} \approx \boxed{200 \text{ MeV}}$

61. Strategy Find the change in mass to determine the energy released by the reaction.

Solution The reaction is $4p + 2e^- \to {}^4\text{He} + 2\nu$.

The change in mass is

$\Delta m = (4.002\ 603\ 2 \text{ u} - 2 \times 0.000\ 548\ 6 \text{ u}) - (4 \times 1.007\ 276\ 5 \text{ u} + 2 \times 0.000\ 548\ 6) = -0.028\ 697\ 2$.

The energy released is $E = 0.028\ 697\ 2 \text{ u} \times 931.494 \text{ MeV/u} = \boxed{26.7313 \text{ MeV}}$.

65. (a) Strategy The nucleon number A is the sum of the total number of protons Z and neutrons N.

Solution The reaction is $_{88}^{226}\text{Ra} \to _{86}^{222}\text{Rn} + _2^4\alpha$.

There will be $226 - 4 - 86 = \boxed{136 \text{ neutrons}}$ and $88 - 2 = \boxed{86 \text{ protons}}$.

(b) Strategy The activity is related to the number of nuclei N and the time constant τ by $R = N/\tau$. The time constant is related to the half-life by $T_{1/2} = \tau \ln 2$.

Solution Calculate the activity of the radon.

$$R = \frac{N}{\tau} = \frac{N \ln 2}{T_{1/2}} = \frac{1.0 \times 10^7 \cdot \ln 2}{3.8 \text{ d} \times 24 \text{ h/d} \times 3600 \text{ s/h}} = 21 \text{ decays/s}$$

So, $\boxed{21 \text{ alpha particles per second}}$ are emitted due to the decaying radon nuclei.

69. Strategy Isotopes have the same atomic number.

Solution Compare the atomic numbers of the unknown nuclides.

$_{71}^{175}(?)$ and $_{71}^{167}(?)$ have the same atomic number 71, so they are isotopes of each other.

$_{74}^{175}(?)$ and $_{74}^{180}(?)$ have the same atomic number 74, so they are isotopes of each other.

Lutetium (Lu) has atomic number 71 and Tungsten (W) has atomic number 74.

To summarize: $\boxed{_{71}^{175}(\text{Lu}) \text{ and } _{71}^{167}(\text{Lu})}$ are isotopes of each other and $\boxed{_{74}^{175}(\text{W}) \text{ and } _{74}^{180}(\text{W})}$ are isotopes of each other.

73. Strategy The energies of the photons correspond to the differences in energy levels for the transitions.

Solution From left to right, the energies are:

$492 \text{ keV} - 0 = \boxed{492 \text{ keV}}$, $472 \text{ keV} - 0 = \boxed{472 \text{ keV}}$, $40 \text{ keV} - 0 = \boxed{40 \text{ keV}}$, $492 \text{ keV} - 40 \text{ keV} = \boxed{452 \text{ keV}}$,

$472 \text{ keV} - 40 \text{ keV} = \boxed{432 \text{ keV}}$, and $327 \text{ keV} - 40 \text{ keV} = \boxed{287 \text{ keV}}$.

77. Strategy Use Eq. (29-23).

Solution

(a) Calculate N/N_0 for a time $t = 4.5 \times 10^9$ yr and a half-life of 4.468×10^9 yr.

$$N = N_0 \left(\frac{1}{2} \right)^{t/T_{1/2}}, \text{ so } \frac{N}{N_0} = \left(\frac{1}{2} \right)^{t/T_{1/2}} = \left(\frac{1}{2} \right)^{\frac{4.5 \times 10^9 \text{ yr}}{4.468 \times 10^9 \text{ yr}}} = \boxed{0.50}.$$

(b) $\dfrac{N}{N_0} = \left(\dfrac{1}{2} \right)^{\frac{4.5 \times 10^9 \text{ yr}}{7.04 \times 10^8 \text{ yr}}} = \boxed{0.012}$

Since ^{235}U decays at a significantly faster rate than ^{238}U, the fact that there are more than 100 times as many ^{238}U atoms as ^{235}U in the Earth today seems reasonable. The answer is $\boxed{\text{yes}}$.

81. (a) Strategy The activity is related to the number of nuclei N and the time constant τ by $R = N/\tau$. The number of nuclei is related to the mass by $N = mN_A/M$, where N_A is Avogadro's number and M is the molar mass. The time constant is related to the half-life by $T_{1/2} = \tau \ln 2$.

Solution Compute the mass of the sample of iodine.

$$m = \frac{NM}{N_A} = \frac{R\tau M}{N_A} = \frac{RT_{1/2}M}{N_A \ln 2}$$

$$= \frac{(64.5 \times 10^{-3} \text{ Ci})(3.7 \times 10^{10} \text{ Bq/Ci})(8.0252 \text{ d})(86,400 \text{ s/d})(130.906\,124\,6 \text{ g/mol})}{(6.022 \times 10^{23} \text{ mol}^{-1}) \ln 2} = \boxed{5.2 \times 10^{-7} \text{ g}}$$

(b) Strategy Use Eq. (29-21).

Solution Find the activity of the iodine 4.5 days later.

$$R = R_0 e^{-t/\tau} = R_0 e^{-t \ln 2/T_{1/2}} = (64.5 \text{ mCi}) e^{-(4.5 \text{ d}) \ln 2/(8.0252 \text{ d})} = \boxed{44 \text{ mCi}}$$

85. (a) Strategy The activity is given by $R = R_0 \left(\frac{1}{2}\right)^{t/T_{1/2}}$.

Solution Compute a ratio of the activities.

$$\frac{R}{R_0} = \left(\frac{1}{2}\right)^{10/87.7} = 0.92$$

The percentage decrease is $100\% - 92\% = \boxed{8\%}$.

(b) Strategy The number of nuclei is related to the mass by $N = mN_A/M$, where N_A is Avogadro's number and M is the molar mass. The activity is related to the number of nuclei N and the time constant τ by $R = N/\tau$. The time constant is related to the half-life by $T_{1/2} = \tau \ln 2$. The power output is equal to the rate of energy produced by the decay times the efficiency.

Solution Initially, the number of ^{238}Pu atoms is

$$N_0 = \frac{mN_A}{M} = \frac{1.0\times10^{-3}\ \text{g}}{238\ \text{g/mol}} \times 6.022\times10^{23}\ \text{atoms/mol} = 2.53\times10^{18}\ \text{atoms}$$

Find the initial activity.

$$R_0 = \frac{N_0}{\tau} = \frac{N_0 \ln 2}{T_{1/2}} = \frac{2.53\times10^{18}\ \text{atoms} \cdot \ln 2}{87.7\ \text{yr} \times 3.156\times10^{7}\ \text{s/yr}} = 6.34\times10^{8}\ \text{Bq}$$

Find the initial power output.

$$P_0 = (\text{decays per sec})(\text{energy per decay})(\text{efficiency}) = 6.34\times10^{8}\ \text{Bq} \times 5.6\times10^{6}\ \text{eV} \times 1.6\times10^{-19}\ \text{J/eV}$$
$$= 5.7\times10^{-4}\ \text{W} = \boxed{0.57\ \text{mW}}$$

The power output after 10.0 years is 92% of P_0.

$$P(t = 10.0\ \text{yr}) = 0.92 \times 0.57\ \text{mW} = \boxed{0.52\ \text{mW}}$$

Chapter 30

PARTICLE PHYSICS

Conceptual Questions

1. Quarks, electrons, and the exchange particles (gluons, photons, W^+, W^-, Z^0, and possibly gravitons) are the only *fundamental* particles that make up an atom of ordinary matter.

5. There are six particles in the lepton family, including their associated anti-particles. These are the electron, the muon, the tau, and the three corresponding neutrinos.

9. Neutrinos interact with matter extremely weakly. This means that matter is essentially transparent to neutrinos. Although 10^{14} of them pass through our bodies every second, practically none of them interact with the particles in our bodies.

Problems

1. Strategy The energy-time uncertainty principle is $\Delta E\,\Delta t \geq \hbar/2$. Use $\Delta E\,\Delta t = \hbar/2$ to estimate the range of the weak force carried by the Z particle.

Solution The range is approximately $c\Delta t$.

$$\Delta E\,\Delta t = \frac{\hbar}{2}, \text{ so } \Delta t = \frac{\hbar}{2\Delta E}. \text{ The range is } \frac{\hbar c}{2\Delta E} = \frac{(1.055\times10^{-34}\ \text{J}\cdot\text{s})(3.00\times10^8\ \text{m/s})}{2(92\times10^9\ \text{eV})(1.602\times10^{-19}\ \text{J/eV})} = \boxed{1.1\times10^{-18}\ \text{m}}, \boxed{\text{which is}}$$

$$\boxed{\text{approximately the same as } 1\times10^{-17}\ \text{m in Table 30.3}}.$$

5. Strategy Replace each of the three quarks that compose a proton with its corresponding antiquark.

Solution The quarks in a proton are uud. Therefore, the quarks in an antiproton are $\boxed{\overline{u}\,\overline{u}\,\overline{d}}$.

9. Strategy Mass is related to rest energy by $E_0 = mc^2$.

Solution Compute the mass.

$$m = \frac{E_0}{c^2} = 10^{19}\ \text{GeV}/c^2, \text{ so } m = \frac{10^{19}\times10^9\ \text{eV}}{(3\times10^8\ \text{m/s})^2}\times\frac{1.6\times10^{-19}\ \text{J}}{1\ \text{eV}} = \boxed{20\ \text{ng}}.$$

11. Strategy The rest energy of an electron is 0.5110 MeV, so an electron with energy 7.0 TeV is extremely relativistic.

Solution Calculate the de Broglie wavelength.

$$E \approx pc \text{ and } p = \frac{h}{\lambda}, \text{ so } \lambda = \frac{hc}{E} = \frac{1240\times10^{-9}\ \text{eV}\cdot\text{m}}{7.0\times10^{12}\ \text{eV}} = \boxed{1.8\times10^{-19}\ \text{m}}.$$

13. Strategy The pions have the same rest energy, so their momenta will be equal in magnitude and opposite in direction, and their kinetic energies will be the same. The total kinetic energy is equal to the difference of the K^0 meson's rest energy and the sum of the rest energies of the pions.

Solution Find the kinetic energies of the pions.
$$K_{\text{total}} = 497.7 \text{ MeV} - (139.6 \text{ MeV} + 139.6 \text{ MeV}) = 218.5 \text{ MeV}$$

Each pion has a kinetic energy of $\dfrac{218.5 \text{ MeV}}{2} = \boxed{109.3 \text{ MeV}}$.

17. Strategy The rest energy of the pion is 0.135 GeV. Conservation of energy requires that the energy of the pion be equal to the energy of the two photons. Conservation of momentum requires that the momenta of the photons are equal and opposite.

Solution Relate the energies.
$$E_\pi = E_{\gamma_1} + E_{\gamma_2}$$
Relate the wavelengths of the photons.
$$p_{\gamma_1} = \frac{h}{\lambda_1} = p_{\gamma_2} = \frac{h}{\lambda_2}, \text{ so } \lambda_1 = \lambda_2.$$
The photons are identical, so let $E_{\gamma_1} = E_{\gamma_2} = E_\gamma$.
Find the energy of each photon.
$$E_\pi = E_\gamma + E_\gamma = 2E_\gamma, \text{ so } E_\gamma = \frac{E_\pi}{2} = \frac{0.135 \text{ GeV}}{2} = \boxed{67.5 \text{ MeV}}.$$

21. Strategy Mesons contain one quark and one antiquark. Baryons contain three quarks. Use Table 30.1 to determine the quark content of each particle.

Solution The charge of an up quark is $\frac{2}{3}e$ and a strange quark has charge $-\frac{1}{3}e$. Therefore, the quark content of the baryon with charge 0 is $\boxed{\text{uss}}$.

25. Strategy Replace each particle in the decay reaction with its corresponding antiparticle to write the two decay modes.

Solution Since π^+ is the antiparticle of π^-, the decay products of π^+ must be antiparticles of the decay products of π^-. The decay modes of π^+ are then $\boxed{\pi^+ \rightarrow \mu^+ + \nu_\mu \text{ and } \pi^+ \rightarrow e^+ + \nu_e}$.

REVIEW AND SYNTHESIS: CHAPTERS 26–30

Review Exercises

1. **Strategy** Use Eq. (26-5) to find the velocity v_{pE} of the escape pod relative to Earth. Then, use Eq. (26-4) to find how long the pod is as measured by people on Earth. Let the positive direction be toward the Earth.

   Solution The velocity of the starship relative to Earth is $v_{sE} = 0.78c$. The velocity of the pod relative to the starship is $v_{ps} = 0.63c$. Find the velocity of the pod relative to Earth.

   $$v_{pE} = \frac{v_{ps} + v_{sE}}{1 + v_{ps}v_{sE}/c^2} = \frac{0.63c + 0.78c}{1 + (0.63)(0.78)} = 0.945c$$

   Find the length of the pod as measured by people on Earth.

   $$L = \frac{L_0}{\gamma} = L_0\sqrt{1 - v^2/c^2} = (12.0\ \text{m})\sqrt{1 - 0.945^2} = \boxed{3.91\ \text{m}}$$

5. **Strategy** Use Eq. (26-8).

   Solution Find the speed of the electron.

   $$K = (\gamma - 1)mc^2$$
   $$\frac{K}{mc^2} = \gamma - 1$$
   $$1 + \frac{K}{mc^2} = \frac{1}{\sqrt{1 - \frac{v^2}{c^2}}}$$
   $$1 - \frac{v^2}{c^2} = \left(1 + \frac{K}{mc^2}\right)^{-2}$$
   $$v = c\sqrt{1 - \left(1 + \frac{K}{mc^2}\right)^{-2}} = c\sqrt{1 - \left[1 + \frac{1.02 \times 10^{-13}\ \text{J}}{(9.109 \times 10^{-31}\ \text{kg})(2.998 \times 10^8\ \text{m/s})^2}\right]^{-2}} = \boxed{0.895c = 2.68 \times 10^8\ \text{m/s}}$$

9. **Strategy** The width of the central maximum is the distance between the first minimum on either side of the central maximum. Use conservation of energy to find the speed of the electrons. Use the de Broglie wavelength for the wavelength of the electrons.

   Solution Find the wavelength of the electrons.

   $$p = mv = \frac{h}{\lambda} \text{ and } K = \frac{1}{2}mv^2 = \frac{p^2}{2m} = e\Delta V, \text{ so } \lambda = \frac{h}{\sqrt{2me\Delta V}}.$$

   The first minimum on either side of the central maximum is given by $a\sin\theta = \lambda$. Solve for θ.

   $$\theta = \sin^{-1}\frac{\lambda}{a} = \sin^{-1}\frac{h}{a\sqrt{2me\Delta V}} = \sin^{-1}\frac{6.626 \times 10^{-34}\ \text{J}\cdot\text{s}}{(6.6 \times 10^{-10}\ \text{m})\sqrt{2(9.109 \times 10^{-31}\ \text{kg})(1.602 \times 10^{-19}\ \text{C})(8950\ \text{V})}} = 1.12555°$$

   If x is the distance to the first minimum on either side of the central maximum and D is the distance from the slit to the screen, then $\tan\theta = x/D$. Find $2x$, the width of the central maximum.

   $$2x = 2D\tan\theta = 2(2.50\ \text{m})\tan 1.12555° = \boxed{9.8\ \text{cm}}$$

13. **Strategy** The rest energy of the lambda particle not used to create the proton and the pion will become kinetic energy of the proton and the pion. Use conservation of momentum and energy.

   Solution Find the energy not used to create the proton and the pion.
   $$K = 1116 \text{ MeV} - (938 \text{ MeV} + 139.6 \text{ MeV}) = 38.4 \text{ MeV}$$

   Let the magnitudes of the momenta of the proton and pion be p_p and p_π, respectively. Then, according to conservation of momentum, $p_p = p_\pi$. Let the kinetic energies of the proton and pion be K_p and K_π, respectively. Then, according to conservation of energy, $K = K_p + K_\pi$. According to Eq. (26-11) and conservation of momentum, $(p_\pi c)^2 = K_\pi^2 + 2K_\pi E_{\pi 0} = (p_p c)^2 = K_p^2 + 2K_p E_{p0}$, where E_{p0} and $E_{\pi 0}$ are the rest energies of the proton and pion, respectively. Substituting for K_π in $(p_p c)^2 = K_\pi^2 + 2K_\pi E_{\pi 0}$ gives

   $$(p_p c)^2 = (K - K_p)^2 + 2(K - K_p)E_{\pi 0}.$$

   Subtracting $(p_p c)^2 = K_p^2 + 2K_p E_{p0}$ from $(p_p c)^2 = (K - K_p)^2 + 2(K - K_p)E_{\pi 0}$ gives

   $$0 = K^2 - 2KK_p + 2KE_{\pi 0} - 2K_p E_{\pi 0} - 2K_p E_{p0}.$$

   Solving this equation for the kinetic energy of the proton K_p gives

   $$K_p = \frac{K^2 + 2KE_{\pi 0}}{2(K + E_{\pi 0} + E_{p0})} = \frac{(38.4 \text{ MeV})^2 + 2(38.4 \text{ MeV})(139.6 \text{ MeV})}{2(38.4 \text{ MeV} + 139.6 \text{ MeV} + 938 \text{ MeV})} = \boxed{5.5 \text{ MeV}}.$$

   Therefore, the kinetic energy of the pion is $K_\pi = K - K_p = 38.4 \text{ MeV} - 5.46 \text{ MeV} = \boxed{33 \text{ MeV}}$.

17. **(a) Strategy** Use Eqs. (29-21) and (29-22).

   Solution Find the activity of the sample of gold-198 after 8.10 days.
   $$R = R_0 e^{-t/\tau} = R_0 e^{-t \ln 2/T_{1/2}} = (1.00 \times 10^{10} \text{ Bq})e^{-8.10 \ln 2/2.70} = \boxed{1.25 \times 10^9 \text{ Bq}}$$

   (b) Strategy The atomic number Z increases by 1, while the mass number A stays the same, so the isotope undergoes beta-minus decay.

   Solution Since the isotope undergoes beta-minus decay, the particles emitted in this process are $\boxed{\text{an electron and an antineutrino}}$.

21. **Strategy** Use Eq. (28-2).

   Solution Find the order of magnitude of the minimum uncertainty in the momentum of the electron.
   $$\Delta x \Delta p_x \geq \frac{\hbar}{2}, \text{ so } (\Delta p_x)_{\text{min}} = \frac{\hbar}{2\Delta x} \sim \frac{10^{-34} \text{ J} \cdot \text{s}}{(10^{-11} \text{ m})(10^{-19} \text{ J/eV})} = \boxed{10^{-4} \text{ eV} \cdot \text{s/m}}.$$

25. **Strategy** According to Eq. (27-9), The cutoff frequency f_{max} of x-rays produced by bremsstrahlung (braking radiation) is directly proportional to the kinetic energy of the incident electrons. Since $\lambda_{min} \propto 1/f_{max}$, the minimum wavelength of the x-rays is inversely proportional to the kinetic energy of the electrons. Now, by conservation of energy, an electron accelerated through a potential difference is given kinetic energy equal to $e\Delta V$. Therefore, the minimum wavelength of the x-rays is inversely proportional to the potential difference through which the electrons are accelerated.

    Solution Find the ratio of the minimum wavelength of x-rays in tube A to the minimum wavelength in tube B.

    $$\frac{\lambda_A}{\lambda_B} = \frac{\Delta V_B}{\Delta V_A} = \frac{40 \text{ kV}}{10 \text{ kV}} = \frac{4}{1}; \text{ the ratio is } \boxed{4:1}.$$

29. **Strategy** The absorbed dose of ionizing radiation is the amount of radiation energy absorbed per unit mass of tissue.

    Solution Find the equivalent dose of energy due to the x-rays.

    $$E = (20 \times 10^{-3} \text{ rad}) \times \frac{0.01 \text{ Gy}}{\text{rad}} \times \frac{1 \text{ J/kg}}{\text{Gy}} \times \frac{65 \text{ kg}}{3} = 0.0043 \text{ J}$$

    Find the energy absorbed by the patient's body.

    $$\text{energy absorbed} = \frac{\text{biologically equivalent dose of energy}}{\text{QF}} = \frac{4.3 \text{ mJ}}{0.90} = \boxed{4.8 \text{ mJ}}$$

MCAT Review

1. **Strategy** The rest energy of an alpha particle is approximately 3.7 GeV. This is much greater than the kinetic energy of the alpha particle, so $K = \frac{1}{2}mv^2$ is a reasonable approximation of the kinetic energy of the particle.

    Solution Compute the approximate speed of the alpha particle.

    $$K = \frac{1}{2}mv^2, \text{ so } v = \sqrt{\frac{2K}{m}} = \sqrt{\frac{2(4.8 \times 10^6 \text{ eV})(1.602 \times 10^{-19} \text{ J/eV})}{(4 \text{ u})(1.66 \times 10^{-27} \text{ kg/u})}} = 1.5 \times 10^7 \text{ m/s}.$$

    The correct answer is $\boxed{\text{C}}$.

2. **Strategy** For each alpha emitted, the nucleus loses two protons and two neutrons. For each beta emitted, the nucleus gains a proton and loses a neutron. Count the number of protons, neutrons, and betas emitted.

    Solution A total of $1+4+1 = 6$ alphas are emitted, so the nucleus loses 12 protons and 12 neutrons. A total of $2+1+1 = 4$ betas are emitted, so the nucleus gains 4 protons and loses 4 neutrons. The new atomic number is $Z = 90 - 12 + 4 = 82$ and the new nucleon number is $A = 232 - 12 - 12 + 4 - 4 = 208$. The correct answer is $\boxed{\text{A}}$.

3. **Strategy** There are three protons and $7 - 3 = 4$ neutrons in the nucleus of lithium-7. Use Eq. (29-8) to find the binding energy.

    Solution Compute the approximate binding energy.

    $$E_B = (\Delta m)c^2 = [(3 \times 1.0073 \text{ u} + 4 \times 1.0087 \text{ u}) - 7.014 \text{ u}] \times 931 \text{ MeV/u} = 40.0 \text{ MeV}$$

    The correct answer is $\boxed{\text{D}}$.

4. **Strategy** The external magnetic field only exerts forces on moving charged particles.

   Solution Alphas are positively charged, betas are negatively charged, and gammas have no charge. Therefore, gamma rays travel straight and alpha and beta rays are bent in opposite directions. The correct answer is $\boxed{\text{D}}$.

5. **Strategy** The nucleus decreases by $|\Delta m| = E/c^2$, where E is the energy of the gamma ray.

   Solution Compute the decrease in mass of the nucleus.
   $$|\Delta m| = \frac{E}{c^2} = \frac{(2.5\times10^6 \text{ eV})(1.602\times10^{-19} \text{ J/eV})}{(2.998\times10^8 \text{ m/s})^2} = 4.5\times10^{-30} \text{ kg}$$
   The correct answer is $\boxed{\text{C}}$.

6. **Strategy** Use Eqs. (29-21) and (29-22).

   Solution Find the original activity R_0 of the sodium-24 sample.
   $$R = R_0 e^{-t/\tau} = R_0 e^{-t\ln 2/T_{1/2}}, \text{ so } R_0 = Re^{t\ln 2/T_{1/2}} = (100 \text{ mCi})e^{24\ln 2/15} = 300 \text{ mCi.}$$
   The correct answer is $\boxed{\text{B}}$.

7. **Strategy** Subtract the mass equivalent of the energy required to break the nucleus of the neon-20 atom into its constituent parts from the masses of $Z = 10$ protons and $N = A - Z = 20 - 10 = 10$ neutrons to find the atomic mass of the atom.

   Solution Find the atomic mass of the atom.
   $$10\times1.0073 \text{ u} + 10\times1.0087 \text{ u} - 0.173 \text{ u} = 19.987 \text{ u}$$
   The correct answer is $\boxed{\text{A}}$.

8. **Strategy** The intermediate product uranium-234 has $238 - 234 = 4$ fewer neutrons than does uranium-238. Count the number of neutrons lost in each sequence.

   Solution Beta emission does not effect the nucleon number A, but does increase the atomic number by one, so choice A is not the correct answer. Alpha emission reduces the nucleon number by four and the atomic number by two. Evaluate the remaining three sequences.
   B: $A = 238 - 4 = 234$ and $Z = 92 - 2 + 3 = 93$.
   C: $A = 238 - 4 - 4 = 230$ and $Z = 92 - 2 - 2 + 1 + 1 = 90$.
   D: $A = 238 - 4 = 234$ and $Z = 92 - 2 + 1 + 1 = 92$.
   The correct answer is $\boxed{\text{D}}$.

9. **Strategy** Every 8 months, the sample has been reduced by half.

   Solution Find the fraction of the sample still remaining after 2 years.
   $$2 \text{ yr} = 24 \text{ mo and } \frac{24}{8} = 3, \text{ so the sample has been reduced to } \left(\frac{1}{2}\right)^3 = \frac{1}{8} \text{ of its original amount.}$$
   The correct answer is $\boxed{\text{C}}$.

10. **Strategy and Solution** In gamma decay, the atomic number Z and the mass number A stay the same. The correct answer is $\boxed{B}$.

11. **Strategy** The nucleon number A is the sum of the total number of protons Z and neutrons N.

    Solution The nucleon number of thallium-201 is $A = 201$, which is equal to the total number of protons and neutrons in the nucleus. The atomic number is $Z = 81$, which is equal to the total number of protons in the nucleus. In a neutral atom, the number of electrons is equal to the number of protons; in this case, there are 81 electrons. There are $N = A - Z = 201 - 81 = 120$ neutrons. The correct answer is $\boxed{D}$.

12. **Strategy** The activity R is given by $R = R_0 e^{-t/\tau}$, where R_0 is the initial activity, t is the time elapsed, and τ is the time constant.

    Solution The activity decreases exponentially with time. The correct answer is $\boxed{C}$.

13. **Strategy** The energy of a photon is related to its wavelength by $E = hc/\lambda$.

    Solution Compute the wavelength of the gamma ray photon.
    $$\lambda = \frac{hc}{E} = \frac{(4.15 \times 10^{-15} \text{ eV} \cdot \text{s})(3.0 \times 10^8 \text{ m/s})}{1.35 \times 10^5 \text{ eV}} = \frac{(4.15 \times 10^{-15})(3.0 \times 10^8)}{1.35 \times 10^5} \text{ m}$$
    The correct answer is $\boxed{A}$.

NOTES

NOTES

NOTES

NOTES

NOTES

NOTES